CHEM
KINE

CHEMICAL KINETICS
The Study of Reaction Rates in Solution

Kenneth A. Connors

WILEY-VCH

New York • Chichester • Weinheim • Brisbane • Singapore • Toronto

Kenneth A. Connors
School of Pharmacy
University of Wisconsin—Madison
425 North Charter Street
Madison, Wisconsin 53706

Originally published as ISBN 1-56081-053-X

Published simultaneously in Canada.

Library of Congress Cataloging-in-Publication Data

Connors, Kenneth A (Kenneth Antonio).
    Chemical kinetics: the study of reaction rates in solution/
  Kenneth A. Connors
        p.   cm.
    Includes bibliographical references.
    ISBN 0-471-72020-8
    1. Chemical Reaction, Rate of.   I. Title.
    [DNLM: 1. Toxicology—popular works. QV 600 091d]
  QD502.C.66   1990
  541.3'94—dc20   90-12119
                                           CIP

Printed in the United States of America.

10 9 8 7 6 5 4 3

To Pat Connors

Preface

Many books on chemical kinetics have been published, but few of these are devoted solely or even primarily to solution phase chemical kinetics. Textbooks of physical organic chemistry must deal with solution chemistry, but kinetics is only one part of their subject. From my teaching experience I have concluded that there is no current text that meets the needs, as I interpret them, of the student and practitioner of solution chemical kinetics.

My goal in this book has been to achieve a useful balance of phenomenology, physical theory, and empiricism. This is a textbook, not a research monograph, and this distinction accounts for the style and level of the treatment. I have tried to provide the reader with the knowledge to go into the laboratory and make the basic kinds of kinetic measurements and to go into the library prepared to interpret the more specialized information and arguments to be found in research papers and monographs. Citations to the literature are intended to be consistent with these aims, with the additional purposes of crediting innovators and directing the reader to interesting developments, examples, exceptions, or complications.

After an introductory chapter, phenomenological kinetics is treated in Chapters 2, 3, and 4. The theory of chemical kinetics, in the form most applicable to solution studies, is described in Chapter 5 and is used in subsequent chapters. The treatments of mechanistic interpretations of the transition state theory, structure–reactivity relationships, and solvent effects are more extensive than is usual in an introductory textbook. The book could serve as the basis of a one-semester course, and I hope that it also may be found useful for self-instruction.

My views on this subject obviously have been influenced by the literature of the field, and my citations to this literature can serve as acknowledgment of this debt; less obviously my views have also been influenced by teachers, colleagues, and students, to whom I express my thanks. Where I have strayed, intentionally or in error, from accepted interpretations, I accept responsibility and hope that students may be led to think independently by such deviant notions.

Some material in Chapter 4 first appeared in my books *A Textbook of Pharmaceutical Analysis* (3rd edition, 1982) and *Binding Constants: The Measurement of Molecular Complex Stability* (1987), both published by Wiley-Interscience. This material is used here with the kind permission of John Wiley & Sons, Inc.

Kenneth A. Connors
Madison, Wisconsin

Contents

1. Introduction to Chemical Kinetics   **1**

   1.1   The Field of Chemical Kinetics   1

       Time as a Variable   1
       Uses of Chemical Kinetics   2
       Literature of the Field   2

   1.2   The Study of Reaction Mechanisms   3

       The Definition of Reaction Mechanism   3
       Methods for Studying Mechanisms   6
       The Role of Kinetics   7
       Classification of Mechanisms   8

   1.3   Experimental Chemical Kinetics   10

       The Rate of Reaction   10
       Rate Equations and Stoichiometric Equations   11
       Rate Constants and Reaction Orders   13
       Temperature Dependence of Rate Constants   13

2. Simple Rate Equations   **17**

   2.1   Integrated Rate Equations   17

       Zero-Order Reactions   17
       First-Order Reactions   18
       Second-Order Reactions   20
       Pseudo-Order Reactions   23

   2.2   Determination of Reaction Order   24

       Use of Integrated Equations   24
       The Isolation and Pseudo-Order Techniques   26
       Initial Rate Method   28
       Fractional Time Methods   29

   2.3   Methods of Data Analysis   31

       Calculation of the Rate Constant from Concentrations   31
       Treatment of Instrument Response Data   34
       Methods When the Final Value is Unknown   36
       Propagation of Errors   40

Linear Least-Squares Regression 41
Weighting in Least-Squares Regression 44
Weighting and the Transformation of Variables 45
Variances of the Parameters 46
Nonlinear Least-Squares Regression 49
Accuracy in Rate Constant Measurements 51

3. Complicated Rate Equations **59**

3.1 Integration of the Rate Equations 60

Reversible Reactions 60
Parallel Reactions 62
Consecutive Reactions 66

3.2 Other Methods of Analysis 77

Measurement of Rates 77
Simplification of the Experimental Kinetics 78
Elimination of the Time Variable 79
Replacement of Time with an Area Variable 81
The Laplace Transformation 82
Secular Equations and Eigenvalues 90
The Preequilibrium Assumption 96
The Steady-State Approximation 100
Numerical Integration 105
Monte Carlo Simulation 109
Analog Simulation 114

3.3 The Kinetic Scheme 115

Model Building 115
Kinetic Equivalence of Rate Terms 123
Microscopic Reversibility and Detailed Balance 125

4. Fast Reactions **133**

4.1 Diffusion-Controlled Reactions 134

4.2 Relaxation Kinetics 136

Linearization of Rate Equations 137
Complex Reactions 139
Experimental Methods 143
Applications 146

4.3 Nuclear Magnetic Resonance Spectroscopy 153

Magnetic Properties of Nuclei 153

The Resonance Condition   155
Spin–Lattice Relaxation   157
Spin–Spin Relaxation   159
The Bloch Equations   160
T_1, T_2, and the Correlation Time   164
Chemical Exchange   166
Pulse NMR Measurements   170
Applications   173

4.4   Rapid Mixing Methods   176

Batch Methods   176
Flow Methods   177

4.5   Other Methods   180

Fluorescence Quenching   180
Electrochemical Methods   181
Common Ion Inhibition   183

5. Theory of Chemical Kinetics   **187**

5.1   Theoretical Approaches   187

The Arrhenius Equation   187
Collision Theory   188
Potential Energy Surfaces   191

5.2   Transition State Theory   200

Assumptions of the Theory   201
The Partition Function   201
The Rate Equation   205
Thermodynamic Interpretation   207

5.3   Chemical Interpretations of the Transition State   209

Reaction Coordinate Diagrams   209
The Rate-Determining Step   213
Composition of the Transition State   216
Position and Height of the Energy Barrier   220
Concerted and Stepwise Reactions   230

6. Phenomena for Study   **245**

6.1   Temperature Dependence of Rates   245

The Arrhenius Equation   245
Curved Arrhenius Plots   251
Standard States   253

Complications in Buffered Solutions   256
Uses of Activation Parameters   259

6.2   The Effect of Pressure   261

6.3   Catalysis   263

Definitions and Examples   263
Determination of Catalytic Rate Constants   268
Distinction between General Base and Nucleophilic Catalysis   271

6.4   pH Effects   273

Preliminaries   273
Curves without Inflection Points   274
Sigmoid Curves   277
Bell-Shaped Curves   285

6.5   Kinetic Isotope Effects   292

Primary Isotope Effects   293
Secondary Isotope Effects   298
Solvent Isotope Effects   300

7. Structure–Reactivity Relationships   **311**

7.1   Extrathermodynamic Relationships   311

7.2   Substituent Effects in Aromatic Compounds   315

The Hammett Equation   315
Deviations and Variations   320
The Substituent Constant   323
The Reaction Constant   328
The Ortho Effect   334
Conclusions   337

7.3   Substituent Effects in Aliphatic Compounds   338

Electronic Effects   338
Steric Effects   342

7.4   Substituent Effects in Reagents   344

General Acid-Base Catalysis   344
Nucleophilicity in Acyl Transfers   349
The α Effect   355
Nucleophilicity at Saturated Carbon   357
Intramolecular Reactivity   363

7.5   Related Topics   368

The Isokinetic Relationship   368
The Reactivity-Selectivity Principle   371

8. Medium Effects **385**

8.1 Introduction to Medium Effects 385

Types of Effects 385
Physical Theories 387
Empirical Correlations 388

8.2 Solvent Properties 389

Physical Properties 389
Intermolecular Forces 391
Chemical Interactions 394
Classification of Solvents 397
Solvent Polarity 399
Solvation 401

8.3 Physical Models of Medium Effects 405

Neutral-Neutral Molecule Reactions 405
Ion-Neutral Molecule Reactions 408
Ion-Ion Reactions 410
The Cavity Model 412
Separation of Initial and Transition State Solvent Effects 418

8.4 Empirical Measures and Correlations 425

Thermodynamic Measures 425
Kinetic Measures 427
Spectroscopic Measures 435
Linear Solvent Effect Relationships 442
Statistical Approaches 444

8.5 Strongly Acid Solutions 446

Acidity Functions 447
Mechanisms of Acid Catalysis 453

Appendices **463**

A. Answers to Selected Problems **463**

B. Physical Constants **467**

Index **469**

CHAPTER 1

Introduction to Chemical Kinetics

1.1 THE FIELD OF CHEMICAL KINETICS

Time as a Variable

Chemistry can be divided (somewhat arbitrarily) into the study of structures, equilibria, and rates. Chemical structure is ultimately described by the methods of quantum mechanics; equilibrium phenomena are studied by statistical mechanics and thermodynamics; and the study of rates constitutes the subject of kinetics. Kinetics can be subdivided into physical kinetics, dealing with physical phenomena such as diffusion and viscosity, and chemical kinetics, which deals with the rates of chemical reactions (including both covalent and noncovalent bond changes). Students of thermodynamics learn that quantities such as changes in enthalpy and entropy depend only upon the initial and final states of a system; consequently thermodynamics cannot yield any information about intervening states of the system. It is precisely these intermediate states that constitute the subject matter of chemical kinetics. A thorough study of any chemical reaction must therefore include structural, equilibrium, and kinetic investigations.

The properties of a system at equilibrium do not change with time, and time therefore is not a thermodynamic variable. An unconstrained system not in its equilibrium state spontaneously changes with time, so experimental and theoretical studies of these changes involve time as a variable. The presence of time as a factor in chemical kinetics adds both interest and difficulty to this branch of chemistry.

Although time as a physical or philosophical concept is an extremely subtle quantity, in chemical kinetics we adopt a fairly primitive notion of time as a linear fourth dimension (the first three being spatial dimensions) whose initial value ($t = 0$) can be set by the experimenter (for example, by mixing two reactant solutions) and whose extent is accurately measurable in standard units. The time dimension persists as a variable until the experimenter stops observing the reaction, or until "infinity time" ($t = \infty$), which is taken to mean the time at which the reaction is "essentially complete."

Uses of Chemical Kinetics

What happens in a chemical reaction during the period between the initial (reactant) state and the final (product) state? An answer to this question constitutes a description of the mechanism of the reaction. The study of reaction mechanisms is a major application of chemical kinetics, and most of this book is devoted to this application; an introduction is given in Section 1.2.

Kinetics can also be applied to the optimization of process conditions, as in organic syntheses, analytical reactions, and chemical manufacturing. This last example constitutes an important aspect of chemical engineering. Yet another practical use of chemical kinetics is for the determination and control of the stability of commercial products such as pharmaceutical dosage forms, foods, paints, and metals.

Some further uses of kinetics, less sweeping in their scope than the preceding applications, are for the testing of rate theories; the measurement of equilibrium constants; the analysis of solutions, including mixtures of solutes; and the measurement of solvent properties that depend upon rates. Some of these applications are treated later in the book.

Literature of the Field

The primary literature of chemical kinetics, like that of any other branch of science, is composed of research articles, which in this field are to be found in periodicals devoted to chemistry and related fields such as biochemistry, pharmacy, chemical engineering, and physics. As sources of instruction and reference, however, the secondary literature, consisting of textbooks, review articles, and monographs, is of more immediate value. This literature can be roughly divided into two groups: the members of one of these, well represented by References 1–16, are primarily directed to the physical chemist, being particularly strong in reaction rate theories, gas phase kinetics, and related fundamental matters. The second group is mainly concerned with kinetics as a means to study reaction mechanisms, and these sources are dominated by the outlook of physical organic chemistry; References 17–34 cite many of these books. (Of course this division into two groups is not always clear-cut.) There are, in addition, volumes in continuing series[35] and collections of kinetic data.[36]

The student should not reject a book without consideration simply because it was published one or two decades ago. If it was a good book when it was published, it remains a good book. Of course, the reader of any scientific work must be aware of its publication date so as to place it in the context of current knowledge. One of the advantages of earlier books, especially in quantitative or theoretical subjects, is that their authors tend to give more expansive and detailed treatments than do later writers, and beginners in a field may benefit from the fuller description.

1.2 THE STUDY OF REACTION MECHANISMS

The Definition of Reaction Mechanism

Consider the following hypothetical one-step reaction:

$$AB + C \rightarrow A + BC$$

The components on the left side we consider the initial state or reactant state; those on the right side are the final state or product state. There are many reactions of this general type, and it is a matter of experience that the reactants are not instantly transformed into the products; that is, the reactants display some relative measure of stability. This resistance to chemical reaction arises because energy is required to pass from the initial to the final state. We imagine that the reacting system follows a path defined by the energy of the system and that this path passes through a maximum somewhere between the initial and final states. This position of maximum energy is called the *transition state* of the reaction, and the difference in energy between the transition state and the initial state is the energy barrier to the reaction. Figure 1-1 illustrates this concept.

A one-step reaction has a single transition state; such a process is called an *elementary reaction*. Many observed ("overall") chemical reactions consist of two

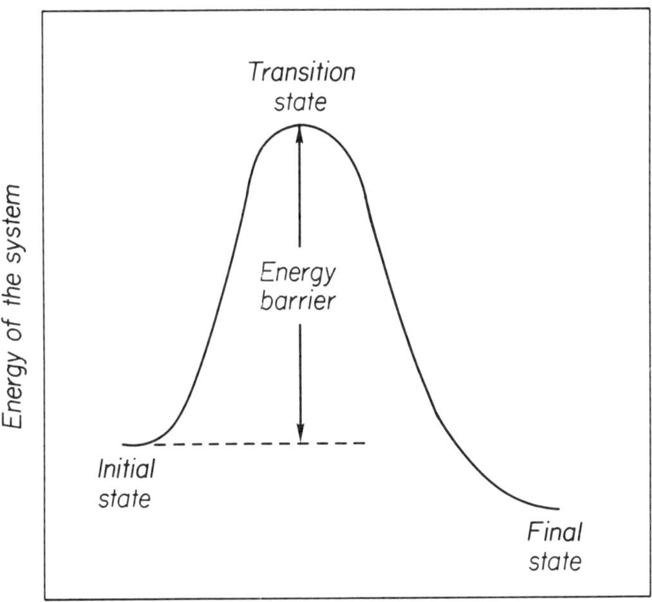

Figure 1-1. Schematic diagram for a hypothetical one-step reaction. (In this example the final state is of lower energy than the initial state, but this need not be so.)

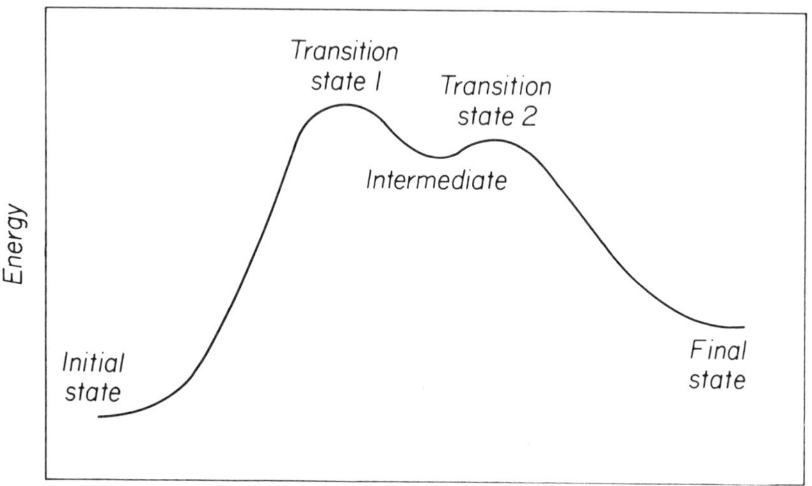

Figure 1-2. Schematic diagram of a complex reaction composed of two elementary reactions, showing an intermediate species.

or more elementary reactions; such a combination of elementary reactions is a *complex* (or *composite*) *reaction*. Figure 1-2 schematically diagrams a complex reaction composed of two elementary reactions. The species characterized by the minimum in this curve is an unstable intermediate; it is not a transition state because it does not occur at a maximum on the reaction path. Alternatively, the intermediate can be viewed as the product of the first elementary reaction and also as the reactant of the second elementary reaction. (The concepts of energy barrier, reaction path, and transition state will be sharpened and elaborated in Chapter 5.)

A postulated reaction mechanism is a description of all contributing elementary reactions (we will call this the *kinetic scheme*), as well as a description of structures (electronic and chemical) and stereochemistry of the transition state for each elementary reaction. (Note that it is common to mean by the term *transition state* both the region at the maximum in the energy path and the actual chemical species that exists at this point in the reaction.)

We illustrate with the example of ester hydrolysis in alkaline solution. The overall reaction is

$$RCOOR' + OH^- \rightarrow RCOO^- + R'OH$$

Much experimental study suggests that this reaction is complex. The essential steps are shown in the following kinetic scheme.

$$RCOOR' + OH^- \rightleftharpoons I$$

$$I \rightarrow RCOOH + R'O^-$$

A third reaction, a fast proton transfer, then generates the equilibrium concentrations of $RCOO^-$ and $R'OH$. In this kinetic scheme I represents an intermediate, whose structure is shown in this amplified description:

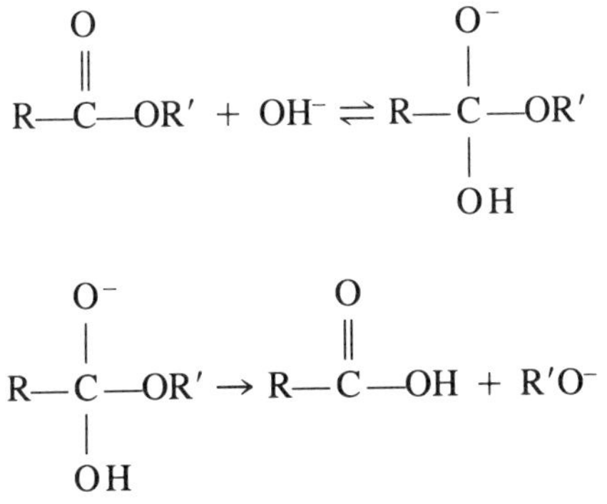

$$R-\underset{\overset{\parallel}{O}}{C}-OR' + OH^- \rightleftharpoons R-\underset{\underset{OH}{\mid}}{\overset{\overset{O^-}{\mid}}{C}}-OR'$$

$$R-\underset{\underset{OH}{\mid}}{\overset{\overset{O^-}{\mid}}{C}}-OR' \rightarrow R-\underset{\overset{\parallel}{O}}{C}-OH + R'O^-$$

This description provides information, via conventional structures, about the constitution of reactants, products, and the intermediate. Transition state structures are more provisional and may attempt to show the electronic distribution and flow in this region of the reaction path. The *curved arrow* symbolism is often used, as shown in structure **1** for the first elementary reaction.

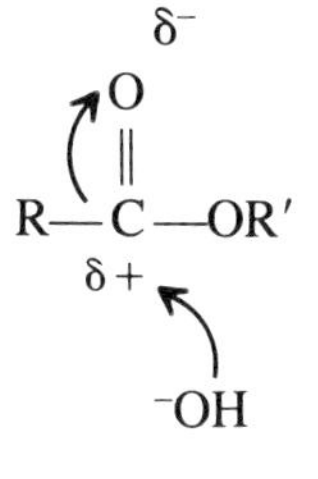

$$R-\underset{\delta+}{\overset{\delta-}{C}}-OR'$$

1

Another convention is to show bonds being formed and broken by dotted lines, as in representation **2** of the transition state for the second elementary reaction.

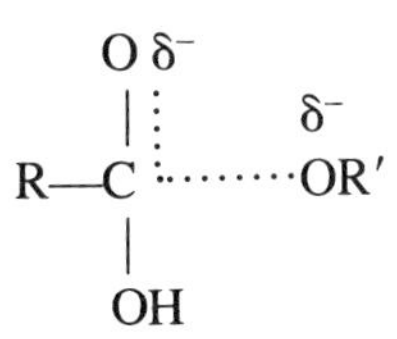

$$R-\underset{\overset{\mid}{OH}}{C}\cdots OR'$$

2

A rough idea of the electronic distribution is conveyed by the requirements that the partial charges must sum correctly and likewise for the partial bonds.

At a deeper level, the reaction mechanism requires a quantitative treatment of the energetics of the reaction path and a quantum mechanical description of the transition state structure.

Methods for Studying Mechanisms

Because the rate of a reaction is related to the height of the energy barrier on the reaction path, chemical kinetics is the preeminent experimental approach to the study of mechanism, but other kinds of information are also useful or even essential.

1. *Reactant and product structures.* Because the transition state structure is normally different from but intermediate to those of the initial and final states, it is evident that the structures of the reactants and products should be known. One should, however, be aware of a possible source of misinterpretation. Suppose the products generated in the reaction of kinetic interest undergo conversion, on a time scale fast relative to the experimental manipulations, to thermodynamically more stable substances; then the observed products will not be the actual products of the reaction. In this case the products are said to be under thermodynamic control rather than kinetic control. A possible example has been given in the earlier description of the reaction of hydroxide ion with ester, when it seems likely that the products are the carboxylic acid and the alkoxide ion, which, however, are transformed in accordance with the relative acidities of carboxylic acids and alcohols into the isolated products of carboxylate salt and alcohol.
2. *Stereochemical course of the reaction.* This kind of information was critical in the elucidation of the S_N1 and S_N2 pathways for nucleophilic substitution at saturated carbon.
3. *Presence and nature of intermediates.* Demonstration of an intermediate shows that a reaction is complex. An intermediate may be sufficiently stable to isolate, but more commonly it can only be detected by physical means (such as some form of spectroscopic observation) or by indirect chemical means ("trapping" it in a side reaction). Despite the instability of some intermediates, these are ordinary chemical species whose properties can, in principle at least, be determined experimentally.
4. *Isotopic substitution.* A classic example is the demonstration by Polanyi and Szabo (37) of acyl-oxygen fission in the alkaline hydrolysis of *n*-amyl acetate. An ester could undergo cleavage at two locations, as indicated in **3**.

$$\mathrm{R-C\!\!\underset{\mid}{\overset{\overset{\textstyle O}{\|}}{C}}\!\!-\!\!O\!\!-\!\!R'}$$

Ac　Al

3

The designation Ac signifies acyl-oxygen fission, whereas Al is alkyl-oxygen fission. When n-amyl alcohol was hydrolyzed in ^{18}O-enriched water, the ^{18}O appeared in the product acid rather than the alcohol, showing that alkyl-oxygen fission could not have occurred. Carpenter[32] gives many examples of isotopic studies.

The Role of Kinetics

The overall reaction stoichiometry having been established by conventional methods, the first task of chemical kinetics is essentially the qualitative one of establishing the kinetic scheme; in other words, the overall reaction is to be decomposed into its elementary reactions. This is not a trivial problem, nor is there a general solution to it. Much of Chapter 3 deals with this issue. At this point it is sufficient to note that evidence of the presence of an intermediate is often critical to an efficient solution. Modern analytical techniques have greatly assisted in the detection of reactive intermediates.[32] A nice example is provided by a study of the pyridine-catalyzed hydrolysis of acetic anhydride.[38] Other kinetic evidence supported the existence of an intermediate, presumably the acetylpyridinium ion, in this reaction, but it had not been detected directly. Fersht and Jencks[38] observed (on a time scale of tenths of a second) the rise and then fall in absorbance of a solution of acetic anhydride upon treatment with pyridine. This requires that the overall reaction be composed of at least two steps, and the accepted kinetic scheme is as follows.

$$P + Ac_2O \rightleftharpoons I^+ + OAc^-$$

$$I^+ + H_2O \rightarrow HOAc + PH^+$$

where P represents pyridine, and structure **4** shows the postulated intermediate, I^+.

$$CH_3\overset{\displaystyle O}{\overset{\displaystyle \|}{C}}-\overset{+}{N}\!\!\bigcirc$$

4

In the process of establishing the kinetic scheme, the rate studies determine the effects of several possible variables, which may include the temperature, pressure, reactant concentrations, ionic strength, solvent, and surface effects. This part of the kinetic investigation constitutes the phenomenological description of the system.

The next level seeks a molecular description, and kinetics again makes a contribution. As will be seen in Chapter 5, the experimental kinetics provides information on both the energetics of the reaction (i.e., the height of the energy barrier on the reaction path) and the atomic composition of the transition state. Any proposed mechanism must therefore be consistent with the kinetic evidence.

Given the initial and final states of an elementary reaction, and therefore a thermodynamic description of the system, there exist a priori an infinite number of paths (i.e., mechanisms) from the initial to the final state. The essential role of kinetics is to eliminate most of these possibilities and to suggest the most likely path. It is often stated that it is only possible to disprove a postulated mechanism; it is not possible to prove that a suggested mechanism is correct. The meaning of this statement is that if a proposed mechanism is inconsistent with the kinetic data, the mechanism *must* be incorrect; whereas if it is consistent with the data, it *may* be correct. Other mechanisms, as yet unimagined, may also be consistent with the data, or later experimental data may rule out the provisionally correct mechanism. We must accept this somewhat discouraging conclusion; let us view it as a challenge.

Classification of Mechanisms

The broadest classification of reactions is into the categories of heterolytic and homolytic reactions. In *homolytic* (free radical) reactions, bond cleavage occurs with one electron remaining with each atom, as in

$$R{:}X \xrightarrow{\text{homolytic}} R{\cdot} \; + \; {\cdot}X$$

Heterolytic (ionic) reactions occur with both electrons remaining with one of the atoms.

$$R{:}X \xrightarrow{\text{heterolytic}} R^{+} \; + \; {:}X^{-}$$

A further classification is into the classes of substitutions (symbolized S), additions (Ad), and eliminations (E).

A widely used system for specifying reaction type or mechanism was introduced by Ingold[18] and has subsequently been modified. In most reactions two reactants are present, and it often is convenient to refer to one of these as the *substrate* and the other as the *reagent*. The distinction is arbitrary and conventional, but it leads to a further classification in terms of reagent type. Reagents are *nucleophiles* (nucleus lovers) if they possess an unshared electron pair and seek electron-deficient sites. Nucleophiles are either bases (in the Bronsted sense) or reducing agents; in a very general interpretation these classes are practically synonymous, for all such reagents function by donating electrons. (However, it is conventional to reserve the term *basicity* to denote equilibrium affinity, usually toward the proton, whereas *nucleophilicity* refers to kinetic reactivity, often toward carbon.) *Electrophiles,* which may be Lewis acids or oxidizing agents, seek sites of high electron density. Reagents may also be classified in terms of their *hardness* or *softness*. A hard acid is small, with high positive charge density and low polarizability. A soft acid is large and polarizable. A hard base has high electronegativity and low polarizability, whereas a soft base is easily polarizable. The hard and soft acid–base (HSAB) concept has been especially useful in understanding substitution reactions of metal ion coordination complexes.[39,40]

Reaction types are classified by specifying the class and the reagent type; thus a nucleophilic substitution (S_N) is a substitution reaction by a nucleophilic reagent, as in Eq. (1-1).

$$I^- + C_2H_5Br \rightarrow C_2H_5I + Br^- \tag{1-1}$$

In this reaction I^- is the nucleophile, and Br^- is called the *leaving group* (or nucleofuge). Beyond this, the classification symbolism may include a designation of the *molecularity* of the reaction. Molecularity is the number of reactant molecules included in the transition state. The above reaction is an S_N2 reaction, because both reactants are present in the transition state. On the other hand, this substitution

$$OH^- + (CH_3)_3CCl \rightarrow (CH_3)_3COH + Cl^- \tag{1-2}$$

is an S_N1 reaction; it is a two-step reaction, and in the first step only the substrate, not the reagent, takes part.

Equation (1-3) is an aromatic electrophilic substitution (S_E), the nitronium ion being the electrophile and the proton the leaving group (electrofuge).

$$\text{benzene} + NO_2^+ \rightarrow \text{nitrobenzene} + H^+ \tag{1-3}$$

Equation (1-4) is an Ad_E reaction.

$$HCl + CH_2 = CH_2 \rightarrow CH_3CH_2Cl \tag{1-4}$$

[Although it is conceivable that the nucleophilic chloride ion initiates the attack, much experience supports the classification of (1-4) as an electrophilic reaction. Of course, if H^+ is the attacking electrophile, the double bond must be functioning as a nucleophile.] Equation (1-5) shows an Ad_N reaction.

$$CN^- + CH_3\overset{\displaystyle O}{\overset{\|}{C}}CH_3 \longrightarrow CH_3\underset{\underset{\displaystyle CN}{|}}{\overset{\overset{\displaystyle O^-}{|}}{C}}CH_3 \tag{1-5}$$

Eliminations often take the form of Eq. (1-6).

$$RCH_2CH_2X \rightarrow RCH = CH_2 + HX \tag{1-6}$$

These reactions are often promoted by a strong base, which assists the departure of the proton. X^- is the leaving group. Both E1 and E2 mechanisms are known, as is a variant designated E1cb, for "unimolecular elimination from the conjugate base of the substrate."

Discussion of acid and ester reaction mechanisms is often carried out in terms of the classification in Table 1-1.[18] This specifies the type of bond fission (Ac or

TABLE 1-1. Classification of Ester Formation and Hydrolysis

	Type of fission	
Conjugate form	Acyl	Alkyl
Acid	$A_{Ac}1$	$A_{Al}1$ (S_N1)
	$A_{Ac}2$	$A_{Al}2$ (unknown)
Base	$B_{Ac}1$ (unknown)	$B_{Al}1$ (S_N1)
	$B_{Ac}2$	$B_{Al}2$ (S_N2)

Al, as shown in structure **3**), the molecularity (1 or 2), and the ionic form of the substrate [A for conjugate acid $RC(OH)OR^+$ and B for conjugate base $RCOOR$]. Note that alkyl-oxygen fission constitutes nucleophilic substitution and is therefore equivalent to the S_N classification:

$$RCO-R + Y^- \longrightarrow RCOO^- + RY$$

$$Al$$

Guthrie[41] and Guthrie and Jencks[42] have proposed an alternative mechanistic symbolism that is capable of more detailed description than the Ingold system, although at the expense of greater complexity. This system may be useful for the computer representation of reaction mechanisms.

1.3 EXPERIMENTAL CHEMICAL KINETICS

The Rate of Reaction

Our interest is in solution kinetics, so we will concern ourselves only with *homogeneous* reactions, which take place in a single phase. (*Heterogeneous* reactions take place, at least in part, at interfaces between phases.) Further, we will mainly work with *closed* systems, those in which matter is neither gained nor lost during the period of observation.

Equation (1-7) is a general representation of any balanced chemical reaction, where A_i represents reactants and products, and v_i is negative for reactants and positive for products.

$$0 = \sum_i v_i A_i \tag{1-7}$$

Let n_i represent amount of substance i in moles. Then the *extent of reaction* ξ is defined by Eq. (1-8)

$$\xi = \frac{n_i - n_i^0}{\nu_i} \tag{1-8}$$

where n_i^0 is the amount of substance at a specified zero time. The *rate of reaction* is then $d\xi/dt$, or

$$\frac{d\xi}{dt} = \frac{1}{\nu_i}\frac{dn_i}{dt} \tag{1-9}$$

In solution kinetics we commonly work with systems at constant volume, and we find it convenient to employ molar concentration units. Dividing both sides of Eq. (1-9) by volume V gives

$$\frac{d(\xi/V)}{dt} = \frac{1}{\nu_i}\frac{dc_i}{dt} = v \tag{1-10}$$

where v is the *rate of reaction per unit volume*[14, p. 13] and c_i is the molar concentration of substance i. Usually v is itself referred to as the rate of reaction, reaction rate, or reaction velocity.

Of course, it is true that the derivative dc_i/dt is also a rate, but omission of the stoichiometric coefficient ν_i may lead to ambiguity. Consider this reaction:

$$A + 2B \rightarrow 3P$$

Applying Eq. (1-10) gives

$$v = -\frac{dc_A}{dt} = -\frac{1}{2}\frac{dc_B}{dt} = \frac{1}{3}\frac{dc_P}{dt}$$

Obviously $dc_B/dt = 2(dc_A/dt)$, because for each molecule of A that reacts, two molecules of B react, so it is necessary to include the stoichiometric coefficient or at least to specify clearly what definition of rate is being used.

Note that the sign convention of ν_i ensures that the reaction rate is a positive quantity.

Rate Equations and Stoichiometric Equations

The rate of reaction nearly always depends upon reactant concentrations and (for reversible reactions) product concentrations. The functional relationship between rate of reaction and system concentrations (usually at constant temperature, pressure, and other environmental conditions) is called the *rate equation*.

The interpretation of kinetic data is largely based on an empirical finding called the Law of Mass Action: In dilute solution the rate of an *elementary* reaction is proportional to the concentrations of the reactants, raised to the powers of their stoichiometric coefficients, and is independent of other concentrations and reactions.[43] Thus if the reaction

$$A + B \rightarrow Y + Z$$

is elementary, its rate equation, according to the law of mass action, is

$$v = kc_A c_B$$

Similarly, if the following reversible reaction is elementary,

$$A \rightleftharpoons B$$

its rate equation is

$$v = k_f c_A - k_r c_B$$

because the forward and reverse reactions independently must conform to the mass action requirement. Thus, for elementary reactions the rate equation can be inferred from the stoichiometric equation.

It is otherwise for complex reactions, for which the rate equation may or may not be simply related to the overall stoichiometric reaction. For example, the rate equation for the alkaline hydrolysis of ethyl acetate, which is a complex reaction (see Section 1.2),

$$CH_3COOC_2H_5 + OH^- \rightarrow CH_3COO^- + C_2H_5OH$$

has the rate equation

$$v = k[CH_3COOC_2H_5][OH^-]$$

(where brackets signify molar concentrations). On the other hand, the rate equation for the imidazole-catalyzed hydrolysis of ethyl acetate is

$$v = k[CH_3COOC_2H_5][\text{imidazole}]$$

although imidazole does not appear in the balanced stoichiometric equation, which is

$$CH_3COOC_2H_5 + H_2O \rightarrow CH_3COOH + C_2H_5OH$$

Another example is the reaction of aryldiazonium ions with nucleophiles in aqueous solution according to

$$Ar - N_2^+ + X^- \rightarrow Ar - X + N_2$$

for which the rate equation is

$$v = k[\text{Ar} - \text{N}_2{}^+]$$

We therefore conclude that for a complex reaction the rate equation cannot be inferred from the stoichiometric equation, but must be determined experimentally. Because we do not know a priori whether a reaction of interest is elementary or complex, we are required to establish the form of all rate equations experimentally.

Note that a rate equation is a differential equation.

Rate Constants and Reaction Orders

Suppose an experimentally determined rate equation has the form

$$v = kc_A^a c_B^b \ldots \tag{1-11}$$

The proportionality constant k is called the *rate constant* (or rate coefficient or specific rate). The rate constant is independent of the concentrations of A, B, . . . , but may depend upon environmental factors such as the temperature and solvent, and of course its magnitude depends on the particular reaction being studied.

The power a is called the *order of reaction* with respect to reactant A, b is the order with respect to B, and the sum $(a + b + \ldots)$ is the *overall order* of the reaction. Many rate equations are of forms different from Eq. (1-11)—for example, concentration terms may appear in the denominator—and then the concept of reaction order is not applicable.

The units of the rate constant depend upon the overall reaction order.

In Section 1.2 we distinguished between *elementary* and *complex reactions*. We now make a distinction between *simple* and *complicated rate equations*. A simple rate equation has the form of Eq. (1-11). A complicated rate equation has a form different from Eq. (1-11); it may be a sum of terms like that in (1-11), or it may have quantities in the denominator. We have seen that there is no necessary relationship between the complexity of the reaction and the form of the experimental rate equation. Simple rate equations are treated in Chapter 2 and complicated rate equations in Chapter 3.

It is important to realize that a rate and a rate constant are different quantities. However, for a simple rate equation, this interpretation can be given to the rate constant: k is the number of moles per liter reacting per unit time when all concentrations are one molar. This interpretation is the basis of the synonym *specific rate* for the rate constant.

Temperature Dependence of Rate Constants

The rates of most reactions are sensitive functions of temperature, and an understanding of the molecular basis for this dependence is an essential goal in theoretical investigations of kinetics. Experimentally it has been found that the rate constant of a reaction can usually be related to the temperature by Eq. (1-12), which is called the *Arrhenius equation*.

$$k = Ae^{-E_a/RT} \qquad (1\text{-}12)$$

In Eq. (1-12) R is the gas constant and T is the absolute temperature. The equation contains two parameters: A is called the *preexponential factor* and E_a is the *experimental activation energy*. A has the units of the rate constant. E_a is commonly expressed in kilocalories or kilojoules per mole, where 1 kJ $\equiv$ 4.184 kcal. Reactions in solution typically have E_a values in the range of 10–30 kcal/mol. These values correspond to the rough rule of thumb that reaction rates increase by a factor of 2 to 3 for each 10°C rise in temperature.

REFERENCES

1. Moelwyn-Hughes, E.A. "The Kinetics of Reactions in Solution", 2nd ed.; Oxford University Press: London, 1947.
2. Amis, E.S. "Kinetics of Chemical Change in Solution"; MacMillan: New York, 1949.
3. Benson, S.W. "The Foundations of Chemical Kinetics"; McGraw-Hill: New York, 1960.
4. Eyring, H.; Eyring E.M. "Modern Chemical Kinetics"; Reinhold: New York, 1963.
5. Boudart, M. "Kinetics of Chemical Processes"; Prentice-Hall: Englewood Cliffs, N.J., 1968.
6. Gardiner, W.C., Jr. "Rates and Mechanisms of Chemical Reactions"; W.A. Benjamin: New York, 1969.
7. Prettre, M.; Claudel, B. "Elements of Chemical Kinetics"; Gordon & Breach Science Publishers: London, 1970.
8. Weston, R.E., Jr.; Schwartz, H.A. "Chemical Kinetics"; Prentice-Hall; Englewood Cliffs, N.J., 1972.
9. Emanuel, N.M.; Knorre, D.G. "Chemical Kinetics: Homogeneous Reactions"; Wiley (Halsted Press): New York, 1973.
10. Skinner, G.B. "Introduction to Chemical Kinetics"; Academic Press, New York, 1974.
11. Hammes, G. G. "Principles of Chemical Kinetics"; Academic Press: New York, 1978.
12. Eyring, H.; Lin, S. H.; Lin, S.M. "Basic Chemical Kinetics"; Wiley-Interscience, New York, 1980.
13. Berry, R.S.; Rice, S.A.; Ross, J. "Physical and Chemical Kinetics", Part III of "Physical Chemistry"; Wiley: New York, 1980.
14. Wilkinson, F. "Chemical Kinetics and Reaction Mechanisms"; Van Nostrand Reinhold: New York, 1980.
15. Moore, J.W.; Pearson, R.G. "Kinetics and Mechanism", 3rd ed.; Wiley-Interscience, New York, 1981.
16. Laidler, K.J. "Chemical Kinetics", 3rd ed.; Harper & Row, New York, 1987.
17. Hammett, L.P. "Physical Organic Chemistry", 2nd ed.; McGraw-Hill: New York, 1970.
18. Ingold, C.K. "Structure and Mechanism in Organic Chemistry", 2nd ed.; Cornell University Press: Ithaca, N.Y., 1969.
19. Frost, A.A.; Pearson, R.G. "Kinetics and Mechanism", 2nd ed.; Wiley: New York, 1961.
20. Hine, J. "Physical Organic Chemistry", 2nd ed.; McGraw-Hill: New York, 1962.
21. Gould, E.S. "Mechanism and Structure in Organic Chemistry"; Holt, Rinehart, & Winston: New York, 1959.
22. Wiberg, K.B. "Physical Organic Chemistry"; Wiley: New York, 1964.
23. Kosower, E.M. "An Introduction to Physical Organic Chemistry"; Wiley: New York, 1968.
24. Jencks, W.P. "Catalysis in Chemistry and Enzymology"; McGraw-Hill: New York, 1969.
25. Bender, M.L. "Mechanisms of Homogeneous Catalysis from Protons to Proteins"; Wiley-Interscience: New York, 1971.

26. Connors, K.A. "Reaction Mechanisms in Organic Analytical Chemistry"; Wiley-Interscience: New York, 1973.
27. Lewis, E.S., Ed. "Techniques of Chemistry", Vol. VI, "Investigation of Rates and Mechanisms of Reactions", Part I, 3rd ed.; Wiley-Interscience: New York, 1974.
28. Ritchie, C.D. "Physical Organic Chemistry: The Fundamental Concepts"; Marcel Dekker: New York, 1975.
29. Lowry, T.H.; Richardson, K.S. "Mechanism and Theory in Organic Chemistry"; Harper & Row: New York, 1976.
30. Drenth, W.; Kwart, H. "Kinetics Applied to Organic Reactions"; Marcel Dekker: New York, 1980.
31. Zuman, P.; Patel, R. "Techniques in Organic Reaction Kinetics"; Wiley-Interscience: New York, 1984.
32. Carpenter, B.K. "Determination of Organic Reaction Mechanisms"; Wiley-Interscience: New York, 1984.
33. Sykes, P. "A Guidebook to Mechanism in Organic Chemistry", 6th ed.; Longman Scientific and Technical (Wiley): Essex, England, 1986.
34. Katakis, D.; Gordon, G. "Mechanisms of Inorganic Reactions"; Wiley-Interscience: New York, 1987.
35. Bamford, C.H.; Tipper, C.F.H., Eds. "Comprehensive Chemical Kinetics"; Elsevier: Amsterdam; Vol. 1., 1969, to Vol. 22, 1980.
36. "Tables of Chemical Kinetics (Homogeneous Reactions)". *Natl. Bur. Stand. (U.S.) Circ.* **1951–1961,** No. 510, Suppls 1, 2, and Monograph 34.
37. Polanyi, M.; Szabo, A.L., *Trans. Faraday Soc.* **1934,** *30,* 508.
38. Fersht, A.R.; Jencks, W.P. *J. Am. Chem. Soc.* **1969,** *91,* 2125.
39. Basolo, F.; Pearson, R.G. "Mechanisms of Inorganic Reactions", 2nd ed.; Wiley: New York, 1967.
40. Pearson, R.G. *J. Chem. Educ.* **1987,** *64,* 561.
41. Guthrie, R.D. *J. Org. Chem.* **1975,** *40,* 402.
42. Guthrie, R.D.; Jencks, W.P. *Acc. Chem. Res.* **1989,** *22,* 343.
43. Hammett, L.P. *J. Chem. Educ.* **1966,** *43,* 464.
44. Laidler, K.J. *Pure Appl. Chem.* **1981,** *53,* 753.

PROBLEMS

1. Classify these reagents as nucleophiles or electrophiles: H^+; OH^-; H_2O; OAc^- (acetate); CH_3NH_2; HPO_4^{2-}; $AlCl_3$; OOH^-.
2. Sketch the reaction progress diagram for the pyridine-catalyzed hydrolysis of acetic anhydride.
3. Postulate a transition state structure for the formation of the acetylpyridinium ion from pyridine and acetic anhydride.
4. Define the rate of this reaction (the oxidation of hydrogen peroxide by permanganate):

$$2KMnO_4 + 3H_2SO_4 + 5H_2O_2 \rightarrow 2MnSO_4 + 8H_2O + 5O_2 + K_2SO_4$$

5. Write the rate equation for Eq. (1-1), assuming that it is an elementary reaction.
6. What are the units of first-order, second-order, and third-order rate constants?
7. Calculate the ratio of rate constants at 30°C and 20°C for a reaction whose activation energy is 20 kcal/mol.
8. From the definition $v = (1/V)(d\xi/dt)$ find the rate of reaction in terms of molar concentration for the case in which the system volume V is not constant (Reference 44).

CHAPTER 2

Simple Rate Equations

In Chapter 1 we defined a rate equation of the form

$$v = kc_A^a c_B^b \ldots \tag{2-1}$$

as a simple differential rate equation. The rate equation of an elementary (single-step) reaction has the form of Eq. (2-1), but even some complex (multistep) reactions may possess simple rate equations. Chapter 2 treats the experimental study of reactions whose kinetics are described by Eq. (2-1).

2.1 INTEGRATED RATE EQUATIONS

Zero-Order Reactions

Improbable as a zero-order reaction may seem on the basis of what has been said thus far, let us consider the possibility of this rate equation:

$$-\frac{dc}{dt} = k \tag{2-2}$$

Integration between the limits of $c = c^0$ when $t = 0$ and $c = c$ when $t = t$ gives the integrated zero-order rate equation.

$$c = c^0 - kt \tag{2-3}$$

Thus, a zero-order reaction yields a linear plot of c vs. t, the slope being equal to $-k$. It is evident that a zero-order rate constant has the units of a rate, for example, moles per liter-second (M s^{-1}).

First-Order Reactions

Equation (2-4) is the stoichiometric equation for an elementary first-order reaction, and Eq. (2-5) is the corresponding differential rate equation.

$$A \xrightarrow{k} Z \tag{2-4}$$

$$-\frac{dc_A}{dt} = kc_A \tag{2-5}$$

Separating the variables and integrating between the limits shown below yields Eqs. (2-6), (2-7), and (2-8) as equivalent forms of the integrated first-order rate equation.

$$\int_{c_A^0}^{c_A} \frac{dc_A}{c_A} = -k \int_0^t dt$$

$$\ln \frac{c_A}{c_A^0} = -kt \tag{2-6}$$

$$\log \frac{c_A}{c_A^0} = -\frac{kt}{2.303} \tag{2-7}$$

$$c_A = c_A^0 e^{-kt} \tag{2-8}$$

For a first-order reaction, therefore, a plot of $\ln c_A$ (or $\log c_A$) vs. t is linear, and the first-order rate constant can be obtained from the slope. A first-order rate constant has the dimension time^{-1}, the usual unit being second^{-1}.

The *half-life* $t_{1/2}$ is defined to be the time required for the reactant concentration to decay to one-half its initial value. To find $t_{1/2}$ for a first-order reaction we use Eq. (2-6) with the substitutions $c_A = c_A^0/2$ and $t = t_{1/2}$, finding

$$t_{1/2} = \frac{\ln 2}{k} = \frac{0.693}{k} \tag{2-9}$$

Note that $t_{1/2}$ is independent of concentration for a first-order reaction.

(Other fractional lives could similarly be defined. For example, the *life-time* τ is the time required for the concentration to decay to $1/e$ its initial value; then we find $\tau = 1/k$. The lifetime is the average time elapsed before a molecule reacts.[1] In pharmaceutics, a *shelf-life* t_{90} is defined to be the time required for c_A to reach the value $0.90\, c_A^0$, giving $t_{90} = 0.105/k$.[2])

Let us define $n = t/t_{1/2}$, so that n is the number of half-lives elapsed. Combining this definition with Eqs. (2-8) and (2-9) gives (2-10), which is a generalized form of the first-order decay curve.

$$\frac{c_A}{c_A^0} = e^{-0.693n} \tag{2-10}$$

Figure 2-1 is a plot of Eq. (2-10) from $n = 0$ to $n = 4$. Note that equal time increments result in equal fractional decreases in reactant concentration; thus in the first half-life c_A/c_A^0 decreases from 1.0 to 0.50; in the second half-life it decreases from 0.50 to 0.25; in the third half-life, from 0.25 to 0.125; and so on. This behavior is implicit in the earlier observation that a first-order half-life is independent of concentration.

For reaction (2-4) we could also write the rate equation in terms of the product concentration as

$$\frac{dc_Z}{dt} = kc_A$$

In view of the mass balance relationship $c_A^0 = c_A + c_Z$, this leads to Eq. (2-11), which can also be obtained from Eq. (2-6).

$$\ln\left(\frac{c_A^0 - c_Z}{c_A^0}\right) = -kt \tag{2-11}$$

Some authors define a *reaction variable x* as the decrease in reactant concentration, which for reaction (2-4) is equal to the increase in product concentration. Equation (2-11) is then often written as Eq. (2-12), where $a \equiv c_A^0$.

$$\ln\left(\frac{a - x}{a}\right) = -kt \tag{2-12}$$

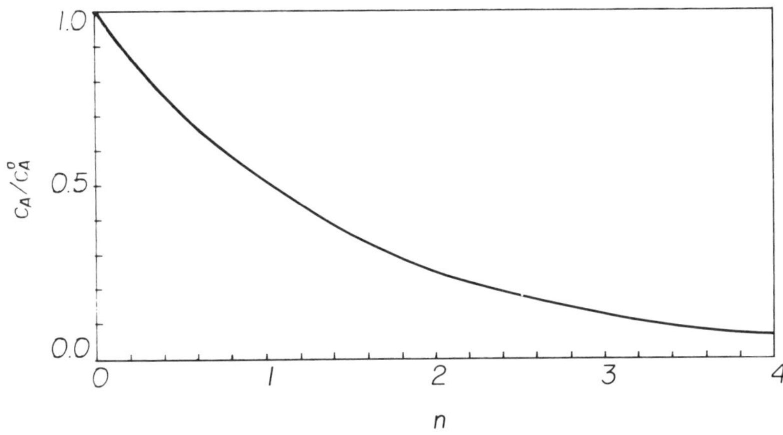

Figure 2-1.   Plot of Eq. (2-10), where c_A/c_A^0 is the fraction of initial reactant concentration remaining after the lapse of n half-lives.

Second-Order Reactions

Special Case

For elementary reaction (2-13),

$$2A \overset{k}{\to} \text{product(s)} \tag{2-13}$$

the differential rate equation is

$$-\frac{dc_A}{dt} = kc_A^2 \tag{2-14}$$

(The stoichiometric coefficient 2 is commonly omitted from the rate expression because there is only one reactant species, so no ambiguity exists.) Separating the variables and integrating between the usual limits gives Eq. (2-15).

$$\frac{1}{c_A} = \frac{1}{c_A^0} + kt \tag{2-15}$$

Therefore, for this type of second-order reaction, a plot of $1/c_A$ vs. t is linear, with the slope equal to k. The usual units of a second-order rate constant are liters per mole-second ($M^{-1}s^{-1}$).

Equation (2-15) also is applicable to the reaction $A + B \to$ products in the special circumstance that $c_A^0 = c_B^0$, which may often be arranged by the experimenter.

Equation (2-15), which emerges from integration of Eq. (2-14), can be rearranged to the functional form $c_A = f(t)$ as in Eq. (2-16).

$$c_A = \frac{c_A^0}{1 + kc_A^0 t} \tag{2-16}$$

Equation (2-15) is one linearized form of Eq. (2-16); Eqs. (2-17) and (2-18) are alternative linear plotting forms.

$$c_A t = -\frac{c_A}{kc_A^0} + \frac{1}{k} \tag{2-17}$$

$$\frac{1}{c_A t} = \frac{1}{c_A^0 t} + k \tag{2-18}$$

Usually the experimental error in the measurement of time is negligible relative to the error in concentration, and the choice of plotting form may be determined by this factor. Another consideration is the extent of reaction that has been followed experimentally. These matters have been treated at length for analogous plotting functions in equilibrium binding studies.[3] Benson[4] recommends Eq. (2-17) (which corresponds to the Scatchard plot).

Dole[5] has rearranged Eq. (2-15) into Eq. (2-19), which is yet another linear plotting form.

$$\frac{t}{c_A^0 - c_A} = \frac{t}{c_A^0} + \frac{1}{c_0^2 k} \tag{2-19}$$

Certain free radical polymerization data gave curves when plotted according to Eq. (2-15) but straight lines according to Eq. (2-19). This apparent paradox was resolved by postulating that some constant portion R of reactant is unreactive and serves to diminish the effective reactant concentration, lowering it to $c_A - R$. The appropriate form of Eq. (2-15) is then

$$\frac{1}{c_A - R} = \frac{1}{c_A^0 - R} + kt$$

which leads to Eq. (2-20) as the form analogous to (2-19).

$$\frac{t}{c_A^0 - c_A} = \frac{t}{c_A^0 - R} + \frac{1}{(c^0 - R)^2 k} \tag{2-20}$$

General Case

Let Eq. (2-21) be an elementary reaction.

$$A + B \xrightarrow{k} \text{product(s)} \tag{2-21}$$

The differential rate equation will be

$$-\frac{dc_A}{dt} = -\frac{dc_B}{dt} = kc_A c_B \tag{2-22}$$

For symbolic convenience we make use of the reaction variable x, which is the decrease in concentration of reactant A in time t. Because of the reaction stoichiometry, x is also the decrease in B concentration. The mass balance expressions are

$$c_A^0 = c_A + x$$

$$c_B^0 = c_B + x$$

so $dx/dt = -dc_A/dt = -dc_B/dt$. Letting $a \equiv c_A^0$ and $b = c_B^0$, Eq. (2-22) becomes

$$\frac{dx}{dt} = k(a - x)(b - x) \tag{2-23}$$

Separating variables gives

$$\int_0^x \frac{dx}{(a - x)(b - x)} = k \int_0^t dt \tag{2-24}$$

This is a standard integral form; the result is Eq. (2-25), for $a \neq b$.

$$\frac{1}{a - b} \ln \frac{b(a - x)}{a(b - x)} = kt \tag{2-25}$$

Equation (2-25) can also be written

$$\ln \frac{(a - x)}{(b - x)} = (a - b)kt + \ln \frac{a}{b} \tag{2-26}$$

showing that a plot of the left side against t should be a straight line for a second-order reaction, when $a \neq b$. If the analytical method yields reactant concentrations rather than the quantity x, the equivalent form is

$$\ln \frac{c_A}{c_B} = \left(c_A^0 - c_B^0 \right)kt + \ln \frac{c_A^0}{c_B^0} \tag{2-27}$$

If $c_A^0 = c_B^0$, the treatment leading to Eq. (2-15) is applicable.

Autocatalysis

Consider this stoichiometric equation.

$$A \rightarrow Z + \ldots$$

having the second-order rate equation

$$-\frac{dc_A}{dt} = kc_A c_Z \tag{2-28}$$

This is an *autocatalytic* reaction, in which a product of the reaction appears in the rate equation for the forward reaction. In this case the mass balance expressions are

$$c_A^0 = c_A + x$$

$$c_Z = c_Z^0 + x$$

so that

$$c_A + c_Z = c_A^0 + c_Z^0 \tag{2-29}$$

Carrying through the integration as in the preceding second-order example gives Eq. (2-30).

$$\ln \frac{c_Z}{c_A} = \left(c_A^0 + c_Z^0 \right) kt + \ln \frac{c_Z^0}{c_A^0} \tag{2-30}$$

An explicit solution for c_Z is found by combining Eqs. (2-29) and (2-30):

$$c_Z = \frac{\left(c_A^0 + c_Z^0 \right)}{1 + \left(c_A^0/c_Z^0 \right) e^{-\left(c_A^0 + c_Z^0 \right) kt}} \tag{2-31}$$

The concentration c_Z is a sigmoid function ("growth curve") of time.[6] Presumably some of product Z must be present at $t = 0$ in order to initiate the reaction.

Pseudo-Order Reactions

Suppose a reaction has this second-order rate equation:

$$v = kc_A c_B \tag{2-32}$$

Now further suppose that c_B is (by means discussed below) held essentially constant throughout the period of experimental observation; then the rate equation can be written as

$$v = k_{obs} c_A \tag{2-33}$$

where $k_{obs} = kc_B$. We say that the second-order reaction has been transformed into a *pseudo-first-order* reaction. The rate constant k_{obs} is a pseudo-first-order rate constant, sometimes symbolized k_{app} (for apparent first-order constant), or k_ψ. If all reactant concentrations are held essentially constant, then a pseudo-zero-order reaction is generated.

This ability to reduce the reaction order by maintaining one or more concentrations constant is a very valuable experimental tool, for it often permits the simplification of the reaction kinetics. It may even allow a complicated rate equation to be transformed into a simple rate equation.

There are several ways to achieve the essential constancy of reactant concentration. Taking Eq. (2-32) as an example, we might set $c_B^0 >>> c_A^0$; then as the concentration of reactant A goes from $c_A = c_A^0$ to $c_A = 0$, the concentration c_B remains essentially constant at c_B^0. For example, if $c_B^0 = 100\ c_A^0$, c_B will decrease only 0.5% by the end of the first half-life of the reaction and only 1% at the completion of the reaction. With ordinary analytical methods, these changes are not significant.

A second way to achieve constancy of a reactant is to make use of a buffer system. If the reaction medium is water and B is either the hydronium ion or the hydroxide ion, use of a pH buffer can hold c_B reasonably constant, provided the buffer capacity is high enough to cope with acids or bases generated in the reaction. The constancy of the pH required depends upon the sensitivity of the analytical method, the extent of reaction followed, and the accuracy desired in the rate constant determination.

A third method, or phenomenon, capable of generating a pseudo reaction order is exemplified by a first-order solution reaction of a substance in the presence of its solid phase. Then if the dissolution rate of the solid is greater than the reaction rate of the dissolved solute, the solute concentration is maintained constant by the solubility equilibrium and the first-order reaction becomes a pseudo-zero-order reaction.

If one of the reactants is the solvent, this reactant is present in large excess, so its kinetic participation will not be observed. Thus a bimolecular hydrolysis reaction commonly follows first-order kinetics. This example shows that the reaction order may not be equal to the reaction molecularity.

Jencks[7] has emphasized a danger in the technique of reducing the reaction order by using an excess concentration of one reactant. If this reactant contains an impurity that itself is very reactive, the impurity concentration may be sufficiently high to lead to spurious results from the unsuspected reaction.

2.2 DETERMINATION OF REACTION ORDER

Use of Integrated Equations

Because the integrated equations for the several integral orders of reaction display different functional dependences of concentration on time, plots of concentration–time data will usually yield the reaction order if a sufficient extent of reaction has been studied. Figures 2-2, 2-3, and 2-4 show plots of simulated data according to Eq. (2-3) (zero-order), Eq. (2-6) (first-order), and Eq. (2-15) (second-order). It is obvious that this reaction conforms to the second-order rate equation. As suggested in the first sentence of this paragraph, a sufficient extent of reaction must be studied to detect curvature if it exists. Over the first few percentage of reaction, plots according to any of these equations may appear to be substantially linear.

An alternative to a graphical display is to calculate the rate constant for each data point using the integrated rate equation, seeking constancy in the calculated k values.

A reaction order determined by plotting the integrated rate equation is sometimes called the *order with respect to time*;[8] this order has an unambiguous meaning only if the order is independent of time, which means that the plotted function is linear

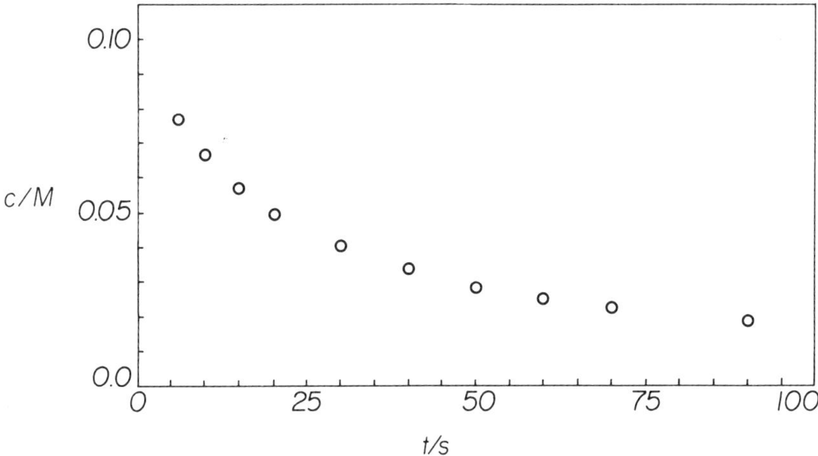

Figure 2-2. Plot of Eq. (2-3) with simulated kinetic data. Evidently the reaction is not zero-order.

over the entire time of reaction. It is very common to observe that a plot according to one of these integrated rate equations is linear for much of the reaction, but shows a tendency to curvature after the first two or three half-lives. One possible cause of such a deviation is a change in the environmental conditions, such as a shift in pH as the reaction proceeds. Another possibility is that a second reaction is taking place; this may be a consecutive reaction in which a product of the first reaction undergoes further change. Whether or not such phenomena are regarded as important and worth further investigation depends upon the particular system and the interests of the investigator. However, in reporting the results of kinetic studies the occurrence and extent of such deviations should be specified.

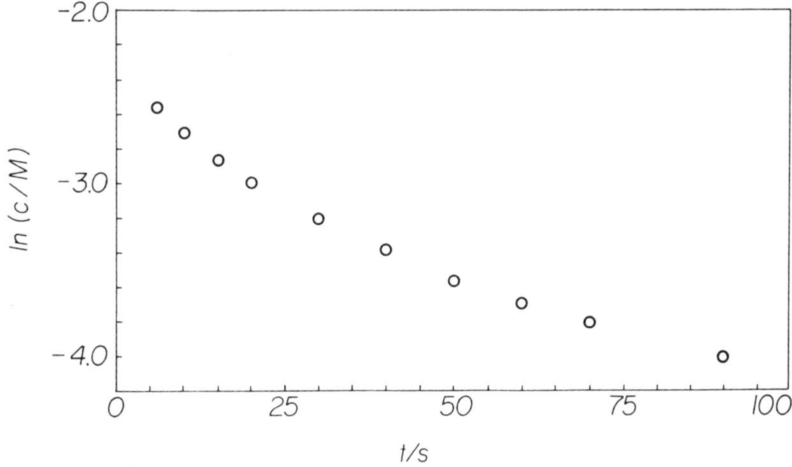

Figure 2-3. Plot of Eq. (2-6) with the same data shown in Fig. 2-2. The reaction is not first-order.

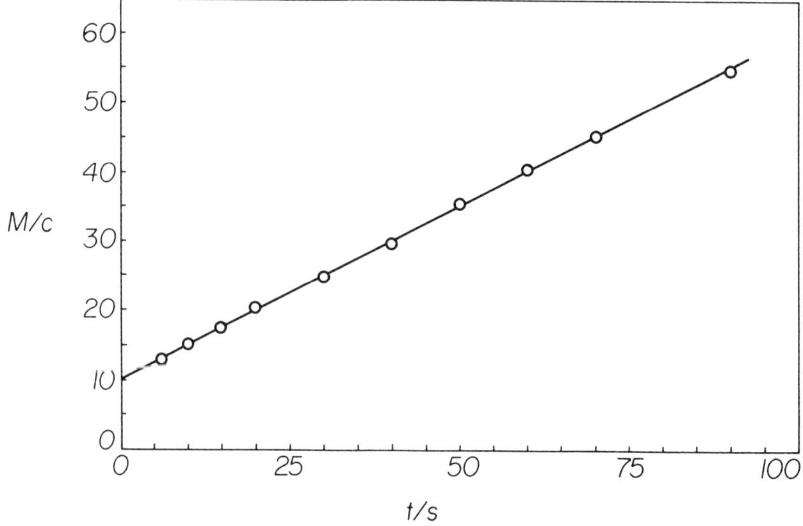

Figure 2-4. Plot of Eq. (2-15) with the data used in Figs. 2-2 and 2-3. The reaction is second-order.

The Isolation and Pseudo-Order Techniques

The isolation experimental design can be illustrated with the rate equation $v = kc_A^a c_B^b$, for which we wish to determine the reaction orders a and b. We can set $c_B^0 >>> c_A^0$, thus establishing pseudo-ath-order kinetics, and determine a, for example, by use of the integrated rate equations, experimentally following c_A as a function of time. By this technique we "isolate" reactant A for study. Having determined a, we may reverse the system and isolate B by setting $c_A^0 >>> c_B^0$ and thus determine b.

More commonly a combination of techniques is used as illustrated by the following study. The reaction is the acylation of allylamine by *trans*-cinnamic anhydride, Eq. (2-34).

$$(C_6H_5CH{=}CHCO)_2O + CH_2{=}CHCH_2NH_2 \xrightarrow{k}$$

$$(2\text{-}34)$$

$$C_6H_5CH{=}CHCONHCH_2CH{=}CH_2 + C_6H_5CH{=}CHCOOH$$

The initial anhydride concentration was about 3×10^{-6} M, and the amine concentration was much larger than this. The reaction was followed spectrophotometrically, and good first-order kinetics were observed; hence, the reaction is first-order with respect to cinnamic anhydride. It was not convenient analytically to use the isolation technique to determine the order with respect to allylamine, because it is easier to observe the cinnamoyl group spectrophotometrically than to follow the loss of amine. Therefore, the preceding experiment was repeated at several amine concentrations, and from the first-order plots the pseudo-first-order rate constants were determined. These data are shown in Table 2-1. Letting A represent

TABLE 2-1. Determination of Reaction Order from Pseudo-First-Order Rate Constants[a]

$10^4[Allylamine]/M$	$10^3\,k_{obs}/s^{-1}$	$k/M^{-1}\,s^{-1}$
1.122	3.75	33.4
2.244	7.56	33.7
4.488	15.3	34.2

[a]For reaction (2-34) at 25.0°C in acetonitrile.

the anhydride and B the amine, the rate equation is now $-dc_A dt = k_{obs}c_A$, where we anticipate $k_{obs} = kc_B^b$. The third column of Table 2-1 lists the quotient k_{obs}/c_B^0, whose constancy shows that $b = 1$.

More complicated behavior is shown by the acylation of n-butylamine with N-$trans$-cinnamoylimidazole, **1.**

$$C_6H_5CH{=}CH{-}\overset{\displaystyle O}{\overset{\displaystyle \|}{C}}{-}N\diagdown\diagup N$$

1

The isolation technique showed that the reaction is first-order with respect to cinnamoylimidazole, but treatment of the pseudo-first-order rate constants revealed that the reaction is not first-order in amine, because the ratio k_{obs}/c_B^0 is not constant, as shown in Table 2-2. The last column in Table 2-2 indicates that a reasonable constant is obtained by dividing k_{obs} by the square of the amine concentration; hence the reaction is second-order in amine. For the system described in Table 2-2, we therefore find that the reaction is overall third-order, with the rate equation

$$v = k[\text{cinnamoylimidazole}][n\text{-butylamine}]^2$$

and rate constant $k = 0.0635\ M^{-2}s^{-1}$.

Data of this type can also be treated graphically. If a pseudo-first-order rate

TABLE 2-2. Determination of Reaction Order: Reaction of n-Butylamine with **1**[a]

$[Amine]/M$	$10^3 k_{obs}/s^{-1}$	$10^3 k_{obs}/[amine]$	$10^2 k_{obs}/[amine]^2$
0.1099	0.768	6.99	6.36
0.1507	1.44	9.56	6.34
0.1952	2.37	12.1	6.20
0.2340	3.56	15.2	6.50
0.2751	4.88	17.7	6.43
0.3277	6.74	20.6	6.29

[a]At 25.0°C in acetonitrile.

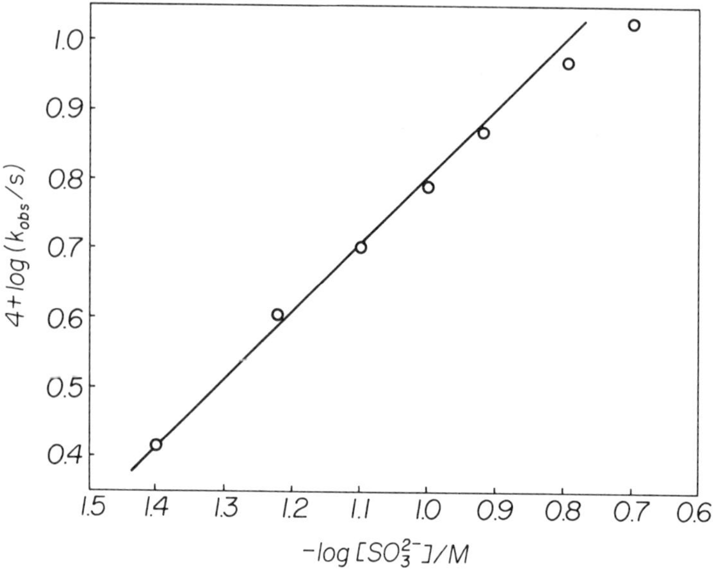

Figure 2-5. Log–log plot according to Eq. (2-36) for the reaction of sulfite with cinnamoylsalicylic acid. The slope of the line is 1.00.

constant is to be tested for the functional form $k_{obs} = kc_B^b$, plots according to Eq. (2-35) or (2-36) can be made.

$$\frac{k_{obs}}{c_B} = kc_B^{b-1} \tag{2-35}$$

$$\log k_{obs} = \log k + b \log c_B \tag{2-36}$$

A plot according to Eq. (2-35) is useful in confirming the absence of another rate term, because the line should pass through the origin. The log–log plot of Eq. (2-36) is valuable because the slope of the line is the order of reaction, for any order. Figure 2-5 is a plot of Eq. (2-36) for the addition of sulfite to the double bond of cinnamoylsalicylic acid. At low concentrations of sulfite the reaction appears to be first-order in sulfite, but at higher sulfite concentrations some deviation is observed.

A reaction order determined by the methods shown in Tables 2-1 and 2-2 and Fig. 2-5 is called an *order with respect to concentration.*[8]

Initial Rate Method

Let the rate equation be $v = kc_A^a$. We initiate the reaction and measure c_A as a function of time during the first few percentage of reaction. The slope of the plot of c_A vs. t at $t = 0$ is the *initial rate* of the reaction, and the rate equation at $t = 0$ is written

$$-\left(\frac{dc_A}{dt}\right)_{t=0} = v_0 = k\left(c_A^0\right)^a$$

which is most usefully expressed in the log–log form:

$$\log v_0 = a \log c_A^0 + \log k \qquad (2\text{-}37)$$

Thus by measuring v_0 at several values of c_A^0, the log–log plot yields the reaction order. This method also works if the rate equation includes the concentrations of other reactants, provided their concentrations are held constant throughout the series of measurements.

Casado et al.[9] have analyzed the error of estimating the initial rate from a tangent to the concentration–time curve at $t = 0$ and conclude that the error is unimportant if the extent of reaction is less than 5%. Chandler et al.[10] fit the kinetic data to a polynomial in time to obtain initial rate estimates.

One advantage of the initial rate method is that it avoids any complications arising from product inhibition or catalysis or from subsequent reactions. Another advantage is that it is applicable to very slow reactions whose study by other methods might be impractical.

The initial rate method yields the reaction order with respect to concentration.

Fractional Time Methods

The functional dependence of the half-life on reactant concentration varies with the reactant order. From the integrated rate equations we obtain these results:

$$\text{Zero-order} \qquad t_{1/2} = \frac{c^0}{2k} \qquad (2\text{-}38)$$

$$\text{First-order} \qquad t_{1/2} = \frac{0.693}{k} \qquad (2\text{-}39)$$

$$\text{Second-order (special case)} \quad t_{1/2} = \frac{1}{c^0 k} \qquad (2\text{-}40)$$

These relationships offer a means for the determination of order by studying the dependence of $t_{1/2}$ on initial concentration. Although this method may not seem to offer an advantage over other procedures, it can provide additional evidence that may be valuable. For example, the observation of a linear first-order plot according to Eq. (2-6) is evidence for a first-order plot. If now the initial concentration is changed substantially and the resulting half-life is unchanged, the conclusion is very sound that the reaction is first-order. If the half-life depends upon concentration, on the other hand, the reaction cannot be first-order. Note that the half-life for any reaction can be obtained from the concentration–time data without recourse to integrated rate equations.

Wilkinson[11] has generalized the fractional time method in the following way. For rate equation $dc/dt = -kc^n$, the integrated equation, for $n \neq 1$, is

$$\frac{1}{c^{n-1}} = \frac{1}{c_0^{n-1}} + (n - 1)kt \tag{2-40}$$

Define the fraction reacted, p, by $p = 1 - c/c_0$. Combination with Eq. (2-40) gives

$$(1 - p)^{1-n} = 1 + (n - 1)kc_0^{n-1}t$$

The left side is expanded in a binomial series, which is truncated after the quadratic term. Combination leads to

$$p = kc_0^{n-1}t - \frac{np^2}{2}$$

At low p, therefore, $p = kc_0^{n-1}t$. This substitution is used to give

$$p = kc_0^{n-1}t - \frac{n}{2} pkc_0^{n-1}t$$

which is rearranged to Eq. (2-41).

$$\frac{t}{p} = \frac{nt}{2} + \frac{1}{kc_0^{n-1}} \tag{2-41}$$

By a different route the same equation can be obtained for $n = 1$. A plot of t/p vs. t leads to an estimate of the order n from the slope.

Because Eq. (2-41) was derived by means of approximations, its use may lead to approximate estimates of order, but its advantage is that it makes use of data taken at many time points. Figure 2-6 is a plot according to Eq. (2-41) with the simulated data used to construct Fig. 2-4. The slope of the line is 1.0, so the order is 2.0, in agreement with Fig. 2-4. [Experience with Eq. (2-41) applied to experimental data for first-order reactions indicates that Eq. (2-41) slightly overestimates the order for these reactions, leading to values $n = 1.2$–1.3.]

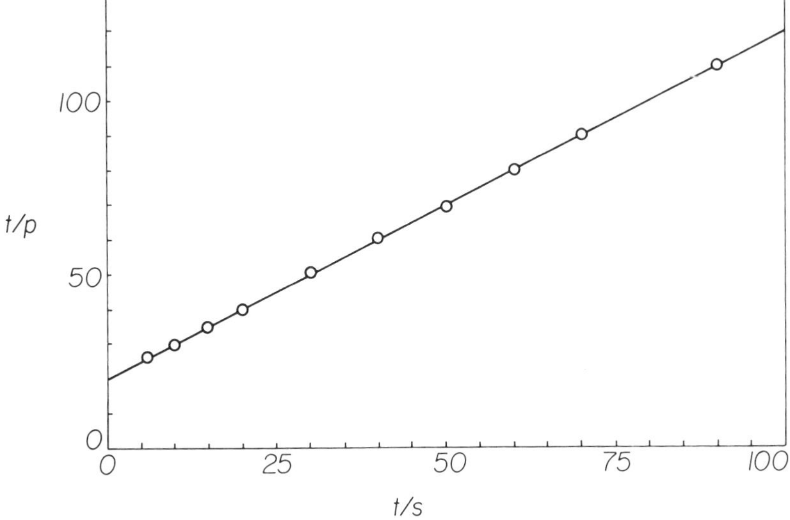

Figure 2-6. Plot according to Eq. (2-41) with the simulated data of Figs. 2-2, 2-3, and 2-4.

2.3 METHODS OF DATA ANALYSIS

Calculation of the Rate Constant from Concentrations

A reading of Section 2.2 shows that all of the methods for determining reaction order can lead also to estimates of the rate constant, and very commonly the order and rate constant are determined concurrently. However, the integrated rate equations are the most widely used means for rate constant determination. These equations can be solved analytically, graphically, or by least-squares regression analysis.

Consider the first-order equation, Eq. (2-6). Writing this for concentrations c_1 and c_2 at times t_1 and t_2 and subtracting gives Eq. (2-42).

$$k = \frac{1}{t_2 - t_1} \ln \frac{c_1}{c_2} \tag{2-42}$$

With Eq. (2-42) the first-order rate constant can be calculated from concentrations at any two times. Of course, usually concentrations are measured at many times during the course of a reaction, and then one has choices in the way the estimates will be calculated. One possibility is to let t_1 be zero time for all calculations; in this case the same value c^0 is employed in each calculation, so error in this quantity is transmitted to each rate constant estimate. Another possibility is to apply Eq. (2-42) to successive time intervals. If, as often happens, the time intervals are all equal, an inefficient mean estimate will result, as shown by Roseveare.[12] Let Δt be the constant time interval; then the mean of n estimates will be

$$\bar{k} = \frac{1}{n}\left(\frac{1}{\Delta t}\ln\frac{c_0}{c_1} + \frac{1}{\Delta_t}\ln\frac{c_1}{c_2} + \ldots + \frac{1}{\Delta t}\ln\frac{c_{n-1}}{c_n}\right)$$

$$\bar{k} = \frac{1}{t_n}\ln\frac{c_0}{c_n}$$

since $n\Delta t = t_n$. Thus the mean value is equal to the rate constant calculated from the first and last data points, all intervening data having been rejected in the averaging process. Roseveare calculated that for a first-order reaction, with negligible error in the time measurement, the best accuracy in calculating k from concentrations at two times is achieved when the time interval constitutes 67% of the total extent of reaction. Usually first-order kinetic data are treated graphically or by linear regression, as discussed subsequently.

Titrimetric analysis is a classical method for generating concentration–time data, especially in second-order reactions. We illustrate with data on the acetylation of isopropanol (reactant B) by acetic anhydride (reactant A), catalyzed by N-methylimidazole. The kinetics were followed by hydrolyzing 5.0-ml samples at known times and titrating with standard base. A blank is carried out with the reagents but no alcohol. The reaction is

$$Ac_2O + (CH_3)CHOH \xrightarrow{\ \ k[NMIM]\ \ } CH_3COOCH(CH_3)_2 + HOAc$$

where $k[NMIM] = k_{obs}$ is the pseudo-second-order rate constant.

In the blank titration, the acetic anhydride is hydrolyzed to give 2 mol of acetic acid per mole of Ac_2O. In a sample titration, each unreacted anhydride molecule likewise yields two of acetic acid, but each reacted Ac_2O molecule yields only one

TABLE 2-3. Kinetics of Acetylation of Isopropyl Alcohol by Acetic Anhydride Catalyzed by N-Methylimidazole[a,b]

t/min	V_S/ml	x/M	c_A/M	c_B/M	k_{obs}/M^{-1} min^{-1}
2	18.70	0.062	0.878	0.308	0.102
6	17.60	0.180	0.760	0.190	0.133
10	17.10	0.234	0.706	0.136	0.125
14	16.70	0.276	0.664	0.094	0.128
18	16.45	0.302	0.638	0.068	0.127

[a] $c_A^0 = 0.94$ M, $c_B^0 = 0.37$ M, $N = 0.5328$, $V_b = 19.30$ ml.
[b] At 45°C in dimethylformamide; [NMIM] $= 0.30$ M; 5.0-ml samples drawn.

of acetic acid. Thus $N(V_b - V_s)$ is the millimoles of alcohol or of acetic anhydride reacted, where N is the normality of the titrant, and

$$x = \frac{N(V_b - V_s)}{\text{sample volume}}$$

is the concentration reacted, x being the reaction variable appearing in Eq. (2-25). The reactant concentrations are then given by $c_A = c_A^0 - x$ and $c_B = c_B^0 - x$. Equation (2-27) is rearranged to

$$k_{obs} = \frac{1}{\left(c_A^0 - c_B^0\right)t} \left(\ln \frac{c_A}{c_B} - \ln \frac{c_A^0}{c_B^0} \right) \tag{2-43}$$

Table 2-3 lists data for a typical kinetic run, with k_{obs} calculated with Eq. (2-43), and Fig. 2-7 shows the plot according to Eq. (2-27).

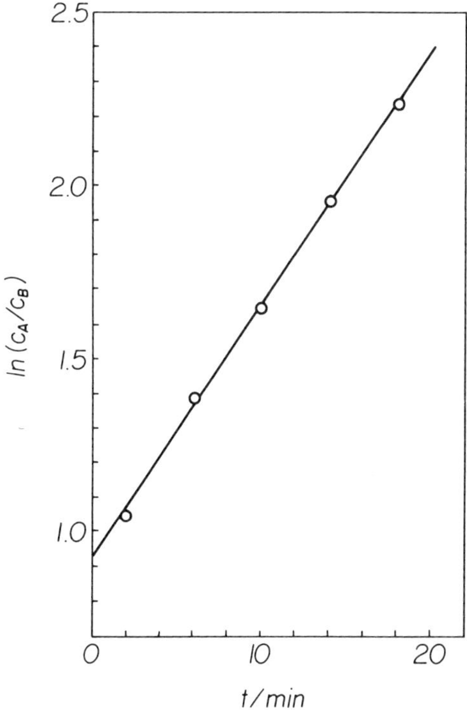

Figure 2-7. Second-order plot of Eq. (2-27) for the acetylation of isopropyl alcohol. Data are from Table 2-3.

Treatment of Instrument Response Data

Equations (2-3), (2-6), and (2-15) are written here in terms of the concentration ratio c/c^0 (or its reciprocal).

$$\text{Zero-order} \qquad \frac{c}{c^0} = 1 - \frac{k}{c^0}\, t \qquad\qquad (2\text{-}44)$$

$$\text{First-order} \qquad \ln\frac{c}{c^0} = -kt \qquad\qquad (2\text{-}45)$$

$$\text{Second-order} \qquad \frac{c^0}{c} = 1 + c^0 kt \qquad\qquad (2\text{-}46)$$

We can reach two useful conclusions from the forms of these equations: First, the plots of these integrated equations can be made with data on concentration *ratios* rather than absolute concentrations; second, a first-order (or pseudo-first-order) rate constant can be evaluated without knowing any absolute concentration, whereas zero-order and second-order rate constants require for their evaluation knowledge of an absolute concentration at some point in the data treatment process. This second conclusion is obviously related to the units of the rate constants of the several orders.

It is often experimentally convenient to use an analytical method that provides an instrumental signal that is proportional to concentration, rather than providing an absolute concentration, and such methods readily yield the ratio c/c^0. Solution absorbance, fluorescence intensity, and conductance are examples of this type of instrument response. The requirements are that the reactants and products both give a signal that is directly proportional to their concentrations and that there be an experimentally usable change in the observed property as the reactants are transformed into the products. We take absorption spectroscopy as an example, so that Beer's law is the functional relationship between absorbance and concentration. Let A be the reactant and Z the product. We then require that $\varepsilon_A \neq \varepsilon_Z$, where ε signifies a molar absorptivity. As initial conditions ($t = 0$) we set $c_A = c_A^0$ and $c_Z = 0$. The mass balance relationship Eq. (2-47) relates c_A^0 and c_Z^∞, where c_Z^∞ is the product concentration at infinity time, that is, when the reaction is essentially complete.

$$c_A^0 = c_A + c_Z = c_Z^\infty \qquad\qquad (2\text{-}47)$$

Applying Beer's law gives

$$A_0 = \varepsilon_A b c_A^0 \qquad\qquad (2\text{-}48)$$

$$A_\infty = \varepsilon_Z b c_Z^\infty \qquad\qquad (2\text{-}49)$$

$$A_t = \varepsilon_A b c_A + \varepsilon_Z b c_Z \qquad\qquad (2\text{-}50)$$

where A_0 is the absorbance at $t = 0$, A_∞ is the absorbance at $t = \infty$, and A_t is the absorbance at time t. Algebraic combination leads to Eq. (2-51), where the form chosen depends upon whether A_0 or A_∞ is the larger quantity.

$$\frac{c_A}{c_A^0} = \frac{A_t - A_\infty}{A_0 - A_\infty} = \frac{A_\infty - A_t}{A_\infty - A_0} \tag{2-51}$$

That is, $(A_0 - A_\infty)$ is proportional to the entire extent of reaction, and $(A_t - A_\infty)$ is proportional to the concentration unreacted at time t. Obviously these quantities can be separated, and the plots made without knowledge of A_0; however, as noted above, A_0 (whose knowledge is equivalent to knowing c^0) is needed to calculate a zero-order or second-order rate constant.

Equation (2-52) is a combination of Eqs. (2-45) and (2-51), and Fig. 2-8 is a plot according to this equation.

$$\log (A_\infty - A_t) = -\frac{k}{2.303}t + \log (A_\infty - A_0) \tag{2-52}$$

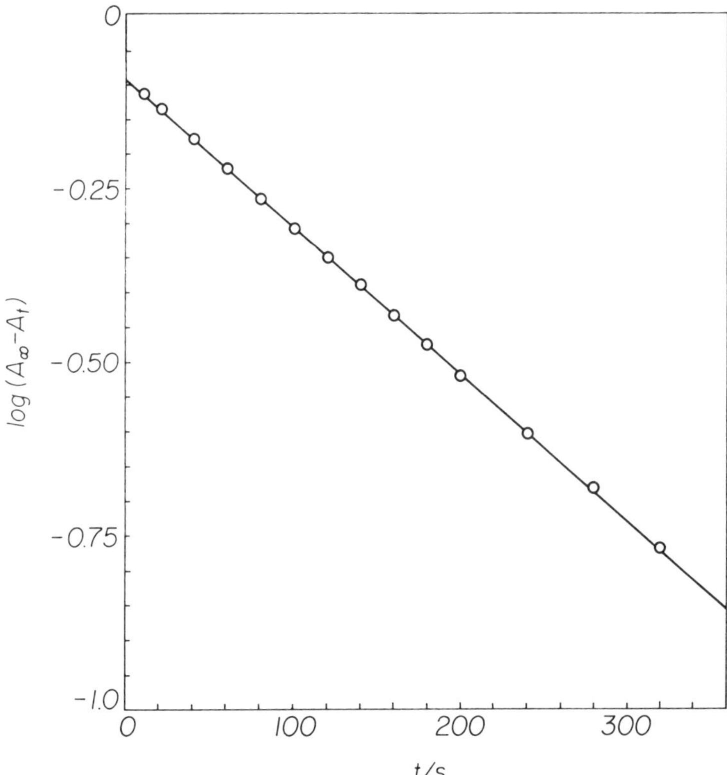

Figure 2-8. First-order plot of the hydrolysis of p-nitrophenyl glutarate at 25°C. Reaction followed spectrophotometrically at 400 nm; $b = 1$ cm, $A_\infty = 0.900$, pH 7.14.

Evidently the reaction shown in Fig. 2-8 is first-order (or pseudo first-order) over the period of time (about 2.5 half-lives) plotted in the figure. From the slope of the line the rate constant is calculated to be 4.90×10^{-3} s^{-1}.

Evidently the measurement of A_∞ should be accomplished with at least the same level of accuracy as the measurement of the A_t values, so the question arises: When does $t = \infty$? That is, when is the reaction essentially complete? For a first-order reaction, we calculate, with Eq. (2-10), that reaction is 99.9% complete after the lapse of 10 half-lives. This would ordinarily be considered an acceptable time for the measurement of A_∞.

Methods When the Final Value is Unknown

Clearly the accurate measurement of the final (infinity time) instrument reading is necessary for the application of the preceding methods, as exemplified by Eq. (2-52) for the spectrophotometric determination of a first-order rate constant. It sometimes happens, however, that this final value cannot be accurately measured. Among the reasons for this inability to determine A_∞ are the occurrence of a slow secondary reaction, the precipitation of a product, an unsteady instrumental baseline, or simply a reaction so slow that it is inconvenient to wait for its completion. Methods have been devised to allow the rate constant to be evaluated without a known value of A_∞; in the process, of course, an estimate of A_∞ is also obtainable.

At the outset it should be clear that, in order to apply these methods, the order of reaction must be known. Most of the following applies to first-order reactions.

The simplest procedure is merely to assume reasonable values for A_∞ and to make plots according to Eq. (2-52). That value of A_∞ yielding the best straight line is taken as the correct value. (Notice how essential it is that the reaction be accurately first-order for this method to be reliable.) Williams and Taylor[13] have shown that the standard deviation about the line shows a sharp minimum at the correct A_∞. Holt and Norris[14] describe an efficient search strategy in this procedure, using as their criterion minimization of the weighted sum of squares of residuals. (Least-squares regression is treated later in this section.)

The most commonly used method for first-order rate constant evaluation in the absence of a final reading is Guggenheim's method.[15] We continue to use spectrophotometric analysis as an example. From Eq. (2-52) we write, for time t,

$$(A_\infty - A_t) = (A_\infty - A_0)e^{-kt} \tag{2-53}$$

For the same reaction, at a later time $(t + \Delta t)$, where Δt is a constant time increment,

$$(A_\infty - A_{t+\Delta t}) = (A_\infty - A_0)e^{-k(t+\Delta t)} \tag{2-54}$$

Subtracting these equations and rearranging gives

$$(A_{t+\Delta t} - A_t) = (A_\infty - A_0)e^{-kt}(1 - e^{-k\Delta t})$$

We take logarithms to obtain Eq. (2-55).

$$\ln(A_{t+\Delta t} - A_t) = -kt + \ln(A_\infty - A_0)(1 - e^{-k\Delta t}) \qquad (2\text{-}55)$$

Because Δt is a constant, Eq. (2-55) is the equation of a straight line, whose slope yields the rate constant. [If $A_0 > A_\infty$, one plots $\ln (A_t - A_{t+\Delta t})$.]

Table 2-4 gives data for the alkaline hydrolysis of phenyl cinnamate under pseudo-first-order conditions, with calculations made in order to apply the Guggenheim method. The plot according to Eq. (2-55) is shown in Fig. 2-9. From the slope the pseudo-first-order rate constant is 3.37×10^{-3} s^{-1}.

The Guggenheim method requires that data be taken at constant time increments equal to Δt. In the past this was often a disadvantage, particularly when the experiment was not designed to be analyzed by this method, but with modern instrumental methods of analysis it is common to acquire a continuous record of instrument response as a function of time, so that data can be taken from this record at any desired times.

The following considerations bear on the selection of the increment Δt. In conventional first-order plotting according to Eq. (2-52), evidently $\Delta t = \infty$, which has the advantage that it minimizes the relative error in the ordinate, which is a difference. If, in the Guggenheim method, Δt is chosen to be very small (relative to the reaction half-life), this relative error will be large. On this basis, then, Δt should be as large as possible. Bacon and Demas[16] used simulated data to find that the relative standard deviation about the Guggenheim line showed a broad minimum in the range $3.6\, t_{1/2} < \Delta t < 5.3\, t_{1/2}$. However, there is another factor to consider. The very reason that an accurate A_∞ cannot be measured experimentally may prevent A_t observations from being made over a very wide extent of reaction, so Δt may be constrained by this limitation. Let T represent the entire time period over which observations are made. Then we have three possibilities:

1. *Choose $\Delta t \approx T/2$.* This choice maximizes Δt subject to the decision not to use any data twice yet to use all data once. This is the procedure used in constructing Table 2-4.

TABLE 2-4. Application of Guggenheim's Method to the Hydrolysis of Phenyl Cinnamate[a]

t/s	A_t	$(t+\Delta t)/s$	$A_{t+\Delta t}$	$(A_t - A_{t+\Delta t})$
0	0.837	400	0.481	0.356
50	0.763	450	0.463	0.300
100	0.699	500	0.447	0.252
150	0.647	550	0.431	0.216
200	0.602	600	0.421	0.181
250	0.563	650	0.411	0.152
300	0.532	700	0.402	0.130
350	0.505	750	0.397	0.108

[a] At 25.0°C and pH 11.80. Initial ester concentration, 8.2×10^{-6} M. Reaction followed at 295 nm in a 5.0-cm cell.

2. *Choose* $\Delta t < T/2$. This sacrifices accuracy in the difference $(A_t - A_{t+\Delta t})$, but produces more data points for the plot by using some data more than once.
3. *Choose* $\Delta t > T/2$. This gives less error in the difference $(A_t - A_{t+\Delta t})$, but it does not make use of all of the data.

The principal source of error in the Guggenheim method appears to lie in taking the difference $(A_t - A_{t+\Delta t})$. This has been circumvented in a method proposed independently by Kezdy et al.,[17] Mangelsdorf,[18] and Swinbourne.[19] Again we write Eqs. (2-53) and (2-54), but now we divide them and rearrange to get Eq. (2-56).

$$A_t = A_{t+\Delta t}\, e^{k\Delta t} + A_\infty(1 - e^{k\Delta t}) \tag{2-56}$$

Thus a plot of A_t vs. $A_{t+\Delta t}$ should be linear with slope equal to $\exp(k\Delta t)$. Figure 2-10 is a plot according to Eq. (2-56) of the data in Table 2-4. Equation (2-56) is obviously very easy to use. It has an unusual feature in that time is not one of the plotting variables; instead both axes consist of variables having comparable experimental error. The statistical analysis of this problem has been given in detail by Schwartz and Gelb[20] and McKinnon et al.[21] Schwartz[22] has generalized these methods, Eqs. (2-55) and (2-56) arising as special cases.

Another approach is possible. We rewrite Eq. (2-53) as

$$A_t = (A_0 - A_\infty)e^{-kt} + A_\infty \tag{2-57}$$

Christensen[23] differentiates Eq. (2-57) with respect to time and then takes logarithms, obtaining Eq. (2-58).

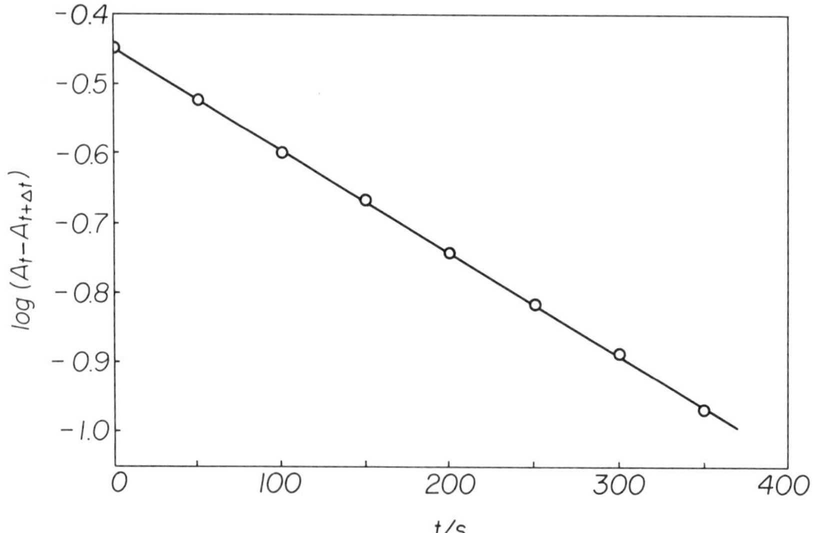

Figure 2-9. Guggenheim plot for the hydrolysis of phenyl cinnamate. Data are from Table 2-4.

$$\ln (dA_t/dt) = -kt + \ln k (A_\infty - A_0) \tag{2-58}$$

dA_t/dt is the tangent to the A_t vs. t curve at time t. A plot of $\ln(dA_t/dt)$ vs. t gives a straight line whose slope is equal to $-k$. Nonlinear least-squares regression analysis, discussed later, is another method that can be applied to Eq. (2-57).[24,25]

Espenson[26] and Livesey[27] have solved the second-order kinetics problem (in the special case) when A_∞ is unknown. Combine Eqs. (2-16) and (2-51) and rearrange to give

$$A_t = A_0 + kc^0 t (A_\infty - A_t) \tag{2-59}$$

Equation (2-59) is also written for time $t + \Delta t$, where Δt is a constant time increment. These equations are subtracted, yielding Eq. (2-60).

$$A_t - A_{t+\Delta t} = kc^0[\Delta t A_{t+\Delta t} - t(A_t - A_{t+\Delta t})] - kc^0 \Delta t A_\infty \tag{2-60}$$

A plot of the left side of the equation against the bracketed term is a straight line with slope equal to kc^0.[26]

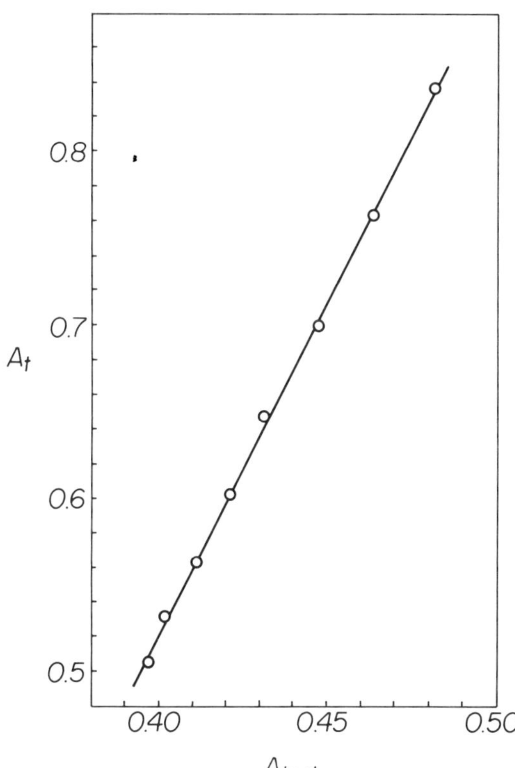

Figure 2-10. Plot of Eq. (2-56) for the hydrolysis of phenyl cinnamate. Data are from Table 2-4.

Propagation of Errors

A parameter such as a rate constant is usually obtained as a consequence of various arithmetic manipulations, and in order to estimate the uncertainty (error) in the parameter we must know how this error is related to the uncertainties in the quantities that contribute to the parameter. For example, Eq. (2-33) for a pseudo-first-order reaction defines k_{obs}, which can be determined by a semilogarithmic plot according to Eq. (2-6). By a method to be described later in this section the uncertainty in k_{obs} (expressed as its variance σ^2) can be estimated. However, k_{obs} is related to the desired parameter k by $k_{obs} = kc_B$, and presumably some uncertainty is associated with c_B. Thus, we need to know how the errors in k_{obs} and c_B are propagated into the rate constant k.

Suppose we have a function $F(x,y)$, and we carry out a Taylor's series expansion about the point (x_0, y_0), thus

$$F(x,y) = F(x_0,y_0) + F_x(x) \cdot (x - x_0) + F_y(y) \cdot (y - y_0)$$

where we truncate the series at the linear term as shown. If we take x close to x_0, y close to y_0, the intervals $\Delta x = x - x_0$, $\Delta y = y - y_0$ will be very small, and the approximation is a reasonable one. In this equation F_x and F_y are partial first derivatives (i.e., $F_x = (\partial F/\partial x)_y$).

Defining $\Delta F = F(x,y) - F(x_0,y_0)$ gives

$$\Delta F = F_x \Delta x + F_y \Delta y \tag{2-61}$$

We now identify the increments Δx and Δy as deviations in x and y. Then Eq. (2-61) reveals, to a good approximation, how these deviations are propagated into the deviation ΔF in the function F.[28] Squaring Eq. (2-61) gives

$$(\Delta F)^2 = (F_x \Delta x)^2 + (F_y \Delta y)^2 + 2F_x F_y \Delta x \Delta y \tag{2-62}$$

Now sum the squared deviations over all possible values of their (small) ranges and take the averages. These quantities can be interpreted as in Eqs. (2-63) and (2-64):

$$\sigma_x^2 = \frac{\Sigma(x - \bar{x})^2}{n}; \qquad \sigma_y^2 = \frac{\Sigma(y - \bar{y})^2}{n} \tag{2-63}$$

$$\sigma_{xy} = \frac{\Sigma(x - \bar{x})(y - \bar{y})}{n} \tag{2-64}$$

where σ_x^2, σ_y^2 are the variances of x and y, and σ_{xy} is the covariance, often designated cov (x,y), which measures the interaction between x and y. Then Eq. (2-62) becomes

$$\sigma_F^2 = F_x^2 \sigma_x^2 + F_y^2 \sigma_y^2 + 2F_x F_y \sigma_{xy} \tag{2-65}$$

This argument obviously can be generalized to any number of variables. Equation (2-65) describes the propagation of mean square error, or the propagation of variances and covariances.

If the errors in the variables are independent, then $\sigma_{xy} = 0$, and the propagation of error equation can be written

$$\sigma_F^2 = F_x^2 \, \sigma_x^2 + F_y^2 \, \sigma_y^2 \qquad (2\text{-}66)$$

To illustrate, we apply Eq. (2-66) to the example $k = k_{obs}/c_B$. Let $k = F$, $k_{obs} = x$, $c_B = y$, so the function is $F = x/y$. Then we find $F_x = \partial F/\partial x = 1/y$ and $F_y = \partial F/\partial y = -x/y^2$. Thus, we obtain

$$\sigma_F^2 = \frac{\sigma_x^2}{y^2} + \frac{x^2\sigma_y^2}{y^4}$$

which may also be written

$$\frac{\sigma_F^2}{F^2} = \frac{\sigma_x^2}{x^2} + \frac{\sigma_y^2}{y^2}$$

Suppose these reasonable results had been obtained: $k_{obs} = 3.0 \times 10^{-3}$ s^{-1}; standard deviation, 6×10^{-5} s^{-1}; $c_B = 0.0150$ M; standard deviation, 0.0002 M. Then we calculate $k = 0.200$ M^{-1} s^{-1}; standard deviation, 0.0046 M^{-1} s^{-1}.

Linear Least-Squares Regression

The most widely used method for fitting a straight line to integrated rate equations is by linear least-squares regression. These equations have only two variables, namely, a concentration or concentration ratio and a time, but we will develop a more general relationship for use later in the book.

It is necessary to make a distinction between two kinds of functions. If a function can be written in the form of

$$f(x_j) = a_0 + a_1x_1 + a_2x_2 + \ldots + a_kx_k \qquad (2\text{-}67)$$

it is said to be linear. The terminology means that it is linear *in the parameters* a_j. The quantities x_j may or may not be linear; for example, we might define $x_j = x^j$; then Eq. (2-67) would be a polynomial in x, but it would remain a linear function in the present sense. If a function is not linear in its parameters, it is a nonlinear function. If a nonlinear function can be transformed into a linear function, it is said to be *intrinsically linear*. Other functions cannot be so transformed; these are called *intrinsically nonlinear functions*. Here we consider linear functions.

This is the general approach:

1. Define the *model function,* Eq. (2-68).

$$\hat{y}_i = \sum_{j=0}^{k} a_j x_{ij} \tag{2-68}$$

2. Define the least-squares *objective function,* Eq. (2-69).

$$G = \sum_{i=1}^{n} (y_i - \hat{y}_i)^2 \tag{2-69}$$

In Eqs. (2-68) and (2-69), the symbols have these meanings:

Observation number	$i = 1, 2, 3, \ldots, n.$
Parameter number	$j = 0, 1, 2, \ldots, k.$
Independent variables	x_{ij} is the ith observation of variable x_j.
Dependent variable	y_i is the ith observation of variable y.
Calculated variable	$\hat{y}_i$ is the ith value of the dependent variable calculated with the model function and the final least-squares parameter estimates.

3. Take partial derivatives of G with respect to each parameter, set each derivative equal to zero, and solve the $k + 1$ simultaneous equations for the parameters.

There are some restrictions that we do not consider here. Our primary requirement is that the y_i are normally distributed (for a given set of x_{ij}) about their mean true values with constant variance. We also, for the present, assume that the errors in the x_{ij} are negligible relative to those in y_i.

To make this specific we take Eq. (2-70) as an example model function.

$$\hat{y}_i = a_0 + a_1 x_{i1} + a_2 x_{i2} + a_3 x_{i3} \tag{2-70}$$

For convenience we omit the observation index i, writing

$$\hat{y} = a_0 + a_1 x_1 + a_2 x_2 + a_3 x_3 \tag{2-71}$$

The objective function is then

$$G = \Sigma[y - (a_0 + a_1 x_1 + a_2 x_2 + a_3 x_3)]^2 \tag{2-72}$$

The necessary partial derivatives are set to zero:

$$\frac{\partial G}{\partial a_0} = 2\Sigma(y - a_0 - a_1 x_1 - a_2 x_2 - a_3 x_3)(-1) = 0$$

$$\frac{\partial G}{\partial a_1} = 2\Sigma(y - a_0 - a_1x_1 - a_2x_2 - a_3x_3)(-x_1) = 0$$

$$\frac{\partial G}{\partial a_2} = 2\Sigma(y - a_0 - a_1x_1 - a_2x_2 - a_3x_3)(-x_2) = 0$$

$$\frac{\partial G}{\partial a_3} = 2\Sigma(y - a_0 - a_1x_1 - a_2x_2 - a_3x_3)(-x_3) = 0$$

Multiplying through and rearranging gives Eqs. (2-73), where we have used the identity $\Sigma a_0 = a_0 n$.

$$a_0 n + a_1\Sigma x_1 + a_2\Sigma x_2 + a_3\Sigma x_3 = \Sigma y \tag{2-73a}$$

$$a_0\Sigma x_1 + a_1\Sigma x_1^2 + a_2\Sigma x_1x_2 + a_3\Sigma x_1x_3 = \Sigma x_1 y \tag{2-73b}$$

$$a_0\Sigma x_2 + a_1\Sigma x_1x_2 + a_2\Sigma x_2^2 + a_3\Sigma x_2x_3 = \Sigma x_2 y \tag{2-73c}$$

$$a_0\Sigma x_3 + a_1\Sigma x_1x_3 + a_2\Sigma x_2x_3 + a_3\Sigma x_3^2 = \Sigma x_3 y \tag{2-73d}$$

The summations in Eqs. (2-73) are over all i. Equations (2-73) are called the *normal regression equations*. With the experimental observations of y_i as a function of the x_{ij}, the summations are carried out, and the resulting simultaneous equations are solved for the parameters. This is usually done by matrix algebra. Define these matrices:

$$A = \begin{bmatrix} a_0 \\ a_1 \\ a_2 \\ a_3 \end{bmatrix}$$

$$Y = \begin{bmatrix} \Sigma y \\ \Sigma x_1 y \\ \Sigma x_2 y \\ \Sigma x_3 y \end{bmatrix}$$

$$X = \begin{bmatrix} n & \Sigma x_1 & \Sigma x_2 & \Sigma x_3 \\ \Sigma x_1 & \Sigma x_1^2 & \Sigma x_1x_2 & \Sigma x_1x_3 \\ \Sigma x_2 & \Sigma x_1x_2 & \Sigma x_2^2 & \Sigma x_2x_3 \\ \Sigma x_3 & \Sigma x_1x_3 & \Sigma x_2x_3 & \Sigma x_3^2 \end{bmatrix}$$

Then Eqs. (2-73) are succinctly written.

$$A\,X = Y$$

Because it is of particular interest in the present context, we now obtain the normal equations for linear regression with a single independent variable. The model function is

$$\hat{y} = a_0 + a_1 x_1 \tag{2-74}$$

Because this is a special case of Eq. (2-70), we can use the earlier results, retaining only the relevant terms. From Eqs. (2-73) we obtain

$$a_0 n + a_1 \Sigma x = \Sigma y$$

$$a_0 \Sigma x + a_1 \Sigma x^2 = \Sigma xy$$

Because there is only one independent variable, the subscript has been omitted. We now note that $\Sigma x/n = \bar{x}$ and $\Sigma y/n = \bar{y}$, so we find Eqs. (2-75) as the normal equations for unweighted univariate least-squares regression.

$$a_0 + a_1 \bar{x} = \bar{y}$$
$$a_0 \Sigma x + a_1 \Sigma x^2 = \Sigma xy \tag{2-75}$$

(A *variable* can take any physically admissible value; a *variate* is a variable that must also satisfy a frequency distribution.) These are called *unweighted normal equations* because each observation is accorded equal weight.

Weighting in Least-Squares Regression

We now consider the case in which, again, the independent variable x_i is considered to be accurately known, but now we suppose that the variances in the dependent variable y_i are not constant, but may vary (either randomly or continuously) with x_i. To show the basis of the method we use the simple linear univariate model, written as Eq. (2-76).

$$\hat{y}_i = a + bx_i \tag{2-76}$$

Let w_i be the weight we wish to assign to point x_i, y_i. The objective function is now defined

$$G = \Sigma[y_i - (a + bx_i)]^2 w_i \tag{2-77}$$

Carrying through the treatment as before yields Eqs. (2-78) as the normal equations for weighted linear univariate least-squares regression.

$$a\Sigma w_i + b\Sigma w_i x_i = \Sigma w_i y_i$$
$$a\Sigma w_i x_i + b\Sigma w_i x_i^2 = \Sigma w_i x_i y_i \tag{2-78}$$

It is conventional to define (for the case being considered) the weight of a measurement y_i to be inversely proportional to the variance of y_i, that is,

$$w_i = \frac{\alpha}{\sigma_i^2} \qquad (2\text{-}79)$$

where σ_i^2 is the variance of y_i corresponding to the independent variable x_i, and α is a proportionality constant. Because α is a constant, it cancels from the normal equations, and its choice is arbitrary. Three conventions are commonly used:

1. *Set $\alpha = 1$.* Then

$$w_i = \frac{1}{\sigma_i^2} \qquad (2\text{-}80)$$

2. *Set $\Sigma w_i = 1$.* Then $\alpha = 1/\Sigma(1/\sigma_i^2)$ and

$$w_i = \frac{(1/\sigma_i^2)}{\Sigma(1/\sigma_i^2)} \qquad (2\text{-}81)$$

3. *Set $\Sigma w_i = n$.* Then $\alpha = n/\Sigma(1/\sigma_i^2)$ and

$$w_i = \frac{n/\sigma_i^2}{\Sigma(1/\sigma_i^2)} \qquad (2\text{-}82)$$

Convention 3 has the feature that the weights add up to n, which they do also in "unweighted" regression, for which the w_i's are all equal to unity.

In order to apply weighting, estimates of the σ_i^2 are needed. These must be obtained experimentally by replicate measurements of the y_i at several x_i.

Weighting and the Transformation of Variables

It frequently happens that we plot or analyze data in terms of quantities that are transformed from the raw experimental variables. The discussion of the propagation of error leads us to ask about the distribution of error in the transformed variables. Consider the first-order rate equation as an important example:

$$c = c_0 e^{-kt}$$

We commonly linearize this by taking logarithms.

$$\ln c = \ln c_0 - kt \qquad (2\text{-}83)$$

Least-squares regression of $\ln c$ on t then gives estimates of $\ln c_0$ and k. Because time is usually measured with much greater accuracy than is concentration, we need only consider the uncertainty in the dependent variable.

We define the transformed variable $c' = \ln c$ and apply the propagation of error argument to find the variance of c'.[28, p. 45] From Eq. (2-66), $\sigma_{c'}^2 = (\partial c'/\partial c)^2 \sigma_c^2$, and because $d \ln u = du/u$, we get

$$\sigma_{c'}^2 = \sigma_c^2/c^2 \tag{2-84}$$

Therefore in applying weighted least-squares analysis to Eq. (2-83), each $c' = \ln c$ should be weighted inversely to σ_c^2/c^2 rather than to σ_c^2.

Variances of the Parameters

It can be argued that the main advantage of least-squares analysis is not that it provides the "best fit" to the data, but rather that it provides estimates of the uncertainties of the parameters. Here we sketch the basis of the method by which variances of the parameters are obtained. This is an abbreviated treatment following Bennett and Franklin.[29] We use the normal equations (2-73) as an example. Equation (2-73a) is solved for a_0:

$$a_0 = \bar{y} - a_1\bar{x}_1 - a_2\bar{x}_2 - a_3\bar{x} \tag{2-85}$$

Equation (2-85) is substituted into Eqs. (2-73b, c, d), multiplied out, and rearranged to give Eqs. (2-86),

$$a_1S(x_1{}^2) + a_2S(x_1x_2) + a_3S(x_1x_3) = S(x_1y)$$

$$a_1S(x_2x_1) + a_2S(x_2{}^2) + a_3S(x_2x_3) = S(x_2y) \tag{2-86}$$

$$a_1S(x_3x_1) + a_2S(x_3x_2) + a_3S(x_3{}^2) = S(x_3y)$$

where

$$S(x_1{}^2) = \Sigma(x_1 - \bar{x}_1)^2$$

$$S(x_1x_2) = \Sigma(x_1 - \bar{x}_1)(x_2 - \bar{x}_2) \tag{2-87}$$

$$S(x_1y) = \Sigma(x_1 - \bar{x}_1)(y - \bar{y})$$

and so on. Thus Eqs. (2-86) represent a model having four parameters, the variables now being expressed relative to their means.

The solution of Eqs. (2-86), which we omit, gives the following equations,

$$a_1 = c_{11}S(x_1y) + c_{12}S(x_2y) + c_{13}S(x_3y)$$

$$a_2 = c_{21}S(x_1y) + c_{22}S(x_2y) + c_{23}S(x_3y) \tag{2-88}$$

$$a_3 = c_{31}S(x_1y) + c_{32}S(x_2y) + c_{33}S(x_3y)$$

where $c_{21} = c_{12}$, $c_{31} = c_{13}$, $c_{23} = c_{32}$, and these $c_{jj'}$ are functions only of the sums involving the x_j. The $c_{jj'}$ are elements of an inverse matrix; they can be obtained from these equations:

$$c_{11}S(x_1^2) + c_{12}S(x_1x_2) + c_{13}S(x_1x_3) = 1$$

$$c_{11}S(x_2x_1) + c_{12}S(x_2^2) + c_{13}S(x_2x_3) = 0 \qquad (2\text{-}89)$$

$$c_{11}S(x_3x_1) + c_{12}S(x_3x_2) + c_{13}S(x_3^2) = 0$$

c_{21}, c_{22}, and c_{23} are found by solving an analogous set of equations with the right sides replaced by 0, 1, 0. c_{31}, c_{32}, and c_{33} are similarly found by replacing the right sides by 0, 0, 1. Thus a_1, a_2, and a_3 are calculated by means of Eqs. (2-88), and a_0 is obtained with Eq. (2-85).

The calculated values $\hat{y}$ of the dependent variable are then found, for x_i corresponding to the experimental observations, from the model equation (2-71). The quantity σ_y^2, the variance of the observations y_i, is calculated with Eq. (2-90), where the denominator is the degrees of freedom of a system with n observations and four parameters.

$$\sigma_y^2 = \frac{\Sigma(y_i - \hat{y}_i)^2}{n - 4} \qquad (2\text{-}90)$$

Finally variance estimates of the parameters are calculated by means of the $c_{jj'}$ elements, as in the following relationships (where *var* and *cov* represent variance and covariance, respectively).

$$\text{var}\,(a_1) = c_{11}\,\sigma_y^2$$

$$\text{var}\,(a_2) = c_{22}\,\sigma_y^2 \qquad (2\text{-}91)$$

$$\text{var}\,(a_3) = c_{33}\,\sigma_y^2$$

$$\text{cov}\,(a_1,a_2) = c_{12}\,\sigma_y^2$$

$$\text{cov}\,(a_1,a_3) = c_{13}\,\sigma_y^2 \qquad (2\text{-}92)$$

$$\text{cov}\,(a_2,a_3) = c_{23}\,\sigma_y^2$$

We still need the variance of a_0. Eliminate a_0 between Eqs. (2-71) and (2-85) to give

$$\hat{y} = \bar{y} + a_1(x_1 - \bar{x}_1) + a_2(x_2 - \bar{x}_2) + a_3(x_3 - \bar{x}_3) \qquad (2\text{-}93)$$

We apply the propagation of errors treatment to Eq. (2-93), where the quantities in parentheses are treated as known constants. The result is

$$\text{var}(\hat{y}) = \text{var}(\bar{y}) + \sum_{j} (x_j - \bar{x}_j)^2 \, \text{var}(a_j) \tag{2-94}$$

Neglecting covariances and using Eqs. (2-91),

$$\text{var}(\hat{y}) = \frac{\sigma_y^2}{n} + \sigma_y^2 \sum (x_j - \bar{x}_j)^2 c_{jj} \tag{2-95}$$

Equation (2-95) gives the variance of $\hat{y}$ at any x_j. With this equation confidence intervals can be estimated, using Student's t distribution, for the entire range of x_j.
In particular, when all $x_j = 0$, $\hat{y} = a_0$, and we find

$$\text{var}\,(a_0) = \sigma_y^2 \left[\frac{1}{n} + \sum \bar{x}_j^2 c_{jj} \right] \tag{2-96}$$

Let us apply these results to the univariate linear model, Eq. (2-74). From Eq. (2-89), $c_{11}s(x_1^2) = 1$, or

$$c_{11} = \frac{1}{\sum (x - \bar{x})^2} \tag{2-97}$$

From Eq. (2-88),

$$a_1 = \frac{\sum (x - \bar{x})(y - \bar{y})}{\sum (x - \bar{x})^2} \tag{2-98}$$

and $a_0 = \hat{y} - a_1 \bar{x}$ from Eq. (2-85). Equation (2-91) gives

$$\text{var}\,(a_1) = \frac{\sigma_y^2}{\sum (x - \bar{x})^2} \tag{2-99}$$

From Eq. (2-95), the variance of $\hat{y}$ at a given x is

$$\text{var}(\hat{y}) = \sigma_y^2 \left[\frac{1}{n} + \frac{(x - \bar{x})^2}{\sum (x - \bar{x})^2} \right] \tag{2-100}$$

Finally, from Eq. (2-100) at $x = 0$,

$$\text{var}\,(a_0) = \sigma_y^2 \left[\frac{1}{n} + \frac{\bar{x}^2}{\sum (x - \bar{x})^2} \right] \tag{2-101}$$

Calculations of the confidence intervals about the least-squares regression line, using Eq. (2-100), reveal that the confidence limits are curved, the interval being smallest at $x_i = \bar{x}$.

Nonlinear Least-Squares Regression

Referring to the earlier treatment of linear least-squares regression, we saw that the key step in obtaining the normal equations was to take the partial derivatives of the objective function with respect to each parameter, setting these equal to zero. The general form of this operation is

$$\frac{\partial G}{\partial a_j} = 2\Sigma(y-\hat{y})\,\frac{\partial \hat{y}}{\partial a_j} = 0$$

Now if the function is linear in the parameters, the derivative $\partial \hat{y}/\partial a_j$ does not contain the parameters, and the resulting set of equations can be solved for the parameters. If, however, the function is nonlinear in the parameters, the derivative contains the parameters, and the equations cannot in general be solved for the parameters. This is the basic problem in nonlinear least-squares regression.

Several methods have been devised to solve this problem.[28, Chap. IV; 30] All involve approximations. The method of Deming is based on linearization of the nonlinear function by a truncated Taylor's series expansion. The treatment will not be given here; very detailed descriptions are available.[3, App. B; 28, Chap. IV] However, it will be useful to show the result for a model function of fairly wide applicability in chemical kinetics. We consider a model having two variables (x, y) and three parameters (a, b, c). Then the linearization procedure leads to this set of normal equations.

$$\Sigma w_i F_a^i F_a^i A \;+\; \Sigma w_i F_a^i F_b^i B \;+\; \Sigma w_i F_a^i F_c^i C \;=\; \Sigma w_i F_a^i F_0^i$$

$$\Sigma w_i F_b^i F_a^i A \;+\; \Sigma w_i F_b^i F_b^i B \;+\; \Sigma w_i F_b^i F_c^i C \;=\; \Sigma w_i F_b^i F_0^i \qquad (2\text{-}102)$$

$$\Sigma w_i F_c^i F_a^i A \;+\; \Sigma w_i F_c^i F_b^i B \;+\; \Sigma w_i F_c^i F_c^i C \;=\; \Sigma w_i F_c^i F_0^i$$

In Eqs. (2-102) the symbols have these meanings: The running index i denotes the observation number (i goes from 1 to n, the total number of observations); $w_i = 1/L_i$, where

$$L_i = \frac{F_{xi}F_{xi}}{w_{xi}} + \frac{F_{yi}F_{yi}}{w_{yi}} \qquad (2\text{-}103)$$

and w_i is the weight of the ith observation; the parameters A, B, and C are defined

$$A = a_0 - a$$

$$B = b_0 - b \qquad (2\text{-}104)$$

$$C = c_0 - c$$

where a_0, b_0, and c_0 are the initial estimates of the parameters, and a, b, and c are the refined estimates, obtained from Eq. (2-104) and the values of A, B, and C found in solving the normal equations. We define

$$F^i(x_i,y_i,a,b,c) = 0 \qquad (2\text{-}105)$$

Then

$$F_0^i = F^i(X_i,Y_i,\ a_0,b_0,c_0) \qquad (2\text{-}106)$$

where X_i and Y_i are the experimental values of the variables in the ith observation. The remaining quantities in Eqs. (2-102) are partial derivatives, defined $F_{xi} = \partial F^i/\partial x_i$, $F_{yi} = \partial F^i/\partial y_i$, $F_a^i = \partial F^i/\partial a$, $F_c^i = \partial F_b^i/\partial b$, $F_c^i = \partial F^i/\partial c$.

To make this more concrete, let us apply Eqs. (2-102) to this simple model function:

$$y = a + bx \qquad (2\text{-}107)$$

Dropping the running index for convenience, we have from Eq. (2-105),

$$F = y - a - bx = 0 \qquad (2\text{-}108)$$

and from Eq. (2-106)

$$F_0 = y - a_0 - b_0x$$

where we understand that the experimental values are taken. The partial derivatives are, from Eq. (2-108), $F_x = -b$, $F_y = 1$, $F_a = -1$, $F_b = -x$. Substituting these into Eq. (2-102) gives

$$\Sigma wA + \Sigma wxB = -\Sigma w(y-a_0 - b_0x)$$

$$\Sigma wA + \Sigma wx^2B = -\Sigma wx(y-a_0 - b_0x)$$

Multiplying out the terms and using Eq. (2-104) gives

$$a\Sigma w + b\Sigma wx = \Sigma wy$$
$$\qquad (2\text{-}109)$$
$$a\Sigma wx + b\Sigma wx^2 = \Sigma wxy$$

which we see are identical to Eqs. (2-78). Notice that the initial parameter estimates a_0 and b_0 do not appear in Eqs. (2-109). When, however, we apply Eqs. (2-102) to a nonlinear model function we find that the initial parameter estimates appear in the normal equations.

The procedure is to use Eqs. (2-102) and the nonlinear model function. Preliminary parameter estimates a_0, b_0, . . . are needed. The resulting parameter values

a, b, . . . are then used as the a_0, b_0, . . . in a second iteration, this calculation being repeated until the estimates converge.

Note that this process has the capability of accounting for uncertainties in both variables through Eq. (2-103). Combining Eqs. (2-79) and (2-103),

$$L_i = \frac{1}{w_i} = \frac{1}{\alpha}\,(F_{xi}^2\sigma_{xi}^2 + F_{yi}^2\sigma_{yi}^2) \qquad (2\text{-}110)$$

There are several possible weighting situations:

1. $\sigma_{xi}^2 <<< \sigma_{yi}^2$; σ_{yi}^2 is independent of i. Then Eq. (2-110) becomes $1/w_i = F_{yi}^2\,\sigma_y^2/\alpha$; because σ_y^2/α is a constant, it cancels from the normal equations, and we have unweighted least-squares analysis.
2. $\sigma_{xi}^2 <<< \sigma_{yi}^2$; $\sigma_{yi}^2 = f(x_i)$. This case is usually described as weighted least-squares analysis. It was discussed earlier.
3. $\sigma_{xi}^2 = \sigma_{yi}^2$. Then the variances can be factored from Eq. (2-110). If they are constant, they cancel from the normal equations. This situation appears to apply to Eqs. (2-17) and (2-56).
4. $\sigma_{xi}^2 \neq \sigma_{yi}^2$, but they are comparable in magnitude. If they are functions of x, both variables must be weighted, and the full Eq. (2-110) must be used.

Equation (2-57) is a typical kinetic example of a nonlinear equation with two variables and three parameters, to which Eqs. (2-102) may be applied.[24,25]

Computer programs capable of carrying out nonlinear least-squares regression are commercially available. The user chooses the model function and provides the data and preliminary estimates of the parameters (which, for the simple rate equations being considered here, are obtainable by linear plotting methods). The program solves the normal equations for the parameters, iterating as required. Variance estimates of the parameters are calculated as described earlier. Kineticists will probably choose to carry out nonlinear regression with such a program rather than to design a program; but it is important to understand the principles of the calculation nevertheless.

Accuracy in Rate Constant Measurements

The experimentalist usually exercises choice in the design of kinetic experiments and in the data analysis. This choice includes the number of replicate experiments and the manner in which replicate data are processed. Phillips et al.[31] argue that least-squares regression using original (nonaveraged) data points gives more reliable variance estimates than does regression using averaged data. For example, if replicate k values are measured at the same concentration levels, fitting the mean k values to a function of the concentration sacrifices information on the variance and leads to a poorer test of goodness of fit (for example, by calculating confidence limits).

Correct weighting procedures for least-squares analysis have been discussed in

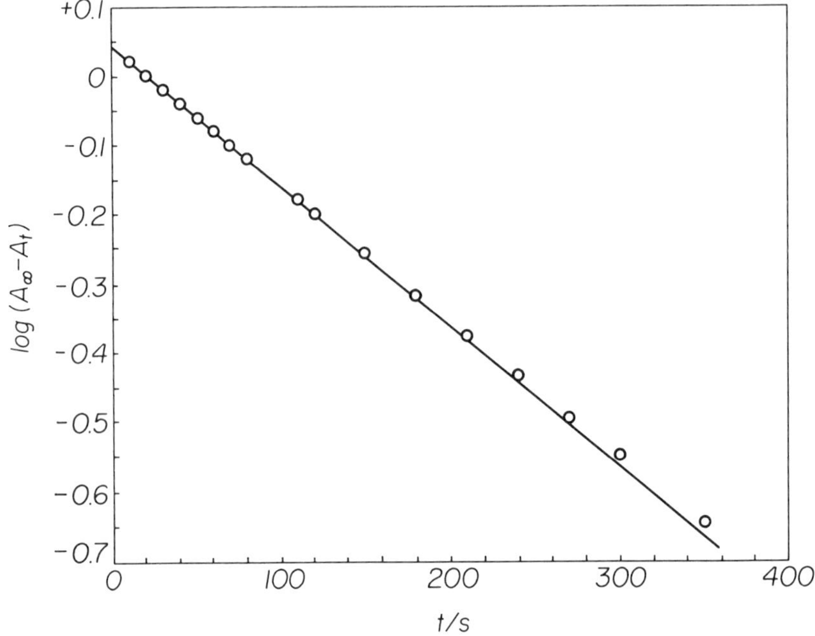

Figure 2-11. Semilogarithmic plot of the hydrolysis of *p*-nitrophenyl glutarate in the presence of *p*-methoxycinnamate ion; 25.0°C, initial pH 7.53, reaction followed at 400 nm. The plot deviates from linearity after the first half-life.

this section, but it must be noted that weighting is probably not employed as often as it could be. Kalantar[32] treated data of the first-order type with both weighted and unweighted least squares, using the linearized semilogarithmic form of the integrated equation. Different kinds of errors were added to the data. The efficiency of the parameter estimates was defined to be the ratio σ^2 (using weights)/σ^2 (not using weights). For a constant relative error (σ_y/y constant) the efficiency was essentially 100%, whereas for a constant absolute error (σ_y constant) the efficiency was less than 100%, decreasing to extremely low values as the range of observations (extent of reaction) increased.

There is a danger in relying on least-squares regression analysis without graphic presentation of the data. The human eye, in combination with chemical knowledge, is a more subtle qualitative judge of data than is regression analysis. Observe Fig. 2-11, a first-order plot of the hydrolysis of *p*-nitrophenyl glutarate in the presence of *p*-methoxycinnamate ion. The straight line has been drawn through the points in the first half-life, and it reveals a distinct curvature in the later portion of the reaction. Least-squares analysis without an accompanying plot might overlook this small but real effect. (In this case the curvature could be easily accounted for: The pH changed by about $+0.03$ unit during the course of the reaction, resulting in erroneous A_t and, especially, A_∞ values, because the proportion of product in the absorbing form is very sensitive to pH at the pH of the experiment.) The availability of computer graphics should be taken advantage of by experimentalists unwilling to plot data manually.

We must ask: How accurate should rate constant measurements be? The answer is that their accuracy should be commensurate with the use to be made of the data. There are instances in which the highest attainable accuracy is required, as in the measurement of kinetic isotope effects, or searches for the temperature dependence of activation energies. Usually, however, it is inefficient to devote excessive time and effort to achieving exceptional accuracy in a rate constant. A single rate constant is, by itself, nearly valueless. Rate constants have meaning primarily in relation to reaction variables such as temperature, pH, catalyst concentration, or solvent composition. It is Jenck's opinion[7, p. 556] that five rate constants having 5% accuracy (and exploring one of the cited reaction variables) are more valuable than a single constant with 1% accuracy, and the labor required for the two outcomes might be very similar.

REFERENCES

1. Wilkinson, F. "Chemical Kinetics and Reaction Mechanism"; Van Nostrand Reinhold: New York, 1980; p 17.
2. Connors, K.A.; Amidon, G.L.; Stella, V.J. "Chemical Stability of Pharmaceuticals", 2nd ed.; Wiley-Interscience: New York, 1986; p 12.
3. Connors, K.A. "Binding Constants: The Measurement of Molecular Complex Stability"; Wiley-Interscience: New York, 1987; pp 60–65, 74, 128–132.
4. Benson, S.W. "The Foundation of Chemical Kinetics"; McGraw-Hill: New York, 1960; p 19.
5. Dole, M. *J. Phys. Chem.* **1987,** *91,* 3117.
6. Frost, A.A.; Pearson, R.G. "Kinetics and Mechanism"; Wiley: New York, 1953; p 19.
7. Jencks, W.P. "Catalysis in Chemistry and Enzymology"; McGraw-Hill: New York, 1969; p 565.
8. Laidler, K.J. *Pure Appl. Chem.* **1981,** *53,* 753.
9. Casado, J.; Lopez-Quintela, M.A.; Lorenzo-Barral, F.M. *J. Chem. Educ.* **1986,** *63,* 450.
10. Chandler, W.D.; Lee, E.J.; Lee, D.G. *J. Chem. Educ.* **1987,** *64,* 878.
11. Wilkinson, R.W. *Chem. Ind.* **1961,** 1395.
12. Roseveare, W.E. *J. Am. Chem. Soc.* **1931,** *53,* 1651.
13. Williams, R.C.; Taylor, J.W. *J. Chem. Educ.* **1970,** *47,* 129.
14. Holt, M.J.J.; Norris, A.C. *J. Chem. Educ.* **1977,** *54,* 426.
15. Guggenheim, E.A. *Phil. Mag.* **1926,** *2,* 538.
16. Bacon, J.R.; Demas, J.N. *Anal. Chem.* **1983,** *55,* 653.
17. Kezdy, F.J.; Jaz, J.; Bruylants, A. *Bull. Soc. Chim. Belg.* **1958,** *67,* 687.
18. Mangelsdorf, P.C. *J. Appl. Phys.* **1959,** *30,* 442.
19. Swinbourne, E.S. *J. Chem. Soc.,* **1960,** 2371.
20. Schwartz, L.M.; Gelb, R.I. *Anal. Chem.* **1978,** *50,* 1592.
21. McKinnon, G.H.; Backhouse, C.J.; Kalantar, A.H. *Int. J. Chem. Kinet.* **1984,** *16,* 1427.
22. Schwartz, L.M. *Anal. Chem.* **1981,** *53,* 206.
23. Christensen, M.K. *Anal. Chem.* **1983,** *55,* 2324.
24. Wentworth, W.E. *J. Chem. Educ.* **1965,** *42,* 96, 162.
25. Moore, P. *J. Chem. Soc. Faraday Trans.* **1972,** *68,* 1890.
26. Espenson, J.H. *J. Chem. Educ.* **1980,** *57,* 160.
27. Livesey, D.L. *Int. J. Chem. Kinet.* **1986,** *18,* 281.

28. Deming, W.E. "Statistical Adjustment of Data"; Wiley: New York, 1943. Reprinted by Dover Publications: New York, 1964; Chapter III.
29. Bennett, C.A.; Franklin, N.L. "Statistical Analysis in Chemistry and the Chemical Industry"; Wiley: New York, 1954; pp 245–250.
30. Draper, N.R.; Smith, H. "Applied Regression Analysis", 2nd ed.; Wiley: New York, 1981; Chapter 10.
31. Phillips, G.R.; Harris, J.M.; Eyring, E.M. *Anal. Chem.* **1982,** *54,* 2053.
32. Kalantar, A.H. *Int. J. Chem. Kinet.* **1987,** *19,* 923.

PROBLEMS

1. Calculate the percentage completion of a first-order reaction for the elapse of 0, 1, 2, . . . ,10 half-lives. Plot percentage completion against the number of half-lives.
2. Find a relationship between the half-life $t_{1/2}$ and the lifetime τ of a first-order reaction.
3. For a reaction that is followed spectrophotometrically, let $v_0' = dA/dt$ be the initial rate in absorbance units per second. Letting ε_R and ε_p be the molar absorptivities of reactant and product, respectively, relate v_0' to v_0, the initial rate in moles per liter per second.
4. Find the integrated rate equation for a third-order reaction having the rate equation $-dc_A/dt = kc_A^3$.
5. These are absorbance–time data for the hydrolysis of *p*-nitrophenyl benzoate in aqueous solution.

t/s	A_t
0	0.168
20	0.218
40	0.265
60	0.303
80	0.342
120	0.402
160	0.451
200	0.488
∞	0.634

The conditions were: pH 7.70; ionic strength, 0.3 M; 25.0°C; initial ester concentration, 8×10^{-6} M; measurements at 400 nm in a 5-cm cell.

(a) Find the first-order rate constant.

(b) Estimate the molar absorptivities of the ester and *p*-nitrophenol under these conditions.

6. These are pseudo-first-order rate constants for the alkaline hydrolysis of ethyl *p*-nitrobenzoate at 25°C.

pH	k/s^{-1}
7.99	5.33×10^{-7}
9.09	6.64×10^{-6}
10.09	6.64×10^{-5}
11.38	1.28×10^{-3}
12.00	5.38×10^{-3}

(a) Find the order with respect to hydroxide ion and estimate the reaction rate constant for hydrolysis.

(b) What is the half-life at pH 11.00?

7. These are initial rates of acetohydroxamic acid formation from acetic acid, catalyzed by nickel chloride.

$$CH_3COOH + NH_2OH \xrightarrow{\text{Ni(II)}} CH_3CONHOH + H_2O$$

$10^8 v_o/M-s^{-1}$	$10^2[CH_3COOH]/M$
3.26	3.45
2.38	2.58
1.65	1.72

The conditions are 90.5°C; pH 5.25; ionic strength, 1.0 M; 0.04 M $NiCl_2$; 0.80 M $NH_2OH \cdot HCl$. Estimate the order with respect to acetic acid.

8. The kinetics of alkaline hydrolysis of phenyl cinnamate were studied at 25°C, in solutions containing 0.8% acetonitrile; ionic strength, 0.3 M; initial ester, 8.19×10^{-6} M; reaction followed spectrophotometrically in 5-cm cells at 295 nm. For studies at three pH values, these absorbance data were obtained. The pH was established with sodium hydroxide of the normality specified in the heading of the table (as titrimetrically determined).

	A_{295}		
t/s	*pH 11.59* *(0.00550 N NaOH)*	*pH 11.69* *(0.00696 N NaOH)*	*pH 11.79* *(0.00872 N NaOH)*
0	0.867	0.865	0.848
50	—	—	0.777
100	0.774	0.751	0.716
150	—	—	0.662
200	0.698	0.662	0.618
250	—	—	0.580
300	0.635	0.592	0.547
350	—	—	0.519
400	0.584	0.538	0.494
450	—	—	0.474
500	0.540	0.496	0.457
550	—	—	0.440

(Continued)

	A_{295}		
t/s	pH 11.59 (0.00550 N NaOH)	pH 11.69 (0.00696 N NaOH)	pH 11.79 (0.00872 N NaOH)
600	0.506	0.465	0.428
650	—	—	0.417
700	0.478	0.438	0.407
750	—	—	0.399
800	0.454	0.418	0.392
850	—	—	0.387
900	0.434	0.403	—
1000	0.419	0.391	—
1100	0.407	0.382	—
1200	0.396	—	—
1300	0.386	—	—

(a) Calculate the pseudo-first-order rate constants for the three studies.

(b) Calculate the second-order rate constant for the hydrolysis on the basis of the titrimetric hydroxide concentration.

(c) Calculate the second-order rate constant for the hydrolysis on the basis of the pH measurements.

9. The acetylation of isopropanol by acetic anhydride is catalyzed by 4-dimethyl-aminopyridine (DMAP). The reaction kinetics can be followed titrimetrically by hydrolyzing samples at known times and titrating them with standard NaOH. A blank is carried out with the reagents but no alcohol.

$$Ac_2O \ + \ (CH_3)_2CHOH \ \xrightarrow{\text{DMAP}} \ HOAc \ + \ CH_3COOCH(CH_3)_2$$

These data were obtained at 45°C in DMF as solvent.

t/min	Sample V_s/ml	Blank V_b/ml
5	7.80	—
10	7.20	—
15	—	9.80
20	6.75	—
30	6.45	—
35	—	9.81
45	6.22	—
50	—	9.80
60	6.10	—

The initial concentrations were: Ac_2O, 0.53 M; ROH, 0.41 M; DMAP, 0.17 M. Portions (5.0 ml) of the sample solution or the blank solution were withdrawn at the specified times and discharged into 20 ml of water. The solutions were titrated with 0.5327 N NaOH, giving the titration volumes shown.

(a) Find the second-order rate constant for the reaction.

(b) Assuming that the reaction is first-order in DMAP, calculate the third-order rate constant.

10. The kinetics of alkaline hydrolysis of an ester can be followed by the "pH-stat" method, in which the pH is held constant by adding a solution of strong alkali to the reacting ester solution. The volume of alkali added in order to keep the pH constant is recorded as a function-time. Find the volume function needed to determine the rate constant; that is, given the volume–time data for this type of experiment, what plot or calculation will yield the rate constant?

11. For a first-order reaction, investigate the properties of a plot of $dc/d(1/t)$ vs. $1/t$.

12. Show how a rate constant can be determined from an initial rate measurement, if the reaction order is known.

13. Suppose an initial rate of reaction v_0 is measured in the first few percentage of a reaction. Calculate the shelf life (t_{90}) if

(a) The reaction is assumed to be zero-order.

(b) The reaction is assumed to be first-order.

14. Derive the linear normal regression equations for the function $y = a_0 + a_1 x + a_2 x^2$.

15. Obtain the weighting function required to carry out weighted least-squares regression analysis of Eq. (2-15).

16. (a) If the standard deviation of absorbance measurements is 0.002 absorbance unit, what is the standard deviation of the quantity $(A_\infty - A_t)$?

(b) If $A_\infty = 0.850$ and $A_t = 0.400$, what is the relative standard deviation of $(A_\infty - A_t)$?

CHAPTER 3

Complicated Rate Equations

In Chapter 1 we distinguished between *elementary* (one-step) and *complex* (multistep reactions). The set of elementary reactions constituting a proposed mechanism is called a *kinetic scheme*. Chapter 2 treated differential rate equations of the form $v = kc_A^a c_B^b\ldots$, which we called *simple* rate equations. Chapter 3 deals with many examples of *complicated* rate equations, namely, those that are not simple. Note that this distinction is being made on the basis of the form of the differential rate equation.

Although it is possible for an elementary reaction to have a complicated rate equation as defined here (our first example below is such an instance), complicated rate equations arise largely in the study of complex reactions. Such kinetic systems can be classified in several ways, perhaps most generally as either parallel reactions or consecutive reactions.[1] In *parallel* reactions a reactant undergoes two or more concurrent reactions to give different products. The grouping might be expanded to include two or more reactants undergoing reaction to give a common product. Parallel reactions are also called *competitive reactions* [although Noyes[1] distinguishes between these classes].

Consecutive reactions are those in which the product of one reaction is the reactant in the next reaction. These are also called *series reactions*. Reversible (opposing) reactions, autocatalytic reactions, and chain reactions can be viewed as special types of consecutive reactions.

There is no general explicit mathematical treatment of complicated rate equations. In Section 3.1 we describe kinetic schemes that lead to closed-form integrated rate equations of practical utility. Section 3.2 treats many further approaches, both experimental and mathematical, to these complicated systems. The chapter concludes with comments on the development of a kinetic scheme for a complex reaction.

3.1 INTEGRATION OF THE RATE EQUATIONS

Reversible Reactions

Scheme I is the simplest reversible reaction.

$$A \underset{k_{-1}}{\overset{k_1}{\rightleftharpoons}} Z$$

Scheme I

[Note the rate constant symbolism denoting the forward (k_1) and backward (k_{-1}) steps.] The differential rate equation is written, according to the law of mass action, as

$$-\frac{dc_A}{dt} = k_1 c_A - k_{-1} c_Z \tag{3-1}$$

That is, the quantity $k_1 c_A$ is the *chemical flux*[2] in the forward direction and $k_{-1} c_Z$ is the flux in the reverse direction, the net rate of change of reactant concentration being their difference.

As the initial conditions we take $c_A = c_A^0$ and $c_Z = 0$; thus the mass balance relationship is $c_A^0 = c_A + c_Z$, which is combined with Eq. (3-1) to give

$$-\frac{dc_A}{dt} = (k_1 + k_{-1})\, c_A - k_{-1} c_A^0 \tag{3-2}$$

At equilibrium $dc_A/dt = 0$; applying this condition to Eq. (3-2) yields $(k_1 + k_{-1})c_A^e = k_{-1} c_A^0$, where c_A^e is the equilibrium value of c_A. This result is used in Eq. (3-2):

$$-\frac{dc_A}{dt} = (k_1 + k_{-1})(c_A - c_A^e) \tag{3-3}$$

After separation of variables, Eq. (3-3) is integrated to give Eq. (3-4) as the integrated rate equation.

$$\ln\left[\frac{c_A - c_A^e}{c_A^0 - c_A^e}\right] = -(k_1 + k_{-1})t \tag{3-4}$$

Evidently simple first-order behavior is predicted, the reactant concentration decaying exponentially with time toward its equilibrium value. In this case a complicated differential rate equation leads to a simple integrated form. The experimentally observed first-order rate constant k is equal to the sum $k_1 + k_{-1}$. From Eq. (3-1) written at equilibrium, when $dc_A/dt = 0$, we get $k_1 c_A^e = k_{-1} c_Z^e$, or $c_Z^e/c_A^e = k_1/k_{-1} = K$, the equilibrium constant for the reaction. Hence, from the measurement of the equilibrium constant and the observed first-order rate constant, the quantities k_1 and k_{-1} can be calculated.

Next consider Scheme II.

$$A + B \underset{k_{-1}}{\overset{k_1}{\rightleftharpoons}} Z$$

Scheme II

If pseudo-first-order conditions can be established, for example, by setting $c_B^0 \gg c_A^0$, then Scheme II collapses to Scheme III,

$$A \underset{k_{-1}}{\overset{k_1 c_B^0}{\rightleftharpoons}} Z$$

Scheme III

which is evidently equivalent to Scheme I. The observed first-order rate constant for Scheme III is $k = k_1 c_B^0 + k_{-1}$, and likewise the equilibrium constant is $c_Z^e/c_A^e = k_1 c_B^0/k_{-1}$.

If pseudo-first-order conditions do not apply, the Scheme II rate equation is

$$-\frac{dc_A}{dt} = k_1 c_A c_B - k_{-1} c_Z \tag{3-5}$$

This can be integrated,[3,4] although the result is not simple. King[5] has shown that Scheme II is more easily described in terms of a quantity Δ defined as the displacement of a concentration from its equilibrium value. Let $c_A = c_A^e + \Delta$; then we also have $c_B = c_B^e + \Delta$ and $c_Z = c_Z^e - \Delta$. The equilibrium constant is $K = k_1/k_{-1} = c_Z^e/c_A^e\, c_B^e$. Algebraic combination of these relationships with Eq. (3-5) gives Eq. (3-6).

$$-\frac{d\Delta}{dt} = k_1 \Delta(Q + \Delta) \tag{3-6}$$

where

$$Q = c_A^e + c_B^e + K^{-1} \tag{3-7}$$

Separation of variables and integration gives the indefinite integral, Eq. (3-8).

$$\ln\left[\frac{\Delta}{Q + \Delta}\right] = -k_1 Q\, t + \text{constant} \tag{3-8}$$

The value of the integration constant is determined by the magnitude of the displacement from the equilibrium position at zero time. King[5] also gives a solution for Scheme IV, and Pladziewicz et al.[6] show how these equations can be used with a measured instrumental signal to estimate the rate constants by means of nonlinear regression.

$$A + B \underset{k_{-1}}{\overset{k_1}{\rightleftharpoons}} Y + Z$$

Scheme IV

This device of Δ, the displacement from equilibrium, is used in the study of very fast reversible reactions by *relaxation kinetics*. We will see, in Chapter 4, that if Δ is very small, all reactions follow first-order kinetics, thus simplifying the interpretation of the kinetics. This approach might be extended to slow reversible reactions.

Parallel Reactions

Scheme V shows a complex reaction in which a single reactant A undergoes independent, concurrent reactions to yield two different products.

$$A \overset{k_1}{\rightarrow} Y$$
$$A \overset{k_2}{\rightarrow} Z$$

Scheme V

The rate equation for the loss of A is

$$-\frac{dc_A}{dt} = k_1 c_A + k_2 c_A \tag{3-9}$$

or

$$-\frac{dc_A}{dt} = k c_A \tag{3-10}$$

where $k = k_1 + k_2$. Thus, the reactant concentration follows a first-order rate equation, the observed rate constant being the sum of the rate constants of the individual reactions. Parallel second-order reactions run under pseudo-first-order conditions will give the analogous result.

The time course of the product formation is interesting. Consider product Y:

$$\frac{dc_Y}{dt} = k_1 c_A = c_A^0 k_1 e^{-kt} \tag{3-11}$$

because, from Eq. (3-10), $c_A = c_A^0 \exp(-kt)$. Integration of Eq. (3-11) gives

$$c_Y = c_Y^0 + \frac{c_A^0 k_1}{k}(1 - e^{-kt}) \tag{3-12}$$

and similarly for product Z:

$$c_Z = c_Z^0 + \frac{c_A^0 k_2}{k}(1 - e^{-kt}) \tag{3-13}$$

Suppose that $c_Y^0 = 0$, $c_Z^0 = 0$, as is often the case. Then the final product concentrations are found by setting $t = \infty$ in Eqs. (3-12) and (3-13); we obtain $c_Y^\infty = c_A^0 k_1/k$ and $c_Z^\infty = c_A^0 k_2/k$. The half-life for the production of Y is then given by Eq. (3-12), setting $c_Y = c_Y^\infty/2$ when $t = t_{1/2}$. We find $t_{1/2} = \ln 2/k$, and the same result is obtained for product Z. Thus, the products are generated in first-order reactions with the same half-life, even though they have different rate constants.

We also find, from Eqs. (3-12) and (3-13), for the case $c_Y^0 = 0$, $c_Z^0 = 0$,

$$\frac{c_Y}{c_Z} = \frac{k_1}{k_2} \tag{3-14}$$

According to this important result, the ratio of product concentrations is equal to the ratio of rate constants, independently of time. Even if the reactions are too fast to follow by conventional techniques, final product analysis will give the rate constant ratio (provided no subsequent reactions introduce artifactual changes).

Scheme VI describes a reaction system in which two different reactants yield a common product.

$$A \xrightarrow{k_1} Z$$

$$B \xrightarrow{k_2} Z$$

Scheme VI

Scheme VII constitutes an equivalent system if the concentration of reagent R is much larger than the reactant concentrations, so that we have pseudo-first-order conditions.

$$A + R \xrightarrow{k_1} Z$$

$$B + R \xrightarrow{k_2} Z$$

Scheme VII

In Scheme VII the reactants A and B compete for reagent R. There may be additional products: the essence of the description is that the analytical method responds identically to the products of the two reactions.

From Scheme VI, the reactants follow first-order kinetics, so $c_A = c_A^0 \exp(-k_1 t)$ and $c_B = c_B^0 \exp(-k_2 t)$. The mass balance expression is

$$c_A^0 + c_B^0 = c_A + c_B + c_z = c_Z^\infty \tag{3-15}$$

Combining these equations gives

$$c_Z^\infty - c_Z = c_A^0 e^{-k_1 t} + c_B^0 e^{-k_2 t}$$

or

$$\ln(c_Z^\infty - c_z) = \ln(c_A^0 e^{-k_1 t} + c_B^0 e^{-k_2 t}) \tag{3-16}$$

Obviously a plot of $\ln(c_Z^\infty - c_Z)$ vs. time will be nonlinear (except for the special case $k_1 = k_2$). However, suppose $k_1 \gg k_2$; then at some time essentially all of reactant A will have reacted, and Eq. (3-16) becomes

$$\ln(c_Z^\infty - c_Z) \approx \ln c_B^0 - k_2 t \tag{3-17}$$

so that k_2 is obtained from the terminal linear portion, which is extrapolated back to $t = 0$ to obtain c_B^0. With c_B^0 and k_2 available, c_B can be calculated and subtracted from $(c_Z^\infty - c_Z)$ to yield c_A. Finally a plot of $\ln c_A$ vs. t gives c_A^0 and k_1.

Figure 3-1 shows calculated plots of Eq. (3-16) for hypothetical systems in which k_1/k_2 has the values 1 and 5. It is evident from the example in which $k_1 = 5k_2$ that the curvature persists well into the reaction and that unambiguous identification of the terminal linear portion may be difficult. The long extrapolation to find c_B^0 is also uncertain. The accuracy of this procedure depends upon the ratios k_1/k_2 and $C_B^0 C_A^0$.

Parallel reactions of the Schemes VI and VII type have attracted much interest because of their analytical utility. If a mixture of two or three reactants can be arranged to undergo parallel reactions, with appropriate rate constant ratios, it may be possible to determine the composition of the initial mixture. Brown and Fletcher[7] introduced the extrapolation technique discussed above for this purpose, and many modifications of the approach have since been made.[8-13]

Next consider the case of parallel first-order and second-order reactions shown in Scheme VIII.

$$A \xrightarrow{k_1} Y$$

$$2A \xrightarrow{k_2} Z$$

Scheme VIII

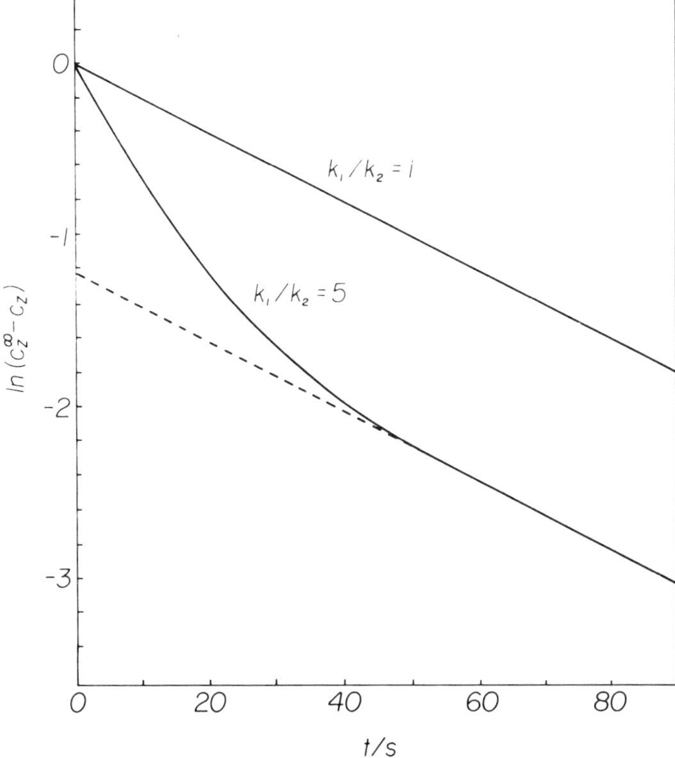

Figure 3-1. Plots of Eq. (3-16) for hypothetical systems having $k_2 = 0.02$ s^{-1} and the indicated values of the ratio k_1/k_2. In both examples $c_A^0 = 0.7$ and $c_B^0 = 0.3$.

The differential rate equation is

$$-\frac{dc_A}{dt} = k_1 c_A + 2k_2 c_A^2 \tag{3-18}$$

Define $r = k_2/k_1$, and use this in Eq. (3-18) to get

$$-\frac{dc_A}{dt} = k_1 c_A(1 + 2rc_A) \tag{3-19}$$

Upon separating variables and integrating, we obtain Eq. (3-20).

$$\ln\left[\frac{c_A}{1 + 2rc_A}\right] = -k_1 t + \ln\left[\frac{c_A^0}{1 + 2rc_A^0}\right] \tag{3-20}$$

Scheme VIII applies also to a system that includes a reagent if the reagent concentration is much larger than c_A^0.

Toby and co-workers[14,15] used Eq. (3-20) by assuming a value for the rate constant ratio r, plotting the left side against t, and calculating the correlation coefficient as a measure of goodness of fit. Then r was varied until the best straight line was generated; this value of r together with the slope gave k_1 and k_2. Under pseudo-order conditions with respect to a reagent, the experiment is repeated at different reagent concentrations in order to determine the orders with respect to the reagent, as described in Section 2-2. Equation (3-20) has been generalized to higher orders.[16]

Consecutive Reactions

Scheme IX is a complex reaction that occurs widely:

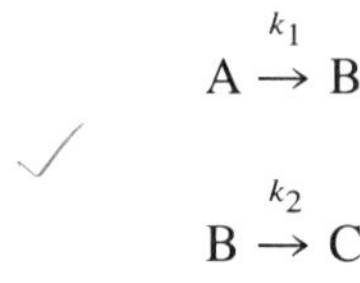

$$A \xrightarrow{k_1} B$$

$$B \xrightarrow{k_2} C$$

Scheme IX

This consists of two consecutive irreversible first-order (or pseudo-first-order) reactions. The differential rate equations are

$$\frac{dc_A}{dt} = -k_1 c_A \tag{3-21}$$

$$\frac{dc_B}{dt} = k_1 c_A - k_2 c_B \tag{3-22}$$

$$\frac{dc_C}{dt} = k_2 c_B \tag{3-23}$$

The reactant A follows a first-order rate equation, hence,

$$c_A = c_A^0 e^{-k_1 t} \tag{3-24}$$

To find the dependence of c_B on time we use the integrating factor method. Write Eq. (3-22) as

$$\frac{dc_B}{dt} + k_2 c_B = k_1 c_A^0 e^{-k_1 t}$$

and multiply both sides by $e^{k_2 t}$, the integrating factor.

$$\left(\frac{dc_B}{dt} + k_2 c_B \right) e^{k_2 t} = k_1 c_A^0 e^{-k_1 t} e^{k_2 t} \tag{3-25}$$

Next notice that

$$\frac{dc_B e^{k_2 t}}{dt} = \left(\frac{dc_B}{dt} + k_2 c_B\right) e^{k_2 t} \tag{3-26}$$

Comparing Eqs. (3-25) and (3-26),

$$\frac{dc_B e^{k_2 t}}{dt} = k_1 c_A^0 \, e^{(k_2 - k_1)t}$$

which is integrated to yield

$$c_B e^{k_2 t} = \frac{c_A^0 k_1}{k_2 - k_1} e^{(k_2 - k_1)t} + \text{constant}$$

The constant is determined by the initial conditions. Suppose $c_B = 0$ at $t = 0$; then we obtain: constant $= -c_A^0 k_1/(k_2 - k_1)$, and the integrated equation becomes

$$c_B = \frac{c_A^0 k_1}{k_2 - k_1} \left[e^{-k_1 t} - e^{-k_2 t} \right] \tag{3-27}$$

For the initial conditions $c_B^0 = 0$, $c_C^0 = 0$, the mass balance relationship is

$$c_A^0 = c_A + c_B + c_C \tag{3-28}$$

Substituting Eqs. (3-24) and (3-27) into Eq. (3-28) and rearranging yields

$$c_C = c_A^0 + \frac{c_A^0}{k_1 - k_2} \left[k_2 e^{-k_1 t} - k_1 e^{-k_2 t} \right] \tag{3-29}$$

Clearly Eqs. (3-27) and (3-29) are inapplicable in the special case $k_1 = k_2$.

Figure 3-2 shows the dependence of c_A, c_B, and c_C on time for a hypothetical system for which $k_1/k_2 = 4$. The rise and then fall of c_B is characteristic; substance B is an intermediate in the overall reaction in which A is transformed into C. Figure 3.3 is the same type of plot for a system having $k_1/k_2 = 1/4$. In this case the concentration of B is at all times less than that in Fig. 3-2; B is a more reactive intermediate in Fig. 3-3. The time at which c_B reaches its maximum value is found by setting $dc_B/dt = 0$; the result is

$$t_B^{\text{max}} = \frac{1}{k_1 - k_2} \ln \frac{k_1}{k_2} \tag{3-30}$$

Equation (3-29) can also be written as in Eq. (3-31).

$$c_C = c_A^0 + \frac{c_A^0}{k_2 - k_1} \left[k_1 e^{-k_2 t} - k_2 e^{-k_1 t} \right] \tag{3-31}$$

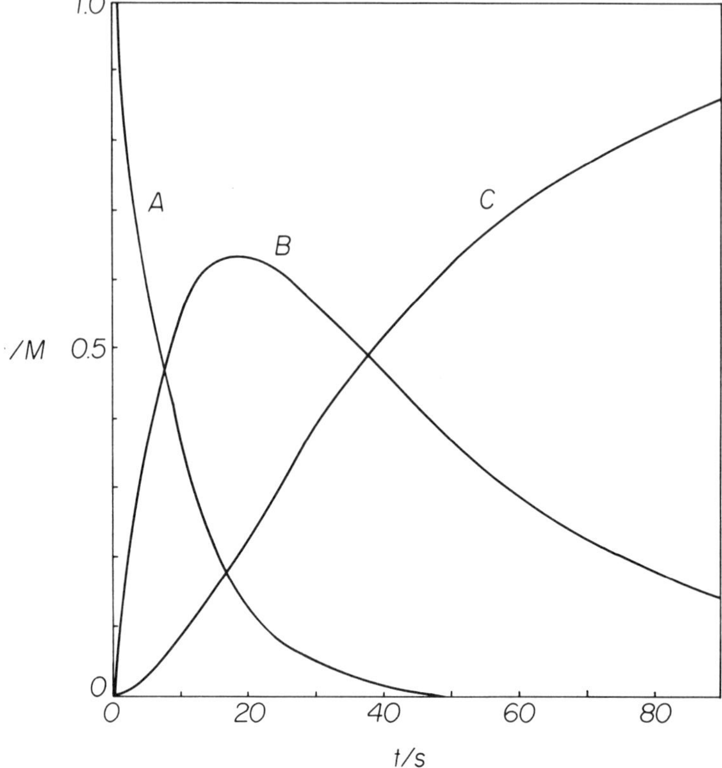

Figure 3-2. Plots of Eqs. (3-24), (3-27), and (3-29) for a hypothetical system having parameters $c_A^0 = 1$, $k_1 = 0.10\ \mathrm{s}^{-1}$, $k_2 = 0.025\ \mathrm{s}^{-1}$. Compare with Fig. 3-3.

Because Eqs. (3-29) and (3-31) are identical, the concentration of C is symmetrical in k_1 and k_2. If two rate constants can be evaluated on the basis of c_C vs. t data alone, it is impossible to establish which is k_1 and which is k_2. This is known as the *slow–fast ambiguity*. In the later stages of reaction, when c_A has decreased to negligible levels, the concentration c_B contains the same kinetic information as does c_C, so a similar conclusion follows; c_B in the later stages is controlled by the smaller rate constant, but one does not know if this is k_1 or k_2. Several authors have discussed the slow–fast ambiguity.[17-20] The problem arises also in pharmacokinetic modeling, when the first step corresponds to absorption of drug into the body and the second step to drug elimination. The data usually consist of drug concentrations in the blood as a function of time, and because absorption is usually more rapid than elimination,[21] the terminal phase is called the *elimination phase,* and the elimination rate constant is evaluated from this phase. However, it is possible for absorption to be slower than elimination, and then the terminal phase kinetics are controlled by the absorption kinetics. This is called the *flip-flop problem.*[22,23]

The legends of Figs. 3-2 and 3-3 reveal that these plots were calculated by simply

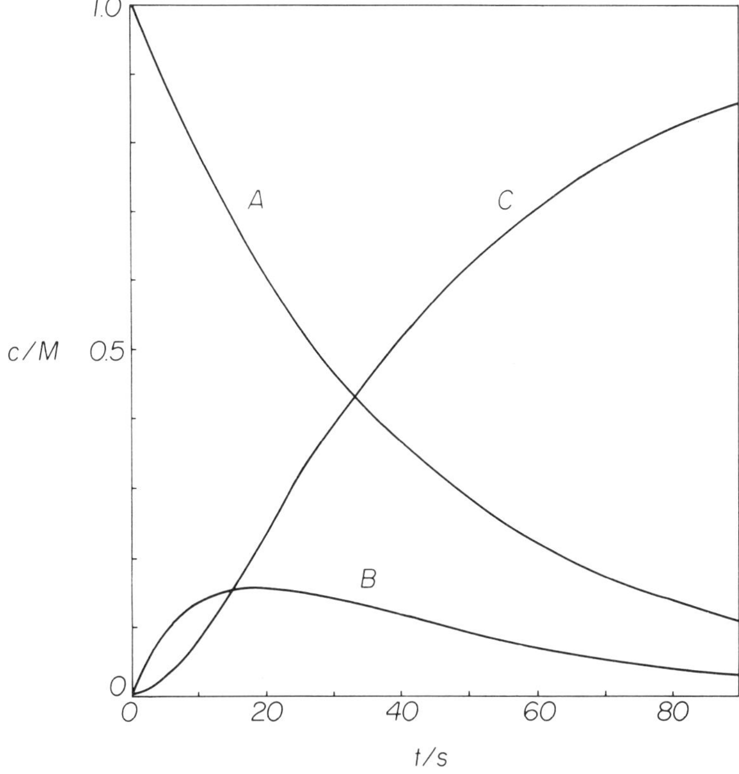

Figure 3-3. Plots of Eqs. (3-24), (3-27), and (3-29) for a system having parameters $c_A^0 = 1$, $k_1 = 0.025$ s^{-1}, $k_2 = 0.10$ s^{-1}. Compare with Fig. 3-2.

interchanging the values of k_1 and k_2. It is for this reason that the time course for the production of C is identical in the two figures. The semilogarithmic plots of the same simulated data, shown in Figs. 3-4 and 3-5, are instructive. An obvious result is that the plot of ln c_A vs. t is linear, because c_A decays according to a first-order rate equation. Moreover, after c_A has decreased to very low levels, the plot of ln c_B vs. t approaches linearity.

The method of evaluation of the rate constants for this reaction scheme will depend upon the type of analytical information available. This depends in part upon the nature of the reaction, but it also depends upon the contemporary state of analytical chemistry. Up to the middle of the 20th century, titrimetry was a widely applied means of studying reaction kinetics. Titrimetric analysis is not highly sensitive, nor is it very selective, but it is accurate and has the considerable advantage of providing absolute concentrations. When used to study the A $\rightarrow$ B $\rightarrow$ C system in which the same substance is either produced or consumed in each step (e.g., the hydrolysis of a diamide or a diester), titration results yield a quantity $F = c_B + 2c_C$. Swain[24] devised a technique, called the *time-ratio method,* to evaluate the rate

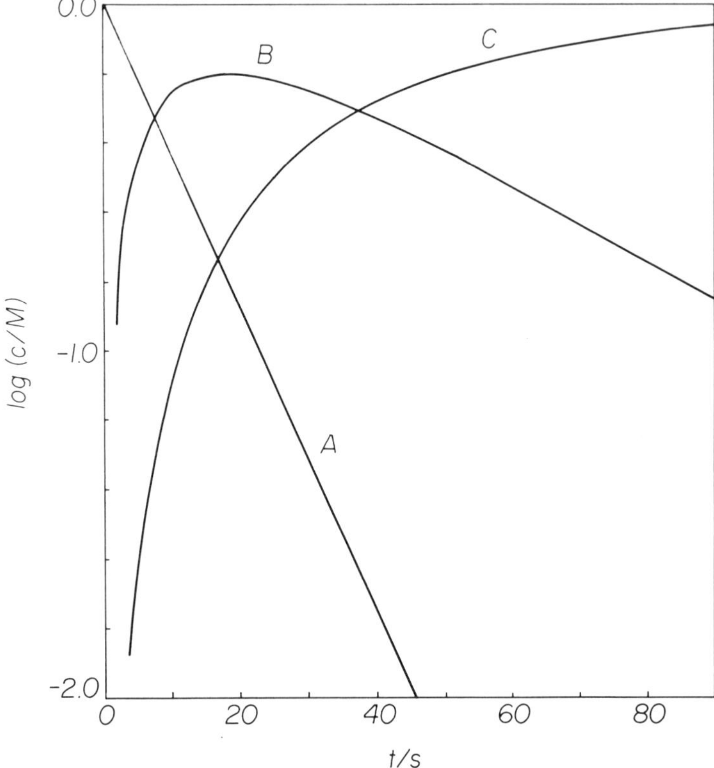

Figure 3-4.   Semilogarithmic plot of the concentration–time curves of Fig. 3-2.

constants when F is the experimental measure of the extent of reaction. Combining Eqs. (3-27) and (3-29) with the definition of F gives

$$F/c_A^0 = 2 - \left[\frac{k_1 - 2k_2}{k_1 - k_2}\right] e^{-k_1 t} - \left[\frac{k_1}{k_1 - k_2}\right] e^{-k_2 t} \qquad (3\text{-}32)$$

Define $r = k_2/k_1$ and $d = k_1 t$. Then Eq. (3-32) becomes

$$F/c_A^0 = 2 - \left[\frac{2r - 1}{r - 1}\right] e^{-d} + \left[\frac{1}{r - 1}\right] e^{-rd} \qquad (3\text{-}33)$$

For a given extent of reaction, Eq. (3-33) is an equation with the two unknowns r and d. The procedure, in essence, is to measure F at two times and to solve the two simultaneous equations. In practice the problem is more difficult than this because an analytical solution cannot be obtained; moreover d is itself dependent upon time. Swain[24] constructed tables of d (and of log d) as a function of r for three different extents of reaction. Curves of log d vs. log r are plotted. The curve

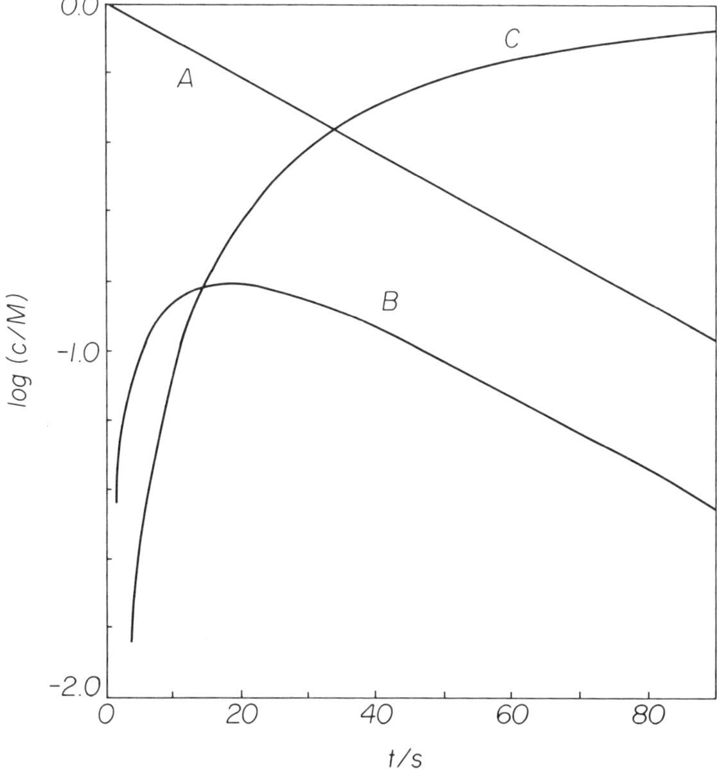

Figure 3-5. Semilogarithmic plot of the concentration–time curves of Fig. 3-3.

for one extent of reaction is superimposed on that for a different extent of reaction, the curves being displaced along the log d axes by a distance corresponding to the logarithm of the experimental time ratio for the two different extents of reaction. From the point of intersection of the two curves the solution is found, and the rate constants are evaluated. Kalonia and Simonelli[25] have critically analyzed the Swain method and have shown that it cannot distinguish between the cases $k_1 \approx 2k_2$ and $k_2 \gg k_1$.

Instrumental methods of analysis provide different measures of the progress of reaction. Consider this kinetic system as observed by absorption spectroscopy. Beer's law applied to the system gives

$$A_\lambda = \varepsilon_A b c_A + \varepsilon_B b c_B + \varepsilon_C b c_C \tag{3-34}$$

where A_λ is the solution absorbance at a single wavelength.

Combination of Eq. (3-34) with Eqs. (3-24), (3-27), and (3-29) provides an equation giving A_λ as a function of time and the parameters c_A^0, ε_A, ε_B, ε_C, k_1, and k_2. Usually ε_A and ε_C can be measured independently, and c_A^0 is known, so the equation has two variables and three unknown parameters. This problem can be solved by nonlinear-least squares regression, as outlined in Section 2.3. Alcock et al.[17] found that the solution includes the slow–fast ambiguity, with two different values for ε_B providing equivalent curve fits. The correct solution could be found by varying k_2/k_1 (for example, by changing reactant structure or temperature) to learn which solution yielded an invariant ε_B. Alcock et al.[17] conclude that a rapid increase and slow decline in absorbance does not necessarily mean that the first reaction is fast and the second one is slow. However, once the correct ε_B has been identified, it is possible to resolve the ambiguity.[18]

Absorption spectroscopy provides an opportunity to follow concentrations of individual species with time by observing the system at more than one wavelength. An example is the dehydration of prostaglandin E_1 methyl ester, in which the essential chemistry is shown as follows:

$$
\text{(A)} \xrightarrow{k_1} \text{(B)} \xrightarrow{k_2} \text{(C)} \tag{3-35}
$$

In this system the product of the first reaction possesses an absorption maximum at 222 nm and the final product has $\lambda_{max} = 288$ nm. The initial reactant is essentially nonabsorbing at these wavelengths. Hence, spectrophotometric observation at 222 and 288 nm allowed two simultaneous equations to be written, and thus c_B and c_C were determined as functions of time. From the known quantity c_A^0, the concentration c_A was calculated with Eq. (3-28). The rate constant k_1 was then found from the plot of ln c_A vs. time. An estimate of rate constant k_2 was obtained from a plot of ln c_B vs. time in the late stages of the reaction, and this value was refined by curve-fitting the c_B and c_C data. Figure 3-6 shows the data and final curve fits.

If the data consist of c_B as a function of time, another approach can be used. As above, the smaller rate constant (say k_2) is estimated from a semilogarithmic plot of c_B at later times when c_A is negligible. This plot is extrapolated back to $t = 0$. This line is described by the equation [from Eq. (3-27)],

$$
\ln c_B^{ext} = \ln [c_A^0 k_1/(k_1 - k_2)] - k_2 t \tag{3-36}
$$

Combining Eqs. (3-36) and (3-27),

$$
\ln (c_B^{ext} - c_B) = \ln [c_A^0 k_1/(k_1 - k_2)] - k_1 t \tag{3-37}
$$

Graphically Eq. (3-37) represents the logarithm of the differences between the experimental c_B values at early times and the values extrapolated from late times. The plots of Eqs. (3-36) and (3-37) should have the same intercepts and their slopes

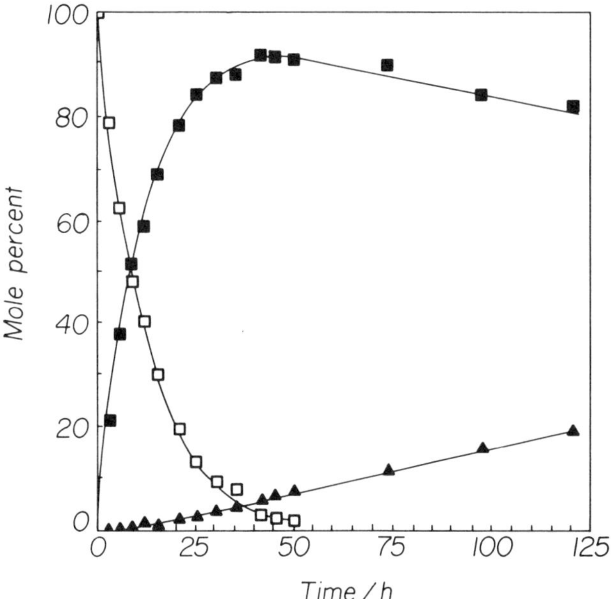

Figure 3-6. Concentration–time behavior of Eq. (3-35) at pH 7.66 and 60°C. The curves were drawn with Eqs. (3-24), (3-27), and (3-29) and the parameters $k_1 = 0.087$ h^{-1}, $k_2 = 0.0020$ h^{-1}. The concentrations are expressed relative to the initial reactant concentration.

yield estimates of the rate constants. Figure 3-7 shows this technique for the hypothetical system of Figs. 3-2 and 3-4.

In pharmacokinetics this plotting technique is called the *method of residuals,* or the *feathering technique;* c_B corresponds to the drug concentration in the blood. Wagner and Metzler[21] have analyzed the errors in the rate constants that result from this method of data analysis. Typically the smaller rate constant (estimated from the final stage of the reaction) is underestimated; as a consequence the larger rate constant is overestimated. These authors suggest that the method of residuals be used to obtain preliminary estimates for a nonlinear least-squares regression analysis.

Scheme X constitutes an extension of the preceding discussion:

$$A \xrightarrow{k_1} B$$

$$B \xrightarrow{k_2} C$$

$$C \xrightarrow{k_3} D$$

Scheme X

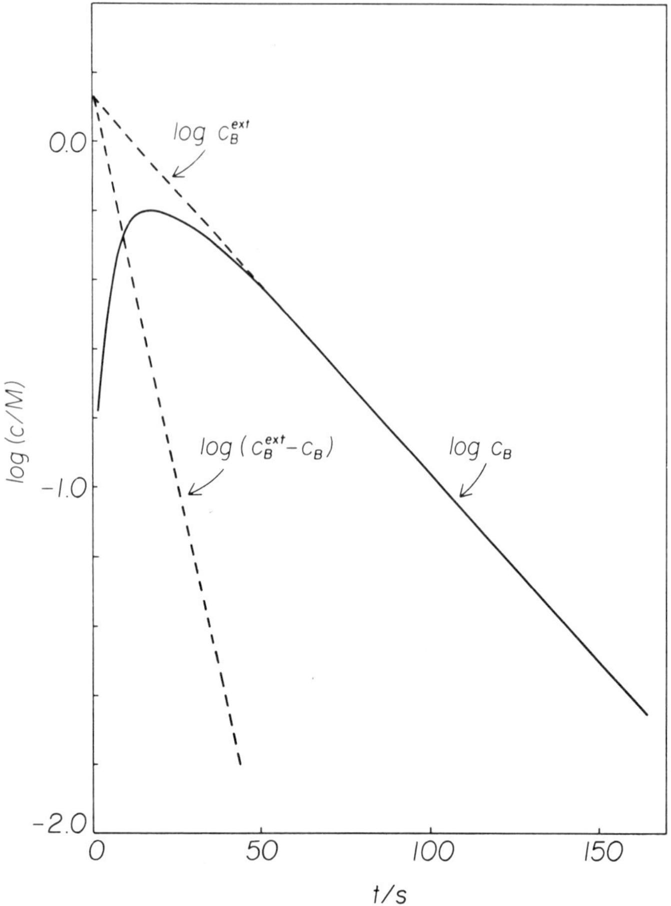

Figure 3-7. The method of residuals applied to the system shown in Figs. 3-2 and 3-4. See the discussion of Eqs. (3-36) and (3-37).

The differential rate equations are

$$\frac{dc_A}{dt} = -k_1 c_A \tag{3-38}$$

$$\frac{dc_B}{dt} = k_1 c_A - k_2 c_B \tag{3-39}$$

$$\frac{dc_C}{dt} = k_2 c_B - k_3 c_C \tag{3-40}$$

$$\frac{dc_D}{dt} = k_3 c_C \tag{3-41}$$

Comparison of Eqs. (3-38) and (3-39) with (3-21) and (3-22) shows that the time dependencies of c_A and c_B are identical in Schemes IX and X, so integrated equations (3-24) and (3-27) apply also to Scheme X. The integration of Eq. (3-40) is carried out analogously to that of Eq. (3-22); the result, for the initial condition $c_C = 0$ at $t = 0$, is

$$c_C = c_A^0 k_1 k_2 \left[\frac{e^{-k_1 t}}{(k_2 - k_1)(k_3 - k_1)} - \frac{e^{-k_2 t}}{(k_2 - k_1)(k_3 - k_2)} + \frac{e^{-k_3 t}}{(k_3 - k_1)(k_3 - k_2)} \right]$$

$$(3\text{-}42)$$

The concentration of D is then found from the mass balance,

$$c_A^0 = c_A + c_B + c_C + c_D \tag{3-43}$$

$$c_D = c_A^0 - \frac{c_A^0 k_2 k_3 e^{-k_1 t}}{(k_2 - k_1)(k_3 - k_1)} + \frac{c_A^0 k_1 k_3 e^{-k_2 t}}{(k_2 - k_1)(k_3 - k_2)} - \frac{c_A^0 k_1 k_2 e^{-k_3 t}}{(k_3 - k_1)(k_3 - k_2)} \tag{3-44}$$

Figure 3-8 is a plot of c_A, c_B, c_C, and c_D for a hypothetical system of the Scheme X type. An interesting feature is the time delay after the start of the reaction before the final product, D, appears in significant concentrations. This delay in product appearance is called an *induction period* or *lagtime*. In order to observe an induction period it is only necessary that the system include several relatively stable intermediates, so that the bulk of the material balance is temporarily "stored" in these prior forms. An experimental measurement of the induction period requires an arbitrary definition of its length.

Generalization of Scheme X to any number of consecutive irreversible first-order reactions is obviously possible, although the equations quickly become very cumbersome. However, Eqs. (3-42) and (3-44) reveal patterns in their form, and Westman and DeLury[26] have developed a systematic symbolism that allows the equations to be written down without integration.

Consecutive reactions involving one first-order reaction and one second-order reaction, or two second-order reactions, are very difficult problems. Chien[27] has obtained closed-form integral solutions for many of the possible kinetic schemes, but the results are too complex for straightforward application of the equations. Chien[27] recommends that the kineticist follow the concentration of the initial reactant A, and from this information rate constant k_1 can be estimated. Then families of curves plotted for the various kinetic schemes, making use of an abscissa scale that is a function of $c_A^0 k_1 t$, are compared with concentration–time data for an intermediate or product, seeking a match that will identify the kinetic scheme and possibly lead to additional rate constant estimates.

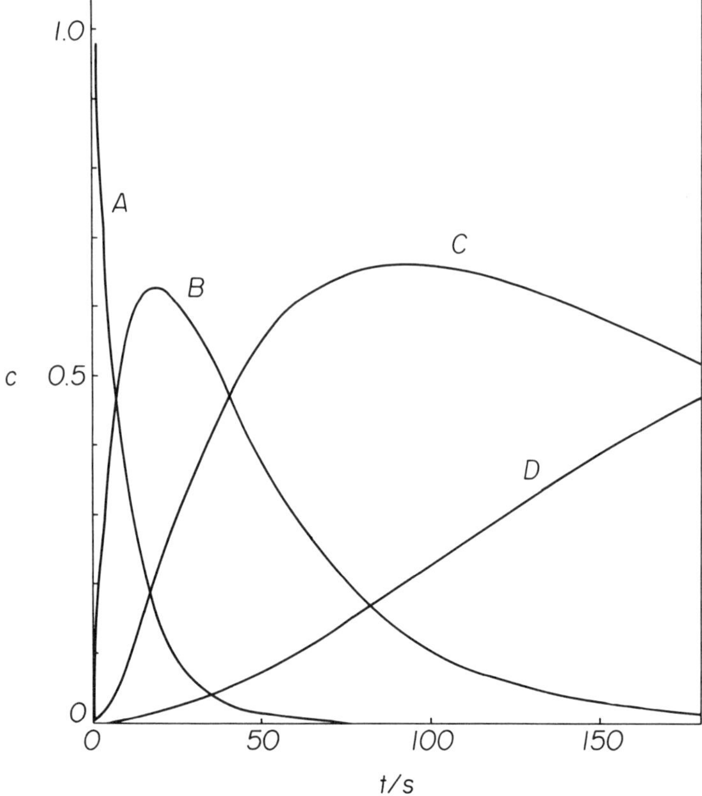

Figure 3-8. Plot of concentration–time data for Scheme X calculated with Eqs. (3-24), (3-27), (3-42), and (3-44) and the parameters $c_A^0 = 1$, $k_1 = 0.10$ s^{-1}, $k_2 = 0.025$ s^{-1}, $k_3 = 0.005$ s^{-1}.

Kinetic scheme XI has been discussed by many authors.

$$A + R \xrightarrow{k_1} B + Z$$

$$B + R \xrightarrow{k_2} C + Z$$

Scheme XI

In Scheme XI, R is a common reagent. The alkaline hydrolysis of a diester is an example of Scheme XI, with the diester $ROOC-(CH_2)_n-COOR$ being A, the half-ester $ROOC-(CH_2)_n-COO^-$ being B, and OH^- serving as R. Frost and Schwemer[28,29] developed a time-ratio method analogous to Swain's method.[24] Such methods are applicable to kinetic data obtained by titrimetric analysis under second-order conditions, in which c_A^0 and c_R^0 are comparable. The kinetic problem is greatly simplified if pseudo-first-order conditions ($c_R^0 \gg c_A^0$) are employed, for then Scheme XI becomes equivalent to Scheme IX. With modern analytical techniques this is the preferred approach to solving this class of kinetic problem.

Paventi[30] has discussed Scheme XI in the following terms. First, consider the simple second-order reaction

$$A + R \xrightarrow{k} B + Z$$

The differential rate equation is $-dc_R/dt = kc_A c_R$, and the mass balance equation is $c_A^0 - c_A = c_R^0 - c_R$. Eliminating c_A between these equations and integrating gives the usual second-order integrated equation, which can be written in this form:

$$c_R = (c_A^0 - c_R^0)[(c_A^0/c_R^0)\exp(c_A^0 - c_R^0)kt - 1]^{-1} \tag{3-45}$$

Paventi[30] then obtains a differential equation for Scheme XI and by analogy writes Eq. (3-46) as a particular solution, where k is a function of k_1 and k_2.

$$c_R = (2c_A^0 - c_R^0)[(2c_A^0/c_R^0)\exp(2c_A^0 - c_R^0)kt - 1]^{-1} \tag{3-46}$$

Paventi[30] argues that, because Eq. (3-46) can describe the experimental c_R, t data with the single rate constant k, it is not possible to evaluate two rate constants from this set of data. The parameter k is the smaller of the two constants k_1 and k_2.

3.2 OTHER METHODS OF ANALYSIS

Measurement of Rates

The differential rate equations of a complex reaction, expressing rates as functions of concentrations, are usually simpler in form than are the corresponding integrated equations, which express concentrations as functions of time; moreover, it is always possible to write down the differential rate equations for a postulated kinetic scheme, whereas it may be difficult or impossible to integrate them. Of course, we usually measure concentration as a function of time. If, however, we can measure rates, we may use the differential equations directly.

Consider Scheme XII, consisting of parallel first-order and second-order reactions (e.g., concurrent S_N1 and S_N2 reactions).

$$A \xrightarrow{k_1} Y + Z$$

$$A + B \xrightarrow{k_2} Y + Z$$

Scheme XII

The differential rate equation for the loss of A is

$$-\frac{dc_A}{dt} = k_1 c_A + k_2 c_A c_B \qquad (3\text{-}47)$$

which can be written

$$v(c_A) = k_1 c_A + k_2 c_A c_B$$

the symbolism $v(c_A)$ emphasizing that the rate depends upon the concentration. Rearranging gives

$$\frac{v(c_A)}{c_A} = k_1 + k_2 c_B \qquad (3\text{-}48)$$

Thus, if c_A and c_B can both be measured as functions of time, a plot of v/c_A vs. c_B allows the rate constants to be estimated. (If it is known that B is also consumed in the first-order reaction, mass balance allows c_B to be easily expressed in terms of c_A.) The rate $v(c_A)$ is the tangent to the curve $c_A = f(t)$ at concentration c_A. This can be determined graphically, analytically, or with computer processing of the concentration–time data. Mata-Perez and Perez-Benito[31] show an example of this treatment for parallel uncatalyzed and autocatalyzed reactions.

Simplification of the Experimental Kinetics

It may be possible to convert a very complex reaction into a less complex system by appropriate choice of reaction conditions or analytical methods. Here are some possibilities.

1. Use of the isolation or pseudo-order technique. This approach is discussed in Chapter 2, where it was shown how a second-order reaction could be converted to a pseudo-first-order reaction by maintaining one of the reactant concentrations at an essentially constant level. The same method may be usefully applied to complex reactions. In this way, for example, Scheme XI can be studied under conditions such that it functions as Scheme IX. A corollary that must be kept in mind is that a reaction system that is observed to behave in accordance with (as an example) Scheme IX may actually be more complex than it appears to be, if an unsuspected reactant is present under pseudo-order conditions.
2. Study of reversible reactions close to equilibrium. This possibility was discussed in connection with Scheme II and is further treated in Chapter 4. It turns out that if the displacement from equilibrium is small, the kinetics approach first-order behavior.
3. Neglect of the terminal stages of the reaction. Perhaps the kinetics become complicated only later in the reaction, when a slow process leading to a by-product begins to become significant. If this last reaction is of no immediate interest, data collection may be terminated before it complicates the kinetic

behavior. This is commonly done in studying enzyme kinetics, in which initial rates are usually measured.

4. Neglect of the initial stages of the reaction. It may be possible to study the final reaction in a series without excessive interference from earlier processes. This is done when estimating the smaller of the two rate constants in the A → B → C series reaction (Scheme IX) by following c_B late in the reaction.

5. Use of an intermediate as a starting reactant. Again taking Scheme IX as an example, if intermediate B is stable enough to be isolated or synthesized, its reaction to give C can be studied independently of its formation from A. (This requires that the A → B step be irreversible). For example, in the overall hydrolysis of a diester, the intermediate is the half-ester, which usually can be synthesized to serve as a reactant in a separate study.

6. Mimicking the reactant with a derivative or analog that cannot undergo the full complexity of the reaction, yet preserves some features of the reaction of interest. Thus an *ortho*-disubstituted benzene derivative may undergo an intramolecular reaction in addition to an intermolecular reaction; the corresponding *para*-di-substituted compound, which is incapable of undergoing the intramolecular process, may be a good model of the intermolecular reaction.

7. Change of reaction conditions to minimize kinetic complications. For example, if two parallel reactions have substantially different activation energies, their relative rates will depend upon the temperature. The reaction solvent, pH, and concentrations are other experimental variables that may be manipulated for this purpose.

8. Choice of initial conditions. To give a very obvious example, in Chapter 2 we saw that a second-order reaction A + B → products could be run with the initial conditions $c_A^0 = c_B^0$, thus permitting a very simple plotting form to be used. For complex reactions, it may be possible to obtain a usable integrated rate equation if the initial concentrations are in their stoichiometric ratio.

9. Exploitation of analytical selectivity. We have seen, in our discussion of the A → B → C series reaction (Scheme IX), that access to the concentration of A as a function of time is valuable because it permits k_1 to be easily evaluated. Modern analytical methods, particularly chromatography, constitute a powerful adjunct to kinetic investigations, and they render nearly obsolete some very difficult kinetic problems. For example, the freedom to make use of the pseudo-order technique is largely dependent upon the high sensitivity of analytical methods, which allows us to set one reactant concentration much lower than another. An interesting example of analytical control in the study of the Scheme IX system is the spectrophotometric observation of the reaction solution at an isosbestic point of species B and C, thus permitting the A to B step to be observed.[32]

Elimination of the Time Variable

For parallel reactions the following technique is often helpful. We illustrate with Scheme V, writing for the products

$$dc_Y/dt = k_1 c_A; \qquad dc_Z/dt = k_2 c_A$$

Dividing one of these equations by the other gives $dc_Y/dc_Z = k_1/k_2$, the time variable canceling. Integration yields

$$\frac{c_Y}{c_Z} = \frac{k_1}{k_2} \tag{3-49}$$

which we had obtained earlier by a different argument. By measurement of the product ratio at any time the rate constant ratio can be calculated.

Scheme XIII is a more complex example. Suppose that reactant B is consumed only in the second-order reaction.

$$A \xrightarrow{k_1} Y$$

$$A + B \xrightarrow{k_2} Z$$

Scheme XIII

The differential rate equations are

$$-\frac{dc_A}{dt} = k_1 c_A + k_2 c_A c_B \tag{3-50}$$

$$-\frac{dc_B}{dt} = k_1 c_A c_B \tag{3-51}$$

Dividing (3-50) by (3-51),

$$\frac{dc_A}{dc_B} = \frac{k_1}{k_2 c_B} + 1 \tag{3-52}$$

Integration gives Eq. (3-53).

$$c_A^0 - c_A = c_B^0 - c_B + \frac{k_1}{k_2} \ln (c_B^0/c_B) \tag{3-53}$$

which relates c_B to c_A for this reaction scheme: Rearrangement gives an expression for the rate constant ratio:

$$\frac{k_1}{k_2} = \frac{(c_A^0 - c_A) - (c_B^0 - c_B)}{\ln (c_B^0/c_B)} \tag{3-54}$$

From Eq. (3-54) and measurements of c_A and c_B, the ratio k_1/k_2 can be obtained. Solution of Eq. (3-53) for c_A, substitution into Eq. (3-51), and integration gives

$$k_2 t = \int_{c_B}^{c_B^0} \frac{dc_B}{c_B[c_A^0 - c_B^0 + c_B - (k_1/k_2)\ln(c_B^0/c_B)]} \tag{3-55}$$

Numerical treatment is necessary. From a measurement of c_B at time t the definite integral in Eq. (3-55) is evaluated, and this gives a value for k_2. Emanuel' and Knorre[33] show an example of this calculation. Benson[3, pp. 43–6] treats other examples of the elimination of the time variable.

Replacement of Time with an Area Variable

In 1950 French[34] and Wideqvist[35] independently described a data treatment that makes use of the area under the concentration–time curve, and later authors have discussed the method.[36,37] We introduce the technique by considering the second-order reaction of A and B, for which the differential rate equation is

$$\frac{dc_A}{dt} = -kc_A c_B \tag{3-56}$$

Now define a variable θ by Eq. (3-57).

$$\theta = \int_0^t c_B dt \tag{3-57}$$

so that $d\theta = c_B dt$. Combining this equality with Eq. (3-56),

$$\frac{dc_A}{d\theta} = -kc_A \tag{3-58}$$

which integrates to Eq. (3-59).

$$c_A = c_A^0 e^{-k\theta} \tag{3-59}$$

Thus, the technique consists of a transformation from the time differential dt to the area differential $d\theta$, and the essential effect of this transformation is a reduction by one of the apparent order of the reaction. The variable θ is the area under the curve of c_B vs. time from $t = 0$ to time t. With modern computer techniques for integrating experimental curves, this method should be attractive.

Applications have been made to consecutive reactions,[34-40] with several methods being developed to extract the rate constants. Consider Scheme XIV.

$$A \underset{k_{-1}}{\overset{k_1}{\rightleftharpoons}} B$$

$$B \overset{k_2}{\rightarrow} C$$

Scheme XIV

The differential rate equations are

$$\frac{dc_A}{dt} = -k_1 c_A + k_{-1} c_B \tag{3-60}$$

$$\frac{dc_B}{dt} = k_1 c_A - (k_{-1} + k_2) c_B \tag{3-61}$$

$$\frac{dc_C}{dt} = k_2 c_B \tag{3-62}$$

The treatment will depend upon the chemistry of the system and upon which concentrations can be followed. Suppose c_B is accessible. Using the definition of Eq. (3-57) converts Eq. (3-62) to

$$\frac{dc_C}{d\theta} = k_2 \tag{3-63}$$

so that the rate of production of C is zero-order when followed with respect to the variable θ. Another approach[37] is to write Eq. (3-60) as

$$- \int_{c_A^0}^{c_A} \frac{dc_A}{c_A} = k_1 \int_0^t dt - k_{-1} \int_0^\theta c_B dt$$

Again using the definition (3-57),

$$- \ln \frac{c_A}{c_A^0} = k_1 t - k_{-1}\theta \tag{3-64}$$

Now two simultaneous equations of the form of (3-64) are written at points $(c_{A,1}, t_1, \theta_1)$ and $(c_{A,2}, t_1, \theta_2)$ and are solved for the two rate constants. If the concentration of c_A is available, a similar development can be made with Eq. (3-61).

The Laplace Transformation

Linear differential equations with constant coefficients can be solved by a mathematical technique called the *Laplace transformation*. Systems of zero-order or first-order reactions give rise to differential rate equations of this type, and the Laplace transformation often provides a simple solution.

Let $y = F(t)$ be a function of t (in our systems t represents time). Then the *Laplace transform* of $F(t)$ is defined by Eq. (3-65).

$$f(s) = \int_0^\infty F(t)e^{-st} dt \tag{3-65}$$

The Laplace transformation converts a function of t, $F(t)$, into a function of s, $f(s)$, where s is the transform variable. The quantity $f(s)$ is called the *Laplace transform of F(t)*. Equation (3-66) shows several equivalent symbolic representations of the Laplace transform of the function $y = F(t)$.

$$f(s) = L[F(t)] = L(y) = \bar{y} \tag{3-66}$$

Some authors use the letter p instead of s for the transform variable.

We will obtain some Laplace transforms by applying the definition, Eq. (3-65). Suppose $F(t) = a$, where a is a constant. Then

$$f(s) = \int_0^\infty ae^{-st}\, dt = \left. -\frac{a}{s} e^{-st} \right]_0^\infty = \frac{a}{s}$$

Next consider the exponential function, which is important in kinetics. Let $F(t) = e^{-at}$.

$$f(s) = \int_0^\infty e^{-at}e^{-st}dt = \int_0^\infty e^{-(s+a)t}dt = \frac{1}{s+a}$$

In this way many Laplace transforms can be found. Table 3-1 gives a small selection of transforms.

Application of the definition shows that the Laplace transform is a linear operator;[41] this property is represented in Eqs. (3-67) and (3-68).

$$L[F_1(t) + F_2(t)] = L[F_1(t)] + L[F_2(t)] \tag{3-67}$$

$$L[aF(t)] = aL[F(t)] \tag{3-68}$$

The Laplace transform of a derivative dy/dt is found by application of Equation (3-65) and integration by parts:

$$f(s) = \int_0^\infty e^{-st} \frac{dy}{dt}\, dt = \left. ye^{-st} \right]_0^\infty -(-s) \int_0^\infty ye^{-st}dt$$

$$f(s) = -y_0 + s\bar{y} \tag{3-69}$$

where y_0 is the value of y when $t = 0$, and $\bar{y}$ is the transform of y.

To take the inverse Laplace transform means to reverse the process of taking the transform, and for this purpose a table of transforms is valuable. To illustrate, we consider a simple first-order reaction, whose differential rate equation is

$$\frac{dc_A}{dt} = -kc_A$$

TABLE 3-1. Some Laplace Transforms[a]

No.	$F(t)$	$f(s)$
1	a	$\dfrac{a}{s}$
2	t	$\dfrac{1}{s^2}$
3	$\dfrac{t^{n-1}}{(n-1)!}$	$\dfrac{1}{s^n}$
4	e^{at}	$\dfrac{1}{s-a}$
5	te^{at}	$\dfrac{1}{(s-a)^2}$
6	$\dfrac{1}{a-b}[e^{at} - e^{bt}]$	$\dfrac{1}{(s-a)(s-b)}$
7	$\dfrac{1}{a-b}[ae^{at} - be^{bt}]$	$\dfrac{s}{(s-a)(s-b)}$
8	$-\dfrac{(b-c)e^{at} + (c-a)e^{bt} + (a-b)e^{ct}}{(a-b)(b-c)(c-a)}$	$\dfrac{1}{(s-a)(s-b)(s-c)}$
9	$\sin at$	$\dfrac{a}{s^2 + a^2}$
10	$\cos at$	$\dfrac{s}{s^2 + a^2}$
11	dy/dt	$s\bar{y} - y_0$

[a]The quantities a, b, and c are constants; n is a positive integer; y_0 is the value of y when $t = 0$.

Applying Eq. (3-68) to the right side and Eq. (3-69) to the left side provides the Laplace transform:

$$s\bar{c}_A - c_A^0 = -k\bar{c}_A$$

We solve this for the transform $\bar{c}_A$:

$$\bar{c}_A = \frac{c_A^0}{s + k}$$

We now take the inverse transform, which converts $\bar{c}_A$ to c_A. From transform No. 4 in Table 3-1, with $a = -k$, we obtain

$$c_A = c_A^0 e^{-kt}$$

which we know to be the correct result.

Thus the Laplace transformation constitutes a method of integration, and a table of Laplace transforms plays a role in this process that is analogous to a table of integrals. It sometimes happens, of course, that a transform that arises in the solution to a problem cannot be found in a table of transforms. In such a case it is often possible, by algebraic manipulation, to place the transform in a form that permits the inverse transform to be taken. For the kinds of functions that arise in kinetics problems the following theorem, which we refer to as the *general partial fraction theorem*,[42] is very useful. This theorem (which we state without proof) applies to a transform that can be written.

$$\bar{y} = \frac{H(s)}{G(s)} \tag{3-70}$$

where $H(s)$ and $G(s)$ are polynomials in s, the degree of $H(s)$ being less than that of $G(s)$. If $G(s)$ can be written in the form

$$G(s) = (s - a_1)(s - a_2) \cdots (s - a_n) \tag{3-71}$$

where the $a_1, a_2, \ldots, a_n$ must all be different, then

$$\bar{y} = \sum_{r=1}^{n} \frac{1}{(s - a_r)} \cdot \frac{H(a_r)}{(a_r - a_1) \cdots (a_r - a_{r-1})(a_r - a_{r+1}) \cdots (a_r - a_n)} \tag{3-72}$$

In Eq. (3-72), $H(a_r)$ is the function $H(s)$ with s set equal to a_r (in each term of the sum). Similarly, in the denominator, s is set equal to a_r in $G(s)$. Note, however, that the factor $(s - a_r)$ is missing in the denominator (this factor would be zero). Hence, one evaluates the rth term in the sum *omitting* the factor $(s - a_r)$ in the denominator.

We illustrate with a numerical example. Suppose

$$\bar{y} = \frac{s + 2}{s^2 + 5s}$$

which can be written

$$\bar{y} = \frac{s + 2}{s(s + 5)}$$

This transform does not appear in Table 3-1. Because it has the form of Eqs. (3-70) and (3-71), with $a_1 = 0$, $a_2 = -5$, we apply Eq. (3-72):

$$\bar{y} = \frac{1}{(s - 0)} \cdot \frac{(0 + 2)}{(0 + 5)} + \frac{1}{(s + 5)} \cdot \frac{(-5 + 2)}{(-5)}$$

$$\bar{y} = \frac{2}{5s} + \frac{3}{5(s + 5)}$$

By taking inverse transforms of the separate terms, using Nos. 1 and 4 of Table 3-1, we find y:

$$y = \frac{1}{5}(2 + 3e^{-5t})$$

We now consider the solution of differential equations by means of Laplace transforms. We have already solved one equation, namely, the first-order rate equation, but the technique is capable of more than this. It allows us to solve simultaneous differential equations.

This is the procedure: From the postulated kinetic scheme we write the differential rate equations. Take the Laplace transforms of the differential equations. Solve the resulting set of algebraic equations for the transforms of the concentrations. Then take the inverse transforms to obtain the concentrations as functions of time.

As an example, we take kinetic scheme IX, for which the differential rate equations are

$$\frac{dc_A}{dt} = -k_1 c_A \tag{3-73}$$

$$\frac{dc_B}{dt} = k_1 c_A - k_2 c_B \tag{3-74}$$

Taking the transforms of these equations gives

$$s\bar{c}_A - c_A^0 = -k_1 \bar{c}_A \tag{3-75}$$

$$s\bar{c}_B = k_1 \bar{c}_A - k_2 \bar{c}_B \tag{3-76}$$

where the initial conditions are $c_A = c_A^0$ and $c_B = 0$ at $t = 0$. Solve Eq. (3-75) for $\bar{c}_A$, place this in Eq. (3-76), and solve for $\bar{c}_B$:

$$\bar{c}_B = \frac{k_1 c_A^0}{(s + k_1)(s + k_2)} \tag{3-77}$$

The inverse transform is given by No. 6 in Table 3-1:

$$c_B = \frac{k_1 c_A^0}{k_2 - k_1}\left[e^{-k_1 t} - e^{-k_2 t}\right] \tag{3-78}$$

We had earlier obtained this result by a different method.

Notice that this method of solving differential equations yields the desired particular solution, the initial conditions being introduced early in the procedure.

We next consider Scheme XIV, for which, in the preceding subsection, a partial solution was obtained. The scheme is rewritten here for convenience.

$$A \underset{k_{-1}}{\overset{k_1}{\rightleftharpoons}} B$$

$$B \overset{k_2}{\rightarrow} C$$

Scheme XIV

The differential rate equations are

$$\frac{dc_A}{dt} = -k_1 c_A + k_{-1} c_B \tag{3-79}$$

$$\frac{dc_B}{dt} = k_1 c_A - k c_B \tag{3-80}$$

where $k = k_{-1} + k_2$. Taking transforms gives Eqs. (3-81) and (3-82).

$$s\bar{c}_A + k_1 \bar{c}_A - k_{-1} \bar{c}_B = c_A^0 \tag{3-81}$$

$$s\bar{c}_B - k_1 \bar{c}_A + k \bar{c}_B = 0 \tag{3-82}$$

Solving Eq. (3-81) for $\bar{c}_A$ gives

$$\bar{c}_A = \frac{c_A^0 + k_{-1} \bar{c}_B}{s + k_1}$$

This is substituted into Eq. (3-82), which is solved for $\bar{c}_B$:

$$\bar{c}_B = \frac{k_1 c_A^0}{s^2 + (k_1 + k_{-1} + k_2)s + k_1 k_2} \tag{3-83}$$

Equation (3-83) can be placed in the form

$$\bar{c}_B = \frac{k_1 c_A^0}{(s + \alpha)(s + \beta)} \tag{3-84}$$

where

$$\alpha\beta = k_1 k_2 \tag{3-85}$$

$$\alpha + \beta = k_1 + k_{-1} + k_2 \tag{3-86}$$

With No. 6 of Table 3-1, the inverse transformation yields

$$c_B = \frac{k_1 c_A^0}{\beta - \alpha} \left[e^{-\alpha t} - e^{-\beta t} \right] \tag{3-87}$$

This is a very interesting result. The time course is identical in form with that given by Eq. (3-78) for Scheme IX, but in Eq. (3-87) the rate parameters α and β are not elementary rate constants; instead they are composite quantities defined by Eqs. (3-85) and (3-86).

To find c_A we substitute Eq. (3-84) into Eq. (3-82) and find $\bar{c}_A$:

$$\bar{c}_A = \frac{(s + k)c_A^0}{(s + \alpha)(s + \beta)} \tag{3-88}$$

We apply the general partial fraction theorem, Eq. (3-72), to Eq. (3-88), with $a_1 = -\alpha$, $a_2 = -\beta$.

$$\bar{c}_A = \frac{(k - \alpha)c_A^0}{(\beta - \alpha)(s + \alpha)} + \frac{(k - \beta)c_A^0}{(\alpha - \beta)(s + \beta)} \tag{3-89}$$

Using No. 4 of Table 3-1,

$$c_A = \frac{c_A^0}{\beta - \alpha} \left[(k - \alpha)e^{-\alpha t} - (k - \beta)e^{-\beta t} \right] \tag{3-90}$$

[The transform $\bar{c}_A$ can be found by alternative algebraic routes, and it will appear to be different from Eq. (3-89), and the inverse transform will not appear to be identical to Eq. (3-90), but these differences in appearance result because the parameters are composite quantities.]

Evidently for any system of linear differential equations with constant coefficients the Laplace transform solution is possible in principle. In practice there are two stages of potential difficulty: The first of these is the solution of the simultaneous algebraic equations for the transforms, and the second is the inverse transformation. The algebraic equations can be solved by any standard method. In the preceding solution the method of elimination was used, and this is effective if there are not more than three unknowns. For larger systems solution by matrix algebra may be preferable. The inverse transformation step does not appear to be a serious problem in kinetics applications, because first-order reactions will give rise to sums of exponentials, and the general partial fraction theorem permits the transform to be expressed in a form convenient for the inverse transformation.

Although a closed-form solution can thus be obtained by this method for any system of first-order equations, the result is often too cumbersome to lead to estimates of the rate constants from concentration–time data. However, the reverse calculation is always possible; that is, with numerical values of the rate constants, the concentration–time curve can be calculated. This provides the basis for a curve-fitting approach to the evaluation of rate constants. The availability of the concentration–time function also allows nonlinear regression analysis to be applied.

One further system will be solved by the transform method. Scheme XV constitutes two consecutive reversible first-order reactions.

$$A \underset{k_{-1}}{\overset{k_1}{\rightleftharpoons}} B$$

$$B \underset{k_{-2}}{\overset{k_2}{\rightleftharpoons}} C$$

Scheme XV

The differential rate equations are

$$\frac{dc_A}{dt} = -k_1 c_A + k_{-1} c_B \tag{3-91}$$

$$\frac{dc_B}{dt} = k_1 c_A - (k_{-1} + k_2) c_B + k_{-2} c_C \tag{3-92}$$

$$\frac{dc_C}{dt} = k_1 c_B - k_{-2} c_C \tag{3-93}$$

With the initial conditions $c_A^0 = c_A^0$, $c_B^0 = 0$, $c_C^0 = 0$, the mass balance relationship is $c_A^0 = c_A + c_B + c_C$. This is used in Eq. (3-92) to eliminate c_C; then the Laplace transforms of (3-91) and (3-92) are taken:

$$s\bar{c}_A - c_A^0 = -k_1 \bar{c}_A + k_{-1} \bar{c}_B \tag{3-94}$$

$$s\bar{c}_B = (k_1 - k_{-2})\bar{c}_A - (k_{-1} + k_2 + k_{-2})\bar{c}_B + \frac{k_{-2} c_A^0}{s} \tag{3-95}$$

Equation (3-94) is solved for $\bar{c}_A$, and this is substituted into (3-95), which is solved for $\bar{c}_B$,

$$\bar{c}_B = \frac{k_1 c_A^0(s + k_{-2})}{s[s^2 + (k_1 + k_{-1} + k_2 + k_{-2})s + (k_1 k_2 + k_1 k_{-2} + k_{-1} k_{-2})]}$$

which can be written

$$\bar{c}_B = \frac{k_1 c_A^0(s + k_{-2})}{s(s + \alpha)(s + \beta)} \tag{3-96}$$

where

$$\alpha + \beta = k_1 + k_{-1} + k_2 + k_{-2} \tag{3-97}$$

$$\alpha\beta = k_1 k_2 + k_1 k_{-2} + k_{-1} k_{-2} \tag{3-98}$$

Applying the general partial fraction theorem, Eq. (3-72), to Eq. (3-96) and then taking inverse transforms gives Eq. (3-99).

$$c_B = k_1 c_A^0 \left[\frac{k_{-2}}{\alpha\beta} + \frac{(k_{-2} - \alpha)}{\alpha(\alpha - \beta)} e^{-\alpha t} - \frac{(k_{-2} - \beta)}{\beta(\alpha - \beta)} e^{-\beta t} \right] \tag{3-99}$$

To find c_A solve Eq. (3-94) for $\bar{c}_B$, substitute it into Eq. (3-95), and solve for $\bar{c}_A$. The result is

$$\bar{c}_A = c_A^0 \left[\frac{k_{-1} k_{-2} + s(s + m)}{s(s + \alpha)(s + \beta)} \right] \tag{3-100}$$

where $m = k_{-1} + k_2 + k_{-2}$, and α and β have the meanings assigned by Eqs. (3-97) and (3-98). Applying Eq. (3-72) and taking the inverse transforms,

$$c_A = c_A^0 \left[\frac{k_{-1} k_{-2}}{\alpha\beta} + \frac{k_{-1} k_{-2} - \alpha(m - \alpha)}{\alpha(\alpha - \beta)} e^{-\alpha t} + \frac{k_{-1} k_{-2} - \beta(m - \beta)}{\beta(\beta - \alpha)} e^{-\beta t} \right] \tag{3-101}$$

Equation (3-101) is a valid expression for c_A, but some simplification is possible by combining the definition of m with Eqs. (3-97) and (3-98):

$$c_A = c_A^0 \left[\frac{k_{-1} k_{-2}}{\alpha\beta} + \frac{k_1(\alpha - k_2 - k_{-2})}{\alpha(\alpha - \beta)} e^{-\alpha t} + \frac{k_1(\beta - k_2 - k_{-2})}{\beta(\beta - \alpha)} e^{-\beta t} \right] \tag{3-102}$$

The concentration c_C is easily found by substituting Eqs. (3-99) and (3-102) into the mass balance expression:

$$c_C = k_1 k_2 c_A^0 \left[\frac{1}{\alpha\beta} + \frac{1}{\alpha(\alpha - \beta)} e^{-\alpha t} - \frac{1}{\beta(\alpha - \beta)} e^{-\beta t} \right] \tag{3-103}$$

Lowry and John[43] studied Scheme XV and discussed the nature of the concentration–time curves, noting that the concentration of B will pass through a maximum if $k_2 < k_1$, whereas if $k_2 > k_1$, it will not display a maximum.

Secular Equations and Eigenvalues

Systems of reversible first-order reactions lead to sets of simultaneous linear differential equations with constant coefficients. A solution may be obtained by means of a matrix formulation that is widely used in quantum mechanics and vibrational spectroscopy.[44] The method was applied to kinetics problems by Zwolinski and Eyring[45] and Matsen and Franklin.[46] It has been generalized by Benson,[3, pp. 39–42] Ritchie,[47] and Carpenter.[48] When the method is expressed in its full mathematical generality, the connection between the mathematics and experiment may be difficult to discern, so here we will develop the method by treating some specific examples in order that the physical content will be more readily evident.

First, we consider Scheme I, a single reversible first-order reaction; Eyring et al.[49] treated this case.

$$A \underset{k_{-1}}{\overset{k_1}{\rightleftharpoons}} Z$$

Scheme I

The differential rate equations are

$$c_A' = -k_1 c_A + k_{-1} c_Z \tag{3-104}$$

$$c_Z' = k_1 c_A - k_{-1} c_Z \tag{3-105}$$

where $c_i' = dc_i/dt$. In matrix notation these equations become

$$\begin{pmatrix} c_A' \\ c_Z' \end{pmatrix} = \begin{pmatrix} -k_1 & k_{-1} \\ k_1 & -k_{-1} \end{pmatrix} \begin{pmatrix} c_A \\ c_Z \end{pmatrix}$$

the parentheses denoting matrices; more succinctly,

$$\mathbf{c'} = \mathbf{kc} \tag{3-106}$$

Equation (3-106) is a simple first-order equation, whose solution is

$$\mathbf{c} = \mathbf{J}e^{-\mathbf{k}t}$$

where boldface symbols represent matrices.

This result shows that we can write particular solutions to Eqs. (3-104) and (3-105) as

$$c_A = J_A\, e^{-\lambda t}; \qquad c_Z = J_Z\, e^{-\lambda t} \tag{3-107}$$

where λ is a parameter that is, at present, an unknown function of the elementary rate constants. General solutions are then constructed as linear combinations of particular solutions; for this problem,

$$c_A = J_{A1}e^{-\lambda_1 t} + J_{A2}e^{-\lambda_2 t}$$

$$c_Z = J_{Z1}e^{-\lambda_1 t} + J_{Z2}e^{-\lambda_2 t} \tag{3-108}$$

From Eqs. (3-107) we obtain

$$c_A' = -\lambda J_A e^{-\lambda t}; \qquad c_Z' = -\lambda J_Z e^{-\lambda t} \qquad (3\text{-}109)$$

Substituting Eqs. (3-107) and (3-109) into (3-104) and (3-105) yields the simultaneous homogeneous algebraic equations (3-110).

$$(k_1 - \lambda)J_A - k_{-1}J_Z = 0$$
$$-k_1 J_A + (k_{-1} - \lambda)J_Z = 0 \qquad (3\text{-}110)$$

These equations are satisfied if $J_A = 0$, $J_Z = 0$, but this trivial solution is of no interest. To ensure a nontrivial solution, it is sufficient to require that the determinant of the coefficients of J_A and J_Z be equal to zero, namely,

$$\begin{vmatrix} k_1 - \lambda & -k_{-1} \\ -k_1 & k_{-1} - \lambda \end{vmatrix} = 0 \qquad (3\text{-}111)$$

Equation (3-111) is the *secular equation* for this problem. [Pauling and Wilson[50] discuss the meaning of the term *secular* in this context.] Expansion of the determinant gives a polynomial in λ that is called the *characteristic equation:*

$$\lambda^2 - (k_1 + k_{-1})\,\lambda = 0 \qquad (3\text{-}112)$$

Evidently the roots of this characteristic equation are

$$\lambda_1 = 0$$

$$\lambda_2 = k_1 + k_{-1}$$

The roots of the characteristic equation are called the *eigenvalues* of the secular equation.

To generalize our findings thus far: If the system contains n chemical states, there will be n differential rate equations, of which $n - 1$ are independent. The secular equation is an nth-order determinant, and the characteristic equation is an nth degree polynomial in λ. The roots $\lambda_1, \lambda_2, \ldots, \lambda_n$ are the n eigenvalues of the determinant. One of the eigenvalues will be zero (corresponding to one of the rate equations being not independent); the remaining $n - 1$ eigenvalues are functions of the elementary rate constants. These statements apply to reversible reactions.

We continue by substituting the eigenvalues, in turn, into the algebraic equations (3-110). Because these equations are not independent, it is not possible to solve uniquely for the individual J_A, J_Z values; only ratios can be obtained, as follows:

$$\lambda_1 = 0: \qquad\qquad k_1 J_{A1} = k_{-1}J_{Z1} \qquad (3\text{-}113)$$

$$\lambda_2 = k_1 + k_{-1}: \qquad -J_{A2} = J_{Z2} \qquad (3\text{-}114)$$

Equations (3-113) and (3-114) are substituted into the general solutions, Eqs. (3-108):

$$c_A = J_{A1}e^{-\lambda_1 t} + J_{A2}e^{-\lambda_2 t}$$

$$c_Z = \frac{k_1}{k_{-1}} J_{A1}e^{-\lambda_1 t} - J_{A2}e^{-\lambda_2 t}$$

(3-115)

We have one further piece of information, namely, the initial conditions. Let $c_A = c_A^0, c_Z = 0$ at $t = 0$; using these conditions allows J_{A1} and J_{A2} to be evaluated,

$$J_{A1} = \frac{c_A^0}{1 + K}; \qquad J_{A2} = \frac{Kc_A^0}{1 + K}$$

(3-116)

where $K = k_1/k_{-1}$, the equilibrium constant for the reaction. The solution is, therefore,

$$c_A = \frac{c_A^0}{1 + K} + \frac{Kc_A^0}{1 + K} e^{-(k_1 + k_{-1})t}$$

$$c_Z = \frac{Kc_A}{1 + K} - \frac{Kc_A}{1 + K} e^{-(k_1 - k_{-1})t}$$

(3-117)

To put this into a more familiar form, let $t = \infty$; then $c_A = c_A^e$, $c_A^e = c_A^0/(1 + K)$, and

$$c_A - c_A^e = \frac{Kc_A^0}{1 + K} e^{-(k_1 + k_{-1})t}$$

When $t = 0$, $c_A - c_A^e = c_A^0 - c_A^e = Kc_A^0/(1 + K)$, giving

$$c_A - c_A^e = (c_A^0 - c_A^e) e^{-(k_1 + k_{-1})t}$$

(3-118)

which is identical to Eq. (3-4).

Next we apply this method to Scheme XV.

$$A \underset{k_{-1}}{\overset{k_1}{\rightleftharpoons}} B$$

$$B \underset{k_{-2}}{\overset{k_2}{\rightleftharpoons}} C$$

Scheme XV

Frost and Pearson[51] treated Scheme XV by the eigenvalue method, and we have solved it by the method of Laplace transforms in the preceding subsection. The differential rate equations are

$$c_A' = -k_1 c_A + k_{-1} c_B$$

$$c_B' = k_1 c_A - (k_{-1} + k_2)c_B + k_{-2} c_C$$

$$c_C' = \qquad\qquad k_2 c_B \qquad - k_{-2} c_C$$

Following exactly the procedure applied in the earlier example, these differential equations are transformed into algebraic equations.

$$(k_1 - \lambda)J_A - k_{-1}J_B = 0$$

$$-k_1 J_A + (k_{-1} + k_2 - \lambda)J_B - k_{-2}J_C = 0 \tag{3-119}$$

$$-k_2 J_B + (k_{-2} - \lambda)J_C = 0$$

The secular equation is, therefore,

$$\begin{vmatrix} k_1 - \lambda & -k_{-1} & 0 \\ -k_1 & k_{-1} + k_2 - \lambda & -k_{-2} \\ 0 & -k_2 & k_{-2} - \lambda \end{vmatrix} = 0 \tag{3-120}$$

The characteristic equation is

$$\lambda^3 - \lambda^2(k_1 + k_{-1} + k_2 + k_{-2}) + \lambda(k_1 k_2 + k_1 k_{-2} + k_{-1} k_{-2}) = 0 \tag{3-121}$$

The roots of the equation are

$$\lambda_1 = 0$$

$$\lambda_2 = \tfrac{1}{2}[M + (M^2 - 4N)^{1/2}] \tag{3-122}$$

$$\lambda_3 = \tfrac{1}{2}[M - (M^2 - 4N)^{1/2}]$$

where

$$M = k_1 + k_{-1} + k_2 + k_{-2}$$
$$\tag{3-123}$$
$$N = k_1 k_2 + k_1 k_{-2} + k_{-1} k_{-2}$$

We digress briefly to find how the eigenvalues are related to the rate constants. From Eqs. (3-122) and (3-123),

$$\lambda_2 + \lambda_3 = k_1 + k_{-1} + k_2 + k_{-2} \tag{3-124}$$

$$\lambda_2\lambda_3 = k_1k_2 + k_1k_{-2} + k_{-1}k_{-2} \tag{3-125}$$

Comparison of Eqs. (3-124) and (3-125) with Eqs. (3-97) and (3-98) shows that $\lambda_2 + \lambda_3 = \alpha + \beta$ and $\lambda_2\lambda_3 = \alpha\beta$, where α and β are the rate parameters that arise in the Laplace transform method of solution.

Continuing, we successively substitute λ_1, λ_2, and λ_3 into Eqs. (3-119) (actually into the first and third of these, for simplicity) and express the J_{Bi} and J_{Ci} in terms of the J_{Ai} (where $i = 1,2,3$). We find these results:

$$J_{Bi} = \left[\frac{k_1 - \lambda_i}{k_{-1}}\right] J_{Ai}$$

$$J_{Ci} = \left[\frac{k_2(k_1 - \lambda_i)}{k_{-1}(k_{-2} - \lambda_i)}\right] J_{Ai} \tag{3-126}$$

Now Eqs. (3-126) are used in the general solutions, which can be summarized as

$$c_X = \sum_{i=1}^{n} J_{Xi}e^{-\lambda_i t} \tag{3-127}$$

where $X = $ A, B, C and $i = 1, 2, 3$. This gives

$$c_A = J_{A1} + J_{A2}e^{-\lambda_2 t} + J_{A3}e^{-\lambda_3 t} \tag{3-128a}$$

$$c_B = \frac{k_1}{k_{-1}} J_{A1} + \left(\frac{k_1 - \lambda_2}{k_{-1}}\right) J_{A2}e^{-\lambda_2 t} + \left(\frac{k_1 - \lambda_3}{k_{-1}}\right) J_{A3}e^{-\lambda_3 t} \tag{3-128b}$$

$$c_C = \frac{k_1k_2}{k_{-1}k_{-2}} J_{A1} + \left[\frac{k_2(k_1 - \lambda_2)}{k_{-1}(k_{-2} - \lambda_2)}\right] J_{A2}e^{-\lambda_2 t}$$

$$+ \left[\frac{k_2(k_1 - \lambda_3)}{k_{-1}(k_{-2} - \lambda_2)}\right] J_{A3}e^{-\lambda_3 t} \tag{3-128c}$$

Next the initial conditions are imposed; let $c_A = c_A^0$, $c_B = 0$, $c_C = 0$ at $t = 0$. Substitution into Eqs. (3-128) gives three equations in the three unknowns J_{A1}, J_{A2}, J_{A3}.

$$c_A^0 = J_{A1} + J_{A2} + J_{A3} \tag{3-129a}$$

$$0 = \frac{k_1}{k_{-1}} J_{A1} + \left(\frac{k_1 - \lambda_2}{k_{-1}}\right) J_{A2} + \left(\frac{k_1 - \lambda_3}{k_{-1}}\right) J_{A3} \tag{3-129b}$$

$$0 = \frac{k_1k_2}{k_{-1}k_{-2}} J_{A1} + \left[\frac{k_2(k_1 - \lambda_2)}{k_{-1}(k_{-2} - \lambda_2)}\right] J_{A2} + \left[\frac{k_2(k_1 - \lambda_3)}{k_{-1}(k_{-2} - \lambda_3)}\right] J_{A3} \tag{3-129c}$$

Equations (3-129) are solved [with the use of Eqs. (3-124) and (3-125)] to give

$$J_{A1} = \frac{c_A^0 k_{-1} k_{-2}}{\lambda_2 \lambda_3}$$

$$J_{A2} = \frac{c_A^0 k_1 (\lambda_2 - k_2 - k_{-2})}{\lambda_2 (\lambda_2 - \lambda_3)} \tag{3-130}$$

$$J_{A3} = \frac{c_A^0 k_1 (k_2 + k_{-2} - \lambda_3)}{\lambda_3 (\lambda_2 - \lambda_3)}$$

Then Eqs. (3-130) are substituted into Eqs. (3-128), giving c_A, c_B, and c_C as functions of time. The final expressions are not written here because we have already derived them by the Laplace transform method; they are Eqs. (3-99), (3-101), and (3-103), with λ_2 and λ_3 replacing α and β.

In the context of chemical kinetics, the eigenvalue technique and the method of Laplace transforms have similar capabilities, and a choice between them is largely dependent upon the amount of algebraic labor required to reach the final result. Carpenter[48] discusses matrix operations that can reduce the manipulations required to proceed from the eigenvalues to the concentration–time functions. When dealing with complex reactions that include irreversible steps by the eigenvalue method, the system should be treated as an equilibrium system, and then the desired special case derived from the general result. For such problems the Laplace transform method is more efficient.

The Preequilibrium Assumption

Scheme XVI is more difficult to analyze than are any of the kinetic schemes treated earlier in this chapter, because it includes a second-order reaction.

$$A + B \underset{k_{-1}}{\overset{k_1}{\rightleftharpoons}} C$$

$$C \overset{k_2}{\rightarrow} Z$$

Scheme XVI

The rate of reaction, expressed as the rate of product formation, is

$$v = k_2 c_C \tag{3-131}$$

The problem can be greatly simplified if k_1 and k_{-1} are much larger than k_2, for then it will be reasonable to assume that the reaction of A and B to form C is essentially always at equilibrium. That is, with this equilibrium assumption the concentration of C is considered to be determined solely by the equilibrium, being unperturbed by the k_2 step of the scheme.

First, consider the nature of equilibrium from the kinetic viewpoint. If the reaction

$$A + B \underset{k_{-1}}{\overset{k_1}{\rightleftharpoons}} C$$

is at thermodynamic equilibrium, the rate of the forward reaction is equal to the rate of the reverse reaction, or

$$k_1 c_A c_B = k_{-1} c_C \tag{3-132}$$

Because the equilibrium constant K is given by $c_C/c_A c_B$, we obtain

$$K = \frac{k_1}{K_{-1}} \tag{3-133}$$

Now the equilibrium assumption is applied to Scheme XVI. From the above considerations, $c_C = K c_A c_B$, which, combined with Eq. (3-131), gives

$$v = k_2 K c_A c_B \tag{3-134}$$

showing that the rate of reaction is described by a simple second-order rate equation. A great simplification has been provided by the equilibrium assumption.

A system of this type is commonly said to possess a *fast preequilibrium* step. Proton transfers constitute a very important class of fast preequilibria, as illustrated by Scheme XVII for acid-catalyzed ester hydrolysis.

$$
\begin{array}{ccc}
\overset{\displaystyle O}{\underset{\displaystyle \|}{}} & & \overset{\displaystyle +OH}{\underset{\displaystyle \|}{}} \\
R\!-\!C\!-\!OR' + H^+ & \underset{\text{Fast}}{\rightleftharpoons} & R\!-\!C\!-\!OR' \\[2em]
\overset{\displaystyle +OH}{\underset{\displaystyle \|}{}} & & \overset{\displaystyle +OH}{\underset{\displaystyle \|}{}} \\
R\!-\!C\!-\!OR' + H_2O & \xrightarrow{\text{Slow}} & R\!-\!C\!-\!OH + R'OH
\end{array}
$$

Scheme XVII

The terms *fast* and *slow* are relative. (The hydrolysis step is followed by another fast proton transfer, but this has no effect on the rate of hydrolysis.)

In order to obtain a better sense of the range of validity of the equilibrium

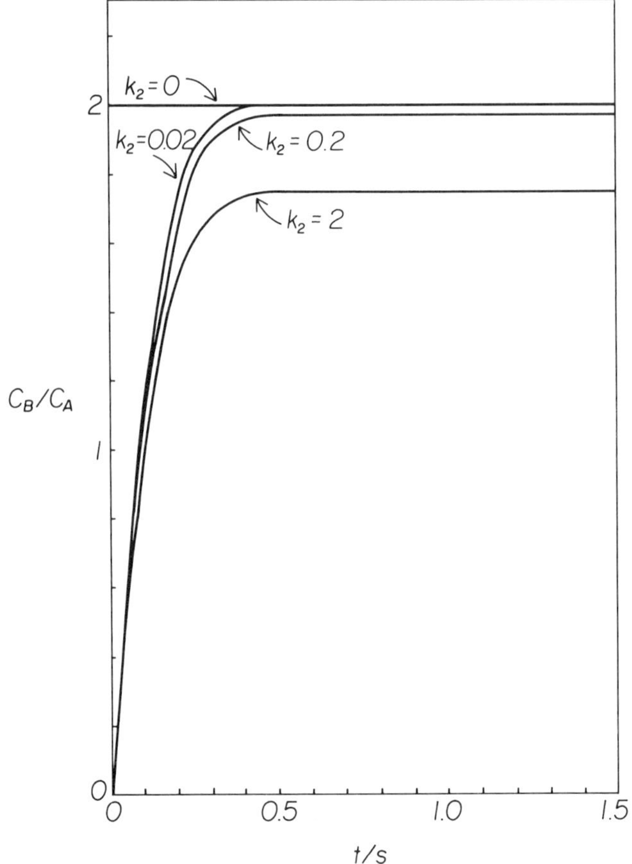

Figure 3-9. Comparison of exact solutions and the preequilibrium assumption for Scheme XIV. For all of the calculations the values $k_1 = 10$ s^{-1}, $k_{-1} = 5$ s^{-1} were used. The k_2 values are given in the figure. The exact solutions were obtained with Eq. (3-135) and the approximate solution (corresponding to $k_2 = 0$) with Eq. (3-136).

assumption, consider Scheme XIV, for which a quantitative description was achieved earlier by means of Laplace transforms (p. 87).

$$A \underset{k_{-1}}{\overset{k_1}{\rightleftharpoons}} B$$

$$B \overset{k_2}{\rightarrow} C$$

Scheme XIV

By combining Eqs. (3-87) and (3-90), Eq. (3-135) is obtained as an exact solution for the ratio c_B/c_A.

$$\frac{c_B}{c_A} = k_1 \left[\frac{e^{-\alpha t} - e^{-\beta t}}{(\beta - k_1)e^{-\alpha t} - (\alpha - k_1)e^{-\beta t}} \right] \tag{3-135}$$

[From Eqs. (3-85) and (3-86), $\alpha\beta = k_1 k_2$ and $\alpha + \beta = k_1 + k_{-1} + k_2$.] The equilibrium assumption gives for this ratio of concentrations

$$\frac{c_B}{c_A} = \frac{k_1}{k_{-1}} \tag{3-136}$$

Figure 3-9 shows plots of Eqs. (3-135) and (3-136) for some hypothetical systems. Obviously the equilibrium approximation is poor in the early stages of the reaction, but in the later stages the assumption can be quite good. The preequilibrium assumption, applied to Scheme XIV, amounts to the statement that k_2 is negligible relative to k_1 and k_{-1}.

The "time scale" of a phenomenon whose characteristic quantity has the units seconds^{-1} is given by the reciprocal of that quantity. In the present context, therefore, we can state that the preequilibrium assumption is valid if the A $\rightleftharpoons$ B reaction is very fast on the time scale $(1/k_2)$ of the B $\rightarrow$ C reaction.

Scheme XVIII introduces another feature, for this is a cyclic kinetic scheme.

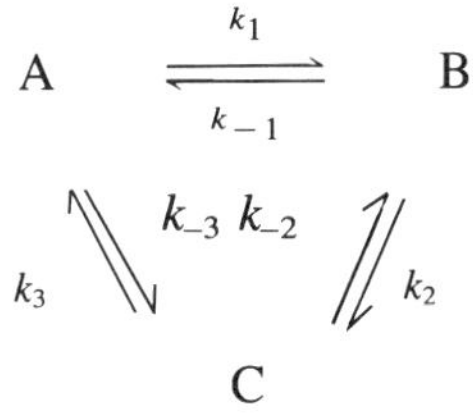

$$A \underset{k_{-1}}{\overset{k_1}{\rightleftharpoons}} B$$

$$k_3 \diagdown \quad k_{-3} \; k_{-2} \quad \diagup k_2$$

$$C$$

Scheme XVIII

This system is capable of attaining thermodynamic equilibrium with respect to all states. [This statement is amplified in Section 3.3, p. 125.] The equilibrium constants are defined

$$K_1 = k_1/k_{-1} = c_B/c_A$$

$$K_2 = k_2/k_{-2} = c_C/c_B$$

$$K_3 = k_3/k_{-3} = c_A/c_C$$

It follows that $K_1 K_2 K_3 = 1$, and, therefore,

$$k_1 k_2 k_3 = k_{-1} k_{-2} k_{-3} \tag{3-137}$$

Equation (3-137) is a general result for this kinetic scheme; it means that only five of the six rate constants are independent.

An exact solution of Scheme XVIII can be found by either the Laplace transform or the eigenvalue method. However, if one of the three equilibria is achieved much more rapidly than the other two, the preequilibrium assumption leads to a simple result. Suppose that k_1 and k_{-1} are much larger than the other rate constants, so that the A, B, pair can be assumed to be in equilibrium throughout the reaction. The rate of change of c_C is

$$\frac{dc_C}{dt} = k_{-3}c_A + k_2c_B - (k_{-2} + k_3)c_C$$

and we have, from the equilibrium assumption, $c_B = K_1c_A$. Let the initial conditions be $c_A = c_A^0$, $c_B = 0$, $c_C = 0$; then the mass balance relationship is $c_A^0 = c_A + c_B + c_C$. Therefore, c_A and c_B can be eliminated from the rate equation to give

$$\frac{dc_C}{dt} = a + bc_C \tag{3-138}$$

where

$$a = -\left[\frac{k_{-3} + k_2K_1}{1 + K_1} + k_{-2} + k_3\right]$$

$$b = c_A^0\left[\frac{k_{-3} + k_2K_1}{1 + K_1}\right]$$

Equation (3-138) is easily integrated, giving a monoexponential time dependence.

Further discussion of the preequilibrium assumption is given in the next subsection.

The Steady-State Approximation

Again we make use of Scheme XIV. Letting $c_i' = dc_i/dt$, the differential rate equations are

$$c_A' = -k_1c_A + k_{-1}c_B \tag{3-139}$$

$$c_B' = k_1c_A - (k_{-1} + k_2)c_B \tag{3-140}$$

$$c_C' = k_2c_B \tag{3-141}$$

Now suppose that the rate constants have values such that the rate of change of c_B is very small relative to the rates of change of other concentrations; then in the conventional formulation it is stated that c_B is at "steady state," and the assumption $c_B' = 0$ is made. With this assumption and Eq. (3-140) we find

$$c_B = \frac{k_1c_A}{k_{-1} + k_2} \tag{3-142}$$

Using Eq. (3-142) in the rate equations for c_A' and c_C' yields

$$c_A' = -c_C' = -\frac{k_1 k_2 c_A}{k_{-1} + k_2} \qquad (3\text{-}143)$$

The reaction displays simple first-order kinetics, with the observed first-order rate constant being equal to $k_1 k_2/(k_{-1} + k_2)$.

This procedure constitutes an application of the *steady-state approximation* [also called the *quasi-steady-state approximation*, the *Bodenstein approximation*,[52] or the *stationary-state hypothesis*]. It is a powerful method for the simplification of complicated rate equations, but because it is an approximation, it is not always valid. Sometimes the inapplicability of the steady-state approximation is easily detected; for example, Eq. (3-143) predicts simple first-order behavior, and significant deviation from this behavior is evidence that the approximation cannot be applied. In more complex systems the validity of the steady-state approximation may be difficult to assess. Because it is an approximation in wide use, much critical attention has been directed to the steady-state hypothesis.

Consider further Scheme XIV and rate equations (3-139) to (3-141). Evidently c_B will be small relative to $(c_A + c_C)$ if $(k_{-1} + k_2) \gg k_1$. Then B plays the role of a "reactive intermediate" in the overall reaction $A \rightarrow C$. This is the usual condition that is taken as a warrant for the application of the steady-state approximation. If c_B is small, it is reasonable that c_B' will be small throughout most the reaction, so it is set equal to zero. As Wong (53) has pointed out, however, the condition $c_B' = 0$ is a sufficient but unnecessary condition for Eq. (3-142) to hold. From Eq. (3-140) we obtain

$$c_B = \frac{k_1 c_A - c_B'}{k_{-1} + k_2}$$

The sufficient and necessary condition is therefore $c_B' \ll k_1 c_A$. As a consequence of imposing the more restrictive condition, which is obviously not correct throughout most of the reaction, it is possible for mathematical inconsistencies to arise in kinetic treatments based on the steady-state approximation. (The condition $c_B' = 0$ is exact only at the moment when c_B passes through an extremum and at equilibrium.)

When reactions other than first-order processes are included in the kinetic scheme, reactant concentrations may appear in the denominator of the rate equation. Scheme XIX is an example.

$$A + B \underset{k_{-1}}{\overset{k_1}{\rightleftharpoons}} C$$

$$C + D \overset{k_2}{\rightarrow} Z$$

Scheme XIX

The rate equations are

$$c'_A = -k_1 c_A c_B + k_{-1} c_C$$

$$c'_C = k_1 c_A c_B - k_{-1} c_C - k_2 c_C c_D$$

$$c'_Z = k_2 c_C c_D$$

Applying the steady-state approximation to C gives

$$c_C = \frac{k_1 c_A c_B}{k_{-1} + k_2 c_D} \tag{3-144}$$

Equation (3-144) substituted in the rate equations for A and Z yields

$$c'_Z = -c'_A = \frac{k_1 k_2 c_A c_B c_D}{k_{-1} + k_2 c_D} \tag{3-145}$$

A study of this system often is carried out with pseudo-order conditions relative to D. Then the apparent second-order rate constant is given by Eq. (3-146).

$$k_{app} = \frac{k_1 k_2 c_D}{k_{-1} + k_2 c_D} \tag{3-146}$$

Observe that k_{app} shows a hyperbolic dependence on c_D; when c_D is very small, k_{app} is a linear function of c_D, whereas at high values of c_D, k_{app} is independent of c_D.

The simplest kinetic scheme that can account for enzyme-catalyzed reactions is Scheme XX, where E represents the enzyme, S is the substrate, P is a product, and ES is an enzyme–substrate complex.

$$E + S \underset{k_{-1}}{\overset{k_1}{\rightleftharpoons}} ES$$

$$ES \overset{k_2}{\rightarrow} P + E$$

Scheme XX

Proceeding in the usual way to apply the steady-state approximation to the complex ES:

$$\frac{d[ES]}{dt} = k_1[E][S] - (k_{-1} + k_2)[ES] = 0$$

$$[ES] = \frac{k_1[E][S]}{k_{-1} + k_2} \tag{3-147}$$

The mass balance expression for the enzyme is

$$E_t = [E] + [ES] \tag{3-148}$$

which, combined with Eq. (3-147) to eliminate [E], gives

$$[ES] = \frac{k_1 E_t [S]}{k_{-1} + k_2 + k_1 [S]} \tag{3-149}$$

The rate of reaction is defined $v = d[P]/dt = k_2[ES]$, or

$$v = \frac{k_1 k_2 E_t [S]}{k_{-1} + k_2 + k_1 [S]}$$

which is usually written

$$v = \frac{V_m [S]}{K_m + [S]} \tag{3-150}$$

where

$$V_m = k_2 E_t$$

$$K_m = \frac{k_{-1} + k_2}{k_1} \tag{3-151}$$

Equation (3-150) is the *Michaelis–Menten equation,* V_m is the maximum velocity (for the enzyme concentration E_t), and K_m is the Michaelis constant.

Several features of this treatment are of interest. Compare the denominators of Eqs. (3-147) and (3-149); Miller[54] has pointed out that the form of Eq. (3-147) is usually seen in chemical applications of the steady-state approximation, whereas the form of Eq. (3-149) appears in biochemical applications. The difference arises from the manner in which one uses the mass balance expressions, and this depends upon the type of system being studied and the information available.

The Michaelis–Menten equation is, like Eq. (3-146), a rectangular hyperbola, and it can be cast into three linear plotting forms. The double-reciprocal form, Eq. (3-152), is called the Lineweaver–Burk plot in enzyme kinetics.[55]

$$\frac{1}{v} = \frac{K_m}{V_m [S]} + \frac{1}{V_m} \tag{3-152}$$

Usually initial rates are measured in enzyme kinetics so as to avoid problems arising from kinetic complications such as product inhibition.

The Michaelis constant has the units of a dissociation constant; however, the dissociation constant of the enzyme–substrate complex is k_{-1}/k_1, which is not equal to K_m unless $k_{-1} \gg k_2$.

The quantitative description of enzyme kinetics has been developed in great detail by applying the steady-state approximation to all intermediate forms of the enzyme. Some of the kinetic schemes are extremely complex, and even with the aid of the steady-state treatment the algebraic manipulations are formidable. Kineticists have, therefore, developed ingenious schemes for writing down the steady-state rate equations directly from the kinetic scheme without carrying out the intermediate algebra.[56-59]

One way to examine the validity of the steady-state approximation is to compare concentration–time curves calculated with exact solutions and with steady-state solutions. Figure 3-10 shows such a comparison for Scheme XIV and the parameters, $k_1 = 0.01$ s^{-1}, $k_{-1} = 1$ s^{-1}, $k_2 = 2$ s^{-1}. The period during which the concentration of the intermediate builds up from its initial value of zero to the quasi-steady-state when dc_B/dt is very small is called the *pre-steady-state* or *transient stage;* in Fig. 3-10 this lasts for about 2 s. For the remainder of the reaction (over 500 s) the steady-state and exact solutions are in excellent agreement. Because the concen-

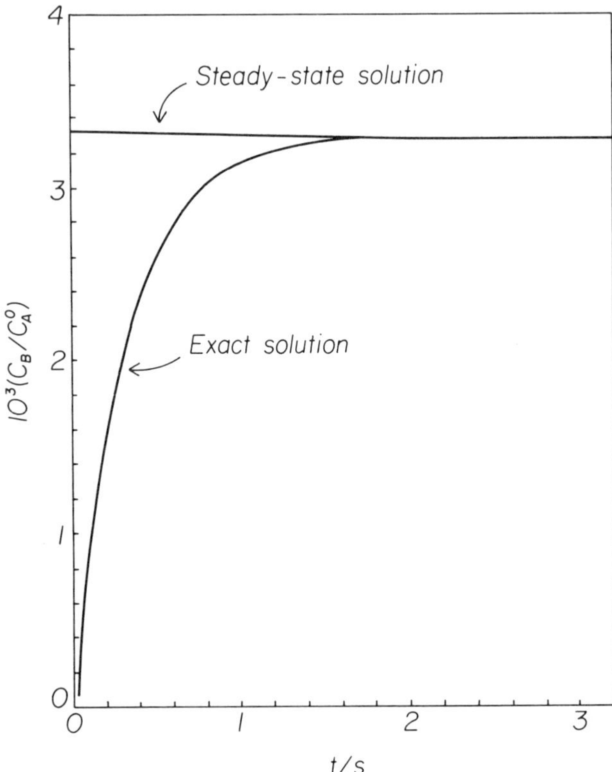

Figure 3-10. Comparison of steady-state and exact solutions for the concentration of intermediate B (relative to c_A^0) for Scheme XIV, where $k_1 = 0.01$ s^{-1}, $k_{-1} = 1$ s^{-1}, $k_2 = 2$ s^{-1}. The exact solution was obtained with Eq. (3-87), the steady-state solution with Eq. (3-142), where c_A was calculated with Eq. (3-90).

tration of intermediate eventually goes to zero, it is obvious that $dc_B/dt \neq 0$ over most of the course of the reaction, but it is certainly very small. Calculations of this type have been published by many authors.[60-63]

In the preceding subsection we described the preequilibrium assumption. Let us now see how that assumption is related to the steady-state approximation. Scheme XIV will serve for the discussion. The equilibrium and steady-state expressions for the intermediate concentration are

$$c_B^{eq} = \frac{k_1 c_A}{k_{-1}} \tag{3-153}$$

$$c_B^{ss} = \frac{k_1 c_A}{k_{-1} + k_2} \tag{3-154}$$

It, therefore, appears that the equilibrium approximation is a special case of the steady-state approximation,[64] namely, the case $k_{-1} \gg k_2$. This may be, but it is possible for the equilibrium approximation to be valid when the steady-state approximation is not. Consider the extreme but real example of an acid–base preequilibrium, which on the time scale of the following slow step is practically instantaneous. Suppose some kind of forcing function were to be applied to c_A, causing it to undergo large and sudden variations; then c_B would follow c_A almost immediately, according to Eq. (3-153). The equilibrium description would be very accurate, but the wide variations in c_B would vitiate the steady-state description. There appear to be three classes of practical behavior, as defined by these conditions:

1. Steady state $\qquad (k_{-1} + k_2) \gg k_1$
2. Preequilibrium $\qquad k_1 \gg k_2 \qquad$ and
   $\qquad\qquad\qquad\quad k_{-1} \gg k_2$
3. Both steady state $\quad (k_{-1} + k_2) \gg k_1 \qquad$ and
   and equilibrium $\qquad k_1 \gg k_2 \qquad\qquad$ and
   $\qquad\qquad\qquad\quad k_{-1} \gg k_2$

(Of course it is also possible for a reaction system not to belong to any of these classes of approximate description.) Only in class III can equilibrium be said to be a special case of the steady-state treatment. Note that, for class III systems, the steady-state concentration of intermediate is very large,[65] whereas for class I it is very small. Zuman and Patel[66] have discussed the equilibrium and steady-state approximations in terms similar to the present treatment.

Numerical Integration

We have seen that a kinetic scheme does not have to be very complex before explicit solutions for concentrations as functions of time become difficult or impossible to obtain. Even with those complex schemes for which solutions are possible, the equations may be too complicated to permit rate constants to be evaluated from experimental concentration–time curves. In these circumstances we must fall back on techniques that are essentially curve-fitting procedures and that involve some degree of approximation.

If an analytical solution is available, the method of nonlinear regression analysis can be applied; this approach is described in Chapter 2 and is not treated further here. The remainder of the present section deals with the analysis of kinetic schemes for which explicit solutions are either unavailable or unhelpful. First, the technique of numerical integration is introduced.

To develop the ideas we take the simple first-order rate equation as an example, namely, $dc/dt = -kc$. Let $c \equiv y$, $t \equiv x$, so

$$\frac{dy}{dx} = -ky \tag{3-155}$$

We approximate differentials with increments, and all subsequent expressions are, therefore, approximate. Equation (3-155) becomes

$$\frac{\Delta y}{\Delta x} = -ky \tag{3-156}$$

A uniform time interval $\Delta x = h$ is taken. The only information needed is an initial value (x_0, y_0) and an estimate of k. Define

$$\Delta y = y_1 - y_0 \tag{3-157}$$

$$\Delta x = x_1 - x_0 = h \tag{3-158}$$

so that y_1 is the value of the dependent variable at the end of the first interval. The procedure consists of making an approximation of $\Delta y/\Delta x$ across the interval h and calculating y_1; then y_1 becomes the initial value for the approximation of y_2 across the next interval, and so on.

From Eqs. (3-156) and (3-158) we have $\Delta y = -hky$, which, combined with (3-156) and (3-157), gives

$$y_1 = y_0 + h\left(\frac{\Delta y}{\Delta x}\right) \tag{3-159}$$

Equation (3-159) is the basic relationship of this method. Several techniques have been developed for the estimation of $\Delta y/\Delta x$. The simplest of these, known as *Euler's method*, is to evaluate $\Delta y/\Delta x$ at x_0. From Eq. (3-156), this gives $(\Delta y/\Delta x)_0 = -ky_0$, which, used in Eq. (3-159), yields

$$y_1 = y_0 - hky_0 \tag{3-160}$$

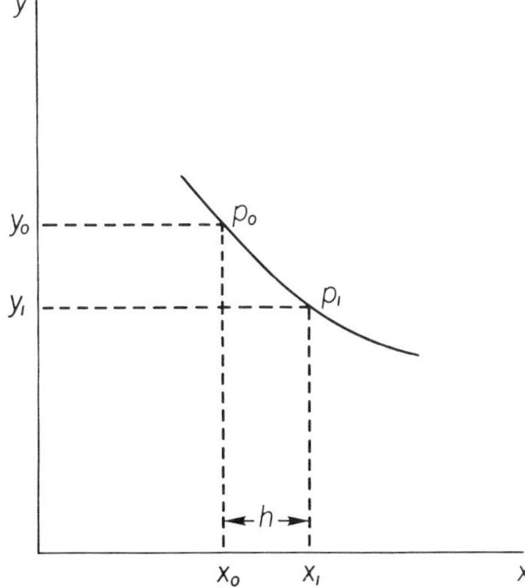

Figure 3-11. An arbitrary function $y = f(x)$, showing the relationship of $\Delta y = y_1 - y_0$ and $\Delta x = x_1 - x_0 = h$.

This method is subject to considerable error, because $y = y_0$ only at the beginning of the interval. It can be improved by taking an average value of y across the interval. The graphical problem is shown in Fig. 3-11. In the Euler method $\Delta y/\Delta x$ is evaluated at point P_0 and applied throughout the interval. We can obtain a better estimate by using as $\Delta y/\Delta x$ in Eq. (3-159) an average of the values at points P_0 and P_1. We have $(\Delta y/\Delta x)_0 = -ky_0$ and $(\Delta y/\Delta x)_1 = -ky_1$, so $(\Delta y/\Delta x)_{av} = -k(y_0 + y_1)/2$, where $(\Delta y/\Delta x)_{av}$ is the mean value.

We know y_0, but y_1 is at present unknown. However, an estimate is available in Eq. (3-160), giving

$$\left(\frac{\Delta y}{\Delta x}\right)_{av} = -ky_0 + \frac{hk^2 y_0}{2} \tag{3-161}$$

Equation (3-161) is then used in Eq. (3-159):

$$y_1 = y_0 - hky_0 + \frac{h^2 k^2 y_0}{2} \tag{3-162}$$

Equation (3-162) should be compared with Eq. (3-160).

This improved procedure is an example of the *Runge-Kutta method* of numerical integration.[67] Because the derivative was evaluated at two points in the interval, this is called a *second-order Runge-Kutta process*. We chose to evaluate the mean derivative at points P_0 and P_1, but because there is an infinite number of points in the interval, an infinite number of choices for the two points could have been made. In calculating the average for such choices appropriate weights must be assigned.

More than two points can be used in the Runge-Kutta method, and the fourth-order Runge-Kutta integration is commonly employed. Obviously computers are used for these calculations. The error in an nth order Runge-Kutta integration is of the order of magnitude h^{n+1}. Schemes have been devised to simplify the arrangement of the calculations and to supply the weighting factors.[68]

Complex reactions require the solution of simultaneous differential equations, and the Runge-Kutta procedure is applicable to these problems. To illustrate the method, Scheme XIV will be used. The rate equations are, in incremental form,

$$\frac{\Delta y}{\Delta x} = -k_1 y + k_{-1} z$$

$$\frac{\Delta z}{\Delta x} = k_1 y - kz \tag{3-163}$$

where $x = t$, $y = c_A$, $z = c_B$, $k = k_{-1} + k_2$.

Proceeding as before, we define $h = \Delta x$, $y_1 - y_0 = \Delta y$, $z_1 - z_0 = \Delta z$, and obtain, as in the earlier development,

$$y_1 = y_0 + h \left(\frac{\Delta y}{\Delta x}\right)$$

$$z_1 = z_0 + h \left(\frac{\Delta z}{\Delta x}\right) \tag{3-164}$$

A second-order treatment will be given, the derivatives being evaluated at (x_0, y_0, z_0) and (x_1, y_1, z_1). From Eqs. (3-163),

$$(\Delta y/\Delta x)_0 = -k_1 y_0 + k_{-1} z$$

$$(\Delta z/\Delta x)_0 = k_1 y_0 - kz_0 \tag{3-165}$$

and

$$(\Delta y/\Delta x)_1 = -k_1 y_1 + k_{-1} z_1$$

$$(\Delta z/\Delta x)_1 = k_1 y_1 - kz_1 \tag{3-166}$$

The quantities y_1 and z_1 are evaluated with Eqs. (3-164) at P_0. These are used in Eqs. (3-166). Then the means are calculated with

$$(\Delta y/\Delta x)_{av} = \tfrac{1}{2}[(\Delta y/\Delta x)_0 + (\Delta y/\Delta x)_1]$$

$$(\Delta z/\Delta x)_{av} = \tfrac{1}{2}[(\Delta z/\Delta x)_0 + (\Delta z/\Delta x)_1] \tag{3-167}$$

Equations (3-167) are used in Eqs. (3-164), yielding

$$y_1 = y_0 - h(k_1 y_0 + k_{-1} z_0) + \frac{h^2}{2}(k_1{}^2 y_0 + k_1 k_{-1} y_0 - k_1 k_{-1} z_0 - k k_1 z_0)$$

$$z_1 = z_0 + h(k_1 y_0 - k z_0) - \frac{h^2}{2}(k_1{}^2 y_0 + k k_1 y_0 - k_1 k_{-1} z_0 - k^2 z_0)$$

$$(3\text{-}168)$$

The procedure, in analyzing kinetic data by numerical integration, is to postulate a reasonable kinetic scheme, write the differential rate equations, assume estimates for the rate constants, and then to carry out the integration for comparison of the calculated concentration–time curves with the experimental results. The parameters (rate constants) are adjusted to achieve an acceptable fit to the data. Carpenter[48, pp. 76–81] shows some numerical calculations. Farrow and Edelson[61] and Porter and Skinner[62] used numerical integration to test the validity of the steady-state approximation in complex reactions.

It is possible for numerical integration to fail, in the sense of yielding physically meaningless results, for certain combinations of coupled equations. This can occur when different reactions in a kinetic scheme have greatly different time scales. An important example is a reaction system to which the steady-state approximation is applicable, the rate of change of an intermediate concentration being essentially zero as a consequence of two very large opposing fluxes. Then very small changes in initial conditions may result in grossly different numerical results. Such equations are said to be "stiff." Curtiss and Hirschfelder[69] called attention to the problem of stiffness in numerical integration, and several computer programs capable of integrating stiff equations are available.[70]

Monte Carlo Simulation

The Monte Carlo or stochastic method for constructing concentration–time curves is based on the statistics of random events applied to large numbers. Schaad[71] described the method for complex kinetic schemes, and it has since been adapted for teaching purposes using paper-and-pencil,[72] pegs-and-pegboard,[73] and computer techniques.[74] The digital computer is the ideal tool for carrying out Monte Carlo simulations, but for the purpose of outlining the method here a manual technique is very satisfactory.

To illustrate the technique we will treat Scheme IX:

$$A \xrightarrow{k_1} B$$

$$B \xrightarrow{k_2} C$$

Scheme IX

For manual calculations 100 molecules of reactant is convenient. Our "reaction flask" is a 10 × 10 grid or matrix, the spaces (cells) in the grid being numbered from 00 to 99. For Scheme IX we need three grids, one for each species:

$$\boxed{A} \xrightarrow{k_1} \boxed{B} \xrightarrow{k_2} \boxed{C}$$

As the initial conditions choose $[A]_0 = 100$, $[B]_0 = 0$, $[C]_0 = 0$, the brackets representing mole percent, which is numerically equal to the number of molecules in a grid, because each grid contains 100 spaces. At zero time we load 100 "molecules" of A in grid A, by writing an A in each of the 100 cells of the grid.

A reaction is simulated by making random selections from the grid. From a table of random numbers, a two-digit random number is selected. If the cell corresponding to this number is occupied by an A, the A is crossed off (it "reacts") and a B is written in the corresponding space in the B grid. If the random number identifies a cell that does not contain an A or that contains a crossed-off A, no reaction occurs and no action is taken.

We need these definitions:

n_1 Number of selections on grid A per cycle.
n_2 Number of selections on grid B per cycle.
m Number of cycles.

In order to begin, we choose values for n_1 and n_2, which is equivalent to defining the rate constant ratio k_1/k_2, namely, $k_1/k_2 = n_1/n_2 = r$. For this illustrative example the values $n_1 = 9$, $n_2 = 3$ were chosen. The operation is to make nine selections

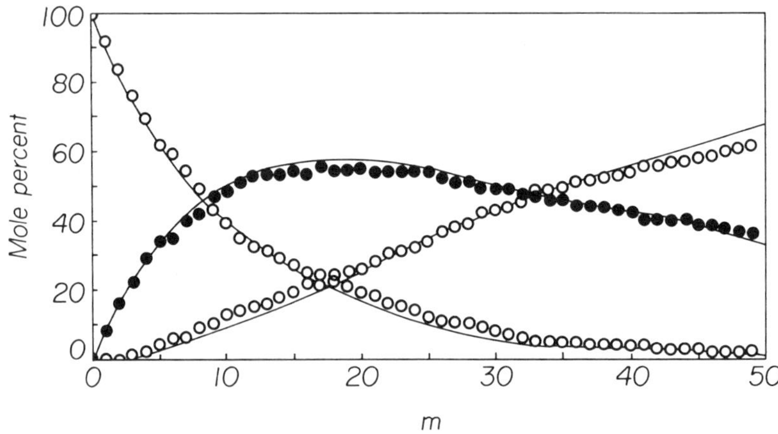

Figure 3-12. Monte Carlo simulation of Scheme IX with $n_1 = 9$, $n_2 = 3$, $[A]_0 = 100$, $[B]_0 = 0$, $[C]_0 = 0$. This is the average of two runs. The smooth lines were calculated with Eqs. (3-173) and (3-174).

on grid A followed by three selections on grid B. This constitutes one cycle. At the conclusion of each cycle the numbers of A, B, and C molecules are counted and tabulated. The process is repeated until the course of the reaction has been adequately defined. A plot is made of numbers of molecules per grid against m, the number of cycles.

Figure 3-12 shows such a plot for the simulation described above. (The fluctuations can be reduced by averaging identical simulations, and the figure shows the average of two simulations.)

To see the connection between this stochastic process and a chemically reacting system, consider the first step of Scheme IX. Each (real) molecule of A has an equal and constant probability of reacting in time t. In the simulation, each position in the grid has an equal and constant probability (p) of being selected. For this first-order reaction, the chemical system is described by

$$\frac{dc_A}{dt} = -k_1 c_A; \qquad \ln \frac{c_A}{c_A^0} = -k_1 t \tag{3-169}$$

The analogous equations in the simulation system are

$$\frac{d[A]}{dm} = -pn_1[A]; \qquad \ln \frac{[A]}{[A]_0} = -pn_1 m \tag{3-170}$$

Supposing that the simulation mimics the chemistry, then,

$$k_1 t = pn_1 m \tag{3-171}$$

Figure 3-13 is the semilogarithmic plot according to Eq. (3-170) for the reaction of A in Fig. 3-12. The slope of the plot is -0.090, in agreement with the values $p = 0.01$ and $n_1 = 9$.

In the same way we write for the second step of the reaction

$$k_2 t = pn_2 m \tag{3-172}$$

With Eqs. (3-171) and (3-172), and the definition $r = k_1/k_2$, Eqs. (3-24) and (3-27) for c_A and c_B become, for the simulation,

$$\frac{[A]}{[A]_0} = e^{-pn_1 m} \tag{3-173}$$

$$\frac{[B]}{[A]_0} = \frac{r}{1-r}\left[e^{-pn_1 m} - e^{-pn_2 m} \right] \tag{3-174}$$

The smooth lines in Fig. 3-12 were calculated with Eqs. (3-173) and (3-174), the values of $[C]/[A]_0$ being obtained from the mass balance. The fit is quite reasonable for such a simple procedure.

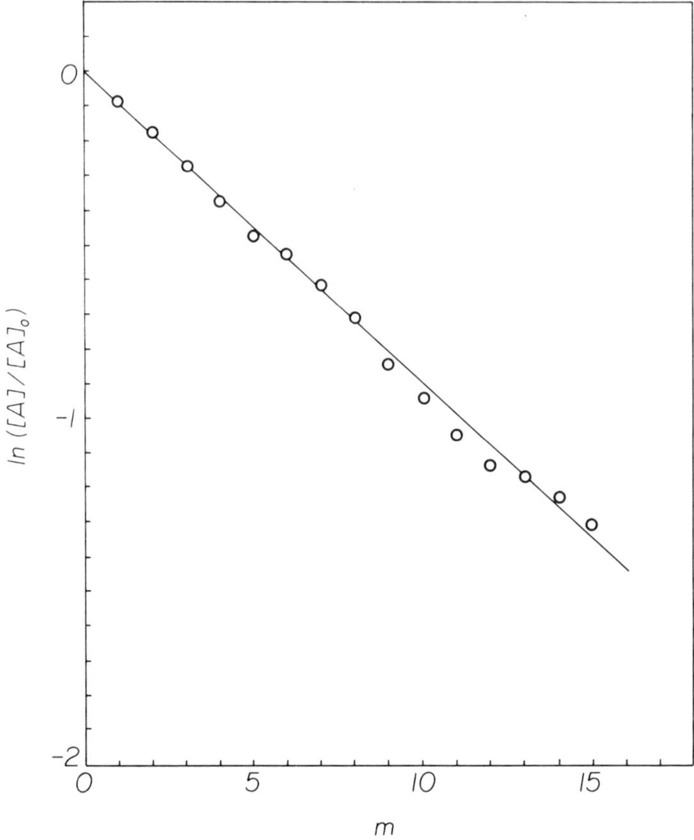

Figure 3-13.   First-order plot of simulated data for reactant A in Fig. 3-12.

The simulation may be expressed in reaction time (t) rather than cycles (m) by assigning absolute values of n_1, n_2, k_1, and k_2, consistent with Eqs. (3-171) and (3-172). Then these equations relate t to m. In this way rate constant estimates may be obtained by curve fitting the simulations to experimental results.

Any combination of first-order reactions can be simulated by extension of this procedure. Reversible reactions add only the feature that reacted species can be regenerated from their products. Second-order reactions introduce a new factor, for now two molecules must each be independently selected in order that reaction occur; in the real situation the two molecules are in independent motion, and their collision must take place to cause reaction. We load the appropriate numbers of molecules into each of two grids. Now randomly select from the first grid, and then, separately, randomly select from the second grid. If in *both* selections a molecule exists at the respective selected sites, then reaction occurs and both are crossed out; if only one of the two selections results in selection of a molecule, no reaction occurs. (Of course, if pseudo-first-order conditions apply, a second-order reaction can be handled just as is a first-order reaction.)

Kinetic schemes that include both first-order and second-order reactions possess an ambiguity related to the different dimensions of the rate constants. We will use Scheme XXI to examine this.

$$A + B \xrightarrow{k_1} I$$

$$I \xrightarrow{k_2} X$$

$$X \xrightarrow{k_3} P$$

Scheme XXI

In grid form the scheme looks like this:

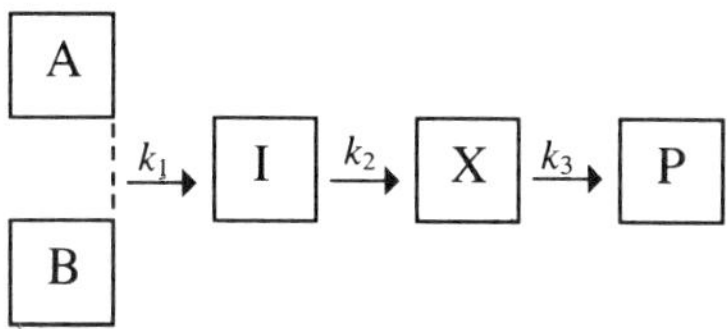

Following Schaad,[71] we write

$$\frac{\text{chemical rate of formation of I}}{\text{chemical rate of formation of X}} = \frac{k_1 c_A c_B}{k_2 c_I}$$

$$\frac{\text{stochastic rate of formation of I}}{\text{stochastic rate of formation of X}} = \frac{n_1[A][B]}{n_2[I]}$$

$$\frac{\text{chemical rate of formation of X}}{\text{chemical rate of formation of P}} = \frac{k_2 c_I}{k_3 c_X}$$

$$\frac{\text{stochastic rate of formation of X}}{\text{stochastic rate of formation of P}} = \frac{n_2[I]}{n_3[X]}$$

The chemical and stochastic concentrations are related by $[A] = \alpha c_A$, etc. Combining these equations and equating the chemical and stochastic rate ratios gives

$$\frac{n_1}{n_2} = \frac{k_1}{\alpha k_2}$$

$$\frac{n_2}{n_3} = \frac{k_2}{k_3}$$

from which is found

$$n_1{:}n_2{:}n_3 = \frac{k_1}{\alpha}{:}k_2{:}k_3 \tag{3-175}$$

Note that $2n_1$ is the total number of selections made on grids A and B in one cycle, n_1 being made on grid A and n_1 on grid B.

If c_A is the mole fraction of A, then $\alpha = 1$. However, if c_A is expressed as a molarity, the value of α is calculated from the initial reactant concentrations. For example, suppose $[A]_0 = 100$, $c_A^0 = 5.0 \times 10^{-3}$ M, $[B]_0 = 50$, and $c_B^0 = 2.5 \times 10^{-3}$ M; then $\alpha = 2 \times 10^4$ M^{-1}.

The relative fluctuations in Monte Carlo simulations are of the order of magnitude $N^{-\frac{1}{2}}$, where N is the total number of molecules in the simulation.[75] The observed error in kinetic simulations is about 1–2% when 10^4 molecules are used.[71,76] In the computer calculations described by Schaad,[71] the grids of the technique shown here are replaced by computer memory, so the capacity of the memory is one limit on the maximum number of molecules. Other programs for stochastic simulation make use of different routes of calculation, and the number of molecules is not a limitation.[76,77] Enzyme kinetics[78] and very complex oscillatory reactions have been modeled.[76,77] These simulations are valuable for establishing whether a postulated kinetic scheme is reasonable, for examining the appearance of extrema or induction periods, applicability of the steady-state approximation, and so on. Even the manual method is useful for such purposes.

A final comment on the interpretation of stochastic simulations: We are so accustomed to writing continuous functions—differential and integrated rate equations, commonly called *deterministic rate equations*—that our first impulse on viewing these stochastic calculations is to interpret them as approximations to the familiar continuous functions. However, we have got this the wrong way around. On a molecular level, events are discrete, not continuous. The continuous functions work so well for us only because we do experiments on very large numbers of molecules (typically 10^{15}–10^{20}). If we could experiment with very much smaller numbers of molecules, we would find that it is the continuous functions that are approximations to the stochastic results. Gillespie[77] has developed the stochastic theory of chemical kinetics without dependence on the deterministic rate equations.

Analog Simulation

Before there were small digital computers, there were small analog computers. An analog computer is a device that can mimic mathematical equations with physical processes. For example, hydrodynamic flow through a capillary can model the chemical kinetics of some reactions.[79] Most analog computers use electrical circuitry, as in the computer designed by Frost and Tamres[80] to solve secular equations. For chemical kinetics applications, the interesting analog computer designs are those that represent differential equations by means of electrical circuits.

These circuits include components that carry out arithmetic operations, differentiation, and integration. They can be assembled to model a postulated kinetic scheme by "patching" together components that build up circuits analogous to the differential rate equations describing the scheme. A voltage is applied linearly with time, and the resulting currents in the circuitry are analogous to concentration. In effect, the computer solves the differential equations, the current–voltage curves representing concentration–time curves. Values of rate constants are supplied by potentiometers, and these are adjusted until satisfactory curve fits are obtained (presuming the kinetic scheme is appropriate to the problem).

The method is quite effective, but is not widely used now because of the ubiquity of digital computers. Zuman and Patel[66, pp. 136–44] show circuit designs for some kinetic schemes. Williams and Bruice[81] made good use of the analog computer in their study of the reduction of pyruvate by 1,5-dihydroflavin. In this simulation eight rate constants were evaluated; variations in these parameters of $\pm 5\%$ yielded discernibly poorer curve fits.

3.3 THE KINETIC SCHEME

Model Building

Sections 3.1 and 3.2 considered this problem: Given a complex kinetic scheme, write the differential rate equations; find the integrated rate equations or the concentration–time dependence of reactants, intermediates, and products; and obtain estimates of the rate constants from experimental data. Little was said, however, about how the kinetic scheme is to be selected. This subject might be dismissed by stating that one makes use of experimental observations combined with chemical intuition to postulate a reasonable kinetic scheme; but this is not very helpful, so some amplification is provided here.

In a general way a kinetic study of reaction mechanism includes these four components:

1. Experimental kinetics.
2. Determination of the rate equation(s).
3. Writing the kinetic scheme.
4. Proposal of transition state structures, stereochemistries, and energetics.

These steps may not proceed in the sequence shown, because a difficult kinetic problem may require cycling of attention among the steps as more is learned about the system, with corrections being made and tests of ideas being applied at each stage. In particular, steps 2 and 3 may be strongly interdependent. Our present concern is with these steps; later chapters deal with step 4. Edwards et al.,[82] Bunnett,[83] and Pearson[84] have formulated provisional rules for proceeding from the rate equation to the mechanism, which includes step 4.

A postulated kinetic scheme is a model or hypothesis. In accord with Occam's razor, the model should be no more complex than is required to account for the experimental observations (which, by extension, include the whole relevant body of chemistry). In the context of the present chapter, we presume that the evidence suggests that the system of interest is not an elementary reaction, although it may possess a simple rate equation. However, in general, we will not yet know the rate equation(s), and the development of the kinetic scheme and the rate equation will proceed concurrently. As Section 3.2 demonstrated, we may be able to write a satisfactory kinetic scheme and yet be unable to obtain a rate equation that explicitly expresses concentration as a function of time.

We now consider some of the kinds of information that will be useful.

1. *Background information on the reasonable and expected chemistry of the system.* This is where chemical experience and intuition play a role. Knowledge of the pertinent literature is valuable. If the type of reaction has been studied earlier, these kinds of issues may be provisionally known:

(a) Probable occurrence and relative stability of intermediates.

(b) Reversibility or essential irreversibility of reaction steps.

(c) Orders of magnitude of anticipated relative rates.

(d) Possible validity of the preequilibrium or steady-state approximations (related to item c).

(e) Anticipated physicochemical properties (e.g., spectroscopic properties) of expected intermediates.

In deciding whether to write an elementary reaction as either a reversible or an irreversible reaction, we take the practical view that if the reverse reaction is negligibly slow on the experimental time scale, the reaction is essentially irreversible. Consider the alkaline hydrolysis of an ester, for which the rate equation is

$$v = k_{OH}[RCOOR'][OH^-]$$

An abundance of mechanistic information in the literature suggests this scheme:

$$RCOOR' + OH^- \underset{k_{-1}}{\overset{k_1}{\rightleftharpoons}} R-\underset{\underset{OH}{|}}{\overset{\overset{O^-}{|}}{C}}-OR'$$

$$R-\underset{\underset{OH}{|}}{\overset{\overset{O^-}{|}}{C}}-OR' \overset{k_2}{\rightarrow} RCOO^- + R'OH$$

Scheme XXII

Among the available pieces of information are the observations that an alkaline solution of the carboxylic acid and alcohol does not generate the ester and that (for most esters) the postulated tetrahedral intermediate cannot be detected. Thus, the overall reaction is irreversible, and the intermediate (if it exists) is present at very low concentration. (Obviously failure to detect an intermediate is not a demonstration of its existence, but in this example there is independent evidence for the tetrahedral intermediate.) These considerations lead to Scheme XXII.

2. *Direct detection of an intermediate.* A nice example, the pyridine-catalyzed hydrolysis of acetic anhydride, was discussed in Chapter 1. Spectroscopic techniques are of great value, because they do not perturb the kinetic system, and because they are selective and sensitive. If the intermediate can be detected, the time course of its appearance and disappearance may be followed.

To show the type of inference that is possible from such observations, we use the example of the acetylation of alcohols or water by acetic anhydride or acetyl chloride, catalyzed by N-methylimidazole.[85] When acetonitrile solutions of acetyl chloride and N-methylimidazole are mixed, the absorbance of the resulting solution is much greater than that calculated for the sum of the reactants. At a given concentration of catalyst and varying concentrations of acetyl chloride, the same absorbance is produced (as long as the acetyl chloride is in excess), showing that the strongly absorbing substance is produced quantitatively. On the other hand, when acetonitrile solutions of acetic anhydride and N-methylimidazole are mixed, no "excess" absorption is observed. However, when water is added, strong UV absorption is seen, similar to that seen for acetyl chloride under anhydrous conditions.

It is inferred that intermediate formation can occur, and these experimental observations suggest Scheme XXIII.

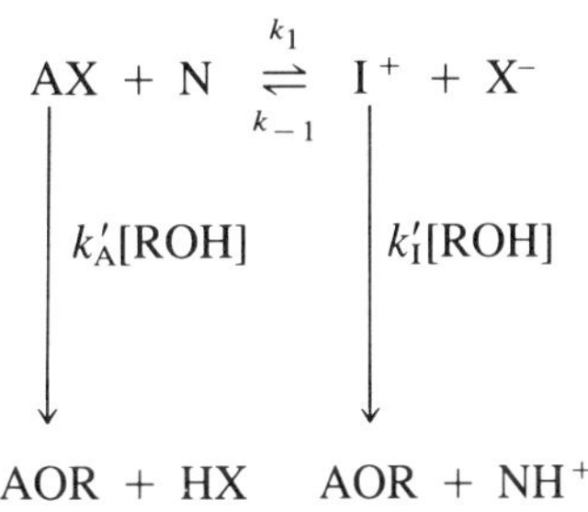

$$AX + N \underset{k_{-1}}{\overset{k_1}{\rightleftharpoons}} I^+ + X^-$$

$$k_A'[ROH] \qquad k_I'[ROH]$$

$$AOR + HX \qquad AOR + NH^+$$

Scheme XXIII

Here AX is the acetyl compound (acetyl chloride or acetic anhydride), N is N-methylimidazole, I^+ is the intermediate (presumably N-acetyl-N'-methylimidazolium ion), X^- is the counterion (chloride or acetate), and ROH is the acetyl acceptor (alcohol or water). A general treatment of Scheme XXIII requires specification of the detailed nature of k_A' and k_I' and is probably too complicated to be of practical use. However, several important special cases may arise from the operation of the ratio k_1/k_{-1}, the behavior of apparent rate constants k_A' and k_I', the relative magnitudes of k_A' and k_I', the relative concentrations of the reactants, the method of observation, and the nature of ROH. These cases are outlined in Scheme XXIV.

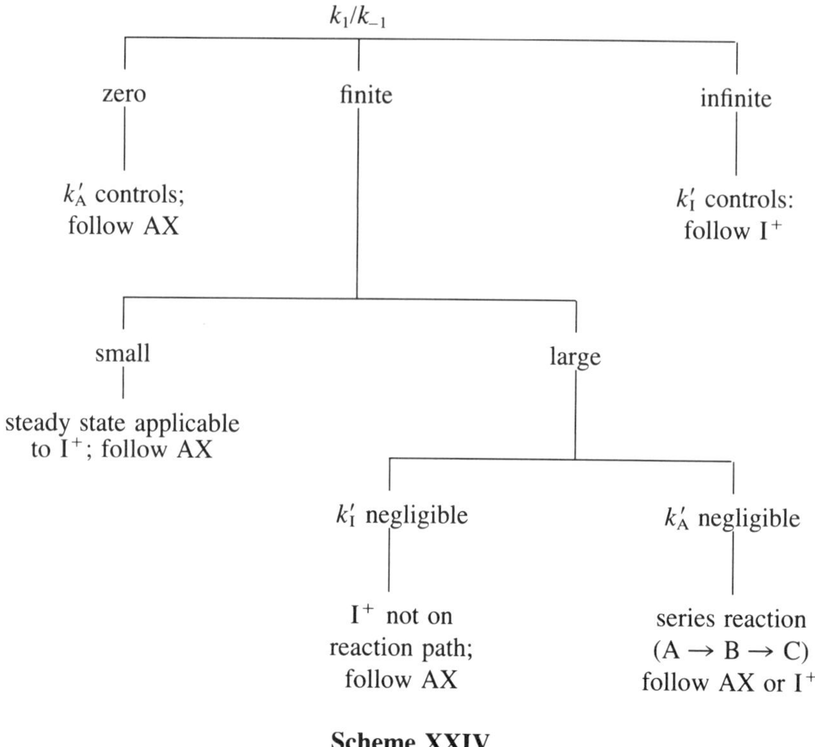

Scheme XXIV

In this study the reactions were followed spectrophotometrically by monitoring the loss of AX or I$^+$ with time. The initial concentration of hydroxy compound was at least 50 times that of the acetylating agent, and pseudo-first-order behavior was observed. This system will be discussed later.

3. *Indirect detection of an intermediate.* The overall reaction of hydroxylamine with a carboxylic acid derivative yields a hydroxamic acid as the product, Eq. (3-176).

$$\underset{\text{RCX}}{\overset{\displaystyle O \atop \displaystyle \|}{}} + NH_2OH \rightarrow \underset{\text{RCNHOH}}{\overset{\displaystyle O \atop \displaystyle \|}{}} + HX \qquad \text{(3-176)}$$

When Jencks[86] reacted hydroxylamine with *p*-nitrophenyl acetate, *p*-nitrophenolate ion was released at a rate faster than that at which acetohydroxamic acid was formed. This "burst effect" is evidence for a two-step reaction. In this case the intermediate is *O*-acetylhydroxylamine, which subsequently reacts with hydroxylamine to form the hydroxamic acid.

$$\overset{\textstyle O}{\overset{\textstyle \|}{RCOAr}} + NH_2OH \rightarrow \overset{\textstyle O}{\overset{\textstyle \|}{RCONH_2}} + ArO^- + H^+$$

$$\overset{\textstyle O}{\overset{\textstyle \|}{RCONH_2}} \xrightarrow{\ NH_2OH\ } \overset{\textstyle O}{\overset{\textstyle \|}{RCNHOH}}$$

Scheme XXV

($R \equiv CH_3$ and $AR \equiv C_6H_4NO_2$.) Actually Scheme XXV and Eq. (3-176) both take place, with some of the hydroxamic acid being formed directly and some via the intermediate. (Note that each of these reactions is itself complex, presumably occurring via a tetrahedral intermediate as shown in Scheme XXII for ester hydrolysis.)

In general, if a reaction leads to two or more products, and the products are not formed at equal rates, there must be an intermediate to account for the material balance. (The converse, of course, is not necessarily true, for an intermediate may be present at vanishingly low concentrations and yet be kinetically important.)

4. *Kinetic evidence for a common intermediate.* Suppose a series of related reactants each separately reacts with a common reagent, and the rates of product formation are the same for each reactant:

$$AX + B^- \xrightarrow{k_X} AB + X^-$$

$$AY + B^- \xrightarrow{k_Y} AB + Y^-$$

$$AZ + B^- \xrightarrow{k_Z} AB + Z^-$$

Scheme XXVI

It is a reasonable inference that, if $k_X = k_Y = k_Z$, these rate constants all describe the same reaction, namely, the reaction, in this example, of B^- with the common intermediate A^+:

$$AX + A^+ + X^-$$

$$A^+ + B^- \xrightarrow{k_X} AB$$

Scheme XXVII

Espenson[87] gives examples from inorganic chemistry. Jencks[88] describes enzyme-catalyzed reactions in which the common intermediate is an acylated enzyme.

5. *Concentration–time curves*. Much of Sections 3.1 and 3.2 was devoted to mathematical techniques for describing or simulating concentration as a function of time. Experimental concentration–time curves for reactants, intermediates, and products can be compared with computed curves for reasonable kinetic schemes. Absolute concentrations are most useful, but even instrument responses (such as absorbances) are very helpful. One hopes to identify characteristic features such as the formation and decay of intermediates, approach to an equilibrium state, induction periods, an autocatalytic growth phase, or simple kinetic behavior of certain phases of the reaction. Recall, for example, that for a series first-order reaction scheme, the loss of the initial reactant is simple first-order. Approximations to simple behavior may suggest justifiable mathematical assumptions that can simplify the quantitative description.

Figure 3-14 shows an induction period in the reaction of carbon suboxide ($O{=}C{=}C{=}C{=}O$) with triethylamine.[89] This reaction is complex and is not yet

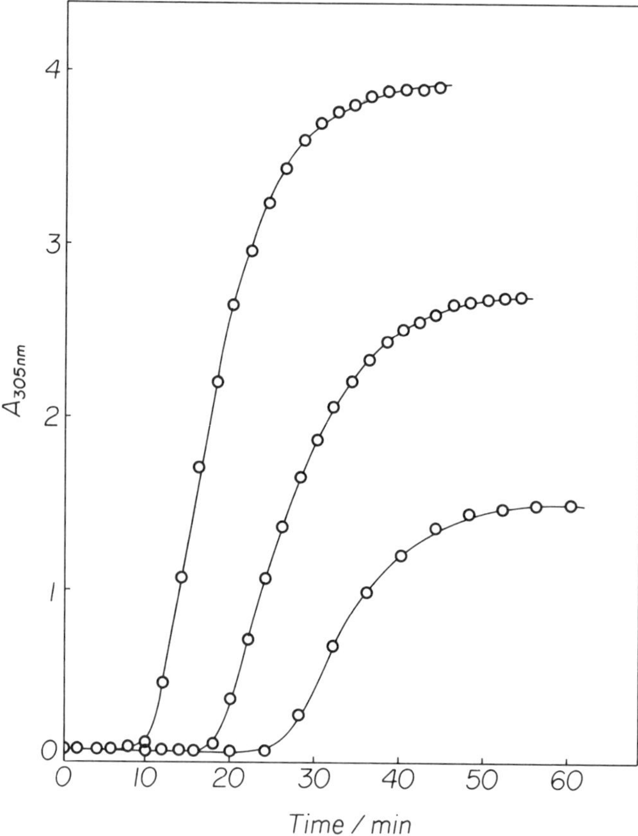

Figure 3-14. Absorbance–time plots for the reaction of carbon suboxide and triethylamine in ether solution in the presence of acetic anhydride. The initial C_3O_2 concentration was 2.03×10^{-2} M; the amine concentrations were 3×10^{-4} M, 5×10^{-4} M, and 7×10^{-4} M.

understood, but clearly this kinetic behavior of a strongly absorbing product is a key observation. The length of the induction period, the maximum rate of product formation, and the final product concentration are all related to the triethylamine concentration.

Considering the attention that we have given in this chapter to concentration–time curves of complex reactions, it may seem remarkable that many kinetic studies never generate a comprehensive set of complicated concentration–time data. The reason for this is that complex reactions often can be studied under simplified conditions constituting important special cases; for example, whenever feasible one chooses pseudo-first-order conditions, and then one studies the dependence of the pseudo-first-order rate constant on variables other than time. This approach is amplified below.

6. *Dependence of apparent constants on concentration.* We continue the consideration of Scheme XXIII by making chemically reasonable tentative selections of the forms of k_A' and k_I'. First, consider the acetyl chloride–alcohol reaction. Because the spectral observations show that intermediate formation is essentially complete, this system belongs to the case in which k_1/k_{-1} may be treated as infinite (Scheme XXIV). The observed reaction is then

$$I^+ + ROH \xrightarrow{k_I'} ROAc + NH^+$$

It is postulated that k_I' is given by Eq. (3-177), where [N] is the concentration of N-methylimidazole.

$$k_I' = k_I + k_{IN}[N] \tag{3-177}$$

The hypothetical rate equation is then

$$\frac{-d[I^+]}{dt} = (k_I + k_{IN}[N])[I^+][ROH]$$

The experimental rate equation was found to be $-d[I^+]/dt = k_{obs}[I^+]$, the plots being linear over the course of the reactions. Therefore,

$$\frac{k_{obs}}{[ROH]} = k_I + k_{IN}[N] \tag{3-178}$$

Equation (3-178) suggests that a plot of $k_{obs}/[ROH]$ vs. [N] will be linear. Because the conversion to the intermediate is quantitative, $[N] = [N]_0 - [AX]_0$. Plots according to Eq. (3-178) were linear, permitting k_I and k_{IN} to be estimated.[85]

Turning to the acetic anhydride–alcohol system, it is inferred that (in the absence of water) k_1/k_{-1} is close to zero (Scheme XXIV). Although the intermediate could not be detected spectrally, its possible presence is admitted in the rate equation for the loss of anhydride:

$$-\frac{d[AX]}{dt} = k_1[AX][N] + k'_A[AX][ROH] - k_{-1}[I^+][X^-]$$

The assumption is made that all of the H^+ produced in the acetylation is accepted by X^- (acetate); then $[X^-] = 0$, so this becomes

$$-\frac{d[AX]}{dt} = k_1[AX][N] + k'_A[AX][ROH]$$

This assumption can be justified. By hypothesis, $k'_A = k_A + k_{AN}[N]$. The experimental rate equation is $-d[AX]/dt = k_{obs}[AX]$, so

$$k_{obs} = k_1[N] + k_A[ROH] + k_{AN}[ROH][N]$$

If in a series of experiments the catalyst concentration is held constant while the alcohol concentration is varied, the plotting form of Eq. (3-179) is used

$$\frac{k_{obs}}{[N]} = k_1 + \left(\frac{k_A}{[N]} + k_{AN}\right)[ROH] \tag{3-179}$$

whereas if [ROH] is kept constant and [N] is varied, Eq. (3-180) can be used.

$$\frac{k_{obs}}{[ROH]} = k_A + \left(\frac{k_1}{[ROH]} + k_{AN}\right)[N] \tag{3-180}$$

Thus, from the slopes and intercepts of these two plots, the constants k_1, k_A, and k_{AN} can be evaluated. It was found that k_1 and k_A were not significantly different from zero.

All of these rate equations can be succinctly expressed in the form of Scheme XXVIII.

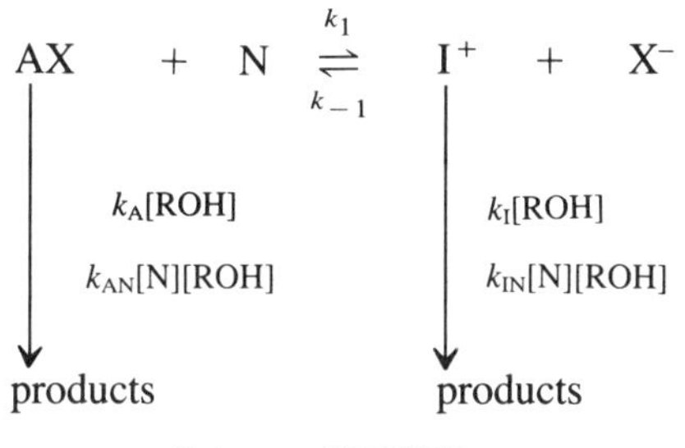

$$\begin{array}{ccccccc}
AX & + & N & \underset{k_{-1}}{\overset{k_1}{\rightleftharpoons}} & I^+ & + & X^- \\
\downarrow & & & & \downarrow & & \\
\end{array}$$

$$k_A[ROH] \qquad\qquad k_I[ROH]$$

$$k_{AN}[N][ROH] \qquad\qquad k_{IN}[N][ROH]$$

$$\text{products} \qquad\qquad\qquad \text{products}$$

Scheme XXVIII

This manner of implicitly including the rate equations in the kinetic scheme is very convenient. It is amplified with the statement that when AX is acetyl chloride, k_1/k_{-1} is very large and the reaction occurs essentially only via the I^+ route. When

AX is acetic anhydride, k_1/k_{-1} is zero, and the reaction occurs entirely via the AX route.

A kinetic scheme that is fully consistent with experimental observations may yet be ambiguous in the sense that it may not be unique. An example was discussed earlier (Section 3.1, Consecutive Reactions), when it was shown that k_1 and k_2 in Scheme IX may be interchanged without altering some of the rate equations; this is the *slow–fast ambiguity*. Additional examples of kinetically indistinguishable kinetic schemes have been discussed.[20,90] The following subsection treats one aspect of this problem.

Kinetic Equivalence of Rate Terms

Suppose an experimental rate equation has the form

$$v = k[\text{HA}] \tag{3-181}$$

where HA is a weak Brønsted acid. Because of the protonic equilibrium

$$\text{HA} \rightleftharpoons \text{H}^+ + \text{A}^-$$

we have the acid dissociation constant $K_{\text{HA}} = [\text{H}^+][\text{A}^-]/[\text{HA}]$, so $[\text{HA}] = [\text{H}^+][\text{A}^-]/K_{\text{HA}}$. Thus Eq. (3-181) can also be written as

$$v = \frac{k[\text{H}^+][\text{A}^-]}{K_{\text{HA}}} = k'[\text{H}^+][\text{A}^-] \tag{3-182}$$

The rate terms $k[\text{HA}]$ and $k'[\text{H}^+][\text{A}^-]$ are said to be *kinetically equivalent* or *kinetically indistinguishable*. There is no purely kinetic basis upon which to make a choice between them; in Chapter 5 we will see why this is so, but a simple interpretation is that the two terms describe equivalent chemical compositions of atoms and charges.

Another example is a second-order term containing the concentration of an acid and a base, say

$$v = k[\text{HA}][\text{B}] \tag{3-183}$$

This is kinetically equivalent to

$$v = k'[\text{A}^-][\text{BH}^+] \tag{3-184}$$

where $k' = kK_{\text{BH}+}/K_{\text{HA}}$. Suppose HA is a carboxylic acid and B is an amine; it may seem most reasonable to write Eq. (3-183) for the reaction, but there is no kinetic reason to prefer this to the reaction of the carboxylate with the protonated amine, Eq. (3-184).

When the elements of the solvent are involved, as in hydrolysis reactions in aqueous media, kinetic equivalence of rate terms may not be so obvious. Let a diprotic acid H_2A be the reactant, with the three forms H_2A, HA^-, A^{2-} being subject to hydrolysis. Suppose uncatalyzed, acid-catalyzed, and base-catalyzed reactions are possible. Then nine reactions can be written, namely,

$$H_2A \xrightarrow{1} P$$

$$H_2A + H^+ \xrightarrow{2} P$$

$$H_2A + OH^- \xrightarrow{3} P$$

$$HA^- \xrightarrow{4} P$$

$$HA^- + H^+ \xrightarrow{5} P$$

$$HA^- + OH^- \xrightarrow{6} P$$

$$A^{2-} \xrightarrow{7} P$$

$$A^{2-} + H^+ \xrightarrow{8} P$$

$$A^{2-} + OH^- \xrightarrow{9} P$$

The elements of water are omitted because the concentration of water cannot be varied and, therefore, is not explicitly included in the rate equation. A complete kinetic scheme requires only reactions 1, 2, 4, 7, and 9; the other reactions are superfluous. For example, the rate term $k_3[H_2A][OH^-]$ is equivalent to $k_4[HA^-]$, as is the term $k_8[H^+][A^{2-}]$.

In these circumstances a decision must be made: which of two (or more) kinetically equivalent rate terms should be included in the rate equation and the kinetic scheme? (It will seldom be justified to include both terms, certainly not on kinetic grounds.) A useful procedure is to evaluate the rate constant using both of the kinetically equivalent forms. Now if one of these constants (for a second-order reaction) is greater than about 10^{10} $M^{-1}s^{-1}$, the corresponding rate term can be rejected. This criterion is based on the theoretical estimate of a diffusion-controlled reaction rate (this is described in Chapter 4). It is not physically reasonable that a chemical rate constant can be larger than the diffusion rate limit.

A weaker but more widely applicable criterion is that the rate constant estimate should be consistent with the body of experimental work on closely related reactions. A third factor is that of style, which is essentially equivalent to the contemporary state of mechanistic chemistry; it may seem more reasonable to write a mechanism for one of the forms than for the alternative. Styles change, however.

Attempts have been made to distinguish kinetically between terms such as those in Eqs. (3-183) and (3-184) on the basis that their different charge types should lead to different salt effects upon the rate. This is futile. A fuller discussion is given in Chapter 8, where it is shown that equations (3-183) and (3-184) have identical salt effects; any differences ascribable to the concentration terms are exactly canceled by the salt effects in the "constants."

Microscopic Reversibility and Detailed Balance

Consider this elementary reversible reaction:

$$A + B \underset{k_{-1}}{\overset{k_1}{\rightleftharpoons}} Y + Z$$

According to the law of mass action the differential rate equation is

$$v = k_1 c_A c_B - k_{-1} c_Y c_Z$$

At equilibrium $v = 0$, the chemical fluxes in the forward and reverse directions are equal, and we write

$$\frac{k_1}{k_{-1}} = \frac{c_Y c_Z}{c_A c_B} = K \tag{3-185}$$

where we call K the equilibrium constant of the reaction. Equation (3-185) requires the conclusion that only two of the three constants k_1, k_{-1}, and K are independent.

Now we write the complex system of Scheme XVIII.

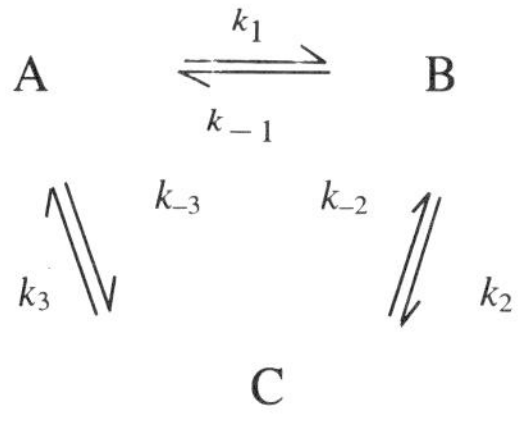

$$A \underset{k_{-1}}{\overset{k_1}{\rightleftharpoons}} B$$

$$C$$

Scheme XVIII

In Section 3.2, p. 99, we had obtained, using the above argument, Eq. (3-186),

$$k_1 k_2 k_3 = k_{-1} k_{-2} k_{-3} \tag{3-186}$$

or $K_1 K_2 K_3 = 1$, where $K_i = k_i/k_{-i}$. Thus only five of the six rate constants are independent.

These are applications of the *principle of detailed balancing,* which can be stated:

> In a system of connected reversible reactions at equilibrium, each reversible reaction is individually at equilibrium.

Detailed balance is a chemical application of the more general *principle of microscopic reversibility,* which has its basis in the mathematical conclusion that the equations of motion are symmetric under time reversal. Thus, any particle trajectory in the time period $t = 0$ to $t = t_1$ undergoes a reversal in the time period $t = -t_1$ to $t = 0$, and the particle retraces its trajectory. In the field of chemical kinetics, this principle is sometimes stated in these equivalent forms:

1. The transition state of an elementary reversible reaction in the forward direction is identical to the transition state in the reverse direction, or
2. The mechanism of a reversible reaction in the rearward direction is identical to the mechanism in the forward direction run backward.

For example, if a proton is donated to a base in the forward direction, the proton must be abstracted from the conjugate acid of the base in the reverse reaction.

These ideas can exclude certain kinetic schemes as impossibilities. Thus, Scheme XVIII is possible, but Scheme XXIX is impossible, because B is formed directly from A, whereas A can only be regenerated from B via C, which is a different mechanism.

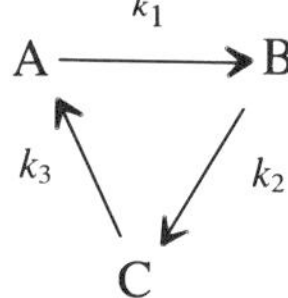

Scheme XXIX

The principle does not rule out the possibility of more than one route (mechanism) between initial and final states, provided that at equilibrium detailed balance applies to all routes.

Aside from qualitative applications such as illustrated in the preceding paragraph, detailed balancing can provide quantitative arguments by means of relationships like Eq. (3-186). We take as an example the hydrolysis of δ-thiovalerolactone (L) catalyzed by imidazole (Im), for which the Scheme XXX was proposed,[91] IH being an intermediate.

$$\begin{array}{ccccc}
 & k_1 & & k_5 & \\
\text{ImH} + \text{L} & \rightleftharpoons & \text{IH} & \rightarrow & \text{P} \\
 & k_{-1} & & & \\
\end{array}$$

$$k_2 \quad k_{-2} \qquad k_{-4} \quad k_4$$

$$\text{Im}^- + \text{L} \underset{k_3}{\overset{k_{-3}}{\rightleftharpoons}} \text{I}^-$$

$$+ \qquad\qquad +$$

$$\text{H}^+ \qquad\qquad \text{H}^+$$

Scheme XXX

The irreversible k_5 step is irrelevant to the following argument, which is based on the equilibrium state. Proceeding to define equilibrium constants as $K_1 = k_1/k_{-1} = [IH]/[ImH][L]$ and so on, we obtain the identity

$$K_1 K_4 = K_2 K_3$$

or, because K_2 and K_4 are acid dissociation constants,

$$\frac{k_1 K_4}{k_{-1}} = \frac{k_3 K_2}{k_{-3}} \tag{3-187}$$

In the original quantitative treatment of the pH dependence of the kinetics,[91] it was necessary to make the assumption that $k_3 K_2 \ll k_1[H^+]$. From Eq. (3-187), this means that

$$\frac{k_1 k_{-3} K_4}{k_{-1}} = k_3 K_2 \ll k_1[H^+]$$

or $k_{-3} K_4 \ll k_{-1}[H^+]$. These two inequalities are inconsistent conditions. Westheimer and Bender[92] pointed this out and proposed an alternative mechanism.

The applications of detailed balancing shown here are typical of the uses to be made of the principle in practical kinetics. At a deeper level these concepts must be justified by quantum mechanical and statistical mechanical arguments. Morrissey[93] and Mahan[94] have discussed these matters.

REFERENCES

1. Noyes, R.M. In "Investigation of Rates and Mechanisms of Reactions", 3rd ed.; Lewis, E.S. Ed.; Wiley-Interscience: New York, 1974; Part I, Chapter IX.
2. Laidler, K.J. *Pure Appl. Chem.* **1981,** *53,* 753.
3. Benson, S.W. "The Foundations of Chemical Kinetics"; McGraw-Hill: New York, 1960; p 29.
4. Moore, J.W.; Pearson, R.G. "Kinetics and Mechanism", 3rd ed.; Wiley-Interscience: New York, 1981; p 305.
5. King, E.L. *Int. J. Chem. Kinet.* **1982,** *14,* 1285.
6. Pladziewicz, J.R.; Lesniak, J.S.; Abrahamson, A.J. *J. Chem. Educ.* **1986,** *63,* 850.
7. Brown, H.C.; Fletcher, R.S. *J. Am. Chem. Soc.* **1949,** *71,* 1845.
8. Mark, H.B., Jr.; Rechnitz, G.A. "Kinetics in Analytical Chemistry"; Wiley-Interscience: New York, 1968; Chapter 5.
9. Mottola, H.A. "Kinetic Aspects of Analytical Chemistry"; Wiley-Interscience: New York, 1988; Chapter 7.
10. Connors, K.A. *Anal. Chem.* **1975,** *47,* 2066.
11. Connors, K.A. *Anal. Chem.* **1976,** *48,* 87.
12. Connors, K.A. *Anal. Chem.* **1977,** *49,* 1650.
13. Connors, K.A. *Anal. Chem.* **1979,** *51,* 1155.
14. Toby, F.S.; Toby, S. *J. Phys. Chem.* **1976,** *80,* 2313.
15. Kaduk, B.A.; Toby, S. *Int. J. Chem. Kinet.* **1977,** *9,* 829.
16. Toby, B.H.; Toby, F.S.; Toby, S. *Int. J. Chem. Kinet.* **1978,** *10,* 417.
17. Alcock, N.W.; Benton, D.J.; Moore, P. *Trans. Faraday Soc.* **1970,** *66,* 2210.

18. Jackson, W.G.; Harrowfield, J.M.; Vowles, P.D. *Int. J. Chem. Kinet.* **1977,** *9,* 535.

19. Carrington, T. *Int. J. Chem. Kinet.* **1982,** *14,* 517.

20. Vajda, S.; Rabitz, H. *J. Phys. Chem.* **1988,** *92,* 701.

21. Wagner, J.G.; Metzler, C.M. *J. Pharm. Sci.* **1967,** *56,* 658.

22. Gibaldi, M., Perrier, D. "Pharmacokinetics"; Marcel Dekker: New York, 1975; pp 21, 22, 35.

23. Niazi, S. "Textbook of Biopharmaceutics and Clinical Pharmacokinetics"; Appleton-Century-Crofts: New York, 1979; p 158.

24. Swain, C.G. *J. Am. Chem. Soc.* **1944,** *66,* 1696.

25. Kalonia, D.S.; Simonelli, A.P. *J. Pharm. Sci.* **1988,** *77,* 1055.

26. Westman, A.E.R.; DeLury, D.B. *Can. J. Chem.* **1956,** *34,* 1134.

27. Chien, J. *J. Am. Chem. Soc.* **1948,** *70,* 2256.

28. Schwemer, W.C.; Frost, A.A. *J. Am. Chem. Soc.* **1951,** *73,* 4541.

29. Frost, A.A.; Schwemer, W.C. *J. Am. Chem. Soc.* **1952,** *74,* 1268.

30. Paventi, M. *Can. J. Chem.* **1987,** *65,* 1987.

31. Mata-Perez, F.; Perez-Benito, J.F. *J. Chem. Educ.* **1987,** *64,* 925.

32. Boehmer, V.; Woersdoerfer, K. *Appl. Spectrosc.* **1977,** *31,* 334.

33. Emanuel', N.M.; Knorre, D.G. "Chemical Kinetics: Homogeneous Reactions"; Wiley (Halsted Press): New York, 1973; p 214.

34. French, D. *J. Am. Chem. Soc.* **1950,** *72,* 4806.

35. Wideqvist, S. *Acta Chem. Scand.* **1950,** *4,* 1216.

36. Svirbely, W.J.; Blauer, J.A. *J. Am. Chem. Soc.* **1961,** *83,* 4115, 4118.

37. Saville, B. *J. Phys. Chem.* **1971,** *75,* 2215.

38. Wideqvist, S. *Arkiv Kemi* **1955,** *8,* 325, 545.

39. Brändström, A. *Acta Pharm Suec.* **1967,** *4,* 139.

40. Svirbely, W.J.; Kundell, F.A. *J. Am. Chem. Soc.* **1967,** *89,* 5354.

41. Boas, M.L. "Mathematical Methods in the Physical Sciences"; Wiley: New York, 1966; p 593.

42. Jaeger, J.C. "An Introduction to the Laplace Transformation"; Methuen: London, 1949; p 7.

43. Lowry, T.M.; John, W.T. *J. Chem. Soc.* **1910,** *97,* 2634.

44. Wilson, E.B., Jr.; Decius, J.C.; Cross, P.C. "Molecular Vibrations"; McGraw-Hill: New York, 1955; p 16 (Dover Publications: New York, 1980, reprint).

45. Zwolinski, B.; Eyring, H. *J. Am. Chem. Soc.* **1947,** *69,* 2702.

46. Matsen, F.A.; Franklin, J.L. *J. Am. Chem. Soc.* **1950,** *72,* 3337.

47. Ritchie, C.D. "Physical Organic Chemistry: The Fundamental Concepts"; Marcel Dekker: New York, 1975; Chapter 1.

48. Carpenter, B.K. "Determination of Organic Reaction Mechanisms"; Wiley-Interscience: New York, 1984; Chapter 4.

49. Eyring, H.; Lin, S.H.; Lin, S.M. "Basic Chemical Kinetics"; Wiley-Interscience: New York, 1980; pp 4–5.

50. Pauling, L.; Wilson, E.B., Jr. "Introduction to Quantum Mechanics"; McGraw-Hill: New York, 1935; p 171.

51. Frost, A.A.; Pearson, R.G. "Kinetics and Mechanism"; Wiley: New York, 1953; pp 160–64.

52. Bodenstein, M. *Z. Physik Chem.* **1913,** *85,* 329.

53. Wong, J.T. *J. Am. Chem. Soc.* **1965,** *87,* 1788.

54. Miller, S.I. *J. Chem. Educ.* **1985,** *62,* 490.

55. Lineweaver, H.; Burk, D. *J. Am. Chem. Soc.* **1934,** *56,* 658.

56. King, E.L.; Altman, C. *J. Phys. Chem.* **1956,** *60,* 1375.

57. Cleland, W.W. *Biochemistry* **1975,** *14,* 3220.

58. Indge, K.J.; Childs, R.E. *Biochem. J.* **1976,** *155,* 567.

59. Gilbert, H.F. *J. Chem. Educ.* **1977,** *54,* 492.

60. Frei, K.; Günthard, H. *Helv. Chim. Acta* **1967,** *50,* 1294.

61. Farrow, L.A.; Edelson, D. *Int. J. Chem. Kinet.* **1974,** *6,* 787.

62. Porter, M.D.; Skinner, G.B. *J. Chem. Educ.* **1976,** *53,* 366.

63. Volk, L.; Richardson, W.; Lau, K.H.; Hall, M.; Lin, S.H. *J. Chem. Educ.* **1977,** *54,* 95.

64. Tardy, D.C.; Cater, E.D. *J. Chem. Educ.* **1983,** *60,* 109.

65. Corio, P.L. *J. Phys. Chem.* **1984,** *88,* 1825.

66. Zuman, P.; Patel, R.C. "Techniques in Organic Reaction Kinetics"; Wiley-Interscience: New York, 1984; pp 109–10.

67. Wylie, C.R.; Barrett, L.C. "Advanced Engineering Mathematics", 5th ed.; McGraw-Hill: New York, 1982; pp 267–73.

68. Beyer, W.H., Ed. "CRC Standard Mathematical Tables", 27th ed.; CRC Press: Boca Raton, Fla., 1984; pp 468–71.

69. Curtiss, C.F.; Hirschfelder, J.O. *Proc. Natl. Acad. Sci. USA*, **1952**, *38*, 235.

70. Warner, D.D. *J. Phys. Chem.* **1977**, *81*, 2329.

71. Schaad, L.J. *J. Am. Chem. Soc.* **1963**, *85*, 3588.

72. Rabinovitch, B. *J. Chem. Educ.* **1969**, *46*, 262.

73. Munson, J.W.; Connors, K.A. *Am J. Pharm. Educ.* **1971**, *35*, 8.

74. Dixon, D.A.; Shafer, R.H. *J. Chem. Educ.* **1973**, *50*, 648.

75. Darvey, I.G.; Ninham, B.W.; Staff, P.J. *J. Chem. Phys.* **1966**, *45*, 2145.

76. Lindblad, P.; Degn, H. *Acta Chem. Scand.* **1967**, *21*, 791.

77. Gillespie, D.T. *J. Phys. Chem.* **1977**, *81*, 2340.

78. Kibby, M.R. *Nature (London)* **1969**, *222*, 298.

79. Davenport, D.A. *J. Chem. Educ.* **1975**, *52*, 379.

80. Frost, A.A.; Tamres, M. *J. Chem. Phys.* **1947**, *15*, 383.

81. Williams, R.F.; Bruice, T.C. *J. Am. Chem. Soc.* **1976**, *98*, 7752.

82. Edwards, J.O.; Greene, E.F.; Ross, J. *J. Chem. Educ.* **1968**, *45*, 381.

83. Bunnett, J.F. In "Investigation of Rates and Mechanisms of Reactions", 3rd ed.; Lewis, E.S., Ed.; Wiley-Interscience: New York, 1974; Part I, pp 399–402.

84. Pearson, R.G. *J. Phys. Chem.* **1977**, *81*, 2323.

85. Pandit, N.K.; Connors, K.A. *J. Pharm. Sci.* **1982**, *71*, 485.

86. Jencks, W.P. *J. Am. Chem. Soc.* **1958**, *80*, 4581, 4585.

87. Espenson, J.H. In "Investigation of Rates and Mechanisms of Reactions", 3rd ed.; Lewis, E.S., Ed.; Wiley-Interscience: New York, 1974; Part I, pp 587–8.

88. Jencks, W.P. "Catalysis in Chemistry and Enzymology"; McGraw-Hill: New York, 1969; pp 50–3.

89. Connors, K.A.; Ifan, A. *J. Pharm. Sci.* **1987**, *76*, 834.

90. Birk, J.P. *J. Chem. Educ.* **1970**, *47*, 805.

91. Bruice, T.C.; Bruno, J.J. *J. Am. Chem. Soc.* **1962**, *84*, 2128.

92. Westheimer, F.H.; Bender, M.L. *J. Am. Chem. Soc.* **1962**, *84*, 4908.

93. Morrissey, B.W. *J. Chem. Educ.* **1975**, *52*, 296.

94. Mahan, B.H. *J. Chem. Educ.* **1975**, *52*, 299.

95. Schubert, W.M.; Lamm, B.; Keeffe, J.R. *J. Am. Chem. Soc.* **1964**, *86*, 4727.

PROBLEMS

1. The benzene anion formed as shown by nucleophilic attack on an aromatic ring is called a Meisenheimer complex.

Give an expression for the pseudo-first-order rate constant for Meisenheimer complex formation.

2. For the hydration of styrene,

$$Ph\text{---}CH{=}CH_2 + H_2O \rightleftharpoons Ph\text{---}CH(OH)CH_3$$

it is reported that $k_{obs} = 5.66 \times 10^{-5}$ s^{-1} and the equilibrium ratio [styrene]/[alcohol] $= 2.3 \times 10^{-2}$. Calculate k_1 and k_{-1}.

3. Suppose a reaction is found to have the experimental rate equation

$$v = k[HA][B][C]$$

where HA is a weak acid with $pK_a = 4.50$, B is a weak base with $pK_b = 8.00$, C is a neutral reactant, and $k = 10$ M^{-2}s^{-1}. Then find the value of k' in the kinetically equivalent form

$$v = k'[A^-][BH^+][C]$$

4. Derive an equation giving c_B^{max}, the value of c_B when $t = t_B^{max}$, for Scheme IX. [*Hint:* Start by deriving $\exp(-k_1 t_B^{max}) = (k_2/k_1)^p$, where $p = k_1/(k_1 - k_2)$, and the analogous equation for k_2.]

5. Define the percentage extent of reaction for Scheme IX if $F = c_B + 2c_C$ is the experimental measure of the progress of the reaction.

6. Show that Eqs. (3-101) and (3-102) are equivalent.

7. What are the units of the Laplace transform variable s when applied to a first-order reaction?

8. For Scheme XIV, and for each of the following sets of rate constants, calculate the exact relative concentration c_B/c_A^0 as a function of time. Also, for each set, calculate the approximate values of c_B/c_A^0 using both the equilibrium assumption and the steady-state approximation.
   (a) $k_1 = 10$ s^{-1}, $k_{-1} = 1$ s^{-1}, $k_2 = 0.01$ s^{-1}.
   (b) $k_1 = 0.01$ s^{-1}, $k_{-1} = 1$ s^{-1}, $k_2 = 10$ s^{-1}.
   (c) $k_1 = 1$ s^{-1}, $k_{-1} = 10$ s^{-1}, $k_2 = 0.01$ s^{-1}.

9. Apply the steady-state approximation to Scheme XXII for ester hydrolysis to find how the experimental second-order rate constant k_{OH} is related to the elementary rate constants.

10. For Scheme XVIII, find the secular equation, the characteristic equation, and the eigenvalues.

11. Two schemes for second-order Runge-Kutta numerical integration can be presented as follows:[68]

	(i)				(ii)		
r	a_r	b_r	w_r	r	a_r	b_r	w_r
1	0	0	1/2	1	0	0	$-1/2$
2	1	1	1/2	2	1/3	1/3	3/2

where $dy/dx = f(x,y)$ and y_0 are given, $x_1 = x_0 + h$, $y_1 = y_0 + K$, and $K = w_1 k_1 + w_2 k_2$, with $k_r = h \cdot f(x_0 + a_r h, y_0 + b_r k_1)$. Show that Schemes i and ii are equivalent to Eq. 3-162.

12. Carry out a manual Monte Carlo simulation of Scheme I, using $[A]_0 = 100$, $[B]_0 = 0$, $n_1 = 7$, $n_{-1} = 3$. Compare the equilibrium constant obtained in the simulation with the true value. Make the semilogarithmic plot according to Eq. 3-40 and interpret its slope.

13. Verify transform No. 6 in Table 3-1 by applying the definition of the Laplace transform, Eq. (3-65).

14. Solve Scheme I using Laplace transforms, obtaining Eq. (3-4) or (3-118), for initial conditions $c_A^0 = c_A^0$, $c_Z^0 = 0$.

15. Equation (3-150) and other equations of the same functional form can be placed in three linear plotting forms, of which the double-reciprocal form is shown as Eq. (3-152). Find the other two linear plotting forms.

CHAPTER 4

Fast Reactions

When we carry out conventional studies of solution kinetics, we initiate reactions by mixing solutions. The time required to achieve complete mixing places a limit on the fastest reaction that can be studied in this way. It is not difficult to reduce the mixing time to about 10 s, so a reaction having a half-life of, say, 10 s is about the fastest reaction we can study by conventional techniques. (See Section 4.4 for further discussion of this limit.) The slowest reaction accessible to study depends upon analytical sensitivity and patience; let us say that the half-life of a graduate student, $2–2\frac{1}{2}$ years, sets an approximate limit. This corresponds to roughly 7×10^7 s. Thus, a range of half-lives of about 10^7 can be studied by conventional techniques.

There are obviously many reactions that are too fast to investigate by ordinary mixing techniques. Some important examples are proton transfers, enzymatic reactions, and noncovalent complex formation. Prior to the second half of the 20th century, these reactions were referred to as *instantaneous* because their kinetics could not be studied. It is now possible to measure the rates of such reactions. In Section 4.1 we will find that the fastest reactions have half-lives of the order 10^{-11} s, so the fast reaction regime encompasses a much wider range of rates than does the conventional study of kinetics.

Despite the great scope for rate studies in the fast reaction field, these still constitute a small fraction of published kinetic studies. In part this is because fast reaction kinetics is still in some respects a specialist's field, requiring equipment (whether commercially purchased or locally fabricated) that is not commonly found in the chemical laboratory's stock of instrumentation. This chapter treats the field at a nonspecialist's level, which is adequate to allow the experimentalist to judge if a certain technique is applicable to a particular problem. Reviews and book-length treatments are available; these should be consulted for more detailed theoretical and experimental descriptions.[1-6]

4.1 DIFFUSION-CONTROLLED REACTIONS

Although the liquid state is obviously less ordered than is the solid state, some order ("structure") exists in the liquid, the extent of order depending upon molecular identity and the temperature. Each molecule of solvent (or of solute, in a dilute solution) is, on average, surrounded by a characteristic number Z of solvent molecules; Z, the *coordination number,* is typically in the range 4 to 12. This sheath of solvent molecules is called a *solvent cage.* A solute molecule enclosed in a solvent cage undergoes many collisions with the solvent molecules surrounding it before it escapes from the cage. This period during which the solute molecule remains in the solvent cage is called an *encounter.*

Consider a dilute solution of two reactant molecules, A and B. Inevitably an A molecule and a B molecule will undergo an encounter, the frequency of such encounters depending upon the concentrations of A and B. If, upon each encounter of A and B, they undergo bimolecular reaction, then the rate of this reaction is determined solely by the rate of encounter of A and B; that is, the rate is not controlled by the chemical requirement that an energy barrier be overcome. One way to find this rate is to treat the problem as one of classical diffusion, and so this maximum possible rate of reaction is often called the *diffusion-controlled rate.* This problem was solved by Smoluchowski.[7] In the following development no provision is made for attractive forces between the molecules.[8]

Consider spherical molecules A and B having radii r_A and r_B and diffusion coefficients D_A and D_B. First, suppose that B is fixed and that the rate of reaction is limited by the rate at which A molecules diffuse to the B molecule. We calculate the flux $J(A \rightarrow B)$ of A molecules to one B molecule. Let n_A and n_B be the concentrations (in molecules/cm^3) of A and B in the bulk, and let r be the radius of a sphere centered at the B molecule. The surface area of this sphere is $4\pi r^2$, so by Fick's first law we obtain

$$J(A \rightarrow B) = 4\pi r^2 D_A \frac{dn_A}{dr} \tag{4-1}$$

where the flux is in molecules per second.

Assume that chemical reaction occurs at contact, namely, when $r = r_A + r_B$; thus, we integrate between limits as shown:

$$J(A \rightarrow B) \int_{r_A + r_B}^{\infty} \frac{dr}{r^2} = 4\pi D_A \int_0^{n_A} dn_A \tag{4-2}$$

The result is

$$J(A \rightarrow B) = 4\pi D_A (r_A + r_B) n_A \tag{4-3}$$

Thus far the single B molecule has been held stationary, and Eq. (4-3) gives the flux of A molecules toward this fixed B. At the same time, however, the B molecule is itself diffusing, so the effective diffusion constant is the sum of D_A and D_B. Moreover, we must multiply our result for the flux to a single B by the total number of B molecules in order to find the total rate leading to diffusion-limited encounters, or

$$v_n = 4\pi(D_A + D_B)(r_A + r_B)n_A n_B \tag{4-4}$$

where v_n is the rate in encounters per cubic centimeter per second.

The conventional second-order rate equation is $v_c = k_D c_A c_B$, where v_c is in moles per liter per second and c_A, c_B are in moles per liter. Because $c = 1000\, n/N_A$ and $v_c = 1000 v_n/N_A$, where N_A is Avogadro's number, we obtain

$$k_D = \frac{4\pi N_A(D_A + D_B)(r_A + r_B)}{1000} \tag{4-5}$$

as the theoretical estimate of k_D, the diffusion-limited second-order rate constant (units, $M^{-1}\ s^{-1}$).

Very commonly Eq. (4-5) is combined with Eq. (4-6), the Stokes–Einstein equation relating the diffusion coefficient to the viscosity η.

$$D = \frac{kT}{6\pi\eta r} \tag{4-6}$$

(k is the Boltzmann constant.) The result is

$$k_D = \frac{2RT}{3000\eta}\left[\frac{(r_A + r_B)^2}{r_A r_B}\right] \tag{4-7}$$

If $r_A = r_B$, as in a dimerization, Eq. (4-7) becomes

$$k_D = \frac{8RT}{3000\eta} \tag{4-8}$$

In Eqs. (4-7) and (4-8) the viscosity is expressed in poise. For reactions in water at 25°C, Eq. (4-8) gives $k_D = 7.4 \times 10^9\ M^{-1}\ s^{-1}$.

This treatment obviously is oversimplified. At the next level of development, it is necessary to incorporate the intermolecular forces between A and B.[9,10] If A and B are ions of opposite charge, it is found that the diffusion-limited rate constant is about $10^{11}\ M^{-1}\ s^{-1}$.

From these considerations we conclude that diffusion-limited bimolecular rate constants are of the order 10^{10}–$10^{11}\ M^{-1}\ s^{-1}$. If an experimentally measured rate constant is of this magnitude, the usual conclusion is, therefore, that it is diffusion limited. For example, this extremely important reaction (in water)

$$H^+ + OH^- \rightleftharpoons H_2O$$

has a measured rate constant in the forward direction of 1.4×10^{11} M^{-1} s^{-1}; hence, the reaction is assumed to be under diffusion control. The corollary to this conclusion is that if the experimental rate constant is substantially smaller than the diffusion-limited value, the reaction rate is chemically controlled; that is, the activation energy is in the range typical of chemical reactions. (The activation energy of a diffusion-controlled reaction is small, typically 2–6 kcal mol^{-1}, and is partly ascribed to the temperature dependence of viscosity.)

Equations (4-5) and (4-7) are alternative expressions for the estimation of the diffusion-limited rate constant, but these equations are not equivalent, because Eq. (4-7) includes the assumption that the Stokes–Einstein equation is applicable. Olea and Thomas[11] measured the kinetics of quenching of pyrene fluorescence in several solvents and also measured diffusion coefficients. The diffusion coefficients did not vary as η^{-1} [as predicted by Eq. (4-6)], but roughly as η^{-2}. Thus Eq. (4-7) is not valid, in this system, whereas Eq. (4-5), used with the experimentally measured diffusion coefficients, gave reasonable agreement with measured rate constants.

In Section 3-3 we discussed the problem of kinetically equivalent rate terms. Suppose one of the rate constants evaluated for such a rate equation were larger than the diffusion-limited value; this is a reasonable basis upon which to reject the formulation of the rate equation leading to this result. Jencks[12] has given examples of this argument.

We have seen that 10^{11} M^{-1} s^{-1} is about the fastest second-order rate constant that we might expect to measure; this corresponds to a lifetime of about 10^{-11} s at unit reactant concentration. Yet there is evidence, discussed by Grunwald,[13] that certain proton transfers have lifetimes of the order 10^{-13} s. These "ultrafast" reactions are believed to take place via quantum mechanical tunneling through the energy barrier. This phenomenon will only be significant for very small particles, such as protons and electrons.

4.2 RELAXATION KINETICS

Consider a reversible chemical reaction at equilibrium. Let a sudden perturbation of the system be made (as by changing the temperature) so as to produce a new equilibrium configuration. Because the actual concentrations are not equal to these new equilibrium values, the system will undergo adjustment until the concentrations have the values required by the new equilibrium position. This process of adjustment of a perturbed system toward its equilibrium configuration is called *relaxation*. The rate of relaxation is a function of the rate constants, which can be determined by studying the relaxation process. Because the perturbation can be produced more rapidly than reactions can be initiated by the conventional mixing of solutions, relaxation kinetics provides a means for studying fast reactions. This approach was developed by Eigen and De Maeyer[14] in the 1950s. Bernasconi[5] has given a thorough treatment of relaxation kinetics.

Linearization of Rate Equations

We begin by considering the simplest reversible reaction (Scheme I).

$$A \underset{k_{-1}}{\overset{k_1}{\rightleftharpoons}} Z$$

Scheme I

The system is initially at equilibrium with concentrations c_A^0 and c_Z^0. Now we rapidly perturb the system so as to alter the magnitude of the equilibrium constant. Let the new equilibrium concentrations, toward which the actual concentrations will relax, be $\bar{c}_A$ and $\bar{c}_Z$. (Clearly one of these will be greater than and one will be less than the initial concentrations.) The concentrations at any time t are c_A and c_Z.

Define a reaction variable x as the concentration by which the actual concentrations differ from the new equilibrium values; thus,

$$x = c_A - \bar{c}_A = \bar{c}_Z - c_Z \tag{4-9}$$

where we have supposed that $c_A^0 > \bar{c}_A$ and $c_Z^0 < \bar{c}_Z$. Figure 4-1 shows how x is related to the initial and final equilibrium concentrations. The differential rate equation is

$$-\frac{dc_A}{dt} = -\frac{dx}{dt} = k_1 c_A - k_{-1} c_Z$$

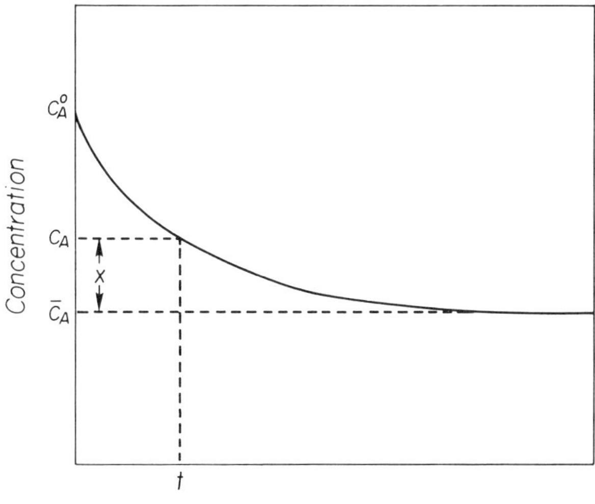

Figure 4-1. Relationship of the reaction variable x to the system concentrations.

which, combined with Eq. (4-9), gives

$$-\frac{dx}{dt} = (k_1 + k_{-1})x + k_1\bar{c}_A - k_{-1}\bar{c}_Z$$

However, the equilibrium constant at the new conditions (after the perturbation) is $K = k_1/k_{-1} = \bar{c}_Z/\bar{c}_A$, so we obtain

$$-\frac{dx}{dt} = (k_1 + k_{-1})x \tag{4-10}$$

Not surprisingly, we find that the relaxation is a first-order process with rate constant $k_1 + k_{-1}$. It is conventional in relaxation kinetics to speak of the *relaxation time* τ, which is the time required for the concentration to decay to $1/e$ its initial value. In Chapter 2 we found that the lifetime defined in this way is the reciprocal of a first-order rate constant. In the present instance, therefore,

$$\tau^{-1} = k_1 + k_{-1} \tag{4-11}$$

The more important equilibrium shown in Scheme II introduces another factor.

$$A + B \underset{k_{-1}}{\overset{k_1}{\rightleftharpoons}} Z$$

Scheme II

The reaction variable is defined by Eq. (4-12).

$$x = c_A - \bar{c}_A = c_B - \bar{c}_B = \bar{c}_Z - c_Z \tag{4-12}$$

and the rate equation is

$$-\frac{dx}{dt} = k_1 c_A c_B - k_{-1} c_Z$$

Substitution leads to

$$-\frac{dx}{dt} = k_1 x^2 + (k_1\bar{c}_A + k_1\bar{c}_B + k_{-1})x + k_1\bar{c}_A\bar{c}_B - k_{-1}\bar{c}_Z$$

However, the equilibrium condition is $K = k_1/k_{-1} = \bar{c}_Z/\bar{c}_A\bar{c}_B$, so

$$-\frac{dx}{dt} = k_1 x^2 + (k_1\bar{c}_A + k_1\bar{c}_B + k_{-1})x \tag{4-13}$$

Equation (4-13) is nonlinear in the concentration x. We, therefore, impose the *linearization condition,* namely, that the perturbation is sufficiently small that the term in x^2 is negligible. The result is

$$-\frac{dx}{dt} = \tau^{-1}x \tag{4-14}$$

where

$$\tau^{-1} = k_1(\bar{c}_A + \bar{c}_B) + k_{-1} \tag{4-15}$$

Equation (4-14) shows that the relaxation is first-order; according to Eq. (4-15), measurements of τ at several values of reactant concentrations allow the rate constants to be estimated.

This treatment illustrates several important aspects of relaxation kinetics. One of these is that the method is applicable to equilibrium systems. Another is that we can always generate a first-order relaxation process by adopting the linearization approximation. This condition usually requires that the perturbation be small (in the sense that higher-order terms be negligible relative to the first-order term). The relaxation time is a function of rate constants and, often, concentrations.

The question of how small the perturbation must be is of practical interest. Let us consider Scheme II and rate equation (4-13), which can be written in this form:

$$-\frac{dx}{dt} = \tau^{-1}x(1 + r) \tag{4-16}$$

where

$$r = x(\bar{c}_A + \bar{c}_B + K^{-1})^{-1}$$

Evidently the perturbation is small enough if $r <<< 1$, so the extent of the acceptable perturbation is governed by the concentrations and the equilibrium constant. For Scheme I it is seen that $r = 0$, so the perturbation in this case is not limited to small values. Brouillard and co-workers[15,16] have analyzed many systems from this point of view.

Complex Reactions

If a reaction system consists of more than one elementary reversible reaction, there will be more than one relaxation time; in general, the number of relaxation times is equal to the number of states of the system minus one. (However, even for multistep reactions, only a single relaxation time will be *observed* if all intermediates are present at vanishingly low concentrations, that is, if the steady-state approximation is valid.) The relaxation times are coupled, in that each relaxation time includes contributions from all of the system rate constants. A system of more than one relaxation time possesses a *relaxation spectrum,* by analogy with infrared spectroscopy, in which an absorption frequency reflects vibrational modes of many bonds in the molecule rather than of just one bond.

We will demonstrate the treatment with the two-step reaction system shown in Scheme III.

$$A + B \underset{k_{21}}{\overset{k_{12}}{\rightleftharpoons}} AB \underset{k_{32}}{\overset{k_{23}}{\rightleftharpoons}} C$$

Scheme III

There are three states, hence, two relaxation times. The differential rate equations are

$$-\frac{dc_A}{dt} = k_{12}c_A c_B - k_{21}c_{AB}$$

$$\frac{dc_{AB}}{dt} = k_{12}c_A c_B + k_{32}c_C - k_{21}c_{AB} - k_{23}c_{AB}$$

(4-17)

(We could have chosen any two of the three possible equations.) Reaction variables are defined as

$$x_A = c_A - \bar{c}_A = c_B - \bar{c}_B$$

$$x_{AB} = c_{AB} - \bar{c}_{AB}$$

(4-18)

$$x_C = c_C - \bar{c}_C$$

Mass conservation requires that the production of one molecule of C result in the loss of one molecule of either A or AB, so

$$x_C = -(x_A + x_{AB})$$

(4-19)

Substitution of Eqs. (4-18) into (4-17) gives

$$-\frac{dx_A}{dt} = k_{12}(x_A^2 + \bar{c}_A x_A + \bar{c}_B x_A + \bar{c}_A \bar{c}_B) - k_{21}(x_{AB} + \bar{c}_{AB})$$

$$\frac{dx_{AB}}{dt} = k_{12}(x_A^2 + \bar{c}_A x_A + \bar{c}_B x_A + \bar{c}_A \bar{c}_B) + k_{23}(\bar{c}_C + x_C)$$

(4-20)

$$- (k_{21} + k_{23})(x_{AB} + \bar{c}_{AB})$$

We apply the linearization condition by neglecting the x^2 terms. The equilibrium conditions are

$$k_{12}\bar{c}_A\bar{c}_B = k_{21}\bar{c}_{AB}$$

$$k_{23}\bar{c}_{AB} = k_{32}\bar{c}_C$$

We also eliminate x_C by means of Eq. (4.19). The result is

$$\frac{dx_A}{dt} = -k_{12}(\bar{c}_A + \bar{c}_B)x_A + k_{21}x_{AB}$$

$$\frac{dx_{AB}}{dt} = [k_{12}(\bar{c}_A + \bar{c}_B) - k_{32}]x_A - (k_{21} + k_{23} + k_{32})x_{AB} \tag{4-21}$$

Equations (4-21) are linear first-order differential equations. We considered in detail the solution of such sets of rate equations in Section 3-2, so it is unnecessary to carry out the solutions here. In relaxation kinetics these equations are always solved by means of the secular equation, but the Laplace transformation can also be used. Let us write Eqs. (4-21) as

$$-\frac{dx_A}{dt} = a_{11}x_A + a_{12}x_{AB}$$

$$-\frac{dx_{AB}}{dt} = a_{21}x_A + a_{22}x_{AB}$$

where the definitions of the coefficients are obvious from Eqs. (4-21). The solutions are found to be

$$\tau_I^{-1} = \tfrac{1}{2}\{(a_{11} + a_{22}) + [(a_{11} + a_{22})^2 - 4(a_{11}a_{22} - a_{12}a_{21})]^{1/2}\}$$
$$\tau_{II}^{-1} = \tfrac{1}{2}\{(a_{11} + a_{22}) - [(a_{11} + a_{22})^2 - 4(a_{11}a_{22} - a_{12}a_{21})]^{1/2}\} \tag{4-22}$$

from which we find

$$\tau_I^{-1}\,\tau_{II}^{-1} = a_{11}a_{22} - a_{12}a_{21}$$

$$\tau_I^{-1} + \tau_{II}^{-1} = a_{11} + a_{22} \tag{4-23}$$

Using the definitions of the coefficients:

$$\tau_I^{-1}\,\tau_{II}^{-1} = k_{12}'\,(k_{23} + k_{32}) + k_{21}k_{32} \tag{4-24a}$$

$$\tau_I^{-1} + \tau_{II}^{-1} = k_{12}' + k_{21} + k_{23} + k_{32} \tag{4-24b}$$

where

$$k'_{12} = k_{12}(\bar{c}_A + \bar{c}_B) \tag{4-24c}$$

Thus, we find that both relaxation times are functions of all four rate constants.

An important special case can be derived from this general result. This is the case of a fast preequilibrium, in which the $A + B \rightleftharpoons AB$ system rapidly equilibrates, the $AB \rightleftharpoons C$ step being much slower. Then the relaxation time for the first step is much shorter than that for the second, and some measure of uncoupling takes place. For such a system $k_{12}, k_{21} \gg k_{23}, k_{32}$, and we obtain $a_{11} = k'_{12}$, $a_{12} = -k_{21}$, $a_{21} \approx -k'_{12}$, $a_{22} \approx k_{21}$. Equations (4-22) then give $\tau_I^{-1} \approx k'_{12} + k_{21}$ and $\tau_{II}^{-1} \approx 0$. Because these are approximations, the result for τ_I is reasonable but that for τ_{II} is not. To reach a reasonable result for τ_{II} we use Eq. (4-24a),

$$\tau_{II}^{-1} = \tau_I [k'_{12}(k_{23} + k_{32}) + k_{21}k_{32}]$$

Using the approximation $\tau_I^{-1} \approx k'_{12} + k_{21}$ yields, after suitable rearrangement,

$$\tau_{II}^{-1} = \frac{K_1(\bar{c}_A + \bar{c}_B)k_{23}}{K_1(\bar{c}_A + \bar{c}_B) + 1} + k_{32} \tag{4-25}$$

where $K_1 = k_{12}/k_{21}$. In the limit that $K_1(\bar{c}_A + \bar{c}_B) \gg 1$, Eq. (4-25) becomes $\tau_{II}^{-1} = k_{23} + k_{32}$, which is the same as Eq. (4-11) for Scheme I; in this extreme limit the two steps are completely uncoupled.

Returning to the general treatment of Scheme III and Eqs. (4-24), and presuming that it is possible experimentally to measure τ_I and τ_{II} as functions of $(\bar{c}_A + \bar{c}_B)$, the four rate constants can be evaluated from plots of $\tau_I^{-1} \tau_{II}^{-1}$ vs. $(\bar{c}_A + \bar{c}_B)$ (plot 1) and of $(\tau_I^{-1} + \tau_{II}^{-1})$ vs. $(\bar{c}_A + \bar{c}_B)$ (plot 2). The plots have these parameters:

$$\text{slope (1)} = k_{12}(k_{23} + k_{32})$$

$$\text{intercept (1)} = k_{21}k_{32}$$

$$\text{slope (2)} = k_{12}$$

$$\text{intercept (2)} = k_{21} + k_{23} + k_{32}$$

Suitable combination allows the four rate constants to be estimated.

We have next to consider the measurement of the relaxation times. Each τ is the reciprocal of an apparent first-order rate constant, so the problem is identical with problems considered in Chapters 2 and 3. If the system possesses a single relaxation time, a semilogarithmic first-order plot suffices to estimate τ. The analytical response is often solution absorbance, or an electrical signal proportional to absorbance or to another physical property. As shown in Section 2.3 (Treatment of Instrument Response Data), the appropriate plotting function is $\ln(A_t - A_\infty)$, where A_t is the signal at time t and A_∞ is the signal when the system has completely relaxed to its final equilibrium position. The slope of the plot of $\ln (A_t - A_\infty)$ vs. t is equal to $-\tau^{-1}$.

If the system has two relaxation times, it is equivalent to two first-order reactions, and the graphical and calculational techniques described in Chapter 3 are appropriate. The analytical signal is a function of two exponentials. If the two relaxation times differ greatly in magnitude, they can be separately measured as if the other were not present, but if they are comparable, the problem may be quite difficult. The *relaxation amplitude*, namely, the maximum value of the analytical signal associated with each relaxation, plays a role in the detectability of the separate relaxation steps. Because the relaxation times depend upon concentrations, it may be possible to alter the relative magnitudes of the relaxation times by changing the concentrations. Other reaction conditions, such as temperature and solvent, may be adjusted to aid in resolving the relaxation times. If the two τ values are fairly similar in magnitude, it may be difficult even to detect the presence of two processes; a semilogarithmic plot may appear to be linear.

Systems having three or more relaxation times are very difficult to analyze, and the concept of a *mean relaxation time* has been developed to describe such reactions; Bernasconi[5] treats this problem.

Experimental Methods

To initiate a chemical relaxation it is necessary to perturb the system from its initial equilibrium position. This is done by applying a *forcing function*, which is an appropriate experimental stress to which the system responds with a shift in equilibrium configuration. Forcing functions can be transient (a sudden, essentially discontinuous "jolt") or periodic (a cyclic stress of constant frequency).

The most widely used transient method is the *temperature-jump* (T-jump) method.[4,5,17] This is based on the van't Hoff equation, which describes the temperature dependence of the equilibrium constant.

$$\left(\frac{\partial \ln K}{\partial T} \right)_P = \frac{\Delta H^0}{RT^2} \tag{4-26}$$

The sensitivity of the equilibrium constant to temperature, therefore, depends upon the enthalpy change ΔH^0. This is usually not a serious limitation, because most reaction enthalpies are sufficiently large and because we commonly require that the perturbation be a small one so that the linearization condition is valid. If ΔH^0 is so small that the T-jump is ineffective, it may be possible to make use of an auxiliary reaction in the following way: Suppose the reaction under study is an acid–base reaction with a small ΔH^0. We can add a buffer system having a large ΔH^0 and apply the T-jump to the combined system. The T-jump will alter the K_a of the buffer reaction, resulting in a *pH jump*. The pH jump then acts as the forcing function on the reaction of interest.

It is obviously desirable that the time required to heat the sample solution be much shorter than the relaxation time(s) of the system. The most commonly used T-jump apparatus is based on Joule heating, which is produced by discharging a high-voltage capacitor through the reaction solution. The time course of the temperature increase is given by

$$\Delta T_t = \Delta T_\infty (1 - e^{-t/\tau_d}) \tag{4-27}$$

where ΔT_t is the temperature at time t, ΔT_∞ is the maximum temperature change, and τ_d is the heating time constant. The heating time constant is given by $\tau_d = RC/2$, where R is the circuit resistance and C is the capacitance. We wish τ_d to be small relative to τ, the chemical relaxation time. The minimum value of the capacitance C is limited by the necessity to store sufficient energy in the capacitor. Therefore τ_d is minimized by reducing R, which can be accomplished by increasing the conductance of the reaction medium. Thus T-jump studies by electric-field discharge must employ polar solvents, often of appreciable ionic strength. Most T-jump studies have been carried out on aqueous solutions. The temperature change is typically 1–10°C.

T-Jumps can also be produced by microwave heating and by laser pulse absorption. These methods remove the restriction to low-resistance solvents; any solvent capable of absorbing energy of the applied frequency may be used. The heating time can be extremely short with laser heating.[18]

The *pressure-jump* (P-jump) method is based on the pressure dependence of the equilibrium constant, Eq. (4-28), where ΔV^0 is the molar volume change of the reaction.

$$\left(\frac{\partial \ln K}{\partial P}\right)_T = -\frac{\Delta V^0}{RT} \tag{4-28}$$

Most reactions in solution have rather small ΔV^0 values (usually ΔV^0 is less than 20 cm^3/mol), so only small perturbations are possible. The pressure change is created by rupturing a diaphragm separating the reaction solution from a pressure vessel. A typical pressure change is about 60 atm.

The *electric field-jump* method is applicable to reactions of ions and dipoles. Application of a powerful electric field to a solution will favor the production of ions from a neutral species, and it will orient dipoles with the direction of the applied field. The method has been used to study metal ion complex formation, the binding of ions to macromolecules, and acid–base reactions.

Concentration-jump methods, such as the pH-jump technique cited earlier, can be used in relaxation kinetics, but this approach is described later (Section 4.4).

Ultrasonic absorption is a so-called stationary method in which a periodic forcing function is used. The forcing function in this case is a sound wave of known frequency. Such a wave propagating through a medium creates a periodically varying pressure difference. (It may also produce a periodic temperature difference.) Now suppose that the system contains a chemical equilibrium that can respond to pressure differences [as a consequence of Eq. (4-28)]. If the sound wave frequency is much lower than $1/\tau$, the characteristic frequency of the chemical relaxation (τ is the

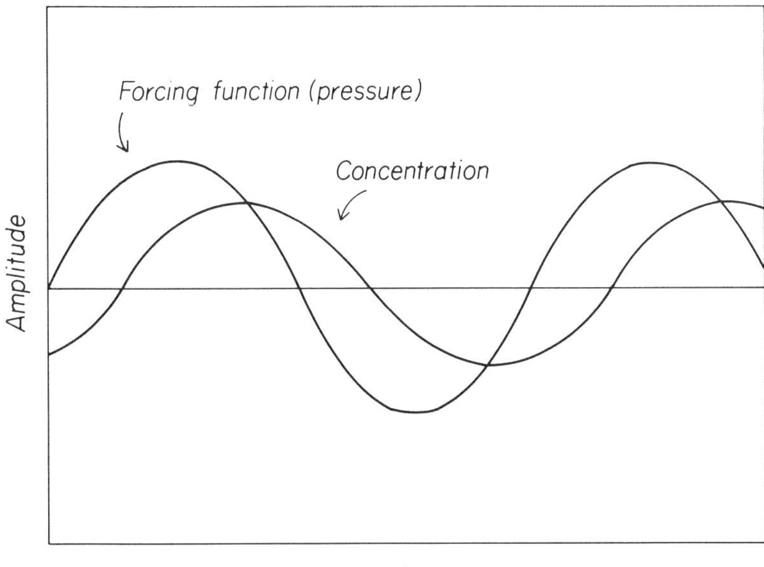

Figure 4-2. The phase lag between concentration and pressure in ultrasonic absorption. The cyclic pressure changes are produced by the sound wave. The cyclic concentration changes are a response to the pressure changes.

relaxation time), then the equilibrium concentrations can respond without significant time delay to the pressure changes. At the other extreme, of extremely high ultrasonic frequency, the chemical system is unable to respond at all to the pressure variations. At frequencies comparable to the chemical relaxation frequency, the chemical system follows the forcing function but with a significant time delay, which manifests itself as a phase lag between the forcing function and the time-dependent concentrations.[19] Figure 4-2 illustrates this concept.

The attenuation of sound intensity I follows Eq. (4-29), which is analogous to Beer's law.

$$I = I_0 e^{-2\alpha x} \tag{4-29}$$

In Eq. (4-29) x is the distance traveled by the wave, and α is the absorption coefficient. Sound absorption can occur as a result of viscous losses and heat losses (these together constitute "classical" modes of absorption) and by coupling to a chemical reaction, as described in the preceding paragraph. The theory of classical sound absorption shows that α is directly proportional to f^2, where f is the sound wave frequency (in Hz), so results are usually reported as α/f^2, for this is, classically, frequency independent.

Writing the total absorption in the form

$$\frac{\alpha}{f^2} = \left(\frac{\alpha}{f^2}\right)_{\text{chemical}} + \left(\frac{\alpha}{f^2}\right)_{\text{classical}}$$

and using the results of theory:

$$\frac{\alpha}{f^2} = \frac{A}{1 + \omega^2\tau^2} + B \tag{4-30}$$

where ω is sound wave frequency in radians per second; thus, $\omega = 2\pi f$.

Equation (4-30) shows that when ω is very small, $\alpha/f^2 = A + B$; absorption occurs by both the chemical and classical modes. In terms of Fig. 4-2, there is no phase-lag between the pressure and concentration curves. When ω is very large, $\alpha/f^2 = B$; the chemical system is unable to absorb energy from the high-frequency forcing function. When ω is comparable with $1/\tau$, α/f^2 is a function of frequency, and α/f^2 passes through an inflection point when $\omega = \tau$; thus τ is found by measuring absorption as a function of frequency.

For sensitivity of measurement, evidently we wish the classical absorption [B in Eq. (4-30)] to be small. Water has $(\alpha/f^2)_{classical} = 22 \times 10^{-17}$ s^2 cm^{-1}, which is a very small value (water does not absorb sound well). Typical values of $(\alpha/f^2)_{chemical}$ are 10–$10^3 \times 10^{-17}$ s^2 cm^{-1}.

The several experimental methods allow a wide range of relaxation times to be studied.[5] T-Jump is capable of measurements over the time range 1 to 10^{-8} s; P-jump, 10 to 5×10^{-5} s; electric field jump, 10^{-4} to 10^{-8} s; and ultrasonic absorption, 10^{-5} to 10^{-11} s. The detection method in the jump techniques depends upon the systems being studied, with spectrophotometry, fluorimetry, and conductimetry being widely used.

Applications

Most proton transfer reactions are fast; they have been carefully studied by relaxation methods. A system consisting of a conjugate acid–base pair in water is a three-state cyclic equilibrium as shown in Scheme IV. [The symbolism is that used by Bernasconi.[5, pp. 52–62]]

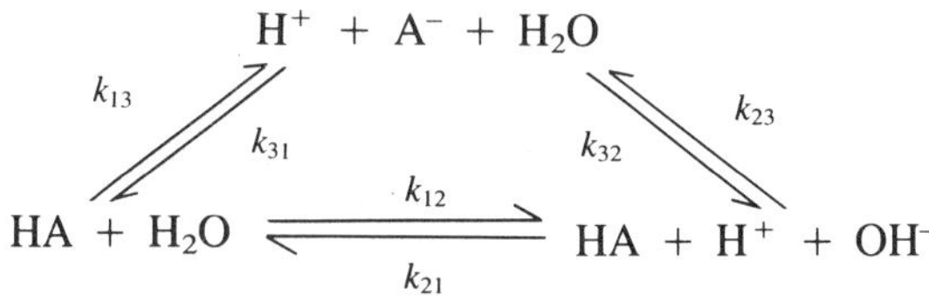

Scheme IV

Because this is a three-state system, it has two independent rate equations and two relaxation times. It is somewhat easier to write the rate equations when Scheme IV is presented in the form of Scheme V.

$$\text{H}_2\text{O} \overset{k_{12}}{\underset{k_{21}}{\rightleftharpoons}} \text{H}^+ + \text{OH}^-$$

$$\text{HA} \overset{k_{13}}{\underset{k_{31}}{\rightleftharpoons}} \text{H}^+ + \text{A}^-$$

$$\text{HA} + \text{OH}^- \overset{k_{23}}{\underset{k_{32}}{\rightleftharpoons}} \text{A}^- + \text{H}_2\text{O}$$

Scheme V

The 12,21 reaction is sometimes called *neutralization,* the 13,31 reaction is *protolysis,* and the 23,32 reaction is *hydrolysis.*

Four rate equations (in the concentrations of HA, A$^-$, H$^+$, OH$^-$) can be written, but only two of these are independent. We choose those in HA and H$^+$, so we will express all reaction variables in terms of these two species. Reaction variables are defined:

$$c_i - \bar{c}_i = x_i \tag{4-31}$$

where c_i is the concentration of species i at time t, and $\bar{c}_i$ is the concentration in the final equilibrium configuration; charges will be omitted for convenience.

The mass balance on solute yields $c_{\text{HA}} + c_{\text{A}} = \bar{c}_{\text{HA}} + \bar{c}_{\text{A}}$, whence

$$x_{\text{HA}} = -x_{\text{A}} \tag{4-32}$$

The mass balance on the solvent gives $c_{\text{H}_2\text{O}} + c_{\text{OH}} = \bar{c}_{\text{H}_2\text{O}} + \bar{c}_{\text{OH}}$, or

$$x_{\text{H}_2\text{O}} = -x_{\text{OH}} \tag{4-33}$$

According to the electroneutrality principle for this system, $c_{\text{H}} = c_{\text{A}} + c_{\text{OH}}$ and $\bar{c}_{\text{H}} = \bar{c}_{\text{A}} + \bar{c}_{\text{OH}}$; subtracting these equations gives

$$x_{\text{H}} + x_{\text{HA}} = x_{\text{OH}} \tag{4-34}$$

From Scheme V we write

$$\frac{dc_{\text{HA}}}{dt} = -k_{13}c_{\text{HA}} + k_{31}c_{\text{H}}c_{\text{A}} - k_{23}c_{\text{HA}}c_{\text{OH}} + k_{32}c_{\text{A}}c_{\text{H}_2\text{O}}$$

$$\frac{dc_{\text{H}}}{dt} = k_{13}c_{\text{HA}} - k_{31}c_{\text{H}}c_{\text{A}} + k_{12}c_{\text{H}_2\text{O}} - k_{21}c_{\text{H}}c_{\text{OH}} \tag{4-35}$$

The c_i in Eqs. (4-35) are replaced by means of Eq. (4-31), reaction variables x_A, x_{OH}, and x_{H_2O} are eliminated by using Eqs. (4-32)–(4-34), the linearization and equilibrium conditions are applied, and we obtain

$$-\frac{dx_{HA}}{dt} = a_{11}x_{HA} + a_{12}x_H$$

$$-\frac{dx_H}{dt} = a_{21}x_{HA} + a_{22}x_H$$

(4-36)

where

$$a_{11} = k_{13} + k_{31}\bar{c}_H + k_{23}(\bar{c}_{OH} + \bar{c}_{HA}) + k_{32}(\bar{c}_{H_2O} + \bar{c}_A)$$

$$a_{12} = k_{23}\bar{c}_{HA} - (k_{31} - k_{32})\bar{c}_A$$

$$a_{21} = k_{12} - k_{13} - (k_{31} - k_{21})\bar{c}_H$$

$$a_{22} = k_{12} + k_{21}(\bar{c}_{OH} + \bar{c}_H) + k_{31}\bar{c}_A$$

(4-37)

Equations (4-36) are identical in form to those developed in the solution of Scheme III, so the solutions for the two relaxation times in terms of the coefficients a_{ij} are given by Eqs. (4-22). Of course, the relaxation times expressed in terms of the rate constants are different in the present problem because the a_{ij} are specific to the kinetic scheme. Equations (4-23) for the sum and the product of the reciprocal relaxation times also apply here, but the results in terms of the rate constants are rather complicated. Many real systems, however, are simpler than the general case because of the relative magnitudes of rate constants and concentrations. For example, a solution of a carboxylic acid in water is essentially describable in terms of the 13,31 equilibrium alone because c_{OH} (which is involved in the 12,21 and 23,32 steps but not in the 13,31 step) is negligible. Thus, it is often possible to work with a system having only a single observable relaxation time. The two relaxation times may be comparable in magnitude (in time units, that is), but the concentration changes associated with one of them may be so small in these special cases that only one relaxation time is detected. Bernasconi[5, pp. 52–62] describes the analysis of these systems in detail.

Another important acid–base process is the transfer of a proton from one solute acid–base pair to a second solute acid–base pair. This can take place via three pathways, shown in Scheme VI.

Direct $\quad\quad$ HA + B $\rightleftharpoons$ BH$^+$ + A$^-$

Protolysis $\quad\quad\quad$ HA $\rightleftharpoons$ H$^+$ + A$^-$

$\quad\quad\quad\quad\quad$ B + H$^+$ $\rightleftharpoons$ BH$^+$

Hydrolysis $\quad$ B + H$_2$O $\rightleftharpoons$ BH$^+$ + OH$^-$

$\quad\quad\quad\quad$ HA + OH$^-$ $\rightleftharpoons$ A$^-$ + H$_2$O

Scheme VI

By constructing a cyclic version as in Scheme VII it is readily seen that this is a four-state system, so it possesses three relaxation times.

$$H^+ + A^- + B + H_2O$$

$$HA + B + H_2O \quad\rightleftharpoons\quad BH^+ + A^- + H_2O$$

$$HA + BH^+ + OH^-$$

Usually special cases of the full scheme are studied so that only one or two relaxation times are observed. Important examples are a solution of an acid–base solute in the presence of an acid–base indicator, and a buffered solution of an acid–base solute.[5, pp. 52–62]

Table 4-1 lists some rate constants for acid–base reactions.[3, Chap. 3] A very simple yet powerful generalization can be made: For "normal" acids, proton transfer in the thermodynamically favored direction is diffusion controlled. Normal acids are predominantly oxygen and nitrogen acids; carbon acids do not fit this pattern. The thermodynamically favored direction is that in which the conventionally written equilibrium constant is greater than unity; this is readily established from the pK_a of the conjugate acid. Approximate values of rate constants in both directions can thus be estimated by assuming a typical diffusion-limited value in the favored direction (most reasonably by inspection of experimental results for closely related

TABLE **4-1.** Rate Constants for Proton Transfer Reactions in Water[a]

Reaction	$k_1/M^{-1}\ s^{-1}$	k_{-1}/s^{-1}
$H^+ + OH^- \rightleftharpoons H_2O$[b]	1.4×10^{11}	2.5×10^{-5}
$H^+ + HCO_3^- \rightleftharpoons H_2CO_3$	4.7×10^{10}	8×10^6
$H^+ + CH_3COO^- \rightleftharpoons CH_3COOH$	4.5×10^{10}	7.8×10^5
$H^+ + C_6H_5COO^- \rightleftharpoons C_6H_5COOH$	3.5×10^{10}	2.2×10^6
$H^+ + NH_3 \rightleftharpoons NH_4^+$	4.3×10^{10}	2.5×10^1
$H^+ + Me_3N \rightleftharpoons Me_3NH^+$	2.5×10^{10}	4
$H^+ + HCO_3^- \rightleftharpoons CO_2 + H_2O$	5.6×10^4	4.3×10^{-2}
$OH^- + NH_4^+ \rightleftharpoons NH_3 + H_2O$	3.4×10^{10}	6×10^5
$OH^- + Me_3NH^+ \rightleftharpoons Me_3N + H_2O$	2.1×10^{10}	1.4×10^6
$OH^- + $ imidazolium ion $\rightleftharpoons$ imidazole $+ H_2O$	2.5×10^{10}	2.5×10^3
$OH^- + EDTAH^{-3} \rightleftharpoons EDTA^{-4} + H_2O$	3.8×10^7	1.1×10^4
$OH^- + CO_2 \rightleftharpoons HCO_3^-$	1.4×10^4	1×10^{-4}

Source: Reference 3 (Chap. 3).

[a]Most of the reactions are at 25°C, ionic strength 0.1 M. k_1 is for the forward reaction as written.

[b]Water concentration is 55.5 M in this reaction; in other reactions water concentration is expressed as mole fraction = 1.

compounds) and combining this with the pK_a. For example, using benzoic acid as a model compound, we might estimate, for this reaction,

$$\text{H}^+ + 4\text{–H}_2\text{N–C}_6\text{H}_4\text{–COO}^- \underset{k_{-1}}{\overset{k_1}{\rightleftharpoons}} 4\text{–H}_2\text{N–C}_6\text{H}_4\text{–COOH}$$

that $k_1 \approx 3.5 \times 10^{10}$ M^{-1} s^{-1}. The pK_a of 4-aminobenzoic acid is 4.92, or $K_a = k_{-1}/k_1 = 1.2 \times 10^{-5}$ M; hence, we calculate $k_{-1} \approx 4.2 \times 10^5$ s^{-1}. The experimental values are $k_1 = 3.5 \times 10^{10}$ M^{-1} s^{-1} and $k_{-1} = 4.4 \times 10^5$ s^{-1}.

Deviations from this generalization may have several sources, including charge repulsion, steric effects, statistical factors, intramolecular hydrogen bonding, and other structural effects that alter electron density at the reaction site. Hague[3, Chap. 3] has discussed these effects.

Metal ion complexation rates have been studied by the T-jump method.[20] Divalent nickel and cobalt have coordination numbers of 6, so they can form complexes ML_n with monodentate ligands L with $n = 1–6$ or with bidentate ligands, $n = 1–3$. The ligands are Bronsted bases, and only the conjugate base form undergoes coordination with the metal ion. The complex formation reaction is then

$$\text{ML}_{n-1} + \text{L} \underset{k_{-n}}{\overset{k_n}{\rightleftharpoons}} \text{ML}_n$$

Scheme VIII

where charges are not specified. We will consider the case in which the only metal species in significant concentration are the ones shown in Scheme VIII; that is, only one of the complex equilibria is significant, which is reasonable if the several complex stability constants differ by several orders of magnitude.

Scheme VIII has the form of Scheme II, so the relaxation time is given by Eq. (4-15)—apparently. However, there is a difference between these two schemes in that L in Scheme VIII is also a participant in an acid–base equilibrium. The proton transfer is much more rapid than is the complex formation, so the acid–base system is considered to be at equilibrium throughout the complex formation. The experiment can be carried out by setting the total ligand concentration comparable to the total metal ion concentration, so that the solution is not buffered. As the base form L of the ligand undergoes coordination, the acid–base equilibrium shifts, thus changing the pH. This pH shift is detected by incorporating an acid–base indicator in the solution.

The rate equation is developed in the usual manner. The reaction variables are related as follows, where these identities arise from mass balance arguments. (We let A represent ML_{n-1} for convenience, and Z represents ML_n).

$$x_\text{A} = x_\text{L} + x_\text{LH} = x_\text{Z} \tag{4-37}$$

$$x_\text{LH} = -x_\text{H} \tag{4-38}$$

The rate equation, developed in terms of x_A, is linearized in the usual way to give

$$-\frac{dx_A}{dt} = \left[k_n\left(\frac{\bar{c}_A}{1 + \alpha} + \bar{c}_L\right) + k_{-n}\right]x_A \qquad (4\text{-}39)$$

where $\alpha = x_{LH}/x_L$. [We are considering only the ligand acid–base equilibrium; a fuller treatment[20] requires consideration of the indicator.] Thus

$$\tau^{-1} = k_n\left(\frac{\bar{c}_A}{1 + \alpha} + \bar{c}_L\right) + k_{-n} \qquad (4\text{-}40)$$

which may be compared with Eq. (4-15).

Letting K_a be the acid dissociation constant of the ligand, $K_a = c_H c_L/c_{LH}$, which is combined with the definitions $x_i = c_i - \bar{c}_i$ and Eq. (4-38) to give

$$\alpha = \frac{\bar{c}_H}{K_a + \bar{c}_B} \qquad (4\text{-}41)$$

Thus, α can be calculated (it is sometimes negligible), and the rate constants are evaluated graphically or by least-squares analysis; the estimates of k_n and k_{-n} must be consistent with the known stability constant.

If two complexes coexist, there will be two relaxation times, and a treatment analogous to the analyses of Schemes III and IV is required. Table 4-2 gives a few rate constants for these reactions.[4, 20] The mechanism of such reactions is believed to consist of at least two steps, shown in simplified form in Scheme IX.

$$M(aq) + L(aq) \overset{K_0}{\rightleftharpoons} M(aq)L$$

$$M(aq)L \underset{k_{-0}}{\overset{k_0}{\rightleftharpoons}} ML(aq) + H_2O$$

Scheme IX

TABLE 4-2. Rate Constants for Metal Ion Complexation at 25°C

Metal ion	Ligand	n	$k_n/M^{-1}\,s^{-1}$	k_{-n}/s^{-1}
Ni(II)	Imidazole	1	5×10^3	2.7
	Imidazole	2	4.3×10^3	8.9
	Imidazole	3	2.4×10^3	17
CO(II)	Imidazole	1	1.3×10^5	507
	Imidazole	2	1.1×10^5	1.3×10^3
Ni(II)	Glycine	1	4.1×10^4	0.057
	Glycine	2	5.6×10^4	0.93
	Glycine	3	4.3×10^4	11
Co(II)	Glycine	1	1.5×10^6	34
	Glycine	2	2.0×10^6	330
	Glycine	3	0.8×10^6	3.3×10^3

In the first step the hydrated ion and ligand form a solvent-separated complex; this step is believed to be relatively fast. The second, slow, step involves the readjustment of the hydration sphere about the complex. The measured rate constants can be approximately related to the constants in Scheme IX by applying the fast preequilibrium assumption; the result is $k_n = K_0 k_0$ and $k_{-n} = k_{-0}$. However, the situation can be more complicated than this.[4,20]

As a final example we consider noncovalent molecular complex formation with the macrocyclic ligand α-cyclodextrin, a natural product consisting of six α-D-glucose units linked 1–4 to form a torus whose cavity is capable of including molecules the size of an aromatic ring. Table 4-3 gives some rate constants for this reaction, where L represents the cyclodextrin and S is the substrate:[21]

$$S + L \underset{k_{-1}}{\overset{k_1}{\rightleftharpoons}} SL$$

The equilibrium binding constant for this 1:1 association is $K_{11} = k_1/k_{-1}$. The K_{11} values were measured spectrophotometrically, and the rate constants were determined by the T-jump method (independently of the K_{11} values), except for substrate No. 6, which could be studied by a conventional mixing technique. Perhaps the most striking feature of these data is the great variability of the rate constants with structure compared with the relative insensitivity of the equilibrium constants. This can be accounted for if the substrate must undergo desolvation before it enters the ligand cavity and then is largely resolvated in the final inclusion complex.[21]

TABLE 4-3. Binding Constants and Rate Constants for Complex Formation between α-Cyclodextrin and Azo Dyes[a,b]

No.	R_1	R_2	K_{11}/M^{-1}	$k_1/M^{-1}\,s^{-1}$	k_{-1}/s^{-1}
1	OH	H	270	1.3×10^7	5.5×10^4
2	O$^-$	H	645	1.7×10^5	2.6×10^2
3	OH	CH$_3$	417	1.2×10^5	3.5×10^2
4	O$^-$	CH$_3$	476	1.5×10^2	0.28
5	OH	CH$_2$CH$_3$	455	6×10^3	19
6	O$^-$	CH$_2$CH$_3$	286	2.8	1×10^{-2}
7	N(CH$_3$)$_2$	H	1010	1.1×10^6	1×10^3

Source: Data from Reference 21.

[a]R_1 and R_2 in

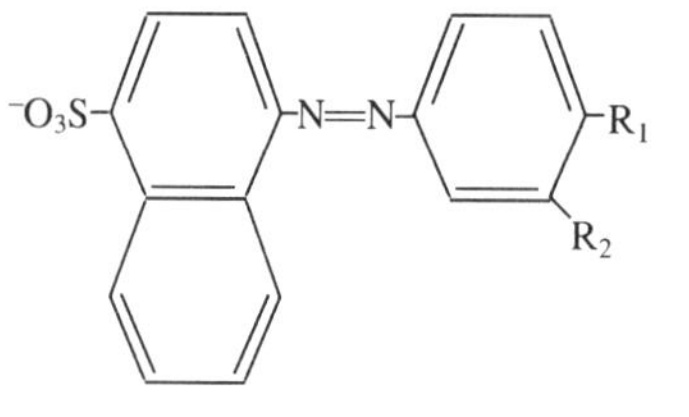

[b]Aqueous solution, 14°C, ionic strength 0.1 M.

4.3 NUCLEAR MAGNETIC RESONANCE SPECTROSCOPY

The connection between spectroscopy and kinetics is frequency, which has the dimension $time^{-1}$. Each type of chemical spectroscopy—ultraviolet, infrared, electron paramagnetic resonance, and nuclear magnetic resonance—has its characteristic frequency range, and, very roughly, the reciprocal of this frequency is the typical lifetime of the state whose transitions are observed by that spectroscopic technique. Nuclear magnetic resonance (NMR) spectroscopy provides valuable chemical kinetic information because of its frequency range, so this section deals with NMR. It is worth noting, however, that electron paramagnetic resonance (EPR) spectroscopy operates on similar principles and samples a different frequency range. EPR is based on transitions induced by the interaction of the electron's magnetic moment with a magnetic field and is a valuable means for studying free radical reactions. NMR is based on transitions induced by the interaction of a nuclear magnetic moment with a magnetic field. Becker[22] and Swift[23] are useful sources on the measurement of rates by NMR.

Magnetic Properties of Nuclei

The nuclei of many isotopes possess an angular momentum, called *spin*, whose magnitude is described by the spin quantum number I (also called the *nuclear spin*). This quantity, which is characteristic of the nucleus, may have integral or half-values; thus $I = 0, \frac{1}{2}, 1, \frac{3}{2}, \ldots$. The isotopes ^{12}C and ^{16}O both have $I = 0$; hence, they have no magnetic properties. ^{1}H, ^{13}C, ^{19}F, and ^{31}P are important nuclei having $I = \frac{1}{2}$, whereas ^{2}H and ^{14}N have $I = 1$.

According to quantum mechanics, the maximum observable component of the angular momentum is $Ih/2\pi$, where h is Planck's constant. A nucleus can assume only $2I + 1$ energy states. Associated with each of these states is a magnetic quantum number m, where m has the values $I, I-1, I-2, \ldots, -I + 1, -I$. For example the proton (^{1}H) has $I = \frac{1}{2}$, so $m = +\frac{1}{2}$ or $-\frac{1}{2}$; the deuteron (^{2}H) has $I = 1$, so $m = +1, 0$, or -1.

A nucleus having spin generates a magnetic moment μ. which is proportional to the angular momentum. Theory is not capable of calculating μ, so it is commonly expressed as Eq. (4-42), where γ is called the *magnetogyric ratio*.

$$\mu = \frac{\gamma I h}{2\pi} \tag{4-42}$$

Thus the experimental manifestation of the nuclear spin is described in terms of either μ or γ.

In the absence of an external magnetic field, the $2I + 1$ energy states of a nucleus are of identical energy (they are said to be degenerate) and, therefore, are equally populated at thermal equilibrium in any assemblage of such nuclei. In the presence of an applied steady field H_0, these $2I + 1$ states will assume different energy levels. This separation of energy levels in a magnetic field is called *nuclear Zeeman splitting*. The energy levels will be equally spaced, and the energy of separation is $\mu H_0/I$.

An intuitive model of the process can be given. Consider the proton, with $I = \frac{1}{2}$; then there are two states, characterized by $m = +\frac{1}{2}$ and $m = -\frac{1}{2}$. In the absence of an applied field, these states are equally populated. The states may be pictured as corresponding to opposite orientations of a tiny bar magnet, which is a crude way of visualizing the magnetic moment vector. Clearly in the absence of an applied field, the orientation of the moment should not affect the energy of the nucleus.

Now let a steady field be applied. The two nuclear states now correspond to orientation of the bar magnet parallel to the field (i.e., N pole to S pole) or antiparallel to the field (N pole to N pole). There will be an energy difference between these states, the orientation with the field (N to S) being of lower energy than the orientation against the field.

Figure 4-3 shows the nuclear splitting effect for the case $I = \frac{1}{2}$. Because the difference in energy levels is $\Delta E = \mu H_0/I$, evidently ΔE is directly proportional to the applied field. The analogy with optical spectroscopy is now obvious. It is possible to induce transitions between adjacent energy states; absorption of energy results in a transition from a lower to a higher energy level, and emission occurs with the reverse process.

The frequency of the electromagnetic radiation that can be absorbed by the nuclear system is easily calculated by equating the energy of a photon and the energy level separation:

$$h\nu = \frac{\mu H_0}{I} \tag{4-43}$$

or $\nu = \gamma H_0/2\pi$ [from Eq. (4-42)]. Thus, for the proton, $I = \frac{1}{2}$ and the experimental value for the moment μ is 1.42×10^{-23} erg G^{-1}. Suppose $H_0 = 10{,}000$ G, a

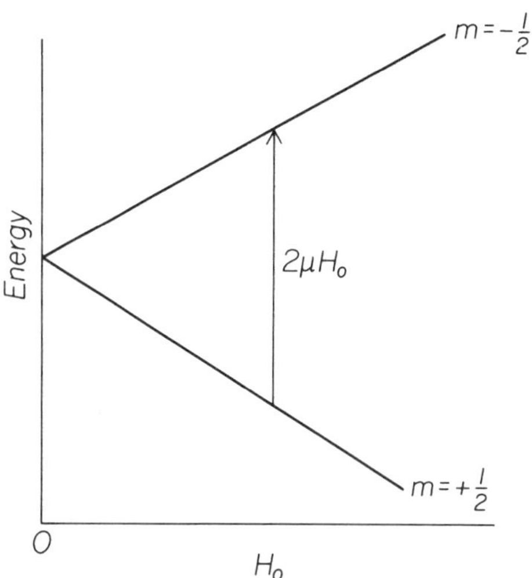

Figure 4-3. Energy level splitting for a nucleus with $I = \frac{1}{2}$ in applied field H_0. The energy separation is proportional to H_0.

reasonable field strength. Then the frequency ν calculated with Eq. (4-43) is 42.6 MHz, which is attainable as radiofrequency (rf) radiation.

Summarizing, nuclei with nonzero spin numbers I can assume $2I + 1$ energy states. In an applied magnetic field H_0, these states are separated by an energy gap of $\mu H_0/I$. Transitions between these states can be excited by rf radiation. A significant difference between this phenomenon and optical spectroscopy is that the frequency of the radiation absorbed by a nuclear system depends upon the strength of the applied magnetic field.

The Resonance Condition

The preceding discussion was presented in terms of quantum mechanics, which is necessary when energies are involved. A classical description is also possible, and this adds useful insights. A nucleus that possesses a magnetic moment ($I \neq 0$) can be considered to be a spinning charged particle. If $I = \frac{1}{2}$ the nucleus behaves as if it were a spinning charged sphere; if I is greater than $\frac{1}{2}$, the nucleus is equivalent to a nonspherical spinning charge (and it possesses a quadrupole moment).

A nucleus with a magnetic moment, when placed in a steady applied magnetic field H_0, will tend to adopt the orientation of the field, but because of its spin will instead assume an angle with the direction of the applied field; that is, the applied field vector and the magnetic moment vector will not be parallel. This situation is shown in Figure 4-4A. The field H_0 tends to decrease the angle θ between the vectors, but because the nucleus is spinning the result is that the moment vector will precess about the field vector (Fig. 4-4A). This action is similar to that of a gyroscope. The angular velocity ω_0 (in rad/s) of this precession is given by Eq. (4-44).

$$\omega_0 = \gamma H_0 \tag{4-44}$$

This velocity is called the *Larmor precessional frequency*.

Now suppose that an additional small magnetic field is applied perpendicular to H_0 in the plane formed by μ and H_0; call this field H_1 (see Fig. 4-4B). Field H_1 will act upon μ to increase the angle θ. If field H_1 is caused to rotate around H_0 at the Larmor precessional frequency of ω_0, the torque produced will steadily act to change the angle θ. On the other hand, if the frequency of rotation of H_1 is not the same as the precessional frequency, the torque will vary depending upon the relative phases of the two motions, and no sustained effect will be produced.

When the frequency of the rotating magnetic field H_1 equals the precessional frequency of the nucleus, energy is absorbed by the nuclear system and a transition occurs from a lower energy level to a higher one; in a sense the magnetic moment vector "flips" from one orientation to the other one. The nuclear absorption of energy from a rotating magnetic field in the presence of a constant magnetic field is called *nuclear magnetic resonance*. The resonance phenomenon occurs only when the angular velocity of the rotating field is equal to the Larmor precessional frequency of the magnetic moment. Because angular velocity ω (in rad s^{-1}) is equal to frequency

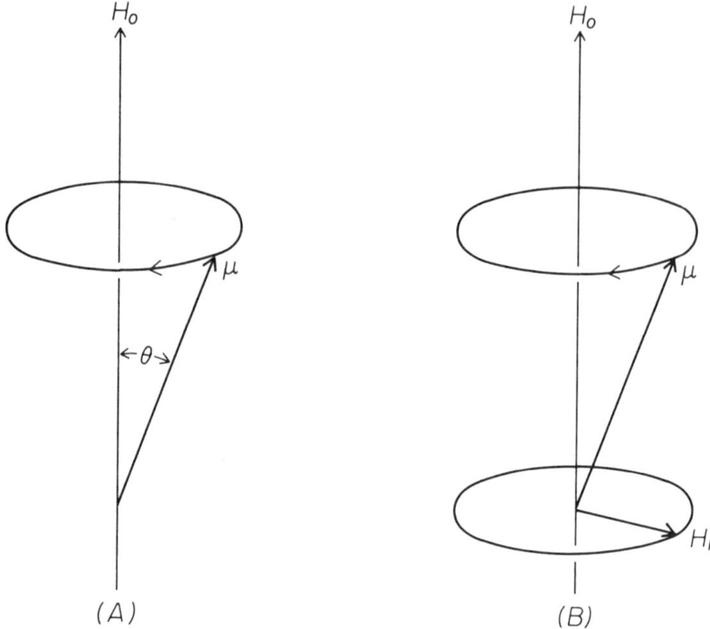

Figure 4-4. (*A*) Magnetic moment vector **μ** precessing about applied field vector H_0. (*B*) Addition of rotating magnetic field H_1 normal to H_0.

ν (in Hz) by $\omega = 2\pi\nu$, comparison of Eqs. (4-43) and (4-44) shows that the Larmor precessional frequency is identical to the transition frequency that was calculated earlier.

If a nucleus (actually a great many nuclei of the same substance, of course) is set precessing in field H_0, and the rotating field H_1 is altered in frequency, sweeping from a frequency lower than the Larmor frequency, through it, to a higher frequency, and if the energy absorbed from the rotating field H_1 is measured, the graph of energy absorbed against frequency of the rotating field is the NMR spectrum. It will consist of a sharp peak in the energy absorbed at the frequency corresponding to the Larmor resonance frequency of the nucleus at the given value of H_0.

It is not easy to apply a rotating magnetic field to the system of precessing nuclei. It is very easy, however, to generate a linearly oscillating field with an rf oscillator. This is perfectly satisfactory, because a linearly oscillating field of total amplitude $2H_1$ can be resolved into two rotating fields, each of amplitude H_1, but rotating in opposite directions. Only the component rotating in the same direction as the precessing nuclei will interact; the other component has no effect.

It was suggested above that the field H_0 is held constant while the H_1 frequency is swept; this mode is called *frequency sweep*. An equivalent result is obtained by holding the rf field at a constant frequency and sweeping the field H_0. NMR spectra obtained by field-sweep measurements are displayed with the field increasing from left to right; spectra obtained by frequency sweep are displayed with the frequency increasing from right to left, so that the two presentations are comparable.

Spin–Lattice Relaxation

Let us consider a set of nuclei (a spin system) with $I = \frac{1}{2}$ at thermal equilibrium in a steady magnetic field H_0. The two nuclear energy levels are separated by $2\mu H_0$, and the state for which $m = +\frac{1}{2}$ is the lower energy state, as in Fig. 4-3. Evidently the tendency is for all of the nuclei to pass to the lower energy level. This tendency, however, is opposed by thermal motion of the nuclei, which tends to produce an equilibrium distribution with identical numbers of nuclei in the two states. The net result is that a small but finite excess of nuclei will, at equilibrium, populate the lower energy level. The ratio of number of nuclei per unit volume in the lower energy level, N_+ (for $m = +\frac{1}{2}$) to the number N_- in the upper level is given by the Boltzmann distribution, Eq. (4-45).

$$\frac{N_+}{N_-} = e^{2\mu H_0/kT} \tag{4-45}$$

Because the energy separation $2\mu H_0$ is very small, the approximation $e^x \approx 1 + x$ is valid, and we have

$$\frac{N_+}{N_-} \approx 1 + \frac{2\mu H_0}{kT} \tag{4-46}$$

showing the excess population of the lower energy state.

The nuclei constituting the spin system may be regarded as embedded in a medium, called the *lattice,* composed of the rest of the sample, that is, of atoms and molecules. Interaction between the spin system and the lattice (that is, transfer of energy) is possible but difficult. Now suppose this spin system is irradiated by an rf field at its resonance frequency, so that absorption of energy occurs. Some nuclei undergo transition from the lower to the higher energy state; thus the absorption of energy has reduced the excess of nuclei in the lower energy state. Equilibrium requires a return to the excess number present before the application of the rf field. Therefore, transitions from the upper energy level to the lower one occur, with transfer of energy from the spin system to the lattice. The return of the spin system to its equilibrium configuration is called *relaxation,* so this mechanism is called *spin–lattice relaxation.* We wish to find the rate of this process.

First, consider the case in which the spin system and the lattice are in equilibrium at temperature T in field H_0. This means that the number of transitions upward must equal the number downward. If W_u is the probability of an upward transition per unit time and W_d is the probability of a downward transition, then $W_u N_+ = W_d N_-$, or $W_d/W_u = 1 + 2\mu H_0/kT$. Combining this with the mean transition probability $W = (W_u + W_d)/2$ gives

$$W_d = W\left(1 + \frac{\mu H_0}{kT}\right) \tag{4-47}$$

$$W_u = W\left(1 - \frac{\mu H_0}{kT}\right) \tag{4-48}$$

Therefore, the probability of a downward transition in a magnetic field is slightly greater than the probability of an upward transition.

Now let energy absorption from an rf field take place. The excess number of nuclei in the lower energy state is denoted n, so $n = N_+ - N_-$. The rate of change of n is

$$\frac{dn}{dt} = 2N_- W_\mathrm{d} - 2N_+ W_\mathrm{u} \tag{4-49}$$

because for each nucleus that makes a transition n changes by 2. Combining Eqs. (4-47)–(4-49),

$$\frac{dn}{dt} = 2W(n_0 - n) \tag{4-50}$$

where $n_0 = N\mu H_0/kT$ and $N = N_+ + N_-$; n_0 is the value of n when the spin system is in equilibrium with the lattice. Integration of Eq. (4-50) gives

$$n_0 - n = (n_0 - n_\mathrm{i})e^{-2Wt} \tag{4-51}$$

where n_i is the initial value of n. A quantity T_1 is defined by $T_1 = 1/2W$, so that Eq. (4-51) expresses the result that the approach to spin–lattice equilibrium is a first-order process characterized by the first-order rate constant $1/T_1$. T_1 is called the *spin–lattice relaxation time*. In liquids, T_1 values are typically in the range 10^{-2}–10^2 s; in solids T_1 may be much larger than in liquids.

In the earlier treatment we reached the conclusion that resonance absorption occurs at the Larmor precessional frequency, a conclusion implying that the absorption line has infinitesimal width. Actually NMR absorption bands have finite widths for several reasons, one of which is spin–lattice relaxation. According to the Heisenberg uncertainty principle, which can be stated

$$\Delta E \cdot \Delta t \approx h/2\pi$$

the uncertainty in energy of a state ΔE times the uncertainty in lifetime of the state Δt is approximately equal to Planck's constant. The spin–lattice relaxation time T_1 is a measure of the mean lifetime of the upper state (more precisely, the mean lifetime cannot exceed T_1), and the energy is related to frequency by $\Delta E = h\Delta\nu$. Thus, we obtain

$$\Delta\nu \approx \frac{1}{2\pi T_1} \tag{4-52}$$

where $\Delta\nu$ is interpreted as the minimum width of the NMR absorption line. This general phenomenon is called *lifetime broadening;* any process that decreases the lifetime of a state results in broadening of the corresponding absorption band.

Spin–Spin Relaxation

We have now to consider the interactions between vicinal nuclei in the spin system. This interaction is of a magnetic dipole–dipole nature. Each nucleus is within the steady field H_0, but is also subjected to a small local magnetic field H_{local}, produced by other nuclei, which are themselves small magnets. Because, in general, the disposition of neighboring nuclei is different for each nucleus, the magnitude of H_{local} will differ at each nucleus. The total magnetic field acting on a nucleus is the vector sum of H_0 and H_{local}. Because this is different at each nucleus, the Larmor precession frequency of each nucleus will be different by a small amount. The range of resonance frequencies for identical nuclei will be of the order $\delta\omega_0 = \gamma H_{local}$. This means that the absorption peak in the NMR spectrum will be a band broadened by the amount $\delta\omega_0$. This broadening is in addition to spin–lattice relaxation broadening. In liquids the molecules are in rapid motion, and this effect of local field variation averages out to a negligible value, but in solids it is a very important contributor to line width.

There is another type of spin–spin interaction. Two adjacent nuclei can interact to cause a transition. For example, the precessing moment of nucleus A sets up an oscillating magnetic field at nucleus B. The oscillating field thus produced will, at some time, be in phase with the precession of B; nucleus B can absorb energy from this field and undergo transition to the upper level. Simultaneously nucleus A must pass to the lower level. Because the precessional frequencies of the two nuclei will differ by about $\delta\omega_0$, the length of time required for the two nuclei to come into phase will be about $1/\delta\omega_0$. This is the approximate lifetime of a nuclear spin state; it is symbolized T_2 and is called the *spin–spin relaxation time*. T_2 incorporates both of the spin–spin relaxation mechanisms described. Note that spin–spin relaxation does not decrease the energy of the spin system, but it does shorten the lifetime of the state, so it leads to line broadening.

We have seen that in a steady field H_0 a small excess, n_0, of nuclei are in the lower energy level. The absorption of rf energy reduces this excess by causing transitions to the upper spin state. This does not result in total depletion of the lower level, however, because this effect is opposed by spin–lattice relaxation. A steady state is reached in which a new steady value, n_s, of excess nuclei in the lower state is achieved. Evidently n_s can have a maximum value of n_0 and a minimum value of zero. If n_s is zero, absorption of rf energy will cease, whereas if $n_s = n_0$, a steady-state absorption is observed. It is obviously desirable that the absorption be time independent or, in other words, that n_s/n_0 be close to unity. Theory gives an expression for this ratio, which is called Z_0, the saturation factor:

$$Z_0 = (1 + \gamma^2 H_1{}^2 T_1 T_2)^{-1}$$

It is, therefore, desirable that the quantity $\gamma^2 H_1{}^2 T_1 T_2$ be as small as possible. The rf field H_1 should be of low power to prevent saturation. The saturation factor appears in the later treatment.

The Bloch Equations

The NMR phenomenon can be quantitatively described in classical terms. This was first done by Bloch.[24] The approach is helpful in developing an understanding of nuclear relaxation processes.

Consider a nucleus with magnetic moment μ in a magnetic field H_0. According to classical mechanics the rate of change of the angular momentum G is the torque T:

$$\frac{dG}{dt} = T$$

The torque is given by the vector cross product of the vectors μ and H_0.

$$T = \mu \times H_0$$

The angular momentum is related to the moment by the magnetogyric ratio.

$$G = \mu/\gamma$$

Therefore $dG/dt = (1/\gamma)(d\mu/dt)$, or

$$(1/\gamma)\frac{d\mu}{dt} = \mu \times H_0$$

If the nuclei in unit volume are summed, the result is the *magnetization M*, so

$$(1/\gamma)\frac{dM}{dt} = M \times H_0 \tag{4-53}$$

Equation (4-53) describes the precession of the magnetization vector about the field vector with angular frequency γH_0, in the absence of the rotating field H_1 (see Fig. 4-4A).

In the presence of a field H_1 rotating at the precessional frequency the nuclear system can absorb energy, following which nuclear relaxation occurs. Thus, the equation of motion must include both the precessional and the relaxation contributions:

$$\frac{dM}{dt} = \frac{\text{precessional}}{\text{motion}} + \frac{\text{relaxational}}{\text{motion}} \tag{4-54}$$

We now adopt a Cartesian coordinate system and write Eq. (4-54) for each of the components of the magnetization. Figure 4-5 shows the coordinate system. The

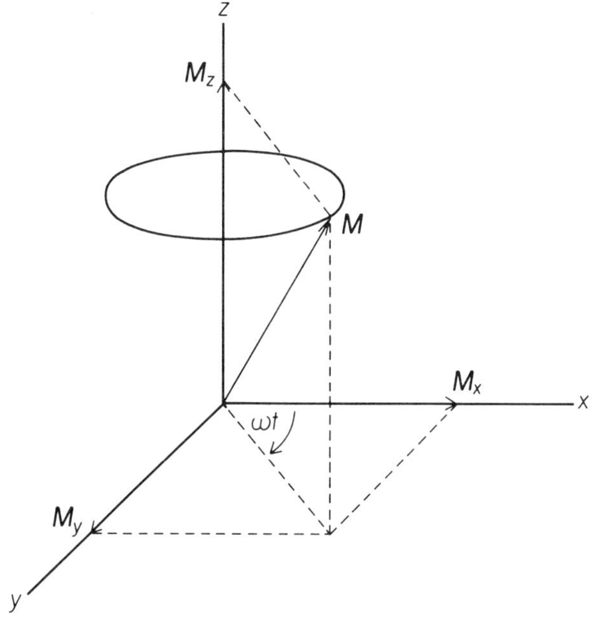

Figure 4-5. Coordinate system for the NMR equation of motion.

applied field H_0 is parallel with the z axis, and the rf field H_1 is in the xy plane. The components of the magnetic field are then

$$H_x = H_1 \cos \omega t \tag{4-55a}$$

$$H_y = -H_1 \sin \omega t \tag{4-55b}$$

$$H_z = H_0 \tag{4-55c}$$

Figure 4-6 illustrates the relaxational contribution to the motion. Figure 4-6A shows moment vectors for a spin system in the absence of the rf field ($H_1 = 0$); the magnetization components are $M_z = M_0$, $M_x = 0$, $M_y = 0$, because in the xy plane the magnetization components cancel. In the presence of the rf field at the resonance frequency the spin system absorbs energy, increasing the angle between H_0 and M and perturbing the thermal equilibrium so that M_x and M_y components are induced and $M_z < M_0$ (Fig. 4-6B). With the passage of time (comparable to the relaxation times T_1 and T_2), relaxation back to the equilibrium configuration takes place, so M_z increases toward M_0, whereas M_x and M_y decrease toward zero as a consequence of the gradual loss of coherence of the moment vectors.

The decay of M_z to M_0 is called *longitudinal relaxation* (because it is parallel with the field H_0); it is identical with the spin–lattice relaxation described earlier. The rate constant for this process is therefore $1/T_1$. The decay of M_x and M_y is

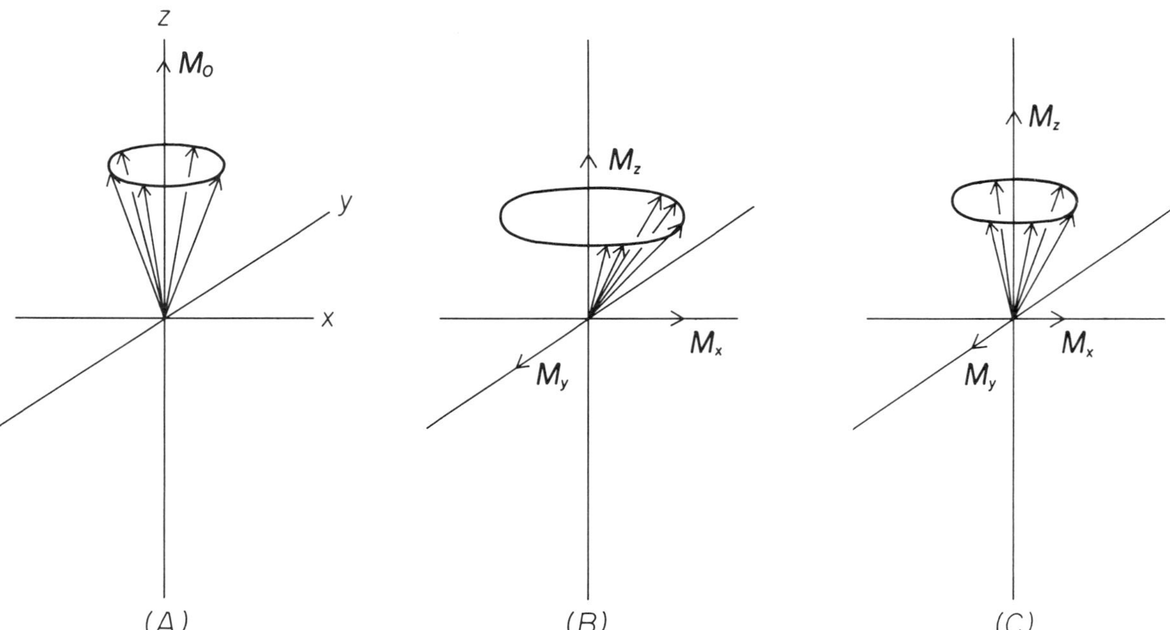

Figure 4-6. Representation of the magnetization components M_x, M_y, and M_z. (A) In presence of field H_0 without field H_1. (B) Immediately after absorption of energy from field H_1. (C) After partial relaxation back to the equilibrium position shown in A.

transverse relaxation or spin–spin relaxation; it is assumed to take place via first-order kinetics with rate constant $1/T_2$. We can now write the differential equations:

$$\frac{dM_x}{dt} = \gamma(\boldsymbol{M} \times \boldsymbol{H})_x - M_x/T_2 \qquad (4\text{-}56a)$$

$$\frac{dM_y}{dt} = \gamma(\boldsymbol{M} \times \boldsymbol{H})_y - M_y/T_2 \qquad (4\text{-}56b)$$

$$\frac{dM_z}{dt} = \gamma(\boldsymbol{M} + \boldsymbol{H})_z + (M_0 - M_z)/T_1 \qquad (4\text{-}56c)$$

The cross product is expanded by Eq. (4-57).

$$\boldsymbol{M} \times \boldsymbol{H} = \begin{bmatrix} i & j & k \\ \boldsymbol{M}_x & \boldsymbol{M}_y & \boldsymbol{M}_z \\ \boldsymbol{H}_1 \cos \omega t & -\boldsymbol{H}_1 \sin \omega t & \boldsymbol{H}_0 \end{bmatrix} \qquad (4\text{-}57)$$

Equations (4-58), the Bloch equations, result:

$$\frac{dM_x}{dt} = \gamma(M_y H_0 + M_z H_1 \sin \omega t) - M_x/T_2 \qquad (4\text{-}58a)$$

$$\frac{dM_y}{dt} = \gamma(M_z H_1 \cos \omega t - M_x H_0) - M_y/T_2 \qquad (4\text{-}58b)$$

$$\frac{dM_z}{dt} = \gamma(-M_x H_1 \sin \omega t - M_y H_1 \cos \omega t) + (M_0 - M_z)/T_1 \qquad (4\text{-}58c)$$

If the rate of sweep through the resonance frequency is small (so-called slow passage), a steady-state solution, in which the derivatives are set to zero, is obtained.[25] The result expresses M_x, M_y, and M_z as functions of ω. These magnetization components are not actually observed, however, and it is more useful to express the solutions in terms of the susceptibility, a complex quantity related to the magnetization. The solutions for the real (χ') and imaginary (χ'') components then are

$$\chi' = \frac{\chi_0 \omega_0 T_2}{2} \left[\frac{(\omega_0 - \omega)T_2}{1 + (\omega_0 - \omega)^2 T_2^2 + \gamma^2 H_1^2 T_1 T_2} \right] \qquad (4\text{-}59a)$$

$$\chi'' = \frac{\chi_0 \omega_0 T_2}{2} \left[\frac{1}{1 + (\omega_0 - \omega)^2 T_2^2 + \gamma^2 H_1^2 T_1 T_2} \right] \qquad (4\text{-}59b)$$

Notice the appearance of the saturation factor in these equations. By working at small values of H_1 these equations become

$$\chi' = \frac{\chi_0 \omega_0 T_2}{2} \left[\frac{(\omega_1 - \omega)T_2}{1 + (\omega_0 - \omega)^2 T_2^2} \right] \tag{4-60a}$$

$$\chi'' = \frac{\chi_0 \omega_0 T_2}{2} \left[\frac{1}{1 + (\omega_0 - \omega)^2 T_2^2} \right] \tag{4-60b}$$

where $\chi_0 = M_0/H_0$. These two equations contain the same information; Eq. (4-60a) describes the dispersion mode, whereas Eq. (4-60b) describes the absorption mode. In practice the absorption mode is always displayed in NMR spectra.

The bracketed term in Eq. (4-60b) describes a Lorentzian line shape for the NMR absorption band. The maximum in the band occurs at the resonance frequency, ω_0. Expressed in units of $\chi_0 \omega_0 T_2/2$, the maximum value of χ'' is 1; at one-half this maximum peak height we find, by substitution, that $(\omega_0 - \omega) = 1/T_2$. Using $\omega = 2\pi\nu$ to convert to frequency (in Hz) gives $(\nu_0 - \nu) = \frac{1}{2}\pi T_2$. However, the peak width is twice this, or

$$\Delta\nu_{1/2} = \frac{1}{\pi T_2} \tag{4-61}$$

where $\Delta\nu_{1/2}$ is the peak width (in Hz) at one-half the maximum height.

Equation (4-61) provides a basis for T_2 measurements (in the absence of chemical exchange, to be discussed subsequently). Magnetic field inhomogeneity also contributes to line width; the field variation is δH_0, and the corresponding resonance frequency range is $\Delta\omega_0 = \gamma\delta H_0$. A typical value for this source is 0.1 Hz, so for accurate T_2 measurement a line should be substantially wider than this. This means that T_2 must be less than (roughly) $1/\pi$ s if the field inhomogeneity contribution is to be less than about 10%.

T_1, T_2, and the Correlation Time

The development thus far can be summarized as follows: A set of nuclei (the spin system) in a steady magnetic field H_0 will absorb energy from an rf field H_1 (normal to H_0) at the Larmor precessional frequency ω_0 of the nuclei. Upon the absorption of energy, the population of nuclear energy levels is altered from the equilibrium Boltzmann distribution appropriate to the sample temperature, so a process occurs of energy transfer from the spin system to the rest of the sample (the lattice), the driving force being the return to thermal equilibrium. This process is spin–lattice relaxation, and it takes place at a rate governed by a first-order rate constant $1/T_1$, where T_1 is the spin–lattice relaxation time. An equivalent viewpoint is that absorption of energy from the rf field generates a component of the spin system magnetization vector in the H_0 direction (the z direction) different from the equilibrium value M_0 in the absence of energy absorption. Relaxation of M_z back to M_0 occurs with a rate $(M_0 - M_z)/T_1$.

There can also be interactions within the spin system. Adjacent nuclei can interact through a dipole–dipole mechanism to cause a transition. The precessing moment of the nucleus 1 sets up an oscillating field at nucleus 2; this field will, at some time, be in phase with the precession of nucleus 2, which can absorb energy from 1. The result is a transfer of energy by a spin–spin relaxation process. Because the precessional frequencies of nuclei differ by a small amount $\delta\omega_0$ because of local induced-field differences, the length of time for two nuclei to come into phase is $1/\delta\omega_0$. This is the approximate lifetime of a nuclear state; it is written T_2, where T_2 is the spin–spin relaxation time. Reverting to the description in terms of magnetization vectors, absorption of energy by the spin system generates components of the magnetization vector in the xy plane (the plane of the H_1 field). These components, M_x and M_y, decay to their equilibrium values of zero with the rate constant $1/T_2$. The spin–spin relaxation phenomenon is reflected in resonance peak widths (the lifetime broadening effect), narrow lines corresponding to long T_2 values.

Both spin–lattice and spin–spin relaxation depend on rates of molecular motion, for relaxation results from the interaction of fluctuating magnetic fields set up by nuclei in the spin system and in the lattice. A quantitative theory of this dependence was given by Bloembergen et al.,[26] who obtained

$$\frac{1}{T_1} \propto \frac{\tau_c}{1 + \omega_0^2 \tau_c^2} \tag{4-62}$$

In Eq. (4-62) ω_0 is the Larmor precessional frequency, and τ_c is the *correlation time,* a measure of the rate of molecular motion. The reciprocal of the correlation time is a frequency, and $1/\tau_c$ may receive additive contributions from several sources, in particular $1/\tau_r$, where τ_r is the rotational correlation time. τ_r is, approximately, the time taken for the molecule to rotate through one radian. Only a rigid molecule is characterized by a single correlation time, and the value of τ_c for different atoms or groups in a complex molecule may provide interesting chemical information.

According to Eq. (4-62), when $\omega_0\tau_0 \ll 1$, T_1 is proportional to $1/\tau_c$, whereas when $\omega_0\tau_c \gg 1$, T_1 is proportional to τ_c. When $\tau_c = \omega_0$, T_1 has its minimum value. Figure 4-7 is a schematic representation of the relationship between T_1 and τ_c. The physical meaning of this relationship is that coupling between the spin system and the lattice is most efficient when the resonance frequency and the frequency of molecular motion are equal. τ_c can be measured by studying the dependence of T_1 on ω_0 (by varying the field strength). For small molecules in solution τ_c is commonly 10^{-12} to 10^{-10} s.

Figure 4-7 also shows the theoretical relationship of T_2 to τ_c. At high frequencies (low τ_c), T_1 and T_2 are equal, as might be expected from the diagrammatic account of Fig. 4-6; but low-frequency phenomena (as in very viscous media or in solids) provide efficient spin–spin coupling and lead to a limiting T_2 value.

The magnetic dipole–dipole interaction is a general mechanism for nuclear re-

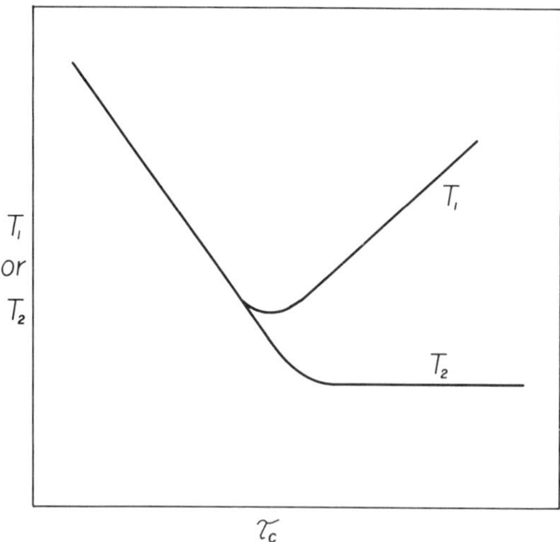

Figure 4-7. Schematic dependence (log–log plot) of T_1 and T_2 as functions of τ_c, the correlation time. The minimum in T_1 occurs at $\tau_c = 1/\omega_0$.

laxation, but there are some special mechanisms that can be more important in certain cases. If a nucleus possesses an electric quadrupole ($I > \frac{1}{2}$), molecular motion results in fluctuating electric fields that can induce relaxation. This can be very efficient, resulting in short relaxation times and broad absorption bands (as in ^{14}N). Another relaxation mechanism is provided by paramagnetic substances, which possess unpaired electrons. Then the dominant contributor to the relaxation is the large electron magnetic moment of the paramagnetic substance. Dissolved oxygen makes an important contribution to relaxation in solutions by this mechanism.

Chemical Exchange

There is arbitrariness in describing phenomena as either *physical* or *chemical*, but in some sense the nuclear relaxation mechanisms we have discussed to this point are physical mechanisms, based as they are on rotational motions of molecules, magnetic dipole–dipole interactions, quadrupolar interactions, and so on. Now we discuss a nuclear relaxation mechanism that is chemical in origin.

Consider a nucleus that can "partition" between two magnetically nonequivalent sites. Examples would be protons or carbon atoms involved in *cis–trans* isomerization, rotation about the carbon–nitrogen atom in amides, proton exchange between solute and solvent or between two conjugate acid–base pairs, or molecular complex formation. In the NMR context the nucleus is said to undergo *chemical exchange* between the sites. Chemical exchange is a relaxation mechanism, because it is a means by which the nucleus in one site (state) is enabled to leave that state.

To make this more specific let us consider the proton transfer from conjugate base A (site A) to base B (site B),

$$\text{HA} \underset{k_B}{\overset{k_A}{\rightleftharpoons}} \text{HB}$$

where k_A and k_B are first-order (or pseudo-first-order) rate constants. From Chapter 3 we have

$$k = k_A + k_B \tag{4-63}$$

where k is an observed first-order rate constant. In NMR it is customary to describe rates in terms of the lifetime τ of a state, where $\tau = 1/k$, so Eq. (4-63) becomes

$$\frac{1}{\tau} = \frac{1}{\tau_A} + \frac{1}{\tau_B} \tag{4-64}$$

or

$$\tau = \frac{\tau_A \tau_B}{\tau_A + \tau_B} \tag{4-65}$$

The fractional occupancy of the sites is

$$x_A = \frac{\tau_A}{\tau_A + \tau_B}; \qquad x_B = \frac{\tau_B}{\tau_A + \tau_B} \tag{4-66}$$

From Eqs. (4-65) and (4-66), $\tau = x_A \tau_B = x_B \tau_A$. For the special case in which $\tau_A = \tau_B$, we have $2\tau = \tau_A = \tau_B$. The total exchange frequency is $1/\tau$; the net exchange frequency is $\frac{1}{2}\tau$, because at equilibrium each time a nucleus passes from A to B, one also passes from B to A.

We are interested in the effect of chemical exchange on line width, its usual manifestation. The total relaxation frequency contributing to line width is

$$
\begin{array}{ccccc}
\text{total} & & \text{spin–spin} & & \text{chemical exchange} \\
\text{relaxation} & = & \text{relaxation} & + & \text{relaxation} \\
\text{frequency} & & \text{frequency} & & \text{frequency}
\end{array}
$$

or

$$\frac{1}{T_2'} = \frac{1}{T_2} + \frac{1}{\tau} \tag{4-67}$$

In Eq. (4-67) T_2 is the spin–spin relaxation time in the absence of chemical exchange (obtainable by reducing the temperature or from model systems lacking the exchange), and T_2' is the spin–spin relaxation time in the presence of exchange. Using Eq. (4-61),

$$\Delta v_{1/2}' = \Delta v_{1/2} + \frac{1}{\pi\tau} \tag{4-68}$$

Equations of the form of (4-67) and (4-68) apply to each site. Equation (4-68) provides a route to the estimation of τ.

The quantitative formulation of chemical exchange involves modification of the Bloch equations making use of Eq. (4-67). We will merely develop a qualitative view of the result.[22, Chap. 11] We adopt a coordinate system that is rotating about the applied field H_0 in the same direction as the precessing magnetization vector. Let ν_A and ν_B be the Larmor precessional frequencies of the nucleus in sites A and B. For simplicity we set $\tau_A = \tau_B$. As the frequency ν_0 of the rotating frame of reference we choose the average of ν_A and ν_B, thus,

$$\nu_0 = \tfrac{1}{2}(\nu_A + \nu_B) \tag{4-69}$$

As a consequence, from the point of view of this rotating frame, a nucleus at site A precesses at frequency $(\nu_0 - \nu_A)$, whereas a nucleus at site B precesses at frequency $(\nu_B - \nu_0)$; that is, the two nuclei (actually their magnetization vectors) precess in opposite directions. We imagine several possible cases.

1. *Very slow exchange.* Slow exchange means that the lifetime $\tau_A = \tau_B$ in each site is very long. Thus, a nucleus in site A precesses many times, at frequency $(\nu_0 - \nu_A)$ in the rotating frame, before it leaves site A, and similarly for a nucleus in site B. Thus, there is time for absorption of energy from the radio-frequency field H_1, and resonance peaks appear at ν_A and ν_B in the laboratory frame.
2. *Moderately slow exchange.* The state lifetime is 2τ; we ask how the absorption band is affected as this becomes smaller. The uncertainty principle argument given earlier is applicable here; lifetime broadening will occur as the state lifetime decreases. Thus, we expect resonance absorption at (or near) frequencies ν_A and ν_B, but the bands will be broader than in the very slow exchange limit. Equation (4-68) is applicable in this regime.
3. *Very fast exchange.* If the lifetimes are very short, a nucleus in site A cannot precess to a significant extent before it leaves A to enter B, where it begins to precess in the opposite direction (in the rotating frame), again enters A, and so on. Therefore from the point of view of the rotating frame, the nucleus is essentially stationary. In the laboratory frame its frequency is ν_0, the frequency of the rotating frame. Thus, according to Eq. (4-69), a single absorption band will be seen at ν_0, the mean of ν_A and ν_B.

At some stage between cases 2 and 3, coalescence into a single broadened band takes place. A full quantitative treatment requires nonlinear regression of the line shape to the theoretical relationship.

In general the lifetimes in the two sites are different, and Eq. (4-69) is replaced by

$$\nu_0 = x_A \nu_A + x_B \nu_B \tag{4-70}$$

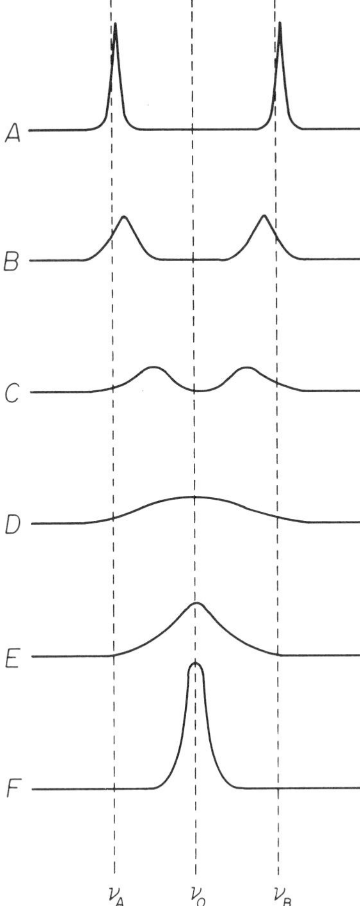

Figure 4-8. NMR absorption by a hypothetical two-identical site system with chemical exchange. (*A*) Slow exchange limit. (*B*) Moderately slow exchange. (*D*) Coalescence. (*F*) Fast exchange limit.

In the slow exchange limit the areas of the absorption bands are proportional to the fractional occupancy of the sites, and in the fast exchange limit the observed resonance frequency is a weighted average of the site frequencies. Because rates are temperature dependent, it is often possible to control the chemical exchange regime of a system by changing its temperature. Figure 4-8 shows the behavior of a hypothetical system.

The preceding discussion did not define the terms *slow* and *fast,* but their meaning is implicit in the imagined experiment with the rotating frame. In this frame the precessional frequencies of the nuclei are $(\nu_0 - \nu_A)$ and $(\nu_B - \nu_0)$, or, using Eq. (4-69), each of these is equal to $(\nu_B - \nu_A)/2$. It is the exchange rate $1/\tau$ relative to this frequency difference that establishes whether an exchange is slow or fast. For proton NMR typical resonance frequency differences would be of the order 10^2 Hz, corresponding to a lifetime of roughly 10^{-2} s. The slow exchange region would require lifetimes much longer than this, the fast exchange limit lifetimes much shorter than this.

Pulse NMR Measurements

To this point we have considered the NMR experiment in the *continuous wave* (cw) form, in which the rf frequency is swept through the resonance frequency, and the signal is displayed as a measure of energy absorbed as a function of frequency; this is the *frequency domain, $F(\omega)$*. However, nearly all modern NMR spectrometers operate in a different mode. The sample, in the applied steady field H_0, is subjected to a brief, very powerful pulse of field H_1, thus rotating the magnetization vector from its initial value M_0 parallel with the z axis to some other direction (often in the xy plane). Upon cessation of this short pulse, the magnetization relaxes back to its equilibrium position in a process called *free induction decay* (FID). The FID signal that is detected by the spectrometer is the magnetization component in the xy plane. The dependence of this signal on time comprises the *time domain, $f(t)$*, of NMR.

The frequency and time domains are related by the Fourier transform,

$$F(\omega) \; = \; \int_{-\infty}^{\infty} f(t)e^{-i\omega t}dt$$

Consequently by collecting the FID signal as a function of time and carrying out a Fourier transformation, the conventional NMR spectrum is obtained. This procedure possesses great advantages, one of which is the ability to time-average successive FIDs so as to improve the signal-to-noise ratio.

The Fourier transform of a pure Lorentzian line shape, such as the function equation (4-60b), is a simple exponential function of time, the rate constant being $1/T_2$. This is the basis of relaxation time measurements by pulse NMR. There is one more critical piece of information, which is that in the NMR spectrometer only magnetization in the xy plane is detected. Experimental design for both T_1 and T_2 measurements must accommodate to this requirement.

Suppose we adopt a rotating frame of reference with coordinates x', y', z' such that the fixed field H_0 lies along the z' axis and the x', y' coordinate system rotates about the z' axis with the frequency of the field H_1. Let H_1 be stationary along the x' axis.

In the presence of H_0 but the absence of H_1, a steady state is established, the magnetization vector having component M_0 along the z' axis, but because of symmetry owing to randomization there is no net magnetization in the $x'y'$ plane. This situation is shown in Fig. 4-9A.

Now let a brief H_1 pulse be applied along the x' axis. The angle through which the magnetization vector is rotated is $\gamma H_1 t_p$, where t_p is the duration of the pulse. The effect is to induce a magnetization component along the y' axis. If t_p is of such magnitude that the magnetization component along the z' axis is reduced to zero, the pulse is called a *90° pulse*. Figure 4-9B illustrates the system immediately after a 90° pulse. Such pulses are typically in the range 1–100 μs. This is so short a time that no significant relaxation occurs during the time the pulse is being applied.

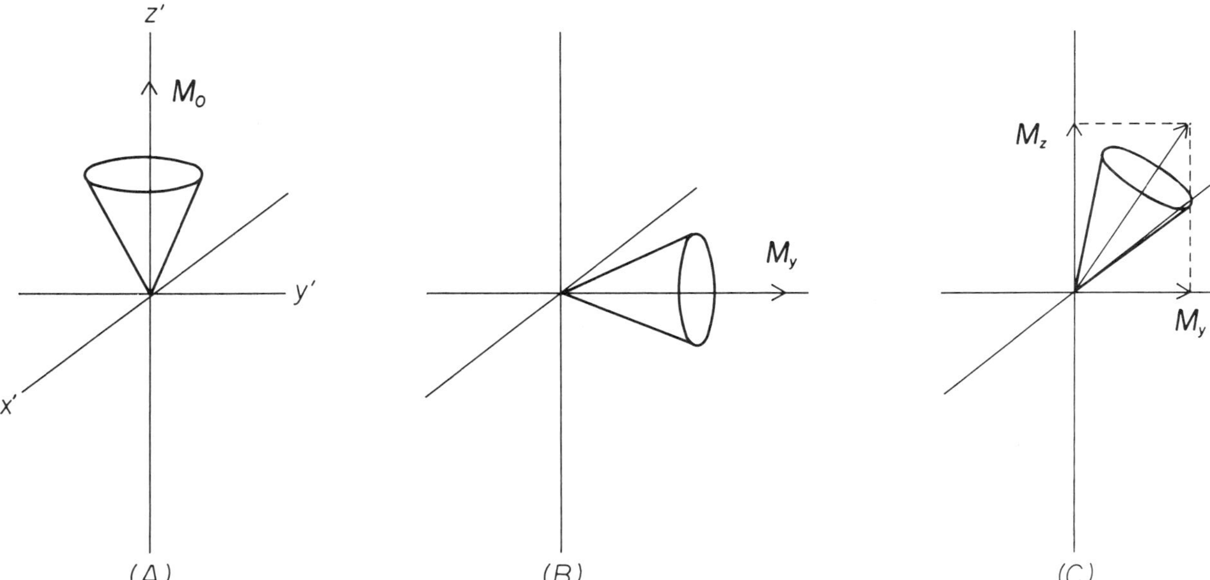

Figure 4-9. (*A*) Precessing moment vectors in field H_0 creating steady-state magnetization vector M_0, with $H_1 = 0$. (*B*) Immediately following application of a 90° pulse along the x' axis in the rotating frame. (*C*) Free induction decay of the induced magnetization showing relaxation back to the configuration in *A*.

Now with H_1 turned off, the induced magnetization must relax to its steady-state value. This is the free induction decay phase. Figure 4-9C shows an intermediate stage in the FID; M_z is increasing from zero toward M_0, and M_y is decreasing toward zero. As we have seen, M_z relaxes with rate constant $1/T_1$, and M_y relaxes with rate constant $1/T_2$.

Because we only detect magnetization in the $x'y'$ plane, it is $1/T_2$ that is measured from the slope of a semilogarithmic first-order plot of the FID signal against time. Actually the effect of magnetic field inhomogeneity, which provides a slight dispersion in resonance frequencies, makes a contribution to the transverse relaxation, and this effect is included in the measured T_2. The field inhomogeneity effect can be eliminated by a multiple pulse technique called the *spin echo method*.[23, Chap. 10; 27]

Because only *xy* magnetization can be detected, a different stratagem is required for T_1 measurements. First, a 180° pulse is applied, which reverses the direction of vector M_z, placing it on the $-z'$ axis. (In this process the higher energy state has been given an excess population.) Immediately following the pulse, longitudinal relaxation begins, with rate constant $1/T_1$. Now, after some time t, a 90° pulse is applied, which induces a magnetization vector on the $-y'$ axis whose magnitude is proportional to the magnitude of the M_z vector at time t.

Next a period of time T ($T >>> T_1$) is allowed for the entire system to relax to its steady-state configuration. Then the pulse sequence is repeated, with a different value for t. In this way the decay of M_z is measured by sampling it via the 90° pulse. The sequence is called a *180°, t, 90° sequence*. $1/T_1$ is found from a semilogarithmic plot.

There is another "hybrid" relaxation time that can be measured by a pulse sequence as follows. First, a 90° pulse is applied along the x' axis, inducing a magnetization along y'. This is immediately followed by a long pulse phase-shifted by 90° so that it is applied along the y' axis. The magnetization vector along y' is said to be *spin-locked,* for its transverse relaxation is restricted by the parallel H_1 field. Instead, its relaxation is analogous to longitudinal relaxation, because it occurs parallel with the H_1 field. For this reason, the corresponding relaxation time is labeled $T_{1\rho}$ for *spin–lattice relaxation in the rotating frame*. $T_{1\rho}$ is closely related to T_2 (as H_1 becomes very small, $T_{1\rho}$ approaches T_2, as can be seen by comparing the spin-lock experiment with the 90° pulse measurement of T_2); but $T_{1\rho}$ is more easily measured.[23]

The exchange rate $1/\tau$ in chemically exchanging systems can be measured by pulse NMR via T_2 measurements. Comparisons of exchange rates measured by continuous wave and pulse techniques sometimes reveal significant differences between the two.[27] Both techniques involve approximations or assumptions that may contribute to systematic errors, so that chemical exchange rates, although chemically very valuable information, should not be treated as highly accurate unless great care has been taken in the design and execution of the study. Typical uncertainties are 10–25%. Examples of chemical exchange rates are given in the following pages.

Applications

The study of chemical exchange by NMR has produced very interesting results. Let us first review the accessible time scales. Let $\Delta\nu_0$ be the frequency difference (chemical shift difference) between the magnetically nonequivalent sites. Then the slow exchange condition is $\tau^{-1} \ll \Delta\nu_0$, and fast exchange signifies that $\tau^{-1} \gg \Delta\nu_0$. The most general cw technique for extracting τ is by total line shape analysis, which clearly is restricted to the range bounded by the slow and fast exchange limits. Thus, we can conclude that $\tau\Delta\nu_0$ must be of the order unity (say $10 \leqslant \tau\Delta\nu_0 \leqslant 0.1$). Because $\Delta\nu_0$ is typically 10–100 Hz, this gives the accessible range in τ as about 10^{-3} s to 1 s. [These estimates are for protons. Other nuclei may have larger $\Delta\nu_0$ values, so correspondingly smaller τ can be measured.][28] If τ is not in this range, it may be ⟨ ⟩ tions to bring it into the measurable range, as b⟨ ⟩ the ⟨ ⟩

Another means is available for studying the ⟨ ⟩ kinetics of second-order reactions—we can adjust a reactant concentration. ⟨ ⟩is may permit the study of reactions having very large second-order rate constants. Suppose the rate equation is $v = kc_A c_B = k_{obs}c_A = \tau^{-1}c_A$, so $kc_B = \tau^{-1}$. For the experimental measurement let us say that we wish τ to be about 10^{-2} s. We can achieve this by adjusting c_B so that the product $kc_B \approx 10^2$ s^{-1} for example, if $k = 10^8$ M^{-1} s^{-1}, we require $c_B = 10^{-6}$ M. This method is possible, because there is no net reaction in the NMR study of chemical exchange.

Much information on proton transfers has been obtained by NMR chemical exchange studies.[1,13] An example is the proton exchange between neopentyl alcohol and acetic acid in acetic acid as the solvent.[29] The reaction is

$$\text{ROH}^\star + \text{HOAc} \rightleftharpoons \text{ROH} + \text{H}^\star\text{OAc}$$

and the rate law was found to be

$$v = k[\text{ROH}]x_{\text{HOAc}} = \tau^{-1}[\text{ROH}]x_{\text{HOAc}}$$

where x_{HOAc} is the mole fraction of acetic acid, which is essentially unity at low alcohol concentrations. The spin echo pulse method was used to measure T_2, and the proton relaxation time τ was obtained from T_2 by means of an approximate relationship relevant to the line-broadening region. The frequency difference $\Delta\nu_0$ was 440 Hz, and τ was found to be 0.15×10^{-5} s, giving $k = 6.7 \times 10^5$ s^{-1}.

Another example is the proton exchange between H_2O and HDO at low water concentrations in organic solvents.[30] τ values were obtained by line shape analysis; τ^{-1} was an approximately linear function of water concentration. At about 1.1 M water, $\tau^{-1}/$s^{-1} was 6.7 (in nitromethane), 0.91 (acetonitrile), 1.0 (acetone), 1.6 (dioxane), 25 (pyridine), 8.3 (dimethylsulfoxide), > 100 (triethylamine). The proton exchange may be catalyzed by H_3O^+ or OH^-, or by solvent lyonium or lyate species, but for the neutral solvents a water trimer (for which some evidence exists) may be involved as shown (S represents the organic solvent):

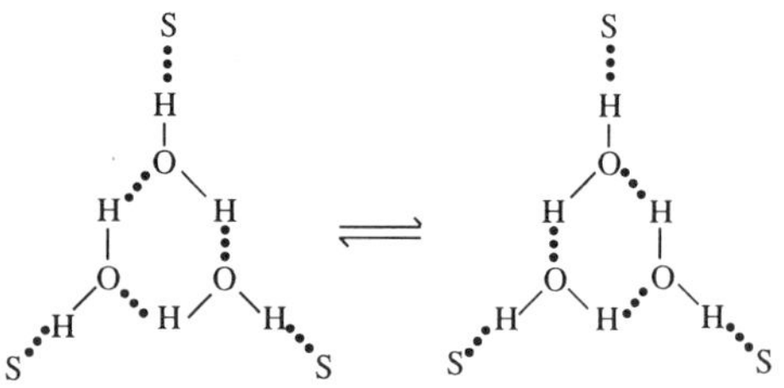

Scheme X

The hydration of acetaldehyde (Scheme XI) constitutes a process in which an oxygen atom exchanges between the solvent and the solute.

$$CH_3CHO + H_2O \rightleftharpoons CH_3CH(OH)_2$$

Scheme XI

This reaction has been studied in ^{17}O-enriched water, the line broadening of the carbonyl and diol ^{17}O resonance lines being analyzed to give the exchange rate τ^{-1}.[31] The exchange is catalyzed by acid, the exchange rate being related to the second-order rate constant by $\tau^{-1} = k[H^+]$. This rate constant was found to be 470 M^{-1} s^{-1}, in agreement with independent measurements of the rate of hydration. Thus, it appears that the oxygen exchange rate can be completely accounted for by the hydration reaction.

Rotation about single bonds and conformational changes can be studied. Amides constitute a classic example. Because of the partial double bond character of the carbon–nitrogen bond as a consequence of the contribution of **2** to the electronic structure, there is an energy barrier to rotation about this bond.

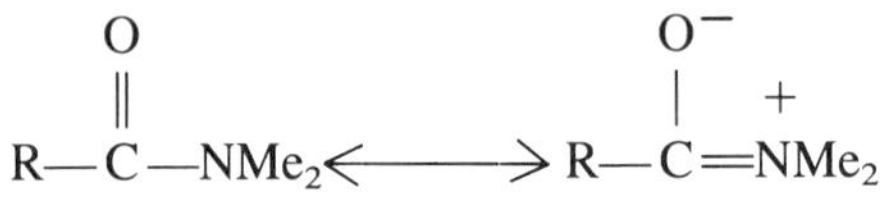

$$R-\underset{\|}{C}-NMe_2 \longleftrightarrow R-\underset{|}{C}=\overset{+}{N}Me_2$$

1   **2**

Consequently rotation about the carbon–nitrogen bond constitutes exchange of the methyl protons between nonequivalent sites, analogous to *cis-trans* isomerization:

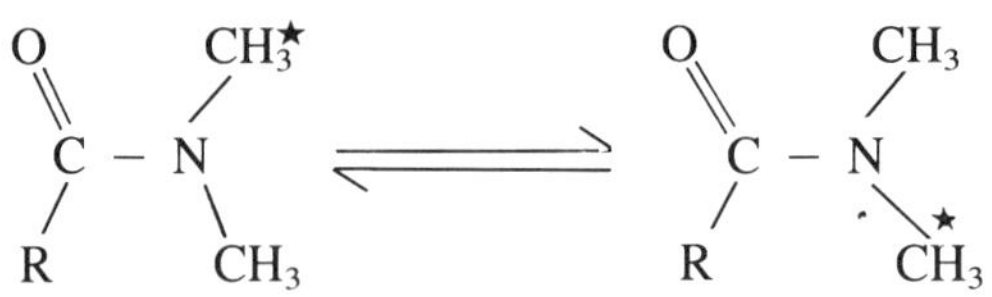

Scheme XII

This reaction has been well studied by NMR.[32] Another important exchange process is the inversion of cyclohexane between equivalent chair forms (Scheme XII), a process in which a proton is exchanged between equatorial and axial positions.

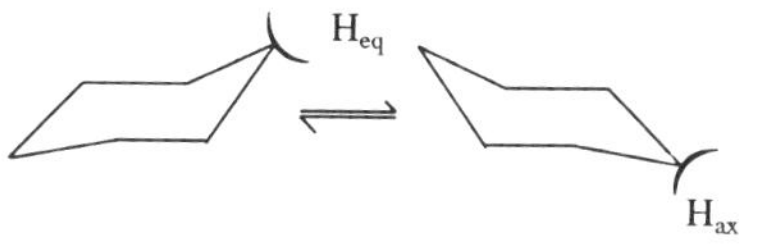

Scheme XIII

This is most readily studied with cyclohexane-d_{11} in which 11 of the 12 protons are replaced with deuterium. The spectrum of cyclohexane-d_{11} resembles the behavior shown in Fig. 4-8; at about $-100°C$ (the slow exchange regime) two sharp lines are seen; these broaden as the temperature is increased, reaching coalescence at $-61.4°C$, and becoming a single sharp line at higher temperatures. (The deuterium nuclei must be decoupled by rf irradiation.) Rate constants τ^{-1} for the conversion were measured over the temperature range $-116.7°C$ to $-24.0°C$ by Anet and Bourne.[33] It is probable that the chair–chair inversion takes place via a boat intermediate.

Turning from chemical exchange to nuclear relaxation time measurements, the field of ^{13}C NMR offers many good examples of chemical information from T_1 measurements. Recall from Fig. 4-7 that T_1 is reciprocally related to τ_c, the correlation time, for high-frequency relaxation modes. For small- to medium-size molecules in the liquid phase, T_1 lies to the left side of the minimum in Fig. 4-7. A larger value of T_1 is, therefore, associated with a smaller τ_c, hence, with a more rapid rate of molecular motion. It is possible to measure T_1 for individual carbon atoms in a molecule, and such results provide detailed information on the local motion of atoms or groups of atoms. Levy and Nelson[34] have reviewed these observations. A few examples are shown here. T_1 values (in seconds) are noted for individual carbon atoms.

$$9.3 \qquad 68 \quad 12.8 \quad 23 \quad 9.8$$

$$(CH_3)_3\ C\text{–}CH_2\text{–}CH\ (CH_3)_2$$

iso-Octane, **3**

$$3.1 \quad 2.2 \quad 1.6 \quad 1.1 \quad 0.84 \quad 0.84 \quad 0.84 \quad 0.77 \quad 0.77 \quad 0.65$$

$$CH_3\ CH_2\ CH_2\ CH_2\ CH_2\ CH_2\ CH_3\ CH_3\ CH_2\ CH_2\ OH$$

n-Decanol, **4**

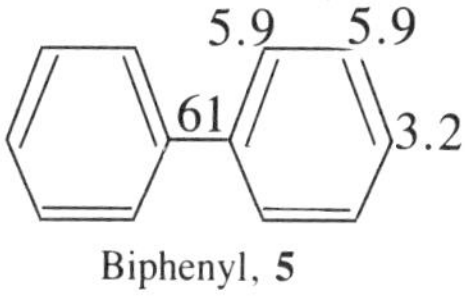

Biphenyl, **5**

Nuclear dipole–dipole interaction is a very important relaxation mechanism, and this is reflected in the relationship between T_1 and the number of protons bonded to a carbon. The motional effect is nicely shown by the T_1 values for *n*-decanol, which suggest that the polar end of the molecule is less mobile than the hydrocarbon tail. Comparison of *iso*-octane with *n*-decanol shows that the entire *iso*-octane molecule is subject to more rapid molecular motion than is *n*-decanol—compare the methyl group T_1 values in these molecules.

The longer T_1 for the *ortho* and *meta* positions in biphenyl relative to the *para* position indicates that rotation about the long axis of the molecule is the dominant motion.

4.4 RAPID MIXING METHODS

Batch Methods

At the beginning of this chapter we pointed out that the rate of mixing of two solutions places a limit on the fastest reactions that can be studied by conventional kinetic methods. In this section we explore the fastest mixing methods that have been devised. These methods therefore constitute a specialized, but otherwise continuous, extension of conventional kinetics into the fast reaction range.

First, let us consider *batch* mixing processes, as exemplified by ordinary laboratory practice in solution kinetics. A portion of one solution (say, of the substrate) is added by pipet to a second solution (containing the reagent) in a flask, the flask is shaken to achieve homogeneity, and then samples are withdrawn at known times for analysis, or the solution is subjected to continuous observation as a function of time, for example, by spectrophotometry. For reactions on a time scale (measured by the half-life) of hours or even several minutes, the time consumed in these operations is a negligible portion of the reaction time, but as the half-life of the reaction decreases, it becomes necessary to consider these preliminary steps. Let us distinguish three stages:

1. The addition of solutions, or their contact, taking a time t_{add} from the beginning to the end of the process.
2. The mixing of the solutions, requiring a time t_{mix} from the beginning of mixing until the solution is essentially homogeneous.
3. The observation or sampling stage, taking time t_{obs} from the completion of mixing until useful quantitative detection begins.

These stages may run consecutively or concurrently; thus we might mix during addition, and we could observe during mixing (although t_{obs} as defined here is measured from the end of mixing). However, when the total time from the beginning of addition to the beginning of data acquisition is significant relative to the half-life, careful analysis of the experiment is required.

One of the problems that arises is specification of *zero time,* the time of initiation of reaction. Some reaction takes place even from the beginning of the addition phase, yet until the solution is well mixed, the extent of reaction is not characteristic of a well-defined system. If the kinetics are first-order, it is not necessary to know the actual zero time; we can arbitrarily set $t = 0$ at any time after quantitative data are first obtained. However, for reactions of other orders or if we wish to extrapolate to zero time so as to estimate initial conditions, we must specify the time of initiation. This is one motivation for reducing t_{add} and t_{mix}.

A more serious problem is that we lose all kinetic information about the system until the data collection begins, and ultimately this limits the rates that can be studied. For first-order reactions we may be able to sacrifice the data contained in the first one, two, or three half-lives, provided the system "amplitude" is adequate; that is, the remaining extent of reaction must be quantitatively detectable. However, this practice of basing kinetic analyses on the last few percentage of reaction is subject to error from unknown side reactions or analytical difficulties.

Discussions of rapid mixing methods make use of the concept of the *dead time*, which is the time from zero time until detection begins. We have already seen that the statement of zero time is ambiguous. It is, moreover, possible that quantitative observation may begin before mixing is complete. The goal of rapid mixing techniques is therefore to reduce t_{add} and t_{mix} to negligible or at least minimum values.

Let us examine some batch results. In trials in which 5 mL of a dye solution was added by pipet (with pressure) to 10 mL of water in a 25-mL flask, which was shaken to mix (as determined visually), and the mixed solution was delivered into a 3-mL rectangular cuvette, it was found that $t_{add} = 3$–5 s, t_{mix} 2–4 s, and t_{obs} 3–5 s. This is characteristic of conventional batch operation. Simple modifications can reduce this dead time. Reaction vessels designed for photometric titrations[35,36] may be useful kinetic tools. For reactions that are followed spectrophotometrically this technique is valuable: Make a flat "button" on the end of a 4-in. length of glass rod. Deliver 3 mL of reaction medium into the rectangular cuvette in the spectrophotometer cell compartment. Transfer 10–100 μL of a reactant stock solution to the button on the rod. Lower this into the cuvette, mix the solution with a few rapid vertical movements of the rod, and begin recording; the dead time will be 3–8 s. A commercial version of the stirrer is available.

None of these procedures constitutes a rapid mixing technique, but they define the approximate limit of conventional kinetics. A fast reaction batch mixing method was devised by Below et al.[37] The reaction vessel is a chamber having quartz windows for spectrophotometric observation. A movable baffle divides the chamber into two tight compartments, which initially contain the two reactant solutions. A spring mechanism quickly raises the baffle, which is designed to produce turbulence and thus to increase the mixing rate. About 90% of complete mixing was achieved in 10–30 ms.

Flow Methods

In flow studies of fast reactions streams of two reactant solutions are forced under pressure to meet in a mixing chamber, from which the mixed solution passes to an observation chamber. These methods have been reviewed by Caldin[1] and Chance.[38] Flow methods function in the time range of (approximately) 1 ms to 10 s, so they very usefully extend the range of conventional batch mixing. Unlike relaxation kinetics (Section 4.2) and NMR (Section 4.3), flow methods are applicable to irreversible reactions.

The *continuous flow* method was developed in 1923 by Hartridge and Roughton[39] in aid of their study of the kinetics of the reaction between hemoglobin and oxygen. The principle is shown schematically in Fig. 4-10. Solutions of reactants A and B are forced by pistons into the mixing chamber, whose design contributes to rapid mixing; the mixed solution flows into an observation tube, where detection by spectroscopy takes place a distance d downstream from the mixer. With continuing injection of reactant solutions a steady state is set up in the observation tube, the concentrations at any point being independent of time. If v is the flow velocity, the distance d is related to reaction time by $t = d/v$. Thus, with a flow velocity of 10 m s^{-1}, a distance of 1 cm corresponds to 1 ms. The time course of the reaction is determined by varying d or v.

Continuous flow devices have undergone careful development, and mixing chambers are very efficient. Mixing is essentially complete in about 1 ms, and half-lives as short as 1 ms may be measured. An interesting advantage of the continuous flow method, less important now than earlier, is that the analytical method need not have a fast response, since the concentrations are at steady state. Of course, the slower the detection method, the greater the volumes of reactant solutions that will be consumed. In 1923 several liters of solution were required, but now reactions can be studied with 10–100 mL.

Two techniques conceptually related to classical continuous flow make use of different injection methods. In one of these[40] a reactant solution formed into a high-speed jet is injected through a sheet or film of the second solution. The jet speed is 40 ms^{-1}, and the mixing time is 1 μs.

In the second technique, two streams of microdroplets (about 100 μm diameter, 40 kHz generation frequency, 15 ms^{-1} velocity) collide to form a single droplet stream, which is observed by Raman spectroscopy.[41] The mixing time is 200 μs.

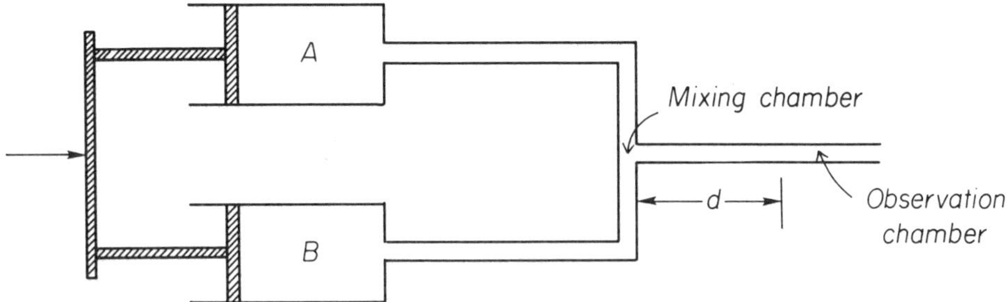

Figure 4-10. Schematic diagram of continuous flow kinetic system. The quantity d is the distance from the mixer to the point of observation.

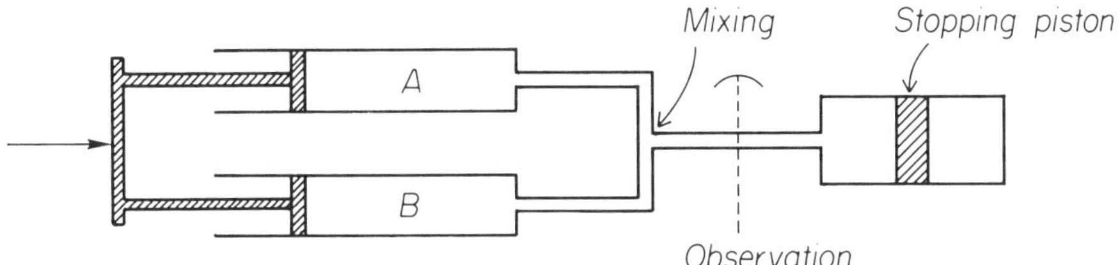

Figure 4-11. Schematic diagram of stopped flow kinetic system.

Stopped flow is probably the most widely used of methods in this chapter for the study of fast reactions, in part because stopped-flow apparatus is commercially available at modest cost. Figure 4-11 shows the principle. As with continuous flow, the reactant solutions are forced from cylinders or syringes, they meet in a mixing chamber, and the mixture flows into an observation tube. However, after a few milliseconds the flow is abruptly stopped by the forcing of a stopping piston against an arrest. Observation by a fast response analytical method (usually spectrophotometry or fluorimetry) is then made as a function of time. The stopped-flow method is, therefore, exactly analogous to conventional batch mixing kinetic studies (although until the stopping action occurs, it is a continuous flow system).

The dead time is typically 3–5 ms, so stopped flow is not quite as fast as continuous flow, but it requires less than a milliliter of each solution per run. Methods have been described for measuring the dead time;[42,43] these are based upon "standard" reactions whose kinetic behavior is well known. The error introduced by collecting data before mixing is complete can be corrected.[43]

Stopped flow is sometimes described as *concentration jump kinetics*. Variations

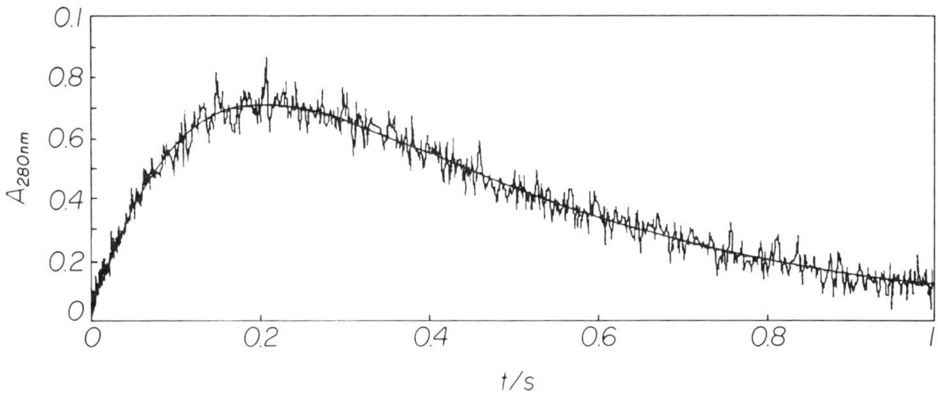

Figure 4-12. Stopped-flow study of the pyridine-catalyzed hydrolysis of acetic anhydride, showing the formation and decay of the acetylpyridinium ion intermediate. Initial concentrations were: 0.087 M pyridine, 2.1×10^{-4} M acetic anhydride; the pH was 5.5; ionic strength, 1.0 M; temperature, 25°C. Five hundred data points (absorbance at 280 nm) were measured in 1 s. The smooth curve is a fit to Eq. (3-27). *Source*: Data of D. Khossravi and S.-F. Hsu, University of Wisconsin.

on the technique include the injection of solvent,[44] the combination of stopped flow
with T-jump,[45] and the use of NMR as the analytical method so as to detect short-
lived intermediates.[46]

Figure 4-12 shows a stopped-flow study of the pyridine-catalyzed hydrolysis of
acetic anhydride. The absorbance–time curve reveals the formation and decay of
the reactive intermediate acetylpyridinium ion.

4.5 OTHER METHODS

This is a very selective survey of additional techniques; Caldin,[1] Hammes,[2] and
Hague[3] give more complete treatments.

The study of relative rates by the competitive method can be useful. The principle
was discussed in Section 3.1 in the context of parallel reactions, for which the ratio
of the product concentrations is equal to the ratio of rate constants (provided the
concentrations are under kinetic control).

Fluorescence Quenching

This is a steady-state competitive method, applicable when a solute is capable of
fluorescing. We consider the simplest case. The solute A undergoes excitation to
the excited singlet state A★ upon absorption of radiation of frequency ν_{ex}.

$$A + h\nu_{ex} \rightarrow A\star$$

The excitation process may generate an excited molecule in any allowed vibrational
state, but the excess vibrational energy is rapidly lost, and the excited state species
may then emit a photon of frequency ν_{em}, this singlet–singlet transition from the
excited to ground state being fluorescence.

$$A\star \xrightarrow{k_F} A + h\nu_{em}$$

Here k_F is the rate constant for this process, and $\nu_{em} < \nu_{ex}$.

A★ may also return to the ground state via a radiationless transition, most commonly by collisional transfer of energy to a solvent molecule.

$$A\star \xrightarrow{k_N} A$$

In the presence of exciting radiation of constant energy, a steady state is established
between the excitation and deexcitation processes.

If a second solute Q is added that is able, in a second-order reaction, to make
available an additional route for return to the ground state, we can write

$$A\star + Q \xrightarrow{k_Q} A + Q'$$

It is possible that Q$'$ is an excited state of Q; if so, we will assume that its emission spectrum does not contribute to the fluorescence intensity at ν_{em}. Q is called a *quencher*, because in its presence the fluorescence intensity of solute A is reduced.

The quantum yield Φ_0 in the absence of quencher is the ratio (No. photons emitted)/(No. photons absorbed), or

$$\Phi_0 = \frac{k_F}{k_F + k_N} \qquad (4\text{-}71)$$

In the presence of Q the quantum yield Φ is smaller because of the additional route:

$$\Phi = \frac{k_F}{k_F + k_N + k_Q c_Q} \qquad (4\text{-}72)$$

In Eq. (4-72), $k_Q c_Q$ is a pseudo-first-order rate constant.

Letting I_{ab} be the intensity of absorbed light and F_0, F the fluorescent intensities, we have $\Phi_0 = F_0/I_{ab}$ and $\Phi = F/I_{ab}$, so

$$\frac{F_0}{F} = \frac{k_F + k_N + k_Q c_Q}{k_F + k_N} \qquad (4\text{-}73)$$

Because in the absence of Q the rate of loss of the excited state is

$$-\frac{dc_{A^*}}{dt} = (k_F + k_N)c_{A^*}$$

we define the mean fluorescence lifetime τ_0 by $\tau_0^{-1} = (k_F + k_N)$. Then Eq. (4-73) becomes

$$\frac{F_0}{F} = 1 + k_Q \tau_0 c_Q \qquad (4\text{-}74)$$

A plot of F_0/F vs. c_Q is called a *Stern–Volmer plot*. From the slope, the quantity $k_Q \tau_0$ is evaluated. This is a relative rate.

In order to obtain the absolute rate constant k_Q, τ_0 must be known. This is a much more difficult measurement, requiring specialized techniques. For singlet–singlet transitions τ_0 is often of the order 10^{-8} s.

This approach can be elaborated to take into account other possible dispositions of the excited state, and it is a valuable means for studying the chemistry of excited state species. Wilkinson has reviewed photochemical kinetics.[47]

Electrochemical Methods

Several electrochemical techniques have been devised for the study of fast reactions.[48,49] These methods require that one of the species involved in the reaction of interest be electroactive, so that the reaction under study is coupled to an electrode reaction. We will use polarography as an example of these methods.[50] Scheme XIV shows the simplest system that can be studied.

$$A \underset{k_{-1}}{\overset{k_1}{\rightleftharpoons}} O$$

$$O + ne \rightarrow R$$

Scheme XIV

In this scheme the reversible conversion of A to O is the reaction whose rate is to be studied, whereas the reduction of O to R is the electrode process. Scheme XIV can also represent a pseudo-first-order formation of O. A specific example is the acid–base equilibrium of pyruvic acid, shown in Scheme XV.

$$CH_3COCOO^- + H^+ \underset{k_{-1}}{\overset{k_1}{\rightleftharpoons}} CH_3COCOOH$$

$$CH_3COCOOH \rightarrow CH_3CHOHCOOH$$

Scheme XV

This is sometimes described as a competitive method, the coupling species O being involved in two separate reactions.

The electrode current depends on the rates of the coupled reactions, but by suitable adjustment of the electrode potential (into the diffusion current region for the electrode reaction) the rate of the reduction reaction can be made so fast that the current depends only on the rate of the prior chemical reaction. The dependence of the observed current on the presence of the chemical reaction is a measure of the rate.

Consider the pyruvic acid system in Scheme XV. Let HA and A^- represent pyruvic acid and pyruvate, respectively, and suppose the system is buffered. At a pH well below the pK_a of HA, a single polarographic wave characteristic of the reduction of HA is observed. At a pH well above the pK_a, a wave (at a much more reducing potential) is observed that is characteristic of the reduction of A^-.

Now at some pH comparable to pK_a, two waves are observed, corresponding to the reduction of both HA and A^-. The currents are proportional to the concentrations of the electroreducible species. Because the pH and pK_a are known, the concentrations of HA and A^- in the bulk solution can be calculated. It is then found that the observed polarographic currents cannot be accounted for on the basis of the known bulk concentrations. It is concluded that the ratio of the concentrations at the electrode surface is different from the ratio of bulk concentrations, and this is a consequence of the coupling between the chemical and electrode processes. In the pyruvic acid system, HA can be converted to the hydroxy acid by the electrode process, and it is also related to A^- by the chemical reaction. A comparison between the polarographic result and that calculated for the bulk concentrations (absence of the chemical reaction) leads to an estimate of the rate constants

The theory of rate measurements by electrochemistry is mathematically quite difficult, although the experimental measurements are straightforward. The techniques are widely applicable, because conditions can be found for which most compounds are electroactive. However, many questionable kinetic results have been reported, and some of these may be a consequence of unsuitable approximations in applying theory. Another consideration is that these methods are mainly applicable to aqueous solutions at high ionic strengths and that the reactions being observed are not bulk phase reactions but are taking place in a layer of molecular dimensions near the electrode surface. Despite such limitations, useful kinetic results have been obtained.[1,48–50]

Common Ion Inhibition

Suppose in Scheme XVI that the steady-state approximation is applicable to the intermediate A^+.

$$AX \underset{k_{-1}}{\overset{k_1}{\rightleftharpoons}} A^+ + X^-$$

$$A^+ \overset{k_2}{\rightarrow} products$$

Scheme XVI

Then the observed rate constant k is given by Eq. (4-75).

$$k = \frac{k_1 k_2}{k_{-1}[X^-] + k_2} \tag{4-75}$$

which can be rearranged to Eq. (4-76), a linear plotting form.

$$\frac{1}{k} = \frac{k_{-1}[X^-]}{k_1 k_2} + \frac{1}{k_1} \tag{4-76}$$

The rate constant k is measured as a function of (added) concentration of the "common ion" X^-, and from the plot according to Eq. (4-76) the ratio $k_{-1}/k_2 = $ slope/intercept is evaluated.

If the intermediate is very unstable, large rate constants may be measured in this way. Thus, Amyes and Jencks[51] studied the hydrolysis of α-azidoethers (N_3^-, the azide ion, being the common ion), finding (because k_{-1} had been independently measured) k_2 values in the range 10^7 to 10^{10} M^{-1} s^{-1} for the reaction with water.

REFERENCES

1. Caldin, E.F. "Fast Reactions in Solution"; Blackwell: Oxford, 1964.
2. Hammes, G.G. *Adv. Protein Chem.* **1968,** *23,* 1.
3. Hague, D.N. "Fast Reactions"; Wiley-Interscience: London, 1971.
4. Hammes, G.G. Ed. "Techniques of Chemistry", Vol. VI, "Investigation of Rates and Mechanisms of Reactions", 3rd ed.; Wiley-Interscience: New York, 1974; Part II.
5. Bernasconi, C.F. "Relaxation Kinetics"; Academic Press: New York, 1976.
6. Kruger, H. *Chem. Soc. Rev.* **1982,** *11*(3), 227.
7. Smoluchowski, M.V. *Z. Phys. Chem.* **1917,** *92,* 129.
8. Gardiner, W.C., Jr. "Rates and Mechanisms of Chemical Reactions"; Benjamin: New York, 1969; pp 165–9.
9. Berry, R.S.; Rice, S.A.; Ross, J. "Physical Chemistry"; Wiley: New York, 1980; Part 3, pp. 1161–5.
10. Weston, R.E., Jr.; Schwartz, H.A. "Chemical Kinetics"; Prentice-Hall; Englewood Cliffs, N.J., 1972; pp 156–61.
11. Olea, A.F.; Thomas, J.K. *J. Am. Chem. Soc.* **1988,** *110,* 4494.
12. Jencks, W.P. "Catalysis in Chemistry and Enzymology"; McGraw-Hill: New York, 1969; pp 190–2.
13. Grunwald, E. *Prog. Phys. Org. Chem.* **1965,** *3,* 317.
14. Eigen, M.; DeMaeyer, L. *Z. Elektrochem.* **1955,** *59,* 986.
15. Brouillard, R. *Faraday Trans. I* **1980,** *76,* 583.
16. Bertigny, J.-P.; Dubois, J.-E., Brouillard, R. *Faraday Trans. I* **1983,** *79,* 209.
17. Yapel, A.F.; Lumry, R. *Methods Biochem. Anal.* **1971,** *20,* 169.
18. Marcandalli, B: Stange, G.; Holzworth, J.F. *Faraday Trans. I* **1988,** *84,* 2807.
19. Stuehr, Jr. In "Techniques of Chemistry", Vol. VI, "Investigation of Rates and Mechanisms of Reactions", 3rd ed.; Hammes, G.G., Ed.; Wiley-Interscience: New York, 1974; Part II, Chapter VII.
20. Hammes, G.G.; Steinfeld, J.I. *J. Am. Chem. Soc.* **1962,** *84,* 4639.
21. Cramer, F.; Saenger, W.; Spatz, H.-Ch. *J. Am. Chem. Soc.* **1967,** *89,* 14.
22. Becker, E.D. "High Resolution NMR", 2nd ed.; Academic Press: Orlando, Fla., 1980.
23. Swift, T.J. In "Techniques of Chemistry", Vol. VI, "Investigation of Rates and Mechanisms of Reactions", 3rd ed.; Hammes, G.G., Ed.; Wiley-Interscience: New York, 1974; Part II, Chapter XII.
24. Bloch, F. *Phys Revs.* **1946,** *70,* 460.
25. Andrew, E.R. "Nuclear Magnetic Resonance"; Cambridge University Press: Cambridge, 1956; Appendix 1.
26. Bloembergen, N.; Purcell, E.M.; Pound, R.V. *Phys Rev.* **1948,** *73,* 679.
27. Boden, N. In "Determination of Organic Structures by Physical Methods"; Nachod, F.C.; Zuckerman, J.J., Ed.; Academic Press: New York, 1971; Vol. 4, Chapter 2.
28. Caldin, E.F. "Fast Reactions in Solution"; Wiley: New York, 1964; p 238.
29. Cocivera, M.; Grunwald, E. *J. Am. Chem. Soc.* **1965,** *87,* 2551.
30. Holmes, J.R.; Kivelson, D.; Drinkard, W.C. *J. Am. Chem. Soc.* **1962,** *84,* 4677.
31. Greenzaid, P.; Luz, Z.; Samuel, D. *J. Am. Chem. Soc.* **1967,** *89,* 756.
32. Allerhand, A.; Gutowsky, H.S.; Jonas, J.; Meinzer, R.A. *J. Am. Chem. Soc.* **1966,** *88,* 3185.
33. Anet, F.A.L.; Bourne, A.J.R. *J. Am. Chem. Soc.* **1967,** *89,* 760.
34. Levy, G.C.; Nelson, G.L. "Carbon-13 Nuclear Magnetic Resonance for Organic Chemists"; Wiley-Interscience: New York, 1972; Chapter 9.
35. Leonard, M.A. In "Comprehensive Analytical Chemistry"; Svehla, G., Ed.; Elsevier: Amsterdam, 1977; Vol. VIII, Chapter III.
36. Sweet, T.R.; Zehner, J. *Anal. Chem.* **1958,** *30,* 1713.
36. Rehm, C.; Bodin, J.I.; Connors, K.A.; Higuchi, T. *Anal. Chem.,* **1959,** *31,* 483.
37. Below, J.F.; Connick, R.E.; Coppel, C.P. *J. Am. Chem. Soc.* **1958,** *80,* 2961.

38. Chance, B. In "Techniques of Chemistry", Vol. VI, "Investigation of Rates and Mechanisms of Reactions", 3rd ed.; Hammes, G.G., Ed.; Wiley-Interscience: New York, 1974; Part II, Chapter II.

39. Hartridge, H.; Roughton, F.J.W. *Proc. Roy. Soc. London Ser. A* **1923**, *104*, 376.

40. Davidovits, P.; Chao, S.-C. *Anal Chem.* **1980** *52*, 2435.

41. Simpson, S.F.; Kincaid, J.R.; Holler, F.J. *Anal. Chem.* **1983,** *55*, 1420.

42. Paul, C.; Kirschner, K.; Haenisch, G. *Anal. Biochem.* **1980,** *101*, 442.

43. Dickson, P.N.; Margerum, D.W. *Anal. Chem.* **1986,** *58*, 3153.

44. Patel, R.C.; Atkinson, G.; Boe, R.J. *J. Chem. Educ.,* **1970,** *47*, 800.

45. Trimm, H.H.; Ushio, H.; Patel, R.C. *Talanta* **1981** *28*, 753.

46. Fyfe, C.A.; Cocivera, M.; Damji, S.W.H. *Acc. Chem. Res.* **1978,** *11*, 277.

47. Wilkinson, F. "Chemical Kinetics and Reaction Mechanisms"; Van Nostrand Reinhold: New York, 1980; Chapter 9.

48. Strehlow, H. In "Techniques of Chemistry", Vol. VI, "Investigation of Rates and Mechanisms of Reactions", 3rd ed.; Hammes, G.G., Ed.; Wiley-Interscience: New York, 1974; Part II, Chapter VIII.

49. Zuman, P.; Patel, R. "Techniques in Organic Reaction Kinetics"; Wiley-Interscience: New York, 1984; Chapter 4.

50. Zuman, P. *Adv. Phys. Org. Chem.* **1967,** 5, 1.

51. Amyes, T.L.; Jencks, W.P. *J. Am. Chem. Soc.* **1989,** *111*, 7888.

52. Grunwald, E.; Jumper, C.F.; Meiboom, S. *J. Am. Chem. Soc.* **1962,** *84*, 4664.

PROBLEMS

1. For the kinetic scheme

$$
\begin{array}{ccccc}
A + B & \rightleftharpoons & C & \rightleftharpoons & D + E \\
\updownarrow & & \updownarrow & & \updownarrow \\
G + H & \rightleftharpoons & I & \rightleftharpoons & K + L
\end{array}
$$

how many relaxation times are there?

2. For the reaction

$$
A + C \underset{k_{-1}}{\overset{k_1}{\rightleftharpoons}} B + C
$$

where C is a catalyst, write the expression for relaxation time in terms of system parameters. (You should be able to do this by inspection.)

3. For the system

$$
A + B \underset{k_{-1}}{\overset{k_1}{\rightleftharpoons}} 2C
$$

derive the relaxation time expression. (Pay attention to the stoichiometric factor 2.)

4. For the kinetic scheme

$$2A \underset{k_{-1}}{\overset{k_1}{\rightleftharpoons}} B$$

find an expression for the relaxation time in terms of the system parameters for a small perturbation of the system.

5. Consider this fast reaction as it would be studied by a small-perturbation chemical relaxation method.

$$A + B \underset{k_{21}}{\overset{k_{12}}{\rightleftharpoons}} C + D$$

Suppose the relaxation time τ is determined under conditions such that reactant B is buffered; that is, essentially no change in the concentration of B occurs during relaxation. Derive an expression for τ in terms of the rate constants and equilibrium concentrations.

6. Show how the four rate constants of Scheme III can be found from measurements of τ_I and τ_{II}.

7. Estimate the rate constants for the acid–base equilibrium of formic acid in water.

8. The relaxation time for this reaction

$$H^+ + OH^- \rightleftharpoons H_2O$$

in water at neutrality and 25°C is 3.5×10^{-5} s. What are the ion recombination and dissociation rate constants?

9. For the rate of proton transfer between the lyonium ion ($MeOH_2^+$) and solvent (MeOH) and the lyate ion (MeO^-) and solvent, Grunwald et al.[52] give this result:

$$\frac{\text{rate of proton}}{\text{transfer/ms}^{-1}} = 8.8 \times 10^{10} [MeOH_2^+] + 1.85 \times 10^{10} [MeO^-]$$

What are the second-order rate constants for these processes?

10. Sketch the 180°, t, 90° pulse sequence for T_1 determination in the style of Fig. 4-9.

11. Explain why, in the study of an acid–base equilibrium, we observe absorption peaks of both species (conjugate acid and base) when using electronic absorption spectroscopy, but only a single peak by NMR.

12. (a) Calculate the NMR peak width at half height in the absence of chemical exchange and field inhomogeneity if $T_2 = 1$ s.

    (b) For $T_2 = 1$ s, calculate the peak width if the system is undergoing chemical exchange with $\tau = 0.1$ s.

13. The ^{13}C T_1 for benzene is 29 s, whereas that for methanol is 15 s. Interpret these results.

CHAPTER 5

Theory of Chemical Kinetics

Earlier chapters have dealt primarily with the phenomenology of chemical kinetics, that is, the observation and description of kinetic phenomena. We now turn to kinetic theory. Theorists have two requirements of a successful theory of rates: that it disclose the factors that control reaction rates and that it permit the calculation of rates from *first principles,* which in this context means from nonkinetic information, such as molecular size and shape, spectroscopic data on vibrational modes, and physical properties like viscosity and density. Experimentalists, among whom we will count ourselves, ask something else from a theory. We need a conceptual framework with which to interpret our kinetic data. We do not require that we be able to calculate a rate constant or an activation energy from first principles; we are willing to measure these quantities, but we wish then to learn how they may be related to system variables or to similar quantities for other reactions. We may even hope for some level of predictive ability by combining experimental findings with theoretical guidance. This chapter is designed to provide a basis for these needs.

Reviews of reaction rate theory by Laidler[1] and Wayne[2] are very helpful. A classic book by Glasstone et al.[3] is still an excellent introduction to the subject. Eyring et al.[4] provide an advanced, detailed treatment of kinetic theory.

5.1 THEORETICAL APPROACHES

The Arrhenius Equation

A large portion of the field of chemical kinetics can be described by, or discussed in terms of, Eq. (5-1), the Arrhenius equation.

$$k = Ae^{-E_a/RT} \tag{5-1}$$

The Arrhenius equation relates the rate constant k of an elementary reaction to the absolute temperature T; R is the gas constant. The parameter E_a is the *activation energy*, with dimensions of energy per mole, and A is the *preexponential factor*, which has the units of k. If k is a first-order rate constant, A has the units seconds^{-1}, so it is sometimes called the *frequency factor*.

The Arrhenius equation is best viewed as an empirical relationship that describes kinetic data very well. It is commonly applied in the linearized form

$$\log k = \log A - \frac{E_a}{2.3RT} \tag{5-2}$$

and an *Arrhenius plot* of $\log k$ against $1/T$ yields E_a from the slope of the straight line. This description implies that A and E_a are temperature independent, an implication that is difficult to test because of the small temperature range usually employed in such studies. We will subsequently comment on the possible temperature dependence of these parameters, but it is reasonable to accept that they are essentially independent of temperature except in unusual circumstances.

Because it is a goal of kinetic theory to calculate rates (or rate constants), this is equivalent to calculating A and E_a. Thus, the aim of theory is to give theoretical expressions for the preexponential factor and the activation energy. A central question for theory is this: If *some* molecules react (in unit time), why do not *all* molecules react? That is, why do some molecules behave differently from others in the same assemblage? The answer given by Arrhenius is that a molecule must possess some threshold energy ε before it can react, and the fraction of molecules that possess this amount of energy is given by the function $\exp(-\varepsilon/kT)$. This idea accounts for the exponential nature of Eq. (5-1), and, in some form, this concept is embodied in all theories of chemical kinetics.

Collision Theory

We are concerned with bimolecular reactions between reactants A and B. It is evident that the two reactants must approach each other rather closely on a molecular scale before significant interaction between them can take place. The simplest situation is that of two spherical reactants having radii r_A and r_B, reaction being possible only if these two particles "collide," which we take to mean that the distance between their centers is equal to the sum of their radii. This is the basis of the hard-sphere collision theory of kinetics. We therefore wish to find the frequency of such bimolecular collisions. For this purpose we consider the relatively simple case of dilute gases.

Suppose particle A moves through space with average speed $\bar{v}$; A will collide with a B particle if their center-to-center distance is less than or equal to $r_A + r_B$. Thus, particle A sweeps out an area $\pi(r_A + r_B)^2\bar{v}$ in which it can collide with B, and the corresponding volume swept out per second is $\pi(r_A + r_B)^2\bar{v}$. If the concentration of B is n_B molecules cm^{-3}, the number of collisions of B particles by this single A particle, per second, is $\pi(r_A + r_B)^2 n_B\bar{v}$. However, the volume also contains A particles at concentration n_A molecules cm^{-3}, so the total number of collisions per second is

$$Z = \pi(r_A + r_B)^2 n_A n_B \bar{v} \tag{5-3}$$

The average molecular speed is[5]

$$\bar{v} = \left(\frac{8kT}{\pi\mu}\right)^{1/2} \tag{5-4}$$

where μ is the reduced mass,

$$\mu = \frac{m_A m_B}{m_A + m_B} \tag{5-5}$$

m_A and m_B being the molecular masses of A and B. Combining Eqs. (5-3) and (5-4),

$$Z = n_A n_B (r_A + r_B)^2 \left(\frac{8\pi kT}{\mu}\right)^{1/2} \tag{5-6}$$

Equation (5-6) can be derived in a more rigorous manner.[6]

Equation (5-6) gives the number of bimolecular collisions per unit time and volume, but not all of these collisions lead to reaction, and so we write rate = collision frequency $\times$ fraction of collisions having energy equal to or greater than that required for reaction, or

$$v = -\frac{dn}{dt} = Ze^{-E/RT} \tag{5-7}$$

where E is the required energy (per mole). To convert this to the usual units of molar concentration c we use the relationship $c = 10^3 n/N_A$, where N_A is Avogadro's number. Then $dc/dt = (10^3/N_A)dn/dt$, or

$$-\frac{dc_A}{dt} = -\frac{dc_B}{dt} = kc_A c_B = \frac{10^6 k n_A n_B}{N_A^2} = -\frac{10^3}{N_A} \cdot \frac{dn}{dt}$$

This gives

$$-\frac{dn}{dt} = \frac{10^3 k n_A n_B}{N_A} \tag{5-8}$$

Comparing Eqs. (5-7) and (5-8),

$$k = \frac{10^{-3} N_A Z e^{-E/RT}}{n_A n_B} \tag{5-9}$$

where Z is given by Eq. (5-6). Equation (5-9) can be written

$$k = Ae^{-E/RT} \qquad (5\text{-}10)$$

which is of the Arrhenius form, the preexponential factor being given by Eq. (5-11).

$$A = 10^{-3}N_A(r_A + r_B)^2(8\pi kT/\mu)^{1/2} \qquad (5\text{-}11)$$

Note that A is predicted by collision theory to be proportional to $T^{1/2}$. For bimolecular reactions A has the units M^{-1} s^{-1} (liter per mole per second).

Let us estimate a typical value for A. Choosing $r_A = r_b = 5$ Å, $\mu = 2 \times 10^{-22}$ g, $T = 300$ K, we find A $\approx 4 \times 10^{11}$ M^{-1} s^{-1}. This is for the gas phase. In solution the situation is somewhat different because of the solvent cage effect described in Section 4.1. During each bimolecular encounter within a solvent cage, several collisions may occur. This results in a predicted A value for liquid solutions somewhat larger than that for gases.[7,8]

The temperature dependence of A predicted by Eq. (5-11) makes a very weak contribution to the temperature dependence of the rate constant, which is dominated by the exponential term. It is, therefore, not feasible to establish, on the basis of temperature studies of the rate constant, whether the predicted $T^{1/2}$ dependence of A is observed experimentally. Uncertainties in estimates of A tend to be quite large because this parameter is, in effect, determined by a long extrapolation of the Arrhenius plot to $1/T = 0$.

Tests of the collision theory consist of comparisons between calculated and experimental values of the preexponential factor, the comparison often being made in terms of a ratio P defined by

$$P = \frac{A \text{ (observed)}}{A \text{ (calculated)}} \qquad (5\text{-}12)$$

Considering the extreme simplicity of the hard-sphere collision theory, it is remarkable that for many reactions, in both the gas and solution phases, P is very close to unity. There are examples of P values, however, that are very much smaller or larger than unity. Table 5-1 gives some results for reactions in solution. [These are drawn from a book by Moelwyn-Hughes,[9] who has been the leading proponent of collision theory applied to solution kinetics.] Several factors may contribute to P values that are far from unity. Perhaps the most obvious of these is that reactant molecules are not spherical. The simple collision theory takes no account of the geometry of approach required for reaction to occur, and because severe steric constraints may exist, very low P values may result from this effect. Another factor is that the collision may not result in the kinetic energy possessed by the reactants being delivered where and when needed in order for reaction to occur. Calculations for reactions in solution are further complicated by the presence of the solvent. It must also be remembered that the theory applies to elementary reactions, and many solution reactions are complex.

TABLE 5-1.   Kinetic Data and P Values for Some Solution Reactions

| | | | 10^{-11} A | | |
Reaction	Solvent	E_a/kcal mol^{-1}	Observed	Calculated	P
$C_2H_5Br + OH^-$	C_2H_5OH	21.4	4.30	3.86	1.11
$C_2H_5O^- + CH_3I$	C_2H_5OH	19.5	2.42	1.93	1.25
$ClCH_2CO_2^- + OH^-$	H_2O	25.9	4.55	2.86	1.59
$C_3H_6Br_2 + I^-$	CH_3OH	25.1	1.07	1.39	0.77
$HOCH_2CH_2Cl + OH^-$	H_2O	19.9	25.5	2.78	9.17
$4\text{-}CH_3C_6H_4O^- + CH_3I$	C_2H_5OH	21.2	8.49	1.99	4.27
$CH_3(CH_2)_2Cl + I^-$	$(CH_3)_2CO$	20.7	0.085	1.57	0.054
$(CH_3)_2SO_4 + CNS^-$	CH_3OH	17.4	0.19	1.91	0.010
$\beta\text{-}C_{10}H_7O^- + C_2H_5I$	CH_3OH	21.0	0.10	2.21	0.045
$(C_2H_5)_3N + C_2H_5Br$	C_6H_6	11.2	—	—	5.3×10^{-10}
$C_5H_5N + CH_3I$	$C_2H_2Cl_4$	13.2	—	—	2.0×10^{-6}
$(H_2N)_2CS + C_2H_5I$	C_2H_5OH	14.6	—	—	1.5×10^{-5}
Lactose $+ H_3O^+$	H_2O	24.7	—	—	1.17
Sucrose $+ H_3O^+$	H_2O	25.8	—	—	5.3×10^3
Melibiose $+ H_3O^+$	H_2O	38.6	—	—	1.5×10^9

Source: Reference 9.

Some P values are greatly in excess of unity, and these may result from very favorable entropic contributions. This factor can be considered quantitatively later in this chapter.

Simple collision theory does not provide a detailed interpretation of the energy barrier or a method for the calculation of activation energy. It also fails to lead to interpretations in terms of molecular structure. The notable feature of collision theory is that, with very simple means, it provides one basis for defining "typical" or "normal" kinetic behavior, thereby directing attention to unusual behavior.

Potential Energy Surfaces

If we were to calculate the potential energy V of the diatomic molecule AB as a function of the distance r_{AB} between the centers of the atoms, the result would be a curve having a shape like that seen in Fig. 5-1. This is a bond dissociation curve, the path from the minimum (the equilibrium internuclear distance in the diatomic molecule) to increasing values of r_{AB} describing the dissociation of the molecule. It is conventional to take as the zero of energy the infinitely separated species.

Most chemical reactions are more complicated than this one, and the system potential energy is a function of more than one variable. Consider this reaction, which is a generalized group-transfer reaction:

$$A + BC \rightarrow AB + C$$

In the simplest case the AB and BC bond axes will remain colinear throughout the reaction; then the potential energy can be expressed as a function of the bond

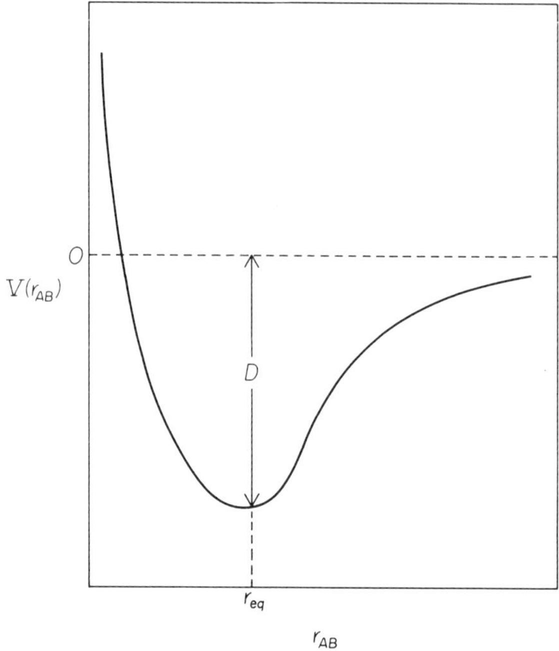

Figure 5-1. Form of a potential energy curve for diatomic molecule AB. $V(r_{AB})$ is the potential energy, r_{AB} is the internuclear distance, r_{eq} is the equilibrium internuclear distance, and D is the bond dissociation energy. (The zero point energy is neglected in the figure.)

distances r_{AB} and r_{BC}, and a three-dimensional figure could be constructed with V on the z axis, and r_{AB}, r_{BC} serving as the x and y coordinates. This figure constitutes a potential energy surface. If the potential energy depends upon more than two variables (e.g., upon r_{AB}, r_{BC}, and the angle A–B–C in this example), a hypersurface in multidimensional space is required to represent the system. Simplification is often possible in order to achieve a convenient form of representation by fixing all but two variables. We will continue the discussion by considering the group transfer with the A–B–C angle being 180°, i.e., the colinear case.

For representation in two dimensions it is convenient to draw a contour map (analogous to a topographic map of geographic features) in which r_{AB} and r_{BC} are the coordinates, and contours of constant potential energy are drawn. A hypothetical potential energy contour diagram is shown in Fig. 5-2. If the r_{BC} distance is very large, the system consists of compound AB (with C far removed), so a section along dashed line *ab* is equivalent to Fig. 5-1.

The potential energy surface consists of two valleys separated by a col or saddle. The reacting system will tend to follow a path of minimum potential energy in its progress from the initial state of reactants (A + BC) to the final state of products (AB + C). This path is indicated by the dashed line from reactants to products in Fig. 5-2. This path is called the *reaction coordinate,* and a plot of potential energy as a function of the reaction coordinate is called a *reaction coordinate diagram.*

Figure 5-3 is the reaction coordinate diagram for Fig. 5-2. Note the region of the maximum potential energy on the reaction coordinate; this region assumes great importance in kinetic theory. At this point the reacting system is unstable with respect to motion along the reaction coordinate. However, at this same point the system possesses minimum energy with respect to motion along dashed line *cd*. This portion of the reaction coordinate is called the *transition state* of the reaction. (This concept was introduced in Fig. 1-1.)

Before discussing the kinds of kinetic information provided by potential energy surfaces we will briefly consider methods for calculating these surfaces, without going into detail, for theoretical calculations are outside the scope of this treatment. Detailed procedures are given by Eyring et al.[4] There are three approaches to the problem.[1] The most basic one is purely theoretical, in the sense that it uses only fundamental physical quantities, such as electronic charge. The next level is the semiempirical approach, which introduces experimental data into the calculations in a limited way. The third approach, the empirical one, makes extensive use of experimental results.

A fully theoretical calculation of a potential energy surface must be a quantum mechanical calculation, and the mathematical difficulties associated with the method require that approximations be made. The first of these is the Born–Oppenheimer approximation, which states that it is acceptable to uncouple the electronic and nuclear motions. This is a consequence of the great disparity in the masses of the electron and nuclei. Therefore, the calculation can proceed by fixing the location

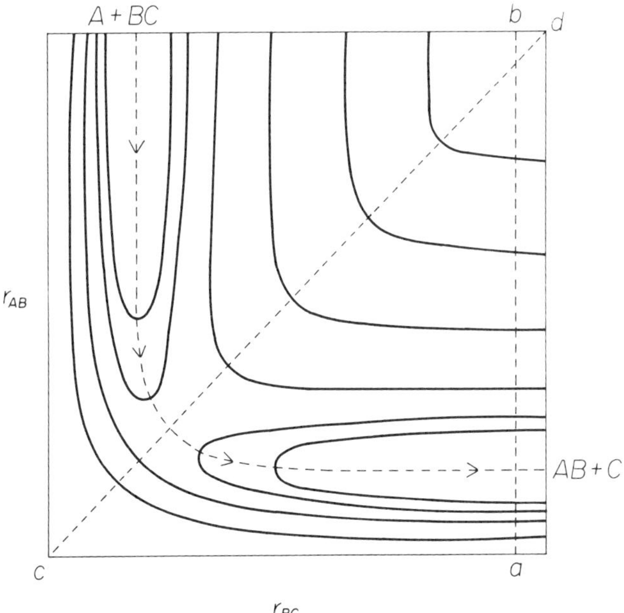

Figure 5-2. A hypothetical potential energy surface for the reaction A + BC → AB + C.

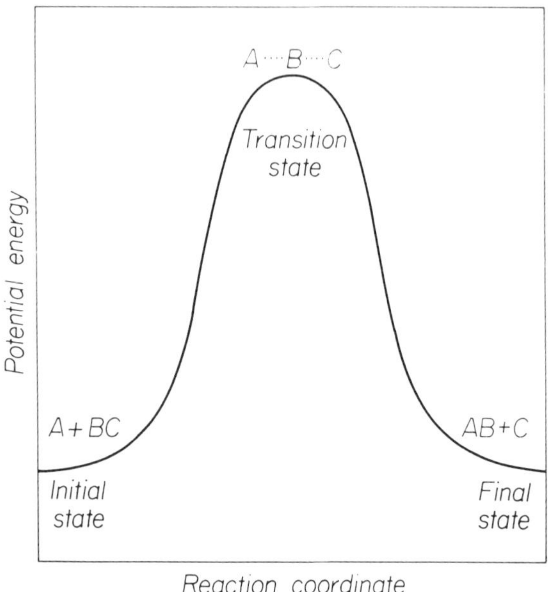

Figure 5-3. Reaction coordinate diagram for the potential energy surface of Fig. 5-2.

of the nuclei and calculating the potential energy due to the electronic configuration for this particular nuclear geometry. Then the nuclear separation can be changed and the calculation repeated, and so on, to map out the entire surface. The second assumption is that the electrons remain in the same electronic energy state (usually the ground state) throughout the reaction. This is the *adiabatic* assumption; it leads to the result that a single potential energy surface describes the entire reaction. (Different electronic states lead to different potential energy surfaces, and it is possible for a reacting system to cross over from one surface to another.)

The most extensive quantum mechanical calculations have been done on this reaction:

$$H + H-H\star \rightarrow H-H + H\star$$

These calculations began in 1927 with Heitler and London's approximate quantum mechanical treatment of the H_2 molecule, which led to Eq. (5-13) for the energy.

$$E = \frac{Q \pm J}{1 \pm S^2} \tag{5-13}$$

Here Q, J, and S are the coulombic, exchange (resonance), and overlap integrals, respectively. Use of the positive signs gives rise to the lower energy (bonding) state, because the integrals are negative. The curve in Fig. 5-1 has the shape of Eq. (5-13) taken with positive signs.

If the overlap integral is neglected, the Heitler–London equation becomes

$$E = Q \pm J \tag{5-14}$$

which happens to reproduce the experimental binding energy for H_2 better than does Eq. (5-13), but this is considered to be a consequence of error cancellation.

London next extended this treatment to the system of three hydrogen atoms, which we may designate A, B, and C. He wrote Eq. (5-15) for this system.

$$E = Q_A + Q_B + Q_C \pm \{ \tfrac{1}{2} [(J_A - J_B)^2 + (J_B - J_C)^2 + (J_C - J_A)^2] \}^{1/2} \tag{5-15}$$

The Q's and J's are coulomb and exchange integrals defined as shown in Table 5-2. Notice that when A is infinitely separated from the B–C pair, Q_B, Q_C, J_B, J_C are all zero, and Eq. (5-15) collapses to $E = Q_A \pm J_A$, as it should. Similarly it gives the appropriate result for the other extreme cases. London did not derive Eq. (5-15), but it has since been derived and is known to apply only to s electrons; moreover it neglects the overlap integrals.

Potential energy surfaces calculated by means of the London equation (5-15) cannot be highly accurate, but the results have been very useful in disclosing the general shape of the surface and the reaction coordinate. The London equation also forms the basis of some semiempirical methods.

A more accurate quantum mechanical approach makes use of the variational method.[3, p. 62] The goal is the solution of the basic wave equation

$$H\psi = E\psi$$

but this is not possible for many-electron problems. The wave function ψ can be expanded in terms of eigenfunctions, ϕ_i, and it can be shown that, no matter how ϕ_i is chosen, the integral

$$\int \bar{\psi} H \psi \, d\tau$$

can never be smaller than the true energy of the system. Thus, by choosing many functions and evaluating the integrals, the lowest result will be closest to the correct

TABLE 5-2. Identification of Coulombic and Exchange
Integrals for the Three-Electron System[a]

Absent	Present	Coulombic	Exchange
A	BC	Q_A	J_A
B	AC	Q_B	J_B
C	AB	Q_C	J_C

[a]See Eq. (5.15).

one. Many functions have been devised for the application of the variational method to the H + H_2 problem,[1] and this system seems to be fairly well understood. Calculations from first principles for more complicated reactions are much more difficult.

The semiempirical methods combine experimental data with theory as a way to circumvent the calculational difficulties of pure theory. The first of these methods leads to what are called London–Eyring–Polanyi (LEP) potential energy surfaces. Consider the triatomic ABC system. For any pair of atoms the energy as a function of intermolecular distance r is represented by the Morse equation, Eq. (5-16),

$$E = D[e^{-2\beta(r-r_{eq})} - 2e^{-\beta(r-r_{eq})}] \tag{5-16}$$

where r_{eq} is the equilibrium internuclear distance, D is the bond dissociation energy (see Fig. 5-1), and β is a constant that can be calculated from spectroscopic data.[4, p. 31] Eyring and Polanyi made the assumption that the ratio of coulombic to total energy is approximately constant (independent of r) at 10–15% (based on earlier theoretical calculations of coulombic and exchange integrals by Sugiura). Thus, taking the BC pair, we have from the London equation $E = Q_A \pm J_A$, from the Morse equation E is known, and with the above assumption Q_A/E is known. Thus, Q_A and J_A can be individually evaluated as functions of r_{BC}. In the same manner Q_B, J_B, Q_C, J_C are found. These quantities are then used in the London equation (Eq. 5-15) to calculate the energy of the triatomic system at various internuclear distances and thus to construct the LEP surface.

The LEP method gives useful estimates of activation energy, but it produces the interesting result (for the H + H_2 system) that there is a shallow basin at the transition state; on a reaction coordinate diagram this would be seen as a slight dip at the top of the energy barrier. This basin implies that the triatomic species at the transition state has some stability relative to motion in all directions—that it is in some sense an intermediate. Quantum mechanical variational calculations do not reveal this basin, and it is probably an artifact of the LEP procedure. To overcome this defect of the LEP method, Sato replaced the assumption of the constant ratio of coulombic to total energy with an alternative route to the estimation of the coulombic and exchange integrals. Other changes were introduced, so the LEPS method is itself quite arbitrary, but it eliminated the basin at the top of the energy barrier. However, the profile along the reaction coordinate of an LEPS surface reveals that the barrier is exceptionally "thin" and, therefore, suggests more quantum mechanical tunneling than seems appropriate. Other modified LEP methods have been devised in order to eliminate the transition state basin. It appears to be necessary that the ratio of coulombic to total energy approach unity as the internuclear distance increases if the basin is to disappear.[4, p. 41]

Empirical methods are of two types: those that permit potential energy surfaces to be calculated and those that only allow activation energies to be estimated. Laidler[1] has reviewed these. A typical approach is to establish a relationship between experimental activation energies and some other quantity, such as heats of reaction, and then to use this correlation to predict additional activation energies. In Section 5.3 we will encounter a different type of empirical potential energy surface.

Sometimes potential energy surfaces are plotted with "skewed" axes; that is, the r_{AB} and r_{BC} axes meet at an angle less than 90°. This is done so that the relative kinetic energy of the three-body system can be represented by the motion of a single point over the surface. In order to achieve this condition it is necessary that the cross-product terms in the kinetic energy drop out. The calculations have been described[3, pp. 100–7; 10] Because our use of potential energy surfaces is qualitative, we will represent them on rectangular axes.

Let us now turn to the surfaces themselves to learn the kinds of kinetic information they contain. First observe that the potential energy surface of Fig. 5-2 is drawn to be symmetrical about the 45° diagonal. This is the type of surface to be expected for a symmetrical reaction like $H + H_2 = H_2 + H$, in which the reactants and products are identical. The corresponding reaction coordinate diagram in Fig. 5-3, therefore, shows the reactants and products having the same stability (energy) and the transition state appearing at precisely the midpoint of the reaction coordinate.

Another feature revealed by these graphical representations (when they result from quantitative calculations) is the "thickness" of the energy barrier that separates the initial and final states. This is important in those systems for which quantum mechanical tunneling may be significant. According to classical mechanics, a reactant system that does not possess sufficient kinetic energy to surmount the barrier cannot cross the barrier and, thus, is prevented from being transformed into the product state. Quantum mechanics makes a different prediction. One way to consider this is to recall that particles may be associated with a corresponding wavelength by the relationship $\lambda = h/mv$, where h is Planck's constant, and m and v are the mass and velocity of the particle. In effect, if the wavelength λ is comparable to the thickness of the barrier, it is conceivable that the particle can "leak" through the barrier by a diffractionlike mechanism. Because λ is inversely proportional to mass, this effect is chemically important only for the very light particles H^+, H, H^-, and the electron.

Let us now consider a chemical reaction whose initial and final states are different. Then the potential energy surface will not be symmetrical. This geological analogy will be helpful: Suppose the valleys are formed by erosion. Then the valley that has eroded faster (or for a longer time) will be both deeper and longer than the less eroded valley, with the necessary consequence that the saddle between the two valleys is shifted toward the shallower valley. Figure 5-4 shows such a surface on which the reactant valley is longer and deeper than the product valley; clearly the transition state is located closer to the final state than to the initial state as a result of this disparity in stabilities.

Figure 5-5 is the corresponding reaction coordinate diagram. This type of surface or behavior is variously described as repulsive, late downhill, late transition state, or productlike transition state. A further interpretation somtimes made is that for the overall reaction $A + BC \rightarrow AB + C$, if the transition state is productlike, the A–B bond is largely formed in the transition state and has properties similar to those it will have in the final state. Therefore, the decrease in potential energy as the system passes over the transition state downward to the final state is largely manifested as an increase in the distance between B and C, that is, as translational energy.

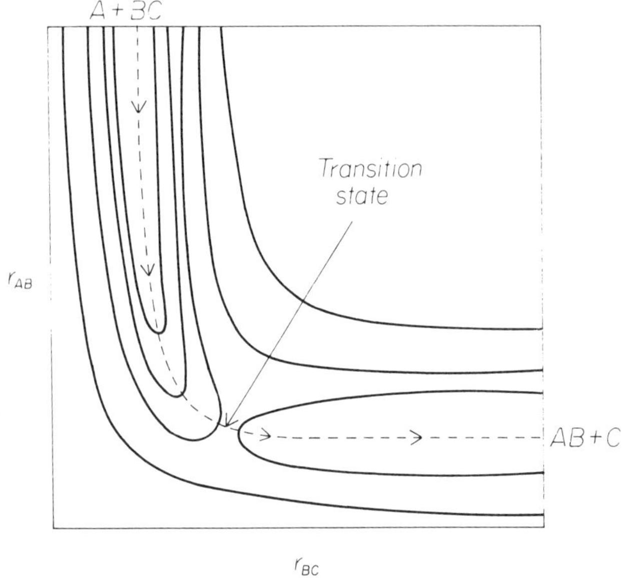

Figure 5-4. Potential energy surface with a late transition state.

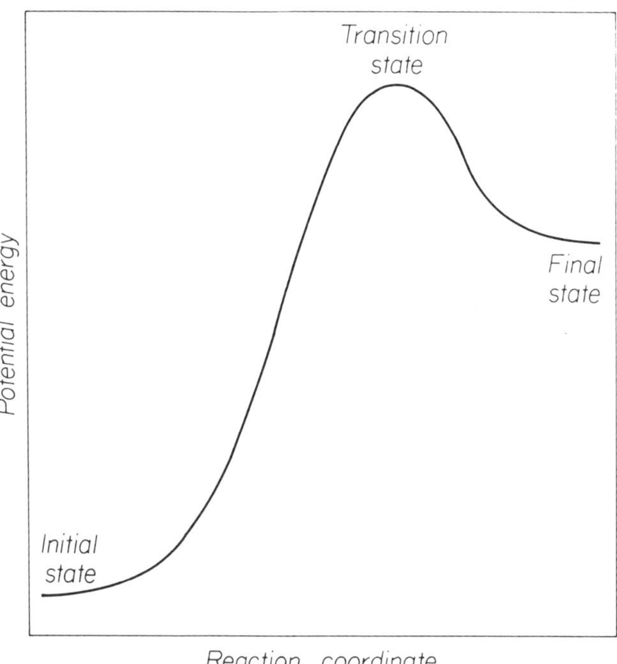

Figure 5-5. Reaction coordinate diagram corresponding to Fig. 5-4, showing that the initial state is more stable than the final state and the transition state is productlike.

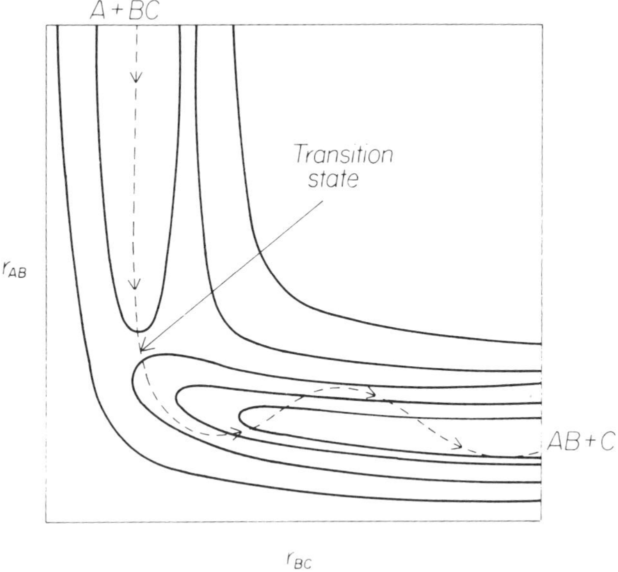

Figure 5-6. Potential energy surface with an early transition state.

Now see Fig. 5-6, in which the product valley is deeper than the reactant valley. This is called an *attractive, early downhill, early transition state,* or *reactantlike transition state surface*. Figure 5-7 is the corresponding reaction coordinate diagram. In this case the transition state is reactantlike, so the A–B distance is much greater than it is destined to be in the final state. Consequently, as the system descends the potential energy surface into the product valley, the A–B bond length undergoes adjustment, which may be taken to mean that part of the translational energy of the initial state is converted to vibrational energy in the final state. The trajectory shown in Fig. 5-6 is intended to convey this meaning.

This question of the location of the transition state on the reaction coordinate is a central issue in the study of reaction mechanisms, and we will return to it in Section 5.3.

A further point is that owing to the kinetic energy of the particle (whose trajectory represents the motion of the chemically reacting system), it is possible for the trajectory to deviate somewhat from the path of minimum potential energy as it crosses the saddle point. The centrifugal force on the particle may carry it on a course that results in the actual energy change being somewhat different from the potential energy barrier height.

Because the accurate calculation of a potential energy surface from first principles for a reaction in solution is—may always be—impracticable, it seems best for our purposes to regard such a construct as a qualitative guide, a heuristic concept, capable of leading us to insights such as the meaning of activation energy and allowing us to think about what happens to a chemically reacting system between

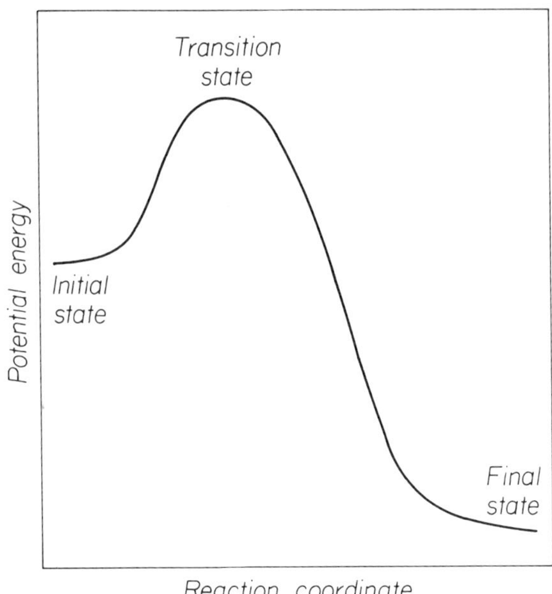

Figure 5-7. Reaction coordinate diagram for Fig. 5-6. The final state is more stable than the initial state, and the transition state is reactantlike.

the initial and final states. In Section 5.3 we explore these uses. Another fruitful result of the potential energy surface concept has been the development of a quantitative theory for the rate constant; this is the subject of Section 5.2.

5.2 TRANSITION STATE THEORY

Shortly after the development of the potential energy surface concept, several authors, most notably Evans, Polanyi, and Eyring, developed quantitative theories of reaction rates based on the idea of the critical role of the transition state in controlling the rate; Glasstone et al.[3] review the early developments. The chemical species existing in the transition state region of the reaction coordinate was named the *activated complex,* so the theory is sometimes called *activated complex theory.* In an alternative terminology, *transition state* is interpreted to mean both the region at the maximum on the reaction coordinate diagram and the species that inhabits this region; thus we have the *transition state theory.* We will use the transition state terminology, despite its imprecision, because the word *complex* tends to be overused in chemistry. Transition state theory has also been called *absolute reaction rate theory.*

Assumptions of the Theory

Transition state theory is based on these three postulates:

1. In passing from the initial state to the final state over the potential energy surface, the reacting system must traverse a region of the reaction path, called the *transition state*, whose potential energy is the highest energy on the path. Figures 5-2 to 5-7 illustrate this idea.
2. The chemical species in the transition state is in equilibrium with the reactant state. This assumption is discussed below.
3. The rate of reaction is equal to the product of the concentration of transition state species formed from the reactant state and the frequency with which this species passes on to the product state.

These statements refer to an elementary reaction, which from this point of view may be defined as a reaction possessing a single transition state. A complex reaction is then a set of elementary reactions, the potential energy surface of the whole being continuous. Thus, for two consecutive reactions the product of the first reaction is the reactant of the second. Each reaction has its own transition state.

Let us examine the equilibrium assumption of transition state theory. Consider a reversible elementary reaction at equilibrium. Because the initial and final states are at equilibrium, assuredly the transition state is in equilibrium with each of these. (It follows that for a reaction at equilibrium, transition state theory is exact insofar as the equilibrium assumption is concerned.)

Now suppose that, from this equilibrium situation, the final state is instantaneously removed. The production of transition state species by the product state will cease. However, the production of transition state species by the reactant state is unaffected by this suppression of the final state, and, according to the third postulate of the theory, the rate of reaction is a function of the transition state concentration formed from the reactant state. This is the usual argument for the equilibrium assumption. Despite its apparent artificiality, the equilibrium assumption is generally considered to be fairly sound, with the possible exception of its application to very fast reactions.[1]

The Partition Function

The derivation of the transition state theory expression for the rate constant requires some ideas from statistical mechanics, so we will develop these in a digression. Consider an assembly of molecules of a given substance at constant temperature T and volume V. The total number N of molecules is distributed among the allowed quantum states of the system, which are determined by T, V, and the molecular structure. Let n_i be the number of molecules in state i having energy ε_i per molecule. Then n_i is related to ε_i by Eq. (5-17), which is known as the *Boltzmann distribution*.

$$n_i = \lambda e^{-\varepsilon_i/kT} \tag{5-17}$$

In Eq. (5-17), k is the Boltzmann constant; k is $R/N_A = 1.380 \times 10^{-16}$ erg $K^{-1} = 1.380 \times 10^{-23}$ J K^{-1}. The parameter λ is the *absolute activity*;[11] it depends upon temperature, but is a constant for all the i quantum states of the system. The absolute activity is related to the chemical potential μ (per molecule) by

$$\lambda = e^{\mu/kT}$$

so $\mu = kT \ln \lambda$.

Figure 5-8 is a plot of Eq. (5-17), showing how the number of molecules having energy ε_i decreases as ε_i increases (at constant T). This plot also reveals that kT is a natural unit of energy on the molecular scale (RT on the molar scale).

Let us now ask what fraction of the total number N of molecules is in the quantum state having energy ε_j. We define this fraction by

$$f_j = \frac{n_j}{N} = \frac{n_j}{\Sigma n_i}$$

or

$$f_j = \frac{e^{-\varepsilon_j/kT}}{\sum_i e^{-\varepsilon_i/kT}} \qquad (5\text{-}18)$$

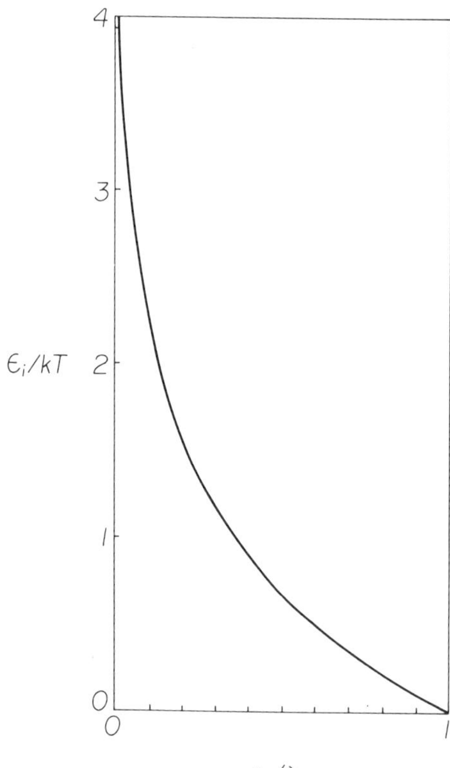

Figure 5-8. Plot of Eq. (5-17), the Boltzmann distribution. n_i is the number of molecules having energy ε_i.

The fraction f_j may also be interpreted as the probability that a molecule will be in the state having energy ε_j.

The denominator in Eq. (5-18) is extremely important, because it represents the distribution of molecules over all of the states available to them. We therefore distinguish it with the symbol Q and the name *partition function*.

$$Q = \sum_i e^{-\varepsilon_i/kT} \tag{5-19}$$

The term *partition function* conveys the idea of distribution over states; the German word is Zustandsumme, sum of states. From the above relationships, we have $N = \lambda Q$ and $\mu = kT \ln (N/Q)$.

We now explore some properties of the partition function. A molecule possesses electronic, vibrational, rotational, and translational energies, and with the assumption that the several contributions to the total energy are independent we will find that the total partition function is factorable.[12,13] To make the argument, suppose the system can be described in terms of two subsystems (e.g., vibrational and rotational) a and b, each having two energy levels, 1 and 2. A molecule may be in the state described by ε_{a1} or ε_{a2}, and at the same time be in the state having energy ε_{b1} or ε_{b2}, so the possible total energy states are $(\varepsilon_{a1} + \varepsilon_{b1})$, $(\varepsilon_{a1} + \varepsilon_{b2})$, $(\varepsilon_{a2} + \varepsilon_{b1})$, $(\varepsilon_{a2} + \varepsilon_{b2})$, and the system partition function is

$$Q = e^{-(\varepsilon_{a1}+\varepsilon_{b1})/kT} + e^{-(\varepsilon_{a1}+\varepsilon_{b2})/kT} + e^{-(\varepsilon_{a2}+\varepsilon_{b1})/kT} + e^{-(\varepsilon_{a2}+\varepsilon_{b2})/kT}$$

This can be written

$$Q = e^{-\varepsilon_{a1}/kT} e^{-\varepsilon_{b1}/kT} + e^{-\varepsilon_{a1}/kT} e^{-\varepsilon_{b2}/kT} + e^{-\varepsilon_{a2}/kT} e^{-\varepsilon_{b1}/kT} + e^{-\varepsilon_{a2}/kT} e^{-\varepsilon_{b2}/kT}$$

or

$$Q = (e^{-\varepsilon_{a1}/kT} + e^{-\varepsilon_{a2}/kT})(e^{-\varepsilon_{b1}/kT} + e^{-\varepsilon_{b2}/kT}) \tag{5-20}$$

Let us define partition functions for the subsystems a and b:

$$q_a = \sum e^{-\varepsilon_{ai}/kT}; \qquad q_b = \sum e^{-\varepsilon_{bi}/kT} \tag{5-21}$$

Comparing Eqs. (5-20) and (5-21),

$$Q = q_a q_b \tag{5-22}$$

This argument can be generalized to any number of subsystems and energy levels. For the case of a molecular system in a given electronic state, the factorization into translational, vibrational, and rotational contributions gives

$$Q = q_{tr} q_{vib} q_{rot} \tag{5-23}$$

Thus far we have specified the energy level relative to some arbitrary zero level. It is common, however, to represent energies as the difference between the ith level and the zeroth level for the molecules. We can write $\varepsilon_i = (\varepsilon_i - \varepsilon_0) + \varepsilon_0$, thus obtaining $\exp(-\varepsilon_i) = \exp(-\varepsilon_0)\cdot\exp[-(\varepsilon_i - \varepsilon_0)]$. Therefore, the quantity $\exp(-\varepsilon_0)$ can be factored out of each term in the partition function, giving

$$Q = e^{-\varepsilon_0/kT} \sum e^{-(\varepsilon_i - \varepsilon_0)/kT} = e^{-\varepsilon_0/kT} Q' \tag{5-24}$$

Q and Q' are both partition functions, but they are referred to different energy levels.

The partition functions q_{tr}, q_{vib}, and q_{rot} for use in Eq. (5-23) can be evaluated by quantum mechanical arguments. We will subsequently require q_{vib}, which is given by Eq. (5-25), where v is the vibrational frequency.

$$q_{vib} = \frac{1}{1 - e^{-hv/kT}} \tag{5-25}$$

The translational partition function is

$$q_{tr} = \left(\frac{2\pi mkT}{h^3}\right)^{3/2} V \tag{5-26}$$

where V is the system volume. We will not require q_{rot}.

Statistical mechanics, via the partition function, provides a route to the calculation of equilibrium constants. Take as an example this reversible reaction:

$$A + B \rightleftharpoons Z$$

The equilibrium constant is defined (on a concentration basis)

$$K = \frac{c_Z}{c_A c_B} = \frac{n_Z V}{n_A n_B} \tag{5-27}$$

where n_A, n_B, n_Z are numbers of molecules contained in volume V. These numbers can be expressed in terms of partition functions by making use of relationships already presented:

$$n_Z = \lambda_Z e^{-\varepsilon_0^Z/kT} Q_Z \tag{5-28a}$$

$$n_A = \lambda_A e^{-\varepsilon_0^A/kT} Q_A \tag{5-28b}$$

$$n_A = \lambda_B e^{-\varepsilon_0^B/kT} Q_B \tag{5-28c}$$

where Q_Z, Q_A, Q_B is each referenced to its zeroth energy level. Combining Eqs. (5-27) and (5-28)

$$K = \frac{\lambda_Z}{\lambda_A \lambda_B} \frac{Q_Z V}{Q_A Q_B} e^{-\Delta\varepsilon_0/kT} \tag{5-29}$$

where $\Delta\varepsilon_0 = \varepsilon_0^Z - \varepsilon_0^A - \varepsilon_0^B$. From Eq. (5-26) for the translational partition function we factor out V from each partition function, obtaining

$$K = \frac{\lambda_Z}{\lambda_A{}^\gamma_B} \frac{Q_Z}{Q_A Q_B} e^{-\Delta\varepsilon_0/kT} \tag{5-30}$$

where we understand that Q_Z, Q_A, Q_B are to be evaluated without inclusion of the volume.[3, p. 182]

Now, the condition for equilibrium is $d\mu = 0$, or, for this reaction, $\mu_A + \mu_B = \mu_Z$. Because $\mu = kT \ln \lambda$, we obtain, at equilibrium, $\lambda_A \lambda_B = \lambda_Z$. This condition applied to Eq. (5-30) gives

$$K = \frac{Q_Z}{Q_A Q_B} e^{-\Delta\varepsilon_0/kT} \tag{5-31}$$

It is more common to express chemical quantities on a molar basis, and Eq. (5-31) becomes

$$K = \frac{Q_Z}{Q_A Q_B} e^{-E_0/RT} \tag{5-32}$$

The Rate Equation

This derivation will be based on a bimolecular reaction, Scheme I.

$$A + B \rightleftharpoons M^\ddagger \rightarrow Z$$

Scheme I

Here Z represents the reaction products. $M^\ddagger$ is the transition state; the double dagger symbol will always signify a quantity or structure relating to the transition state. Scheme I incorporates the equilibrium assumption by writing the conversion of the initial state into the transition state as an equilibrium. This assumption then allows us to apply statistical mechanics to the rate problem; making use of Eq. (5-32), we have

$$K^{TS} = \frac{Q_M}{Q_A Q_B} e^{-E_0/RT} \tag{5-33}$$

The critical portion of the progress of the chemically reacting system over the potential energy surface is in the transition state region. The transition state is regarded as a normal molecule in almost all aspects, but it is very special with respect to motion along the reaction coordinate, for this motion transforms it into product. This special motion is considered to be a bond vibration, which in the transition state becomes an extremely "loose" (i.e., low-frequency) vibration. We will factor this vibrational mode out of the vibrational partition function and take the limit as the frequency goes to zero, thus

$$\lim_{v \to 0} \frac{1}{1 - e^{-hv/kT}} = \frac{1}{1 - (1 - hv/kT)} = \frac{kT}{hv} \tag{5-34}$$

where the approximation $e^x \approx 1 + x$ has been used. Now Q_M is written

$$Q_M = Q^{\ddagger} \left(\frac{kT}{hv} \right) \tag{5-35}$$

$Q^{\ddagger}$ is the partition function of the transition state with this special vibrational mode factored out.

We can combine Eqs. (5-33) and (5-35) and the definition of K^{TS} to obtain

$$K^{TS} = \frac{c^{\ddagger}}{c_A c_B} = \frac{kT}{hv} \cdot \frac{Q^{\ddagger}}{Q_A Q_B} e^{-E_0/RT} \tag{5-36}$$

where $c^{\ddagger}$ is the concentration of transition state species. This equation is rearranged to

$$vc^{\ddagger} = c_A c_B \frac{kT}{h} \frac{Q^{\ddagger}}{Q_A Q_B} e^{-E_0/RT} \tag{5-37}$$

However, one of the postulates of transition state theory is that the rate of reaction is equal to the product of the transition state species concentration and the frequency of their conversion to products, so the theoretical rate equation is

$$v = c_A c_B \frac{kT}{h} \frac{Q^{\ddagger}}{Q_A Q_B} e^{-E_0/RT} \tag{5-38}$$

The experimental rate equation is

$$v = k c_A c_B \tag{5-39}$$

where k is the second-order rate constant. Comparing Eqs. (5-38) and (5-39)

$$k = \frac{kT}{h} \frac{Q^{\ddagger}}{Q_A Q_B} e^{-E_0/RT} \tag{5-40}$$

Equation (5-40) gives the transition state theoretical result for the rate constant.

Equations (5-38) and (5-40) can also be derived by making use of the translational partition function.[1-4]

The quantity kT/h is a fundamental term in transition state theory. The ratio k/h has the value 2.0836×10^{10} $K^{-1}s^{-1}$, and $kT/h = 6.213 \times 10^{12}$ s^{-1} at 25°C.

For gas-phase reactions, Eq. (5-40) offers a route to the calculation of rate constants from nonkinetic data (such as spectroscopic measurements). There is evidence, from such calculations, that in some reactions not every transition state species proceeds on to product; some fraction of transition state molecules may return to the initial state. In such a case the calculated rate will be greater than the observed rate, and it is customary to insert a correction factor κ, called the *transmission coefficient,* in the expression. We will not make use of the transmission coefficient.

If the transition state theory is applied to the reaction of two hard spheres, the result is identical with that of simple collision theory.[3, pp. 16-7] Because transition state theory is an equilibrium theory, it can be inferred that collision theory is also an equilibrium theory.

The preexponential factor of the Arrhenius equation is approximately given by

$$A \approx \frac{kT}{h} \frac{Q^{\ddagger}}{Q_A Q_B} \tag{5-41}$$

Thus, A is apparently temperature dependent, but not highly so, because the partition functions are not very sensitive functions of temperature.

Thermodynamic Interpretation

Let us compare Eqs. (5-33) and (5-40), and define a new equilibrium constant $K^{\ddagger}$ by Eq. (5-42).

$$K^{\ddagger} = \frac{Q^{\ddagger}}{Q_A Q_B} e^{-E_0/RT} \tag{5-42}$$

$K^{\ddagger}$ is a special kind of equilibrium constant because it lacks the partition function contribution for the vibrational motion leading to products. With Eqs. (5-40) and (5-42) we can express transition state theory in its most succinct form:

$$k = \frac{kT}{h} K^{\ddagger} \tag{5-43}$$

Equation (5-43) has the practical advantage over Eq. (5-40) that the partition functions in (5-40) are difficult or impossible to evaluate, whereas the presence of the equilibrium constant in (5-43) permits us to introduce the well-developed ideas of thermodynamics into the kinetic problem. We define the quantities $\Delta G^{\ddagger}$, $\Delta H^{\ddagger}$, and $\Delta S^{\ddagger}$ as, respectively, the standard *free energy of activation, enthalpy of activation, and entropy of activation;* from thermodynamics we now can write

$$\Delta G^{\ddagger} = -RT \ln K^{\ddagger} \tag{5-44}$$

$$\Delta G^{\ddagger} = \Delta H^{\ddagger} - T\Delta S^{\ddagger} \tag{5-45}$$

$$\frac{d \ln K^{\ddagger}}{d(1/T)} = -\frac{\Delta H^{\ddagger}}{R} \tag{5-46}$$

These quantities (which are standard molar quantities) describe the process

$$\text{initial state} \rightleftharpoons \text{transition state}$$

The numerical values of $\Delta G^{\ddagger}$ and $\Delta S^{\ddagger}$ depend upon the choice of standard states; in solution kinetics the molar concentration scale is usually used. Notice (Eq. 5-43) that in transition state theory the temperature dependence of the rate constant is accounted for principally by the temperature dependence of an equilibrium constant.

By combining Eqs. (5-43)–(5.45) we can express the rate constant in these alternative forms:

$$k = \frac{kT}{h} e^{-\Delta G^{\ddagger}/RT} \tag{5-47}$$

$$k = \frac{kT}{h} e^{-\Delta H^{\ddagger}/RT} \cdot e^{\Delta S^{\ddagger}/R} \tag{5-48}$$

The evaluation of the activation parameters $\Delta G^{\ddagger}$, $\Delta H^{\ddagger}$, and $\Delta S^{\ddagger}$ proceeds as follows. From the Arrhenius equation, Eq. (5-1), we have

$$\frac{d \ln k}{d(1/T)} = -\frac{E_a}{R} \tag{5-49}$$

Equation (5-43) gives

$$\ln K^{\ddagger} = \ln k + \ln(1/T) + \ln(h/\mathbf{k})$$

which leads to

$$\frac{d \ln K^{\ddagger}}{d(1/T)} = \frac{d \ln k}{(d(1/T)} + \frac{d \ln(1/T)}{d(1/T)}$$

Therefore,

$$\Delta H^{\ddagger} = E_a - RT \tag{5-50}$$

[This argument neglects the volume change in the process, which makes a negligible contribution in solution reactions.[1, p. 55]] Because E_a is found from an Arrhenius plot, $\Delta H^{\ddagger}$ is accessible with Eq. (5-50). The value of T used in this calculation can be in the middle of the experimental temperature range; this is not a critical matter because the experimental uncertainty in E_a is typically of the order RT (which is 0.6 kcal mol^{-1} at room temperature).

The entropy of activation is obtained via Eq. (5-48):

$$\Delta S^{\ddagger} = \Delta H^{\ddagger}/T - R \ln(T/k) - R \ln(k/h) \qquad (5\text{-}51)$$

where k is the rate constant at temperature T. Finally $\Delta G^{\ddagger}$ can be calculated with Eq. (5-45), or it can be found by means of Eq. (5-47).

In Eq. (5-50) we see that E_a is related to $\Delta H^{\ddagger}$. By combining Eqs. (5-1) and (5-48) we find that A is related to $\Delta S^{\ddagger}$ by Eq. (5-52).

$$\Delta S^{\ddagger} = R \ln A - R \ln(kT/h) - R \qquad (5\text{-}52)$$

The formulation of transition state theory has been in terms of reactant and transition state concentrations; let us now define an equilibrium constant $K_0^{\ddagger}$ in terms of activities.

$$K_0^{\ddagger} = \frac{a^{\ddagger}}{a_A a_A} = K^{\ddagger} \frac{\gamma^{\ddagger}}{\gamma_A \gamma_B} \qquad (5\text{-}53)$$

Then with Eq. (5-43) we get

$$k = k_0 \frac{\gamma_A \gamma_B}{\gamma^{\ddagger}} \qquad (5\text{-}54)$$

where $k_0 = (kT/h)K_0^{\ddagger}$; therefore, k_0 is the value of the rate constant in the reference state of unit activity coefficients. Equation (5-54) provides an interpretation of the nonideal behavior of rates. The activity coefficient $\gamma^{\ddagger}$ applies to the transition state species, and it has the usual properties of an activity coefficient. It is unusual, however, in that it applies to a specific reaction; unlike activity coefficients in equilibrium systems, it cannot be measured in one reaction and applied to another.[4, p. 409] $\gamma^{\ddagger}$ cannot be determined by most of the usual physical methods, though it can be estimated with Eq. (5-54) if γ_A and γ_B are known. Analogy with stable compounds of closely related structure may also provide estimates of $\gamma^{\ddagger}$.[14] The dependence of transition state activity coefficients on solution acidity has been measured for acid-catalyzed reactions.[15]

5.3 CHEMICAL INTERPRETATIONS OF THE TRANSITION STATE

Reaction Coordinate Diagrams

Thus far we have drawn reaction coordinate diagrams with potential energy as the ordinate. However, the free energy is a much more accessible quantity (actually *differences* in free energy, of course), so we will now construct these diagrams using the free energy of the system. The free energy of activation, $\Delta G^{\ddagger}$, is related to the rate constant by Eq. (5-47) and the standard free energy change is related to the equilibrium constant by the thermodynamic result

$$\Delta G^{0} = -RT \ln K \tag{5-55}$$

With these relationships and the appropriate experimental data we can plot reaction coordinate diagrams that are quantitatively useful in displaying the free energy differences between states. Figure 5-9 is an example, the data being drawn from Table 4-3, System 2. For this reversible reaction,

$$A + B \underset{k_{-1}}{\overset{k_{1}}{\rightleftharpoons}} Z$$

the kinetic results are $k_1 = 1.7 \times 10^5$ M^{-1}s^{-1} and $k_{-1} = 2.6 \times 10^2$ s^{-1}, or $K = k_1/k_{-1} = 6.5 \times 10^2$ M^{-1}. With Eq. (5-47) we calculate $\Delta G^{\ddagger}_{1} = 9.91$ kcal mol^{-1} and $\Delta G^{\ddagger}_{-1} = 13.61$ kcal mol^{-1}; Eq. (5-55) gives -3.69 kcal mol^{-1} (which is

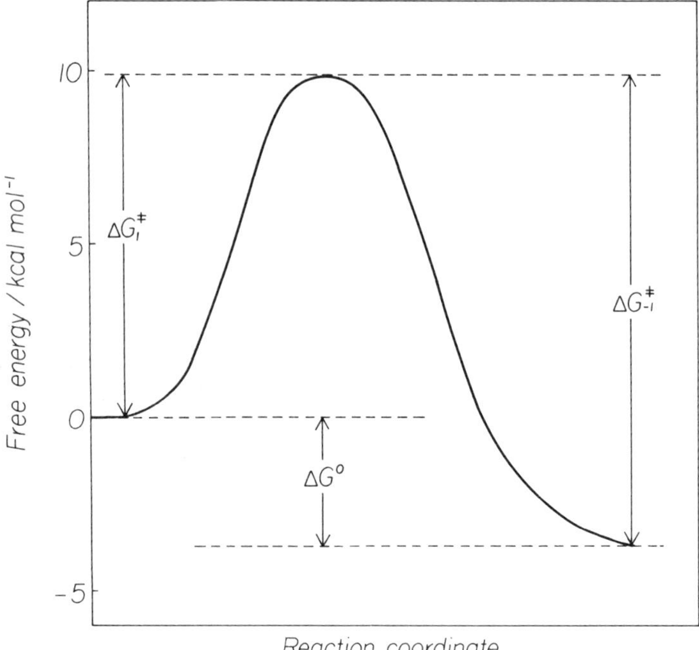

Figure 5-9. Free energy reaction coordinate diagram for System 2 of Table 4-3, the formation of a cyclodextrin inclusion complex.

evidently the difference between $\Delta G_1^{\ddagger}$ and $\Delta G_{-1}^{\ddagger}$). These quantities are plotted in Fig. 5-9, where the zero of free energy is arbitrarily assigned to the initial state.

Several features of this treatment are notable. First, the calculations treat only the vertical axis, and this only at the defined states (initial, transition, final in this example of an elementary reaction). The curve drawn through these three points has a shape suggested by the potential energy surfaces of Section 5.1, but it has no quantitative significance except at the three defined states. This is a consequence of the thermodynamic impossibility of being able to specify the reaction path (i.e., mechanism) from knowledge only of the initial and final states of a system. Thus, the equilibrium constant K yields ΔG^0, but allows an infinite number of paths between the initial and final states. By stepping outside of thermodynamics and introducing kinetic theory we are enabled to circumvent this limitation and thus to learn something about another point (the transition state) on the reaction path. However, in formulating transition state theory we postulated an equilibrium between the initial and transition states, and in doing this we restricted ourselves from knowing what happens between these two states.

A second point is that we, as yet, have no quantitative basis for the placement of the transition state along the horizontal axis. Figure 5-9 shows the transition state located slightly closer to the initial state than to the final state, in accordance with the argument of Section 5.1, "Potential Energy Surfaces." This problem is dealt with later in the present section.

The potential energy is a mechanical concept and is applicable to an individual particle. The free energy is a thermodynamic concept and is applicable to large numbers of particles. The free energy has the disadvantages that it is a composite quantity ($\Delta G = \Delta H - T\Delta S$) and that it is temperature dependent. (At the absolute zero the potential and free energies would be equal.) In general the location of the maximum along the reaction coordinate will be different for the potential energy and the free energy[1, pp. 76–9] because of the entropic contribution. In making these free energy reaction coordinate diagrams it is important to remember that $\Delta G^{\ddagger}$ is *not* the activation energy for the reaction; combining Eqs. (5-45) and (5-50) gives $\Delta G^{\ddagger} = E_a - T(R + \Delta S^{\ddagger})$. One could (with sufficient experimental effort) construct separate reaction coordinate diagrams for $\Delta H^{\ddagger}$ and $\Delta S^{\ddagger}$. Notice that the magnitudes of $\Delta G^{\ddagger}$ and $\Delta S^{\ddagger}$ depend upon the standard state (i.e., the concentration scale used to express the rate constants). This is especially important if the sign of $\Delta S^{\ddagger}$ is to be used as a mechanistic criterion.

Let us now sketch the reaction coordinate diagram for the complex reaction of Scheme II, where R represents the reactant state, P the products, and I an intermediate.

$$\text{R} \underset{k_{-1}}{\overset{k_1}{\rightleftharpoons}} \text{I} \underset{k_{-1}}{\overset{k_2}{\rightleftharpoons}} \text{P}$$

Scheme II

Figure 5-10 shows this diagram. As sketched, the final state is more stable than the initial state. The new feature of Fig. 5-10 is the presence of the intermediate

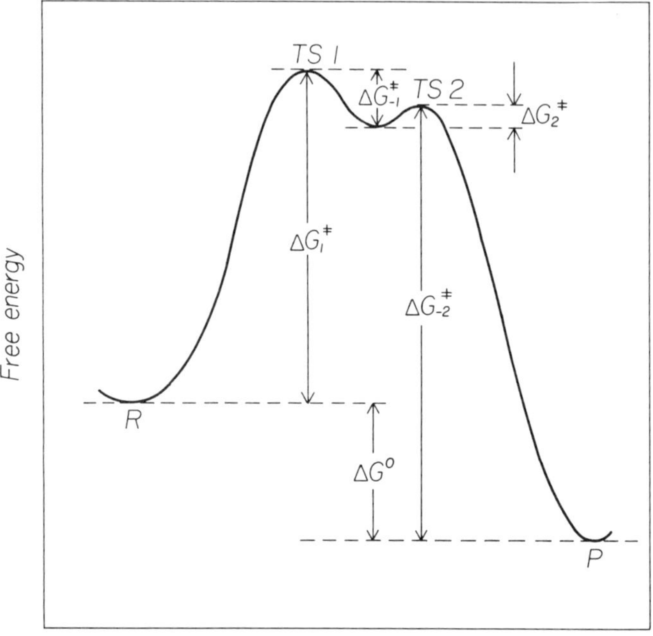

Figure 5-10.   Hypothetical free energy reaction coordinate diagram for Scheme II (TS $\equiv$ transition state).

on the reaction path. This intermediate is unstable with respect to reactants and products, but it is represented by a local energy minimum, which differentiates it from a transition state, which occurs at an energy maximum. This reaction scheme therefore possesses two transition states, one for each elementary reaction. From the lengths of the lines in Fig. 5-10, together with Eq. (5-47), it is seen that $k_2 > k_{-1} > k_1 > k_{-2}$.

The reader may have become unsettled by the treatment given indiscriminately to first-order and second-order rate constants in making these free energy reaction coordinate diagrams, for these rate constants have different units, yet the same equation (5-47) was applied to both. Let us begin with the thermodynamic relationship, Eq. (5-55), which is commonly written in this form. Yet for the example to which it is applied ($K = 650$ M^{-1}), it is not possible to take the logarithm of a unit. We, therefore, create a pure number by algebraic rearrangement, namely, $K/\text{M}^{-1} = 650$, and thus Eq. (5-55) really should be written (for this case) $\Delta G^0 = -RT \ln (K/\text{M}^{-1})$.

To analyze the rate constant problem we start with Eq. (5-43), $k = (kT/h)K^{\ddagger}$. The term (kT/h) has the unit second^{-1}, so consistency is achieved if the concentration units of k and $K^{\ddagger}$ are identical. As before, we pass to pure numbers, writing (for a second-order rate constant)

$$(k/M^{-1}s^{-1}) = (kT/hs^{-1})(K^{\ddagger}/M^{-1})$$

From Eq. (5-44), $\Delta G^{\ddagger} = -RT \ln (K^{\ddagger}/M^{-1})$, so $(K^{\ddagger}/M^{-1}) = \exp(-\Delta G^{\ddagger}/RT)$, a pure number. Taking logarithms of Eq. (5-47),

$$-\Delta G^{\ddagger}/RT = \ln (k/M^{-1}s^{-1}) - \ln (kT/hs^{-1})$$

Therefore Eq. (5-47) is applicable to first-order and to second-order rate constants, it being understood that the arithmetic operations are carried out on pure numbers generated as shown. We have not evaded the requirement of dimensional consistency, which is provided by Eq. (5-43).

This algebraic treatment of units to create pure numbers is also convenient for the unambiguous presentation of data in tables and figures.

The Rate-Determining Step

Consider the series reaction A→B→C. If the first step is very much slower than the second step, the rate of formation of C is controlled by the rate of the first step, which is called the *rate-determining step* (rds), or rate-limiting step, of the reaction. Similarly, if the second step is the slower one, the rate of production of C is controlled by the second step. The slower of these two steps is the "bottleneck" in the overall reaction. This flow analogy, in which the rate constants of the separate steps are analogous to the diameters of necks in a series of funnels, is widely used in illustration of the concept of the rds.

Strictly speaking, the flow analogy is valid only for consecutive irreversible reactions, and it can be misleading if reverse reactions are significant. Even for irreversible reactions the rds concept has meaning only if one of the reactions is much slower than the others. For reversible reactions the free energy reaction coordinate diagram is a useful aid. In Fig. 5-10, for example, the intermediate I is unstable with respect to R and P, and its formation (the k_1 step) is the rds of the overall reaction.

When the overall reaction includes more than two elementary steps, the situation may not be easy to analyze. The product of the nth step is the reactant of the $(n + 1)$st step, but in order for these two states to be represented by the same free energy they must have the same composition; this means that the stoichiometric composition must be constant throughout the entire series of reactions.[16] Suppose that it has been possible to construct the free energy reaction coordinate. Murdoch[17] gives this method for identifying the rds:

1. Label the reactants R, the products P, the intermediates I_1, I_2, . . . in order, and the transition states T_1, T_2, . . . in order.
2. Divide the reaction into sections, the first section beginning with R and ending at the first intermediate (I_i) more stable than R. The second section begins at the end of the first section and ends at the next intermediate more stable than I_i. Continue until P is reached.

3. Calculate the energy difference between the transition state of highest energy in each section and the initial energy of the section.
4. The section having the greatest energy difference contains the rds, which is the step leading to the transition state of highest energy within that section.

Let us apply this method to the hypothetical reaction coordinate diagram of Fig. 5-11, which consists of two sections. The requisite energy differences are for the vertical distances $(T_2 - R)$ and $(T_3 - I_2)$. Because $(T_3 - I_2) > (T_2 - R)$, the second section contains the rds, which must be the $I_2 \rightarrow T_3$ step. Note that T_3 actually has a lower free energy than do T_1 and T_2; it is the change in free energy from the valley at the beginning of the section that determines the rate.

This is an interesting exercise, but we should not become excessively concerned with formal schemes for the identification of the rds. We want to know the rds because it is a piece of information about the reaction mechanism. If we have already acquired so much information about the system that we can construct a reaction coordinate diagram displaying all intermediates and transition states, we probably have no need to specify the rds. As an example of the experimental detection of the rds we will describe Jencks' study of the reaction of hydroxylamine with acetone.[18] The overall reaction is

$$
\underset{\displaystyle CH_3\,C\,CH_3}{\overset{\displaystyle \overset{\textstyle O}{\|}}{}} + NH_2OH \rightarrow \underset{\displaystyle CH_3\,CCH_3}{\overset{\displaystyle \overset{\textstyle N-OH}{\|}}{}} + H_2O
$$

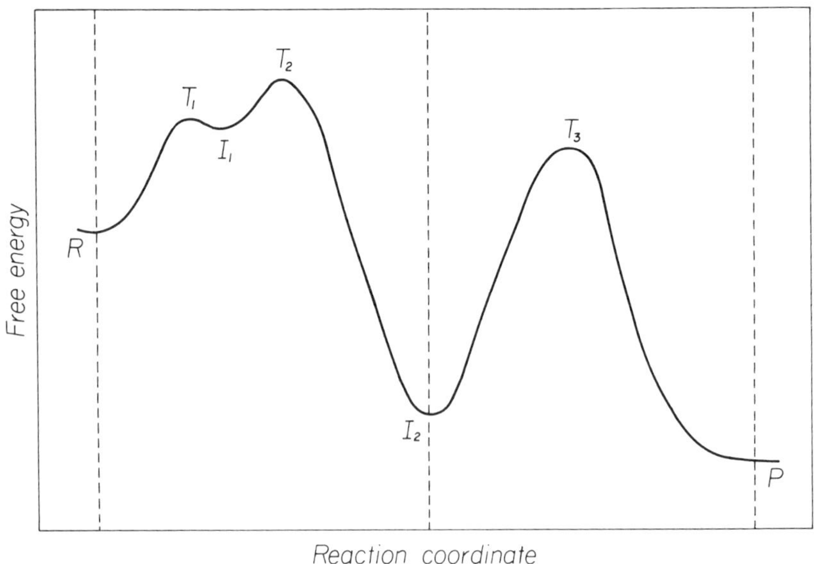

Figure 5-11. Demonstration of the rds concept. The $I_2 \rightarrow T_3$ step is the rds.

The reaction is known to be complex, proceeding through a tetrahedral intermediate called a *carbinolamine*. Generalizing to the reactions of amines with carbonyls, the two-step reaction sequence is

$$\diagup\!\!\!\diagdown C{=}O \;+\; RNH_2 \;\rightleftharpoons\; C\diagup\!\!\!\diagdown \begin{smallmatrix}OH\\NHR\end{smallmatrix} \;\rightleftharpoons\; \diagup\!\!\!\diagdown C{=}NR \;+\; H_2O$$

Scheme III

When acetone is treated with hydroxylamine in aqueous solution near neutral pH, the carbonyl UV absorption intensity decreases very rapidly; this fast spectral change is followed by a much slower absorption increase that is due to the appearance of the oxime product. This suggests that, at such pH values, the initial addition is very rapid and the second step, dehydration of the carbinolamine, is the rds. Figure 5-12 is a plot of the apparent first-order rate constant against pH for this reaction. As the pH is decreased from neutrality, the rate increases, indicating that the rds

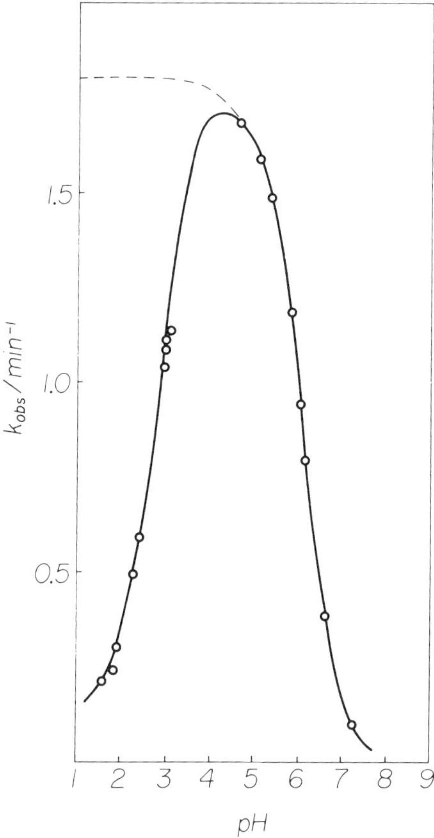

Figure 5-12. pH-rate profile for the reaction of hydroxylamine with acetone in water at 25°C. Dashed line: rate of acid-catalyzed dehydration step; solid line: observed rate.

is acid catalyzed. The rate increase produced by this acid catalysis does not continue indefinitely, however, because it eventually is exactly compensated by the protonation of hydroxylamine, whose conjugate acid is unreactive. Thus, the rate of the second step reaches (at low pH) a limiting value that is independent of pH (the dashed line in Fig. 5-12).

However, as the pH–rate plot shows, at very low pH the observed rate actually decreases. Because, as the preceding argument shows, rate-determining dehydration should result in a pH-dependent rate at low pH, this decreased rate must mean that the rds has changed. This is reasonable, for at pH values well below the pK_a of hydroxylamine, the decreasing proportion of the hydroxylamine in the unprotonated form will decrease the rate of the initial addition. At some pH, then, the rate of the addition step will fall below that of the dehydration step, and the observed rate curve will lie lower than the rate predicted for the dehydration.

Composition of the Transition State

Perhaps the single most important piece of information to be derived from a kinetic study is the composition of the transition state. The basis of the inference can be developed by means of this elementary bimolecular reaction,

$$A + B \rightleftharpoons M^{\ddagger} \rightarrow Z$$

where $M^{\ddagger}$ is the transition state. We define $K^{\ddagger} = c_M{}^{\ddagger}/c_A c_B$, and from transition state theory we have $v = (kT/h)c_M^{\ddagger}$, giving

$$v(\text{theory}) = \left(\frac{kT}{h}\right)K^{\ddagger}c_A c_B \tag{5-56}$$

However, for this elementary reaction the experimental rate equation is

$$v(\text{expt}) = kc_A c_B \tag{5-57}$$

We, therefore, conclude that the concentration dependence of the experimental rate gives the composition of the transition state; in this example the transition state is composed of one molecule of A and one of B, for the experimental rate constant is first-order in each reactant.

This argument can be extended to consecutive reactions having a rate-determining step.[3, p. 200] The composition of the transition state of the rds is given by the rate equation. This composition includes reactants prior to the rds, but nothing following the rds. Thus, the rate equation may not correspond to the stoichiometric equation. We will consider several examples. In Scheme IV a fast acid–base equilibrium precedes the slow rds.

$$\text{A} + \text{H}^+ \overset{\text{fast}}{\rightleftharpoons} \text{AH}^+$$

$$\text{AH}^+ \overset{k_2}{\underset{\text{slow}}{\rightarrow}} \text{Z}$$

Scheme IV

Writing $K = c_{\text{AH}^+}/c_{\text{A}}c_{\text{H}^+}$, *we have*

$$v = k_2 c_{\text{AH}^+} = k_2 K c_{\text{A}} c_{\text{H}^+} \tag{5-58}$$

Thus, the transition state includes one A and one H^+, but the kinetics says nothing about the charge distribution or spatial configuration of these components; this is the basis of the problem of kinetic indistinguishability that we discussed in Section 3.3.

Another problem is revealed by Scheme V.

$$\text{A} + \text{H}^+ \overset{\text{fast}}{\rightleftharpoons} \text{AH}^+$$

$$\text{AH}^+ + \text{H}_2\text{O} \overset{k_2}{\underset{\text{slow}}{\rightarrow}} \text{Z}$$

Scheme V

For this scheme we again obtain Eq. (5-58), because the water (solvent) concentration is essentially constant and so is absorbed into the rate constant. Thus, the rds is bimolecular, but the rate equation is first-order; the role of the solvent in the transition state is not evident from the rate equation.

Kinetic evidence in the hydroxide-catalyzed halogenation of acetone provides definitive mechanistic information. The overall reaction is

$$\text{OH}^- + \text{CH}_3\overset{\overset{\text{O}}{\|}}{\text{C}}\text{CH}_3 + \text{X}_2 \rightarrow \text{CH}_3\overset{\overset{\text{O}}{\|}}{\text{C}}\text{CH}_2\text{X} + \text{X}^- + \text{H}_2\text{0}_2$$

The rate equation is first-order in acetone, first-order in hydroxide, but it is independent of (i.e., zero order in) the halogen X_2. Moreover, the rate is the same whether X_2 is chlorine, bromine, or iodine. These results can only mean that the transition state of the rds contains the elements of acetone and hydroxide, but not of the halogen, which must enter the product in a fast reaction following the rds. Scheme VI satisfies these kinetic requirements.

$$\text{CH}_3\text{COCH}_3 + \text{OH}^- \overset{\text{slow}}{\rightleftharpoons} \text{CH}_3\text{COCH}_2^- + \text{H}_2\text{O}$$

$$\text{CH}_3\text{COCH}_2^- + \text{X}_2 \overset{\text{fast}}{\rightarrow} \text{CH}_3\text{COCH}_2\text{X} + \text{X}^-$$

Scheme VI

Just as the rate equation gives the transition state composition but not its structure, neither does it tell us the order in which the components were assembled. Thus, Schemes VII and VIII both give the same rate equation, $v = kK_1K_2c_Ac_B\,c_Dc_F$, so this equation correctly identifies the rds transition state composition, but it cannot distinguish between these reaction schemes (or others that can be drawn).

$$\text{A} + \text{B} \overset{\text{fast}}{\rightleftharpoons} \text{C}$$

$$\text{C} + \text{D} \overset{\text{fast}}{\rightleftharpoons} \text{E}$$

$$\text{E} + \text{F} \overset{\text{slow}}{\rightarrow} \text{Z}$$

Scheme VII

$$\text{A} + \text{F} \overset{\text{fast}}{\rightleftharpoons} \text{E}$$

$$\text{E} + \text{B} \overset{\text{fast}}{\rightleftharpoons} \text{C}$$

$$\text{C} + \text{D} \overset{\text{slow}}{\rightarrow} \text{Z}$$

Scheme VIII

All of these examples have shown rate equations in which each reactant appears to the first-order. In Section 2.2 we obtained this rate equation from kinetic data:

$$v = k[\text{cinnamoylimidazole}]\,[n\text{-butylamine}]^2$$

We conclude that the rds transition state includes the elements of one cinnamoylimidazole and two butylamine molecules, but we do not know anything about their assembly. However, because a termolecular collision is very improbable, we are justified in supposing that this is a complex reaction, the three molecules having been brought together in stepwise fashion.

When we expand our scope to consider complex reactions with slow reversible steps or an rds that is not of the bottleneck type, the interpretation of the rate equation may be less clear-cut. Suppose Scheme IX is applicable, with the steady-state approximation being justifiable.

$$A + B \underset{k_{-1}}{\overset{k_1}{\rightleftharpoons}} C$$

$$C + D \overset{k_2}{\rightarrow} Z$$

Scheme IX

Then we find

$$c_C = \frac{k_1 c_A c_B}{k_{-1} + k_2 c_D}$$

and the rate equation is

$$v = \frac{k_1 k_2 c_A c_B c_D}{k_{-1} + k_2 c_D} \tag{5-59}$$

The appearance of c_D in the denominator means that D is coupled to a reversible step prior to the rds.[19] If k_1 and k_{-1} were so large that the fast preequilibrium assumption is valid, then the c_D term in the denominator would drop out, and we would have $v = k_2 K c_A c_B c_D$, giving the composition of the rds transition state. If k_2 is very much larger than k_1 and k_{-1}, Eq. (5-59) becomes $v = k_1 c_A c_B$; the first step is now the rds, and the rate equation gives the transition state composition.

Parallel reactions give rate equations having sums of rate terms. Each term provides the transition state composition for a reaction path. For example, some acid-catalyzed reactions have the rate equation

$$v = k_1[S][H^+] + k_2[S][HA]$$

where S is the substrate (reactant) and HA is a weak acid. The k_1 term describes *specific acid catalysis,* and its transition state composition is given by this term. The k_2 term describes *general acid catalysis;* its transition state includes the weak acid molecule, not just the hydrogen ion. These are two separate reactions, proceeding concurrently.

In a special circumstance the rate equation for parallel reactions may be misleading.[20,21] If two parallel reactions are catalyzed by a common catalyst, and if a significant fraction of the catalyst is tied up in the form of intermediates, then the two reactions are not independent, and the rate equation will not give the transition state composition. King has analyzed this case in terms of enzyme-catalyzed reactions.[20]

There is a kinetic source of information about the transition state in addition to the rate equation—this is $\Delta S^{\ddagger}$, the entropy of activation. The entropy change for a process is a measure of the change in disorder or "looseness"; a decrease in entropy reflects a decrease in disorder or motional modes, whereas an entropy increase indicates a corresponding increase in disorder (decrease in constraint) of the system. Therefore, we might expect $\Delta S^{\ddagger}$ for a unimolecular reaction to be approximately zero, because one particle in the initial state is transformed to one particle in the transition state. For a bimolecular reaction, two reactant particles are converted to a single transition state particle, with a loss of disorder or "freedom," so $\Delta S^{\ddagger}$ should be negative. This criterion has been applied to determine the molecularity of reactions, such as solvolyses, in which the rate law provides no information about participation of the solvent in (or prior to) the rds. Schemes IV (unimolecular reaction, the A-1 mechanism of acid catalysis) and V (bimolecular, A-2 mechanism) are examples in which the entropy of activation may be a useful quantity. Schaleger and Long[22] have reviewed this application of $\Delta S^{\ddagger}$. For many of these reactions (whose mechanisms are believed to be known on the basis of other information), the A-1 reactions give $\Delta S^{\ddagger} \approx 0$ to 10 eu, whereas A-2 reactions give $\Delta S^{\ddagger} \approx -15$ to -30 eu. (The *entropy unit,* or eu, is equal to 1 cal K^{-1} mol^{-1}. These $\Delta S^{\ddagger}$ values are on the molar concentration scale.) This criterion is not definitive, however, because it omits consideration of other factors, such as changes in solvation between the initial and transition states, that could affect $\Delta S^{\ddagger}$. Nevertheless, it is usually found that bimolecular reactions involving the solvent in the rds have negative entropies of activation, and unimolecular reactions not involving the solvent have positive or very small negative values of the entropy of activation.[22,23]

The standard state chosen for the calculation of $\Delta S^{\ddagger}$ controls its magnitude and even its sign. The standard state is established when the concentration scale is selected. For most solution kinetic work the molar concentration scale is used, so $\Delta S^{\ddagger}$ values reported by different workers are usually comparable. Nevertheless, an important chemical question is implied: Because the sign of $\Delta S^{\ddagger}$ may depend upon the concentration scale used for the evaluation of the rate constant, which concentration scale should be used when $\Delta S^{\ddagger}$ is to serve as a mechanistic criterion? The same question appears in studies of equilibria. The answer (if there is a single answer) is not known, though some analyses of the problem have been made.[24–26] Further discussion of this issue is given in Section 6.1.

Position and Height of the Energy Barrier

If, for an elementary reaction, we know the location of the transition state along the reaction coordinate and also the height of the energy barrier (relative to the energies of the initial and final states), then we know a great deal about the mechanism of the reaction. Much theoretical attention has, therefore, been directed to establishing the factors that control these measures and to the possibility that they may be related. In 1955 Hammond[27] made the proposal that if two states occurring consecutively on a reaction coordinate have very similar energies, their interconversion will require only small changes in their structures. This *Hammond postulate* is usually applied in this extended form: The transition state of an elementary reaction

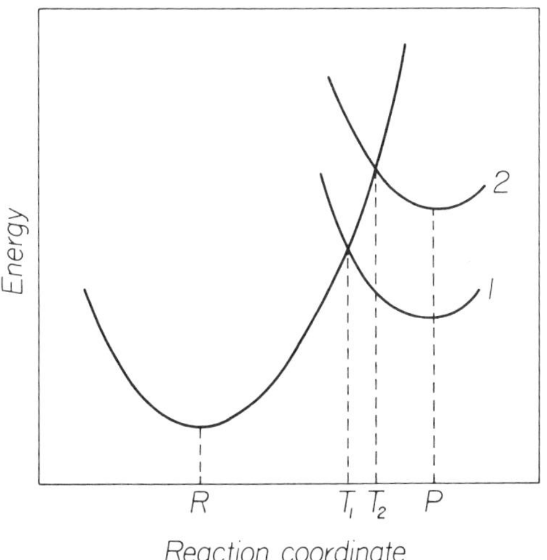

Figure 5-13 Demonstration of the Hammond postulate with harmonic potential wells.

lies closer on the reaction coordinate to that state (initial or final) having the higher energy. The chemical basis of the Hammond postulate is that if two species (related by chemical transformation) have similar energies, they probably have similar structures. This postulate is restricted to adiabatic pathways (see Section 5.1, "Potential Energy Surfaces"), for which it is reasonable to expect structures along the reaction path to be describable as combinations of the initial and final state structures. Hammond also restricted the postulate to potential energy functions, which do not include entropic contributions, but it is commonly applied to free energies.

The Hammond postulate is a valuable criterion of mechanism, because it allows a reasonable transition state structure to be drawn on the basis of knowledge of the reactants and products and of energy differences between the states (i.e., $\Delta G^{\ddagger}$ and ΔG^0). Throughout this chapter we have located transition states in accordance with the Hammond postulate.

Several qualitative demonstrations of the Hammond postulate have been given. In Section 5.1 we used a geological analogy to the potential energy surface to reach the same conclusion as the Hammond postulate. Another viewpoint is shown in Fig. 5-13, which represents the potential energy surfaces in the vicinities of the reactant and product as parabolic wells describing harmonic oscillators. The transition state occurs at the intersection of the two potential functions.[28] Consider, in Fig. 5-13, the intersection of the reactant parabola with product parabola 1; in this case T_1 lies closer to P than to R, because the product has higher energy than the reactant. This is in accord with the postulate. Now suppose the product is destabilized relative to reactant (product parabola 2). Then T_2 moves even closer to P. Thornton[29] showed that the addition of a linear perturbation to a parabolic potential shifts the position of the extremum in the direction expected from the Hammond postulate; Figs. 5-14A and B show this concept. [Kurz[30] has shown that two intersecting parabolic wells are equivalent to a parabolic barrier; see Fig. 5-14C.] Agmon[31] abstracts the reaction progress curve as the straight lines shown in Fig.

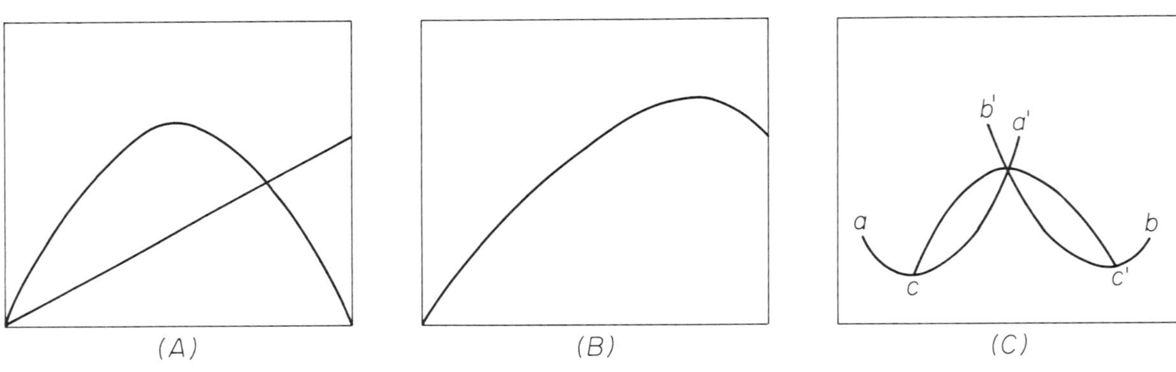

Figure 5-14. (*A*) A parabolic potential barrier and a linear perturbation. (*B*) Sum of the parabolic and linear functions, showing shift in maximum in accord with the Hammond postulate. (*C*) Two parabolic potential wells *aa'* and *bb'* are equivalent to the parabolic barrier *cc'*.

5-15, drawing an analogy from optics to trace the path from R to T to P. The proposition is made that the path of least time is followed, which results in a trajectory such that reflection from the line AB occurs with the angle of incidence equal to the angle of reflection, as shown in Fig. 5-15. Thus the transition state lies closer to the state of higher energy. This optical analogy introduces a symmetry condition.

Thus, we see that if the reaction is exergonic (reactant having higher energy than product), the transition state is early or reactantlike, whereas if the reaction is endergonic (product having higher energy than reactant), the transition state is late or productlike. If the reaction is isergonic (reactant and product of equal energy), we expect the transition state to be equidistant between the initial and final states on the reaction coordinate.

It would be desirable to achieve a quantitative version of the Hammond postulate. For this purpose we need a measure of progress along the reaction coordinate. Several authors have used the bond order for this measure.[31-33] The chemical significance of bond order is that it is the number of covalent bonds between two atoms; thus the bond orders of the C—C, C=C, C≡C bonds are 1, 2, and 3, respectively. For the displacement reaction

$$A + B\text{—}C \rightarrow A\text{—}B + C$$

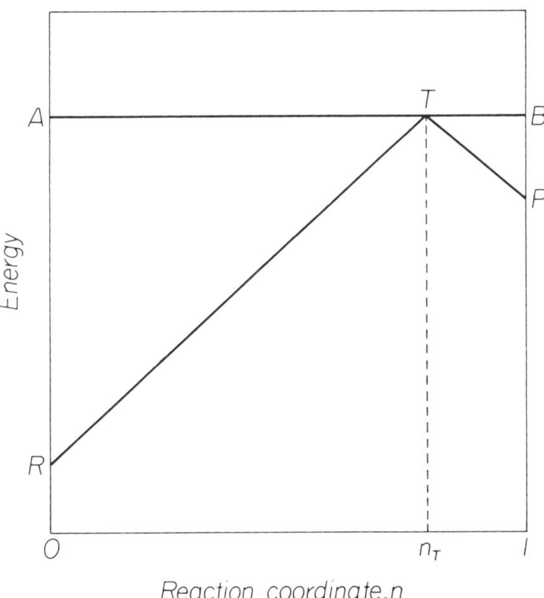

Figure 5-15. Representation of a reaction coordinate diagram by straight line trajectories determined by the condition of minimum time or distance. The reaction coordinate is specified as the bond order of a bond formed in the reaction.

the bond order n_{AB} of the A—B bond changes from 0 (in the initial state) to 1 (in the final state), whereas the order n_{BC} of the B—C bond changes from 1 to 0. The relative ease with which reactions of this type take place (i.e., the reason $\Delta G^{\ddagger}$ is much smaller than bond dissociation energies) is because the energy of breaking the B—C bond is furnished in part by the concurrent formation of the A—B bond. This suggests a *conservation of bond order* property, such that $n_{AB} + n_{BC} = 1$. The bond order n_{AB} is thus a reasonable measure of the reaction coordinate, its value in the transition state, n_T, serving to locate the transition state quantitatively along the reaction coordinate, and describing the extent to which A—B bond formation has proceeded in the transition state.

The simplest argument is that of Agmon[31] based on Fig. 5-15. Because of the reflection symmetry at point T, the triangles ATR and BTP are similar, so we can write AT/AR = TB/PB, or, if we express the ordinate in free energies,

$$\frac{n_T}{\Delta G^{\ddagger}} = \frac{1 - n_T}{\Delta G^{\ddagger} - \Delta G^0}$$

which rearranges to

$$n_T = \frac{1}{2 - (\Delta G^0/\Delta G^{\ddagger})} \tag{5-60}$$

According to this very simple derivation and result, the position of the transition state along the reaction coordinate is determined solely by ΔG^0 (a thermodynamic quantity) and $\Delta G^{\ddagger}$ (a kinetic quantity). Of course, the potential energy profile of Fig. 5-15, upon which Eq. (5-60) is based, is very unrealistic, but, quite remarkably, it is found that the precise nature of the profile is not important to the result provided certain criteria are met, and Miller[34] obtained Eq. (5-60) using an arc length minimization criterion. Murdoch[35] has analyzed Eq. (5-60) in detail. Equation (5-60) can be considered a quantitative formulation of the Hammond postulate. The transition state in Fig. 5-9 was located with the aid of Eq. (5-60).

We will explore further the idea that there may be a relationship between rates and equilibria. Although such a relationship is not required by thermodynamics, neither is it forbidden, and much empirical evidence supports the frequent occurrence of such relationships. Chapter 7 is devoted to this topic; here we restrict attention to correlations of $\Delta G^{\ddagger}$ (or log k) with ΔG^0 (or log K) *of the same reaction.* Such correlations are usually sought within a *reaction series* in which a set of reactants having a common reaction site but different substituent sites are subjected to the same reaction.

We consider the effect of the structural change on the free energy of the transition state. Leffler[36,37] postulated that changes in $G^{\ddagger}$ can be represented as a linear combination of changes in the free energies of reactants and products,

$$\delta G^{\ddagger} = a\delta G_P^0 + b\delta G_R^0$$

where a and b are constants whose values are fixed by the further condition that, although $G^\ddagger$ is greater than either G_R or G_p, changes in $G^\ddagger$ will be intermediate in magnitude to the corresponding changes in G_R or G_p. Thus, $a + b = 1$, and we write

$$\delta G^\ddagger = \alpha\delta G_P^0 + (1 - \alpha)\delta G_R^0 \tag{5-61}$$

where $0 \leq \alpha \leq 1$. (This condition means that we adopt the adiabatic assumption.)

Passing to differential symbolism and defining $\Delta G^\ddagger = G^\ddagger - G_R^0$ and $\Delta G^0 = G_P^0 - G_R^0$, Eq. (5-61) becomes

$$d\Delta G^\ddagger = \alpha d\Delta G^0 \tag{5-62}$$

where we have used the commutation property $\Delta dG = d\Delta G$. Equation (5-62) is a correlation between rates and equilibria within a reaction series. The quantity α is the slope of the plot of $\Delta G^\ddagger$ against ΔG^0; α is often called a *Brønsted coefficient,* but this term is more commonly applied to a different (though related) quantity that arises in studies of general acid–base catalysis (Chapter 7). It may be useful to think of α as a *sensitivity coefficient,* because it expresses the sensitivity of response of $\Delta G^\ddagger$ to changes in ΔG^0. Another interpretation of α is that it is a measure of the extent to which the transition state resembles the product; thus, it is an alternative to n_T, the bond order of the transition state, as a measure of location of the transition state on the reaction coordinate.

We will follow Murdoch's development.[38] Figure 5-16 shows schematic reaction coordinate diagrams for a reaction series of varying product stability. It is evident that α at the transition state varies with ΔG^0 for the reaction. We will assume a linear relationship,

$$\alpha = m\Delta G^0 + b \tag{5-63}$$

Clearly when $\Delta G^0 = 0$ (isergonic reaction), $\alpha = \frac{1}{2}$, so $b = \frac{1}{2}$. Figure 5-16 shows that as G_p increases, ΔG^0 increases, until the transition state becomes so productlike that $\alpha = 1$; further increases in ΔG^0 lead to no further increase in $\Delta G^\ddagger$, so when $\alpha = 1$, we can write $\Delta G^0 = \Delta G_{max}^0 = \Delta G_{max}^\ddagger$. (When $\alpha = 0$, $\Delta G^0 = -\Delta G_{max}^0$). Substituting into Eq. (5-63) we find $m = \frac{1}{2}\Delta G_{max}^0$. Thus, Eq. (5-63) becomes

$$\alpha = \frac{\Delta G^0}{2\Delta G_{max}^0} + \frac{1}{2} \tag{5-64}$$

We substitute (5-64) into (5-62) and integrate over the limits shown.

$$\int_{\Delta G_0^\ddagger}^{\Delta G_{max}^\ddagger} d\Delta G = \int_0^{\Delta G_{max}^0} \left(\frac{\Delta G^0}{2\Delta G_{max}^0} + \frac{1}{2}\right) d\Delta G^0 \tag{5-65}$$

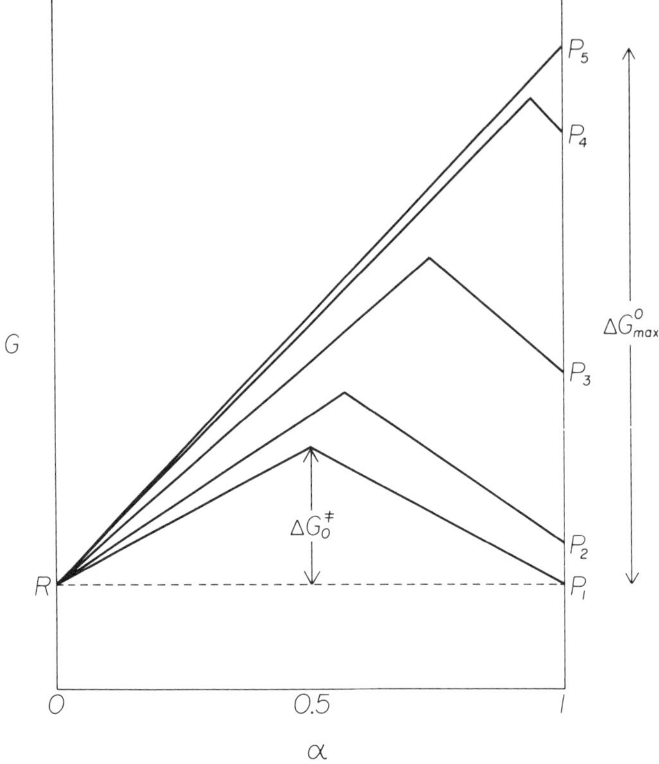

Figure 5-16. Schematic reaction coordinate diagrams of a reaction series, showing $\Delta G_0^{\ddagger}$, the intrinsic barrier, and ΔG_{max}^0, the maximum standard free energy difference.

That is, $\Delta G^{\ddagger} = \Delta G_0^{\ddagger}$ when $\Delta G^0 = 0$ (see Fig. 5-16); moreover $\Delta G_{\text{max}}^{\ddagger} = \Delta G_{\text{max}}^0$, so we obtain

$$\Delta G_0^{\ddagger} = \Delta G_{\text{max}}^0/4 \tag{5-66}$$

$\Delta G_0^{\ddagger}$ is called the *intrinsic barrier* for the reaction series. (Strictly speaking, we should write $\Delta G_0^{\ddagger} = |\Delta G_{\text{max}}^0|/4$, because ΔG^0 can be negative.)

We now use Eq. (5-66) in (5-64) to get

$$\alpha = \frac{\Delta G^0}{8\Delta G_0^{\ddagger}} + \frac{1}{2} \tag{5-67}$$

This is substituted into Eq. (5-62) and integrated:

$$\int_{\Delta G^{\ddagger}}^{\Delta G^{\ddagger}_0} d\Delta G^{\ddagger} = \int_{\Delta G^0}^{0} \left(\frac{\Delta G^0}{8\Delta G_0^{\ddagger}} + \frac{1}{2} \right) d\Delta G^0 \tag{5-68}$$

$$\Delta G^\ddagger = \Delta G_0^\ddagger + \frac{\Delta G^0}{2} + \frac{(\Delta G^0)}{16\Delta G_0^\ddagger} \tag{5-69}$$

or

$$\Delta G^\ddagger = \Delta G_0^\ddagger \left(1 + \frac{\Delta G^0}{4\Delta G_0^\ddagger} \right)^2$$

Equation (5-69) is an important result. It was first obtained by Marcus[39–41] in the context of electron-transfer reactions. Marcus' derivation is completely different from the one given here.[39–42] In electron transfer from one molecule (or ion) to another, no bonds are broken or formed, so the transition state theory does not seem to be applicable. Marcus assumed negligible orbital overlap in the electron-transfer transition state, but he later obtained the same equation for group transfer reactions requiring significant overlap.[43] Many applications have been made to proton transfers[44] and nucleophilic displacements.[45,46]

Equation (5-69) describes rate–equilibrium relationships in terms of a single parameter, the intrinsic barrier $\Delta G_0^\ddagger$, which therefore assumes great importance in interpretations of such data. It is usually assumed that $\Delta G_0^\ddagger$ is essentially constant within the reaction series; then it can be estimated from a plot of $\Delta G^\ddagger$ vs. ΔG^0 as the value of $\Delta G^\ddagger$ when $\Delta G^0 = 0$. Another method is to fit the data to a quadratic in ΔG^0 and to find $\Delta G_0^\ddagger$ from the coefficient of the quadratic term.[47]

Figure 5-17 is a plot of Eq. (5-69) for a hypothetical system having $\Delta G_0^\ddagger = 20$ units. Over a wide range in ΔG^0 the curvature is easily perceived, although it may not be evident if a small range is covered by a set of experimental data. Equation (5-67) gives the slope α as a function of ΔG^0; $\alpha = \frac{1}{2}$ when $\Delta G^0 = 0$, $\alpha > \frac{1}{2}$ for positive values of ΔG^0, and $\alpha < \frac{1}{2}$ for negative values of ΔG^0. Figure 5-18 is a rate–equilibrium plot for the isomerization of 1-substituted 5-aminotriazoles.[48]

$$\underset{\text{R}}{\underset{|}{\text{triazole}}}\text{—NH}_2 \;\rightleftharpoons\; \underset{\text{H}}{\underset{|}{\text{triazole}}}\text{—NHR}$$

From the intercept at $\Delta G^0 = 0$ we find $\Delta G_0^\ddagger = 31.9$ kcal mol^{-1}, and the slope is 0.77. As we have seen, if Eq. (5-69) is applicable, the slope should be 0.5 when $\Delta G^0 = 0$. In this example either the data cover too small a range to allow a valid estimate of the slope to be made or the equation does not apply to this system. Such a simple equation is not expected to be universally applicable. Recall that it was derived for an elementary reaction, so multistep reactions, even if showing simple rate–equilibrium behavior, introduce complications in the interpretation. The simple interpretation of Eq. (5-69) also requires that $\Delta G_0^\ddagger$ be constant within the reaction series, but this condition may not be met.[47] Later pages describe another possible reason for the failure of Eq. (5-69).

It is interesting to note that α (Eq. 5-67) and n_T (Eq. 5-60) are different measures

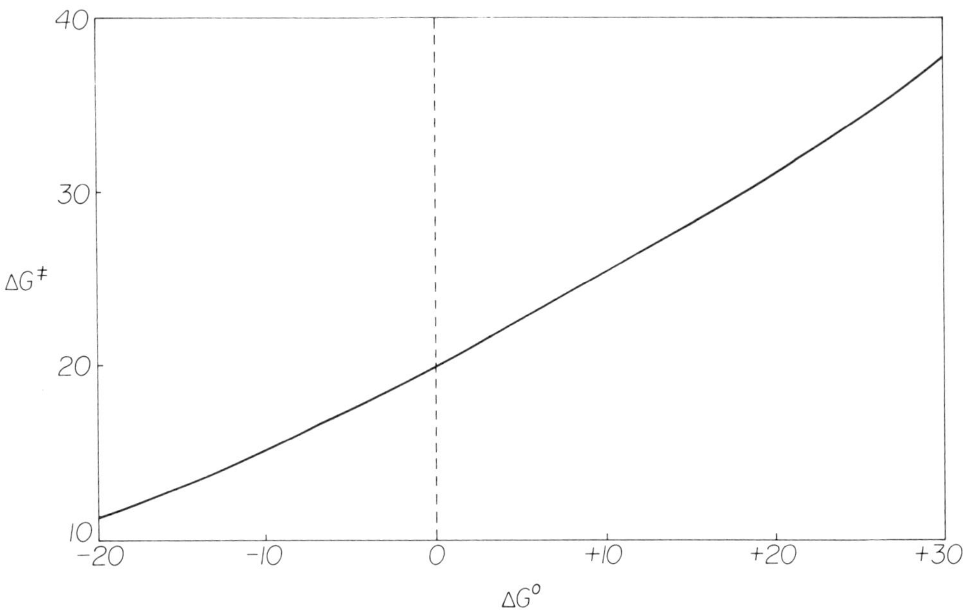

Figure 5-17. Rate–equilibrium relationship according to Eq. (5-69) for a hypothetical system with $\Delta G_0^{\ddagger} = 20$.

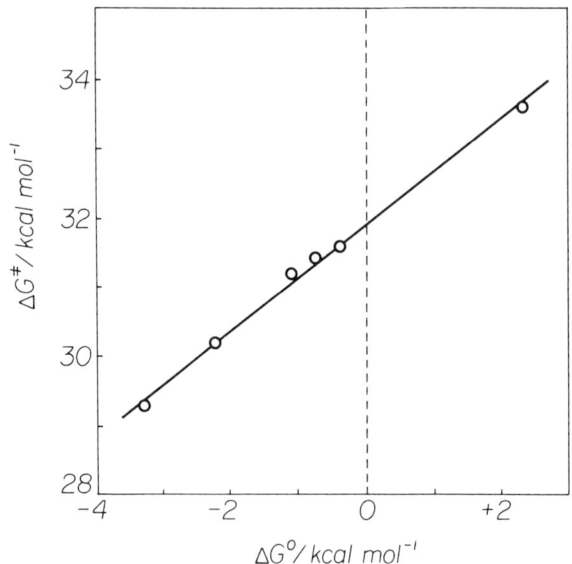

Figure 5-18. Rate–equilibrium plot for the isomerization of substituted 5-aminotriazoles at 423 K.[48]

of the location of the transition state on the reaction coordinate. If Eq. (5-69) applies to the system, $\alpha = n_T$ at $\Delta G^0 = 0$, $\alpha > n_T$ at $\Delta G^0 < 0$, and $\alpha < n_T$ at $\Delta G^0 > 0$. If Eq. (5-69) is not applicable, these relationships are not valid. For the reaction series in Fig. 5-18, $\alpha = 0.77$ (from the slope of the line), whereas $n_T = 0.49$ (from Eq. 5-60 applied to each of the reactants). Thus, caution is necessary in drawing mechanistic conclusions from these parameters.

If the intrinsic barrier $\Delta G_0^{\ddagger}$ could be independently estimated, the Marcus equation (5-69) provides a route to the calculation of rate constants. An additivity property has frequently been invoked for this purpose.[39–43,46,49,50] For the *cross-reaction*

$$A + BC \rightleftharpoons AB + C \tag{5-70}$$

we define two *identity reactions:*

$$A + BA \rightleftharpoons AB + A \tag{5-71}$$

$$C + BC \rightleftharpoons CB + C \tag{5-72}$$

Reactions (5-70), (5-71), and (5-72) have intrinsic barriers $\Delta G_0^{\ddagger}$, $\Delta G_{AA}^{\ddagger}$, and $\Delta G_{CC}^{\ddagger}$, respectively. Then $\Delta G_0^{\ddagger}$ is estimated with Eq. (5-73).

$$\Delta G_0^{\ddagger} = (\Delta G_{AA}^{\ddagger} + \Delta G_{CC}^{\ddagger})/2 \tag{5-73}$$

For the identity reactions, the intrinsic barriers are their free energies of activation, which can be determined by tracer studies[51] or less directly by rate–equilibrium correlations.[46]

There are several equations other than the Marcus equation that describe rate–equilibrium relationships. Murdoch[50] writes all of these equations in the general form

$$\Delta G^{\ddagger} = \Delta G_0^{\ddagger}(1 - g_2) + \tfrac{1}{2} \Delta G^0 (1 + g_1) \tag{5-74}$$

where g_1 and g_2 are functions of ΔG^0. For example, if $g_1 = \Delta G^0/4\Delta G_0^{\ddagger}$ and $g_2 = g_1^2$, Eq. (5-74) becomes Eq. (5-69).

Values of the slope, α, of a plot of $\Delta G^{\ddagger}$ vs. ΔG^0 obviously must lie in the range of 0–1 if the interpretation of α as the fractional resemblance of the transition state to the product is to be valid. Some examples of α values lying outside this range have been observed.[52,53] Thus, Bordwell et al.[52] found $\alpha = 1.31$ for this reaction:

$$ArCHMeNO_2 + H_2O \rightleftharpoons (ArCMeNO_2)^- + H_2O$$

Several explanations have been given for this result.[52,54,55] One possibility is the failure of the adiabatic assumption; in this ionization process a transition state is generated that may not be a combination of the initial and final states, but may include mixing in of an excited state structure. This idea will reappear in the following pages.

Concerted and Stepwise Reactions

As a qualitative concept, the reaction coordinate can represent any number of progress variables, but if we wish to attach quantitative significance to the reaction coordinate, as we did in the preceding pages when we attempted to define the location of the transition state, a single variable must suffice to define the progress of the reacting system. This might be a bond distance, a bond order, or the sensitivity coefficient α. However, when more than one process is involved in the development of the transition state, we need more than one progress variable to specify the extent to which each process has occurred. The types of processes that may take place concurrently, but not necessarily synchronously, are bond formation and bond breaking, solvation and desolvation, proton transfer, and charge delocalization.

A great many reactions in solution can profitably be treated by an extension of the earlier ideas to include more than one progress variable. We will introduce the idea with the general reaction

$$Y^- + G\text{–}X \rightarrow Y\text{–}G + X^-$$

where the group G represents a central structure that remains essentially unchanged, Y^- is the attacking group, and X^- is the leaving group. If the reaction proceeds with concurrent formation of the Y–G bond and cleaving of the G–X bond such that the sum of the bond orders is unity, then a single progress variable suffices to describe the reacting system. However, there are other ways this reaction could take place: Thus G–X might ionize to give $G^+ + X^-$, followed by combination of G^+ with Y^-; or Y^- might attack to yield $(Y\text{–}G\text{–}X)^-$, from which X^- then dissociates to give the products. The species G^+ and $(Y\text{–}G\text{–}X)^-$ are intermediates (or hypothetical intermediates), and reactions via these routes are stepwise reactions. The elementary reaction is a concerted reaction. Let us use this symbolism:

R   Reactant (initial state)

P   Product (final state)

I^+ Cationic intermediate

I^- Anionic intermediate

In the pure concerted reaction there is no need to invoke the cationic or anionic intermediates in describing the transition state, but it now becomes evident that some deviation from this idealized route may be possible, and then we need a way to comment upon and to measure the extent to which the cationic or anionic character is "mixed in" in the transition state. This is now widely accomplished with the aid of energy surfaces of the type shown schematically in Fig. 5-19. Depending on the nature of the surface, the reaction path may follow a route far from the diagonal representing the pure concerted reaction, and the primary goal is to identify the location of the transition state on this surface.

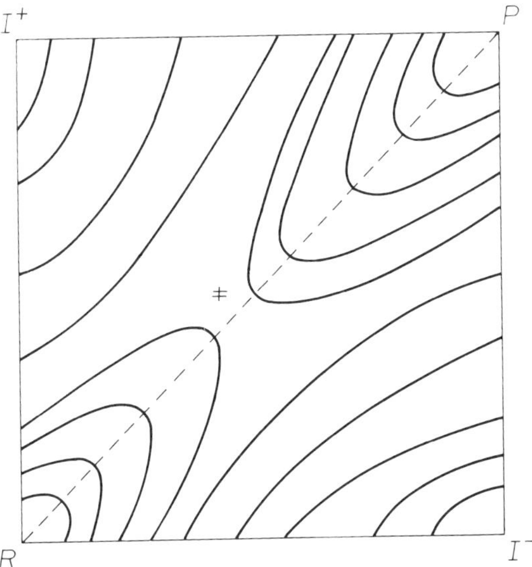

Figure 5-19. Two-dimensional energy surface for generalized reaction R → P showing the cationic (I⁺) and anionic (I⁻) possible intermediates. The dashed line is the reaction path for the pure concerted reaction.

Energy surfaces of this type appear to have been described first by Hughes et al.[56] in 1936. Their current popularity and utility are a consequence of developments by Albery,[57] More O'Ferrall,[58] and Jencks,[59] and they are sometimes called More O'Ferrall plots or More O'Ferrall–Jencks plots. We will refer to them as RIP diagrams, this designation arising from the symbolism in Fig. 5-19.

To make this more specific, Table 5-3 gives examples of several reaction types that fit the RIP pattern. Consider nucleophilic substitution on saturated carbon. The concerted mechanism is the one-step bimolecular S_N2 process:

$$Y^- + RCH_2X \rightarrow [Y^{\delta^-} \cdots \overset{\overset{\displaystyle H \quad\;\; H}{\diagdown \diagup}}{\underset{\underset{\displaystyle R}{|}}{C}} \cdots {}^{\delta^-}X]^{\ddagger} \rightarrow YCH_2R + X^-$$

A stepwise mechanism is also possible; this is the two-step unimolecular S_N1 reaction:

$$R_3CX \overset{slow}{\rightleftharpoons} R_3C^+ + X^-$$

$$R_3C^+ + Y^- \overset{fast}{\rightarrow} R_3CY$$

TABLE 5-3.   Examples of the Concerted/Stepwise Reaction Description

Reaction type	State description[a]			
	R	I^+	I^-	P
Nucleophilic aliphatic substitution	$Y^- R{-}X$	$\underline{Y^- R^+\ X^-}$	$Y^- R^-\ X^+$	$Y{-}R\ X^-$
β-Elimination[b]	$B\ H{-}R{-}R{-}X$	$B\ H{-}R{-}R^+\ X^-$	$\underline{BH^+\ {}^-R{-}R{-}X}$	$BH^+\ R{=}R\ X^-$
Acyl transfer	$Y^- RCOX$	$Y^- RCO^+\ X^-$	$\underline{RC(O^-)XY}$	$RCOY\ X^-$
Electrophilic aromatic substitution	$Y^+\ Ar{-}X$	$\underline{(Y{-}Ar{-}X)^+}$	$Y^+\ Ar^-\ X^+$	$Y{-}Ar\ X^+$
Nucleophilic aromatic substitution	$Y^-\ Ar{-}X$	$Y^-\ Ar^+\ X^-$	$\underline{(Y{-}Ar{-}X)^-}$	$Y{-}Ar\ X^-$

[a]The underlined configuration is the more probable intermediate (more stable intermediate) in most systems.
[b]B represents a base.

In this example the S_N1 mechanism is the I^+ route of Fig. 5-19. The I^- route is clearly much less favored (the I^- state is of higher energy) than the I^+ route for this reaction type.

Now let us consider the possible rate changes that may occur in a reaction series in which we change the relative stabilities of states by making structural changes. Suppose the energy surface is constructed of a flexible material as a model with which energy changes can be made to the surface. Let the product state P be made more stable relative to R. This can be modeled by pulling down the P corner of the surface. The lowering of the energy surface at P will be transmitted in attenuated form across the surface, and the result is that the location of the transition state (marked with a double dagger in Fig. 5-19) will be shifted toward the reactants R. This behavior is in accord with the Hammond postulate. This energetic stress on the system is called a *parallel* displacement because it is exerted parallel to the reaction path.

In contrast, suppose we now make a substituent change that stabilizes intermediate I^+ relative to the reactant state. This is modeled by pulling down the I^+ corner of the surface. The result is to distort the surface such that the transition state moves closer to I^+. This is called *anti-Hammond behavior,* and the stress that it produces it is a *perpendicular* displacement.[29] Thus, both these parallel and perpendicular displacements of the energy surface lower the energy of the transition state, but parallel displacements lead to Hammond behavior and perpendicular displacements to anti-Hammond behavior. [Kurz[60] has shown that in special cases a parallel displacement can lead to anti-Hammond behavior.] An example of a perpendicular displacement is the successive replacement of hydrogen by methyl in the series CH_3Cl, CH_3CH_2Cl, $(CH_3)_2CHCl$, $(CH_3)_3CCl$ in a nucleophilic substitution reaction. Increasing methyl substitution at the reaction site stabilizes the incipient carbocation (carbonium ion) intermediate, I^+, with the result that the location of the transition state shifts from the pure concerted reaction path toward the I^+ corner, and with the fully substituted substrate the reaction becomes stepwise.

An alternative description of these energetic effects has been given by Pross and Shaik.[61,62] Again the S_N reaction is used as an example. A valence bond (VB) description is developed by writing linear combinations of a basis set of configurations. These constitute a reasonable set of configurations (compare with Table 5-3):

$$Y\bar{:} \quad R\cdot \;\cdot X \qquad \text{(R, reactantlike)}$$

$$Y\cdot\cdot R \quad :X^- \qquad \text{(P, productlike)}$$

$$Y\bar{:} \quad R^+ \quad :X^- \qquad (I^+, \text{ carbocationlike})$$

$$Y\bar{:} \quad R\bar{:} \quad X \qquad (I^-, \text{ carbanionlike})$$

The initial state is dominated by configuration R, and the final state by P, though each state includes contributions from all configurations. An S_N2 reaction can be represented as in Fig. 5-20). The R configuration naturally rises in energy because it is energetically unfavorable to maintain the initial state configuration while progressing along the reaction coordinate. Likewise the P configuration decreases in energy as it approaches the final state. At the point where these configurations meet, an *avoided crossing* takes place so that the reacting system follows the path of least energy. Now, suppose the P state were to be stabilized relative to R by a

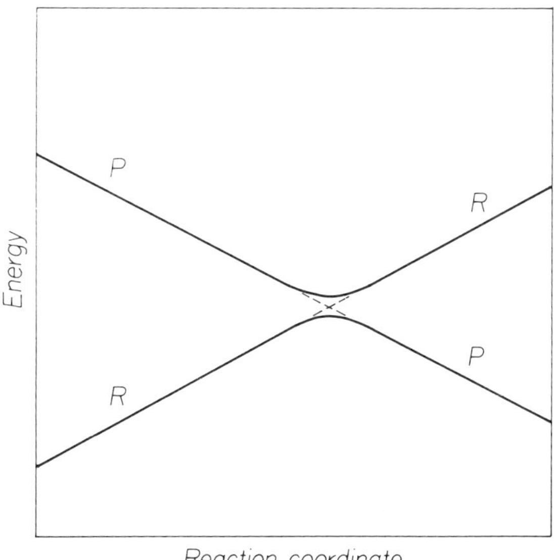

Figure 5-20. Reaction coordinate diagram generated from a valence bond description of initial and final state configurations.

structural change. This would lower the P energy profile in Fig. 5-20, resulting in an earlier transition state; this is Hammond behavior.

This valence bond description leads to an interesting conclusion.[63] Because the transition state occurs at the point where the initial and final state VB configurations cross, the transition state receives equal contributions from each. This is so whether the transition state is early or late. Thus, the nucleophile Y and the leaving group X possess about equal charge densities in the transition state. This conclusion means that an early transition state is not (in this sense) "reactantlike", for a reactantlike transition state should have most of the charge on Y. Similarly, a late transition state is not necessarily productlike. This view is at variance with other interpretations.

Continuing to develop this VB approach, suppose now that the carbocationlike configuration is not an important contributor to the initial or final states, but that it is significant for the midregion of the reaction coordinate. Figure 5-21 shows the concept. Now the transition state is a combination of R, P, and I^+, whereas the initial and final states are not significantly influenced by I^+. This is equivalent to a perpendicular displacement as described earlier, and it also is a possible explanation of anomalous α values, as noted in preceding pages.[64]

We will return to our consideration of RIP diagrams. Figure 5-22 summarizes the possible reaction paths. Recall that an intermediate is a state of minimum energy on the reaction path, so that all four corners may constitute energy minima, but for any given type of reaction it is unlikely that both I^+ and I^- will be of comparable stability. As Table 5-3 indicates, one of these is apt to be the favored intermediate,

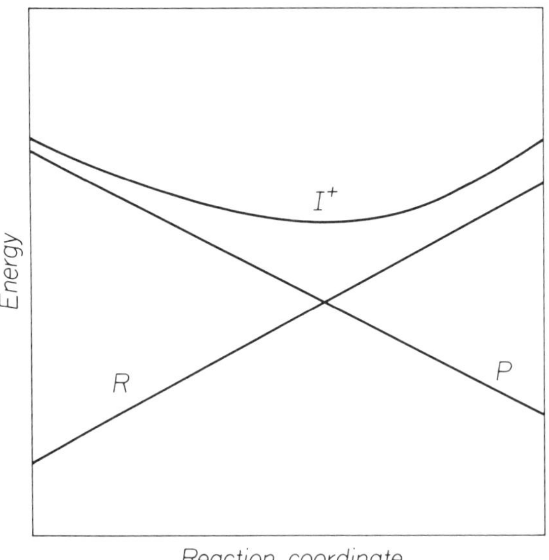

Figure 5-21. Energy profiles of R, P, and I^+ configurations showing how I^+ can mix in to the transition state while not contributing significantly to the initial and final states.

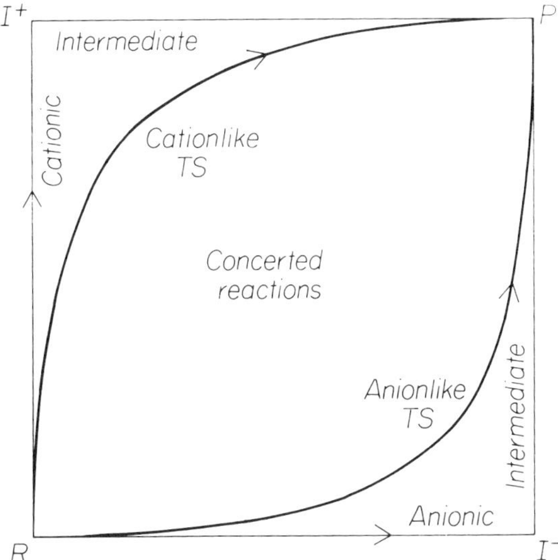

Figure 5-22. Generalized RIP diagram.

if indeed an intermediate exists. The experimental observation of an intermediate or the demonstration of a two-step reaction is obviously very valuable information. It is possible, however, for concerted and stepwise reactions to coexist.

The essential goal is to locate the transition state on the RIP diagram. This involves specifying two coordinates, so two quantitative progress variables are required. One approach, fairly widely applied, can be illustrated with the study by Hill et al.[65] of the general acid-catalyzed addition of substituted anilines to dicyanamide. The overall reaction is

$$ArNH_2 + HN(CN)_2 \rightarrow ArNH\underset{\overset{|}{HA}}{C}\overset{NH_2}{=}N\!\!-\!\!CN$$

Here $ArNH_2$ is an attacking nucleophile and HA is the acid catalyst. Two reaction series can be designed for the present purpose; in one of these $ArNH_2$ is held constant and the structure of HA is varied, and in the second HA is held constant and the structure of $ArNH_2$ is varied. To continue the treatment we must anticipate Chapter 7 by defining these quantities:

α_{HA} The slope of a plot of log k vs. pK_a of HA ($ArNH_2$ held constant).

β_{nuc} The slope of a plot of log k vs. pK_a of $ArNH_3^+$ (HA held constant).

The quantities of α_{HA} and β_{nuc} are sensitivity coefficients closely analogous to the quantity α that we encountered in the Marcus treatment, and their interpretation is similar. α_{HA} represents the fractional extent to which the proton transfer has been accomplished in the transition state of this process

$$HA + {}^-N(CN)_2 \rightarrow A^- + HN(CN)_2$$

(The pK_a of dicyanamide is less than 1, so it exists mainly as the anion under the reaction conditions. α_{HA} is a negative number, so the absolute value is taken.) This equation constitutes the $R \rightarrow I^+$ step. The quantity β_{nuc} describes the fractional extent of the $R \rightarrow I^-$ step, which is

$$ArNH_2 + {}^-N(CN)_2 \rightarrow ArNH_2\overset{+}{C}\begin{smallmatrix} {\nearrow}N^- \\ \\ N{-\!-}CN^- \end{smallmatrix}$$

Figure 5-23 shows the RIP diagram.

Several lines of evidence led to the conclusion that the reaction is concerted. Experimental values of the sensitivity coefficients were $\alpha_{HA} = -0.54$ and $\beta_{nuc} = 0.63$. These values are plotted in Fig. 5-23 to define the position of the transition state, which probably has the following charge distribution, where the net charge is -1, and bond formation from the nucleophile to carbon is well advanced.

$$\left[A_r - \overset{\delta+}{N}H_2 \cdots C{=\!=}N \cdots H \cdots \overset{\delta-}{A} \atop \qquad\qquad\quad N{-\!\!-}CN \right]^{\ddagger}$$

Other examples of this sensitivity coefficient approach are given by Bernasconi[66] on the deprotonation of phenylnitromethane and by Thibblin[67] on elimination reactions. Elaboration of the method takes into account variations in the coefficients within and between reaction series.[66,68,69]

Other measures of the reaction coordinate have been proposed. Jones and Kirby[70] suggest that (within a reaction series) the longer the bond, the more readily it will cleave heterolytically. This idea is based on the premise that a longer covalent bond carries a greater contribution from the charge-separated valence bond structure, as in

$$R\text{–}X\text{–}R' \leftrightarrow R^+ \; {}^-X\text{–}R'$$

Thus, the greater the charge separation in the initial state, the easier the cleavage to the ion pair. Crystallographic bond lengths, therefore, provide some information about the reaction coordinate. Shaik[71] has given a quantitative form to the terms *looseness* and *tightness* of a transition state. Qualitatively a loose transition state has weak, partial, long bonds. Shaik draws a bond dissociation curve (like Fig. 5-1) for a transition state; i.e., the transition state is the species at the minimum of this curve. The completely dissociated species corresponds to the maximum possible

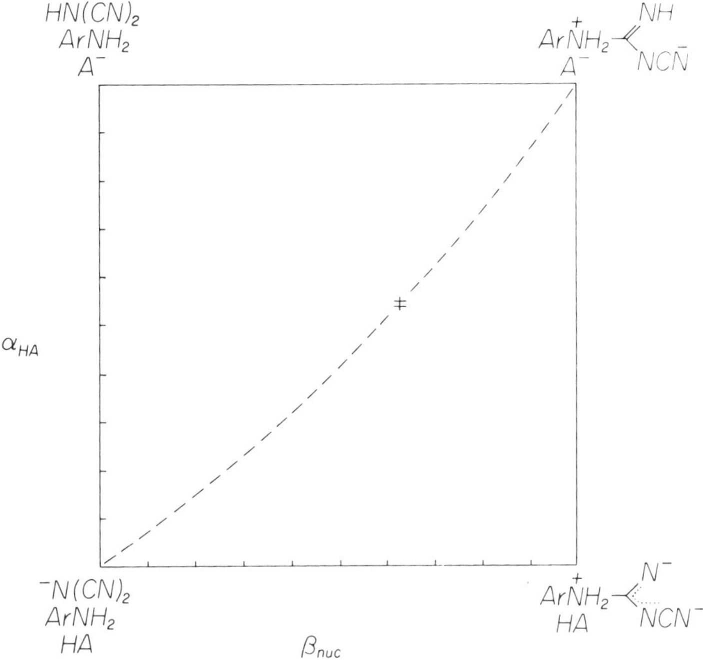

Figure 5-23. RIP diagram for the general acid-catalyzed addition of anilines to dicyanamide.

looseness. The tightness of the transition state is defined $T^{\ddagger} = D^{\ddagger}/D$, where $D^{\ddagger}$ is the binding energy of the transition state (relative to reactants) and D is the bond dissociation energy. The looseness is then $L^{\ddagger} = 1 - T^{\ddagger}$. It is inferred that a small energy barrier corresponds to a tight transition state, and a high barrier to a loose one, because the loose transition state is closer to the dissociated condition.

By locating the transition state within the RIP diagram with the aid of progress variables such as sensitivity coefficients, we have not yet made explicit use of the third dimension of these diagrams, the energy. It has proved possible to construct the energy surface, by completely empirical means, and thus to deduce the reaction path taken by the reacting system over the surface. One needs to know the energies (relative to some state such as the initial state) of several points, for example, the four corners and the transition state, together with the mathematical shape of the surface, which is generally assumed to be a hyperbolic paraboloid. Agmon[72] and Guthrie and Pike[73] show examples of this approach. Murdoch[74] and Grunwald[75,76] have generalized Marcus theory for the case of two progress variables. We will outline Grunwald's method.

The R→P reaction is called the *main reaction,* and the hypothetical process I⁻ → I⁺ is called the *disparity reaction.* Figure 5-24 is an RIP diagram with two coordinate systems. In the type of diagram we have used thus far, the coordinates are u and v; for example in Fig. 5-23, $u = \beta_{nuc}$ and $v = \alpha_{HA}$. The normalized

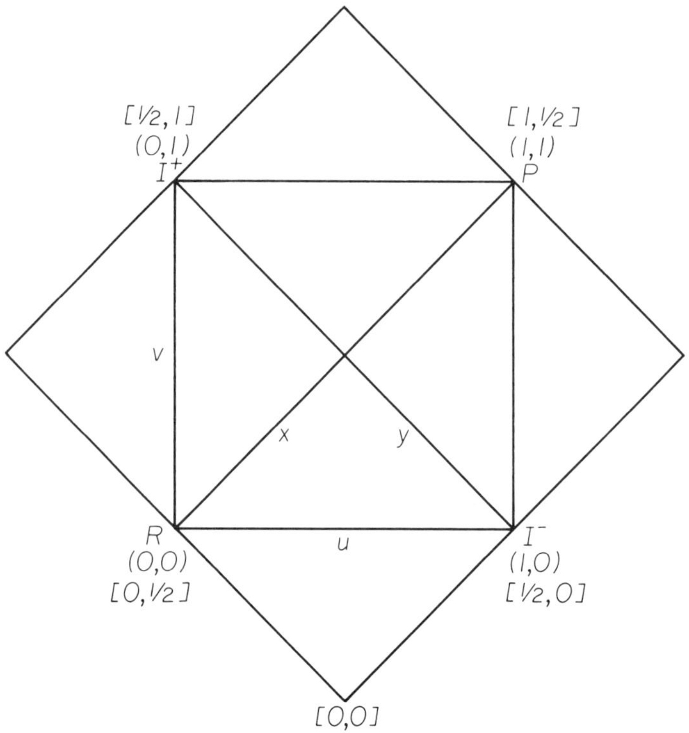

Figure 5-24.   The u,v and x,y coordinate systems of Grunwald's analysis of the RIP diagram. The u,v coordinates are in parentheses and the x,y coordinates are in brackets.

u,v coordinates of the four corners of the RIP diagram are enclosed in parentheses in Fig. 5-24.

A normalized x,y coordinate system is generated by a translation and 45° rotation as shown by the larger square in Fig. 5-24, and the x,y coordinates of the four corners (R, P, I⁻, I⁺) of the RIP diagram are given in square brackets. The u,v, and x,y coordinate systems are related by Eq. (5-75).

$$x = (u + v)/2 \tag{5-75a}$$

$$y = \tfrac{1}{2} + (v - u)/2 \tag{5-75b}$$

Thus, the coordinate x measures the progress of the main reaction from R to P, whereas $(y - \tfrac{1}{2})$ is a measure of the perpendicular displacement from the pure concerted reaction path.

For a system describable with a single progress variable, we derived the Marcus equation, Eq. (5-76).

$$\Delta G^{\ddagger} = \Delta G_0^{\ddagger} + \frac{\Delta G^0}{2} + \frac{(\Delta G^0)^2}{16\Delta G_0^{\ddagger}} \tag{5-76}$$

The position of the transition state on the reaction coordinate, α, is given by

$$\alpha = \frac{1}{2} + \frac{\Delta G^0}{8\Delta G_0^{\ddagger}} \tag{5-77}$$

The extension to two progress variables[74-76] gives Eq. (5-78), which may be compared with Eq. (5-76).

$$\Delta G^{\ddagger} = \Delta G_0^{\ddagger} + \frac{\Delta G^0}{2} + \frac{(\Delta G^0)^2}{16\Delta G_0^{\ddagger}} - \frac{(\Delta G_I^0)^2}{16\Delta G_I^{\ddagger}} \tag{5-78}$$

In Eq. (5-78), $\Delta G_0^{\ddagger}$ is the intrinsic energy barrier of the main reaction, ΔG^0 is the standard free energy change of the main reaction, $\Delta G_I^{\ddagger}$ is the intrinsic energy well of the disparity reaction, and ΔG_I^0 is the standard free energy change of the disparity reaction. Figure 5-25 shows profiles of the hyperbolic paraboloid surface, the main reaction path along the x coordinate passing over the barrier of height $\Delta G_0^{\ddagger}$ (since

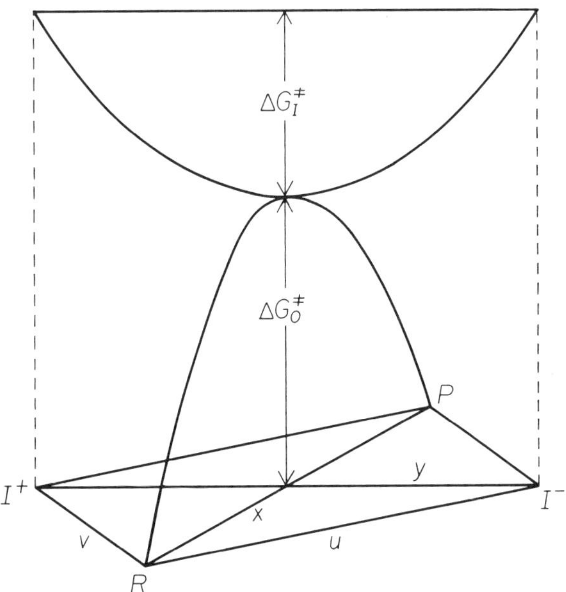

Figure 5-25. Sections through a hyperbolic paraboloid energy surface constructed over an RIP diagram. The intrinsic barrier $\Delta G_0^{\ddagger}$ of the main reaction and the intrinsic well $\Delta G_I^{\ddagger}$ of the disparity reaction are shown.

$\Delta G^0 = 0$ as shown), whereas the perpendicular disparity reaction path along the y coordinate passes through an energy minimum, whose magnitude (taken as a positive number) is $\Delta G_I^{\ddagger}$, because $\Delta G_I^0 = 0$ in this diagram. The coordinates of the transition state are (see Problem 9)

$$x^{\ddagger} = \frac{1}{2} + \frac{\Delta G^0}{8\Delta G_0^{\ddagger}} \tag{5-79a}$$

$$y^{\ddagger} = \frac{1}{2} - \frac{\Delta G_I^0}{8\Delta G_I^{\ddagger}} \tag{5-79b}$$

When the states I^+ and I^- are of equal energy, $\Delta G_I^0 = 0$ and Eq. (5-78) becomes Eq. (5-76); with this condition, also, $x^{\ddagger} = \alpha$ and $y^{\ddagger} = \frac{1}{2}$, describing the pure concerted reaction along the x coordinate with no net perpendicular displacement. An apparent failure of Eq. (5-76) may mean that Eq. (5-78) is applicable.

Grunwald[75,76] has shown applications of Eqs. (5-78) and (5-79) as tests of the theory and as mechanistic criteria. One way to do this, for a reaction series, is to estimate ΔG^0 and ΔG_I^0 from thermodynamic data and from reasonable approximations and then to fit experimental rate data ($\Delta G^{\ddagger}$ values) to Eq. (5-78) by nonlinear regression. This yields estimates of $\Delta G_0^{\ddagger}$ and $\Delta G_I^{\ddagger}$ (which are constants within the reaction series), and these are then used in Eq. (5-79) to obtain the transition state coordinates.

REFERENCES

1. Laidler, K.J. "Theories of Chemical Reaction Rates"; McGraw-Hill: New York, 1969.
2. Wayne, R.P. In "Comprehensive Chemical Kinetics"; Bamford, C.H.; Tipper, C.F.H., Eds.; Elsevier: Amsterdam, 1969; Vol. 2, Chapter 3.
3. Glasstone, S.; Laidler, K.J.; Eyring, H. "The Theory of Rate Processes"; McGraw-Hill: New York, 1941.
4. Eyring, H.; Lin, S.H.; Lin, S.M. "Basic Chemical Kinetics"; Wiley-Interscience: New York, 1980.
5. Moore, W.J. "Physical Chemistry," 3rd ed.; Prentice-Hall: Englewood Cliffs, N.J., 1962; p 238.
6. Berry, R.S.; Rice, S.A.; Ross, J. "Physical Chemistry"; Wiley: New York, 1980; Part III, pp 1058–62.
7. Clark, I.D.; Wayne, R.P. In "Comprehensive Chemical Kinetics"; Bamford, C.H.; Tipper, C.F.H. Eds.; Elsevier: Amsterdam, 1969; Vol. 2, pp 310–1.
8. Moore, J.W.; Pearson, R.G. "Kinetics and Mechanism", 3rd ed.; Wiley-Interscience; New York, 1981; pp 237–41.
9. Moelwyn-Hughes, E.A. "The Kinetics of Reactions in Solution," 2nd ed.; Oxford University Press: London, 1947; Chapter III.
10. Eyring, H.; Eyring, E.M. "Modern Chemical Kinetics"; Reinhold: New York, 1963; pp 22–5.
11. Guggenheim, E.A. "Thermodynamics", 3rd ed.; North-Holland: Amsterdam, 1957; pp 73, 76–7.
12. Hill, T.L. "An Introduction to Statistical Thermodynamics"; Addison-Wesley; Reading, Mass., 1960; p 60.
13. Adamson, A.W. "A Textbook of Physical Chemistry"; Academic Press: New York, 1973; pp 143–4.
14. Weston, R.E., Jr.; Schwartz, H.A. "Chemical Kinetics"; Prentice-Hall; Englewood Cliffs, N.J., 1972; p 176.

15. Yates, K.; Modro, T.A. *Acc. Chem. Res.* **1978,** *11,* 190.

16. Boyd, R.K. *J. Chem. Educ.* **1978,** *55,* 84.

17. Murdoch, J.R. *J. Chem. Educ.,* **1981,** *58,* 32.

18. Jencks, W.P. *J. Am. Chem. Soc.* **1959,** *81,* 475.

19. Espenson, J.H. In "Investigation of Rates and Mechanisms of Reactions", 3rd ed.; Lewis, E.S., Ed.; Wiley-Interscience: New York, 1974; Part I, p 563.

20. King, E.L. *J. Phys. Chem.* **1956,** *60,* 1378.

21. Noyes, R.M. In "Investigation of Rates and Mechanisms of Reactions", 3rd ed.; Lewis, E.S., Ed.; Wiley-Interscience: New York, 1974; Part I, pp 528–9.

22. Schlaeger, L.L.; Long, F.A. *Adv. Phys. Org. Chem.* **1963,** *1,* 1.

23. Jencks, W.P. "Catalysis in Chemistry and Enzymology"; McGraw-Hill: New York, 1969; pp 607–9.

24. Gurney, R.W. "Ionic Processes in Solution"; McGraw-Hill: New York, 1953; Chapters 5 and 6. (Dover, New York, reprint, 1962.)

25. Connors, K.A. "Reaction Mechanisms in Organic Analytical Chemistry"; Wiley-Interscience: New York, 1973; pp 17–21, 59–60.

26. Connors, K.A. "Binding Constants: The Measurement of Molecular Complex Stability"; Wiley-Interscience: New York, 1987; pp 35–42.

27. Hammond, G.S. *J. Am. Chem. Soc.* **1955,** *77,* 334.

28. Salem, L. "Electrons in Chemical Reactions: First Principles"; Wiley-Interscience: New York, 1982; p 50.

29. Thornton, E.R. *J. Am. Chem. Soc.* **1967,** *89,* 2915.

30. Kurz, J. *Chem. Phys. Lett.* **1978,** *57,* 243.

31. Agmon, N. *J. Chem. Soc., Faraday Trans.* **1978,** *74* II, 388.

32. Johnston, H.S., Parr, C. *J. Am. Chem. Soc.,* **1963,** *85,* 2544.

33. Agmon, N., Levine, R.D. *Chem. Phys. Lett.* **1977,** *52,* 197.

34. Miller, A.R. *J. Am. Chem. Soc.* **1978,** *100,* 1984.

35. Murdoch, J.R. *J. Am. Chem. Soc.* **1983,** *105,* 2667.

36. Leffler, J.E. *Science* **1953,** *117,* 340.

37. Leffler, J.E.; Grunwald, E. "Rates and Equilibria of Organic Reactions"; Wiley: New York, 1963; p 157.

38. Murdoch, J.R. *J. Am. Chem. Soc.* **1972,** *94,* 4410.

39. Marcus, R.A. *J. Chem. Phys.* **1956,** *24,* 966.

40. Marcus, R.A. *J. Chem. Phys.* **1963,** *38,* 1858.

41. Marcus, R.A. In "Investigation of Rates and Mechanisms of Reactions", 3rd ed.; Lewis, E.S., Ed.; Wiley-Interscience: New York, 1974; Part I, pp 38–9.

42. Weston, R.E., Jr.; Schwartz, H.A. "Chemical Kinetics"; Prentice-Hall: Englewood Cliffs, N.J.; 1972; pp 205–13.

43. Marcus, R.A. *J. Phys. Chem.* **1968,** *72,* 891.

44. Cohen, A.O.; Marcus, R.A. *J. Phys. Chem.* **1968,** *72,* 4249.

45. Dodd, J.A.; Brauman, J.I. *J. Am. Chem. Soc.* **1984,** *106,* 5356.

46. Harris, J.M.; McManus, S.P., Eds., "Nucleophilicity"; American Chemical Society: Washington, D.C. Advanced Chemical Series 215, Chapters 2 and 3.

47. Murdoch, J.R. *J. Phys. Chem.* **1983,** *87,* 1571.

48. Lieber, E.; Rao, C.N.R.; Chao, T.S. *J. Am. Chem. Soc.* **1957,** *79,* 5962.

49. Murdoch, J.R.; Magnoli, D.E. *J. Am. Chem. Soc.* **1982,** *104,* 3792.

50. Murdoch, J.R. *J. Am. Chem. Soc.* **1983,** *105,* 2159.

51. Lewis, E.S.; Hu, D.D. *J. Am. Chem. Soc.* **1984,** *106,* 3292.

52. Bordwell, F.G.; Boyle, W.J., Jr.; Hautala, J.A.; Yee, K.C. *J. Am. Chem. Soc.* **1969,** *91,* 7224.

53. Satchell, R.S.; Satchell, D.P.N. *Proc. Chem. Soc.* **1964,** 362.

54. Marcus, R.A. *J. Am. Chem. Soc.* **1969,** *91,* 7224.

55. Kresge, A.J. *J. Am. Chem. Soc.* **1970,** *92,* 3210.

56. Hughes, E.D.; Ingold, C.K.; Shapiro, U.G. *J. Chem. Soc.* **1936,** 225.

57. Albery, W.J. *Prog. React. Kinet.* **1967,** *4,* 355.

58. More O'Ferrall, R.A. *J. Chem. Soc.* **1970,** 274.

59. Jencks, W.P. *Chem. Revs.* **1972,** *72,* 705.
60. Kurz, J.L. *J. Org. Chem.* **1983,** *48,* 5117.
61. Pross, A.; Shaik, S.S. *J. Am. Chem. Soc.* **1981,** *103,* 3702.
62. Pross, A.; Shaik, S.S. *Acc. Chem. Res.* **1983,** *16,* 363.
63. Pross, A.; Shaik, S.S. *Tetrahedron Lett.* **1982,** *23,* 5467.
64. Pross, A. *J. Org. Chem.* **1984,** *49,* 1811.
65. Hill, S.V.; Longridge, J.L.; Williams, A. *J. Org. Chem.* **1984,** *49,* 1819.
66. Bernasconi, C.F. *Acc. Chem. Res.* **1987,** *20,* 301.
67. Thibblin, A. *J. Am. Chem. Soc.* **1988,** *110,* 4582.
68. Jencks, D.A.; Jencks, W.P. *J. Am. Chem. Soc.* **1977,** *99,* 7948.
69. Jencks, W.P. *Chem. Revs.* **1985,** *85,* 511.
70. Jones, P.G.; Kirby, A.J. *J. Am. Chem. Soc.* **1984,** *106,* 6207.
71. Shaik, S.S. *J. Am. Chem. Soc.* **1988,** *110,* 1127.
72. Agmon, N. *J. Org. Chem.* **1987,** *52,* 2192.
73. Guthrie, J.P., Pike, D.C. *Can. J. Chem.* **1987,** *65,* 1951.
74. Murdoch, J.R. *J. Am. Chem. Soc.* **1983,** *105,* 2660.
75. Grunwald, E. *J. Am. Chem. Soc.* **1985,** *107,* 125.
76. Grunwald, E. *J. Am. Chem. Soc.* **1985,** *107,* 4710. Grunwald, E. *J. Am. Chem. Soc.* **1985,** *107,* 4715.
77. Jencks, W.P.; Gilchrist, M. *J. Am. Chem. Soc.* **1968,** *90,* 2622.

PROBLEMS

1. Find K, the overall equilibrium constant, for Scheme II as a function of the rate constants.
2. For the oximation of a ketone (see Fig. 5-12 and accompanying text), sketch the free energy reaction coordinate diagram at pH 7 and pH 2.
3. Give an equation for $\Delta S^{\ddagger}$ in calories per mole per degree Kelvin. (This combination of units is called an entropy unit, eu.)
4. Collision theory leads to this equation for the rate constant: $k = A \exp(-E/RT) = A'T^{1/2} \exp(-E/RT)$. Show how the energy E is related to the Arrhenius activation energy E_a (presuming the Arrhenius preexponential factor is temperature independent).
5. Show that the slope of a plot of $\log(k/T)$ against $1/T$ is equal to $-\Delta H^{\ddagger}/2.3R$.
6. What condition is required in order that n_T (Eq. 5-60) and α (Eq. 5-67) be equal for all ΔG^0?
7. Jencks and Gilchrist[77] have shown that $\beta_{nuc} \approx 0.3$ and $\beta_{lg} \approx 0.3$ for the attack of strongly basic anionic oxygen nucleophiles on esters, where β_{nuc} is β for the attacking alkoxide ion and β_{lg} is for the leaving group. Construct an RIP diagram, showing structures of R, I^+, I^-, P, and the location of the transition state.
8. Let x be the normalized progress variable in a system subject to the Marcus equation (Eq. 5-69), so $x^{\ddagger} = \frac{1}{2} + \Delta G^0/8\Delta G_0^{\ddagger}$, where $x^{\ddagger}$ has the significance of α in Eq. (5-67). Then deduce this equation, which describes the energy change, relative to the reactant, over the reaction coordinate:

$$\Delta G = 4\Delta G_0^{\ddagger} x(1 - x) + x\Delta G^0$$

Hint: $x^{\ddagger}$ is the value of x when $d\Delta G/dx = 0$.

9. By analogy with Problem 8, we write for a system having two progress variables

$$G = 4\Delta G_0^{\ddagger}x(1 - x) + x\Delta G^0 - 4\Delta G_I^{\ddagger}y(1 - y) + y\Delta G_I^0 + C$$

where C is a constant. [See Ref. 75 for a justification of this equation.]

(a) Derive Eq. (5-79) for $x^{\ddagger}$ and $y^{\ddagger}$, the coordinates of the transition state.

(b) Derive Eq. (5-78).

CHAPTER 6

Phenomena for Study

Chapters 2, 3, and 4 treat the measurement of rate constants, and Chapter 5 describes physical ideas (theories) useful in providing a conceptual context for the interpretation of kinetic measurements. We are now in a position to explore the kinds of experiments that can yield kinetic information pertinent to chemically interesting problems. The general approach is to modify the reaction system or to apply a stress to it and to observe the response of the system to this stress. The stress may constitute a change in the reaction environment, such factors as temperature, pressure, pH, and solvent composition being useful experimental variables for this type of manipulation; or we may alter the system by introducing catalysts or by making structural changes in the reactants. Chapters 6, 7, and 8 deal with these methods for applying kinetics. Chapter 6 treats a miscellany of useful approaches. Chapter 7 deals with structural changes, and Chapter 8 with changes in the reaction medium. (Salt effects are included in Chapter 8, but pH effects are in Chapter 6; the separation is based on whether or not the kinetic effect can be described by a concentration term in the rate equation.)

6.1 TEMPERATURE DEPENDENCE OF RATES

The Arrhenius Equation

Chemical reaction rates increase with an increase in temperature because at a higher temperature, a larger fraction of reactant molecules possesses energy in excess of the reaction energy barrier. Chapter 5 describes the theoretical development of this idea. As noted in Section 5.1, the relationship between the rate constant k of an elementary reaction and the absolute temperature T is the *Arrhenius equation:*

$$k = Ae^{-E_a/RT} \tag{6-1}$$

In Eq. (6-1), A is called the *preexponential factor* and E_a is the *activation energy*. In this section we are concerned with the experimental evaluation of A and E_a and with their uses.

Usually the Arrhenius equation is placed in the linear form

$$\ln k = \ln A - E_a/RT \qquad (6\text{-}2)$$

and a plot of $\ln k$ against $1/T$ (called an Arrhenius plot) yields estimates of A and E_a from the intercept and slope. The differential form of Eq. (6-2), useful later in this section, is

$$\frac{d \ln k}{(d(1/T)} = -\frac{E_a}{R} \qquad (6\text{-}3)$$

From Eq. (6-1) it is evident that A has the units of k and that E_a has the units energy per mole. For many decades the usual units of E_a were kilocalories per mole, but in the International System of Units (SI) E_a should be expressed in kilojoules per mole (1 kJ = 4.184 kcal). In order to interpret the extant and future kinetic literature, it is essential to be able to use both of these forms.

Figure 6-1 is an Arrhenius plot for the chair–chair conformational inversion of cyclohexane, determined by NMR methods by Anet and Bourn.[1] (Of course logarithms to the base 10 can also be used, with the advantage of providing rapid order-of-magnitude interpretations.) The equation of the straight line is $\ln(k/s^{-1}) = 30.5 - 5600/T$, from which is found $A = 1.76 \times 10^{13}$ s^{-1} and $E_a = 11.2$ kcal mol^{-1}. Although the general appearance of Fig. 6-1 is common to nearly all Arrhenius plots, this example possesses several unusual features. One of these is the low temperatures of the rate studies (-24.0 to $-116.7°C$); most kinetic studies are around room temperature, so the abscissa on an Arrhenius plot is typically of the order $1/T \approx 0.003$. More importantly, the temperature range in Fig. 6-1 is unusually wide, and the study generated a large number of points. It is much more common to see Arrhenius studies of three to five points covering a range of 20–40°C. Notice the excellent linearity of the plot in Fig. 6-1.

The values of A and E_a provide a full description of the kinetic data, but it may be desirable, for mechanistic interpretation, to express the results in terms of the activation parameters $\Delta H^{\ddagger}$ and $\Delta S^{\ddagger}$. We developed equations in Section 5.2 for this purpose; for convenience these are repeated here:

$$\Delta H^{\ddagger} = E_a - RT \qquad (6\text{-}4)$$

$$\Delta S^{\ddagger} = R \ln A - R \ln (kT/h) - R) \qquad (6\text{-}5)$$

In these calculations T is taken in the middle of the experimental range; k is the Boltzmann constant and h is Planck's constant. From Eq. (5-48) it can be seen that a plot of $\ln(k/T)$ against $1/T$ has a slope of $-\Delta H^{\ddagger}/R$; such a graph is called an *Eyring plot*.

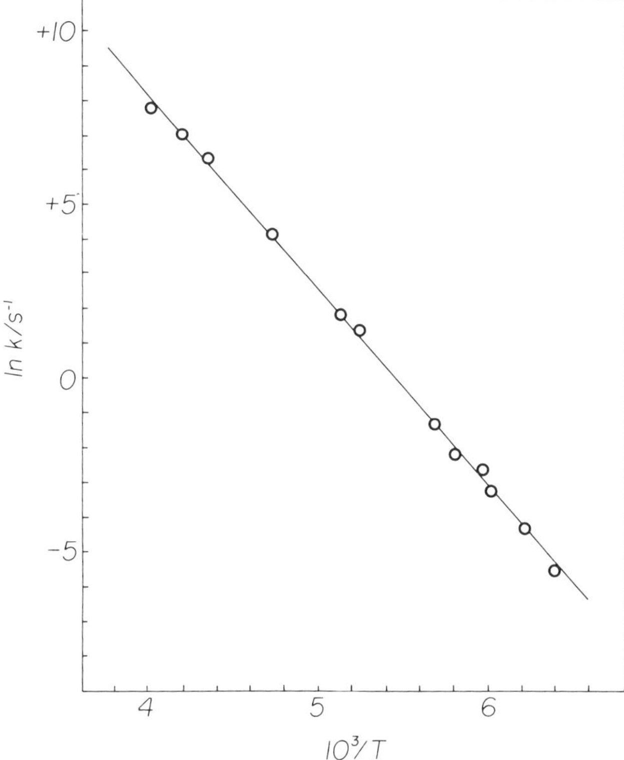

Figure 6-1. Arrhenius plot for the chair–chair ring inversion of cyclohexane.[1]

This is all very straightforward in principle, but in practice there may be difficulties. These arise from two sources: the limited temperature range that is usually studied and the very long extrapolation to $1/T = 0$ that is required to estimate ln A. It is, therefore, advisable to make use of statistical techniques in order to obtain the best parameter estimates, and we now consider this problem. We developed the necessary mathematics in Section 2.3.

We wish to apply weighted linear least-squares regression to Eq. (6-2), the linearized form of the Arrhenius equation. Let us suppose that our kinetic studies have provided us with data consisting of k_i, T_i, and σ_{ki}, for at least three temperatures, where σ_{ki} is the experimental standard deviation of k_i. We will assume that the error in T is negligible relative to that in k. For convenience we write Eq. (6-2) as

$$y = a + bx \qquad (6-6)$$

where $y = \ln k$, $x = 1/T$, $a = \ln A$, $b = -E_a/R$. The observation index i is omitted for simplicity.

In order to carry out the weighting of y values, we need σ_y^2, the variance of y. Applying the propagation of errors treatment (Eq. 2-66) to the function $y = \ln k$, we have

$$\sigma_y^2 = \left(\frac{dy}{dk}\right)^2 \sigma_k^2$$

$$\sigma_y^2 = \frac{\sigma_k^2}{k^2}$$

(6-7)

The weight of the ith observation is inversely proportional to the variance of the observation; we will use Eq. (2-82) for this quantity, n being the number of observations.

$$w_y = \frac{n/\sigma_y^2}{\Sigma(1/\sigma_y^2)}$$

(6-8)

Finally, from Eq. (2-78), we have the normal equations for weighted linear least-squares regression:

$$a\Sigma w + b\Sigma wx = \Sigma wy$$

(6-9a)

$$a\Sigma wx + b\Sigma wx^2 = \Sigma wxy$$

(6-9b)

We will apply these results to data on the addition of m-nitrophenylazide to norbornene in ethyl acetate:[2]

Table 6-1 lists the experimental quantities, k, T, σ_k, the transformed variables x, y, and the weights w. (It is necessary, in least-squares calculations, to carry many more digits than are justified by the significant figures in the data; at the conclusion, rounding may be carried out as appropriate.) The sums required for the solution of the normal equations are

$$\Sigma w = 3$$

$$\Sigma wx = 9.943214 \times 10^{-3}$$

$$\Sigma wx^2 = 32.977 \times 10^{-6}$$

$$\Sigma wy = -15.99700444$$

$$\Sigma wxy = -53.169043 \times 10^{-3}$$

Inserting these quantities into Eq. (6-9) and solving yields the parameters a and b.

We also want estimates of the uncertainties of the parameters. Equations (2-99) and (2-101), repeated here, provide these,

$$\sigma_a^2 = \sigma_y^2 \left[\frac{1}{n} + \frac{\bar{x}^2}{\Sigma(x - \bar{x})^2} \right] \tag{6-10}$$

$$\sigma_b^2 = \frac{\sigma_y^2}{\Sigma(x - \bar{x})^2} \tag{6-11}$$

where σ_y^2 measures the scatter of the experimental points about the least-squares line:

$$\sigma_y^2 = \frac{\Sigma(y_{\text{obs}} - y_{\text{calc}})^2}{n - 2} \tag{6-12}$$

Table 6-2 lists the results of the analysis. [To calculate σ_A^2 we use the relationship $\sigma_A^2 = A^2\sigma_a^2$; see Eq. 6-7).] The regression analysis was also carried out without weighting, that is, by setting $w = 1$ for each point; these results are included in Table 6-2.

Examination of Table 6-1 reveals how the weighting treatment takes into account the reliability of the data. The intermediate point, which has the poorest precision, is severely discounted in the least-squares fit. The most interesting features of Table 6-2 are the large uncertainties in the estimates of A and E_a. These would be reduced if more data points covering a wider temperature range were available; nevertheless it is common to find the uncertainty in E_a to be comparable to RT. The uncertainty of A is a consequence of the extrapolation to $1/T = 0$, which, in effect, is how $\ln A$ is determined. In this example, the data cover the range 0.003 23 to 0.003 41 in $1/T$, and the extrapolation is from 0.003 23 to zero; thus about 95% of the line constitutes an extrapolation over unstudied tempertures. Estimates of A and E_a are correlated, as are their uncertainties.[3]

TABLE 6-1. Data for Weighted Linear Least-Squares Arrhenius Analysis

T	$10^3 k/M^{-1}\ min^{-1}$	$10^3 \sigma_k/M^{-1}\ min^{-1}$	x	y	w
293.55	2.54	0.02	0.003 406 575	$-5.975\ 591\ 2$	1.343 109 485
298.25	3.99	0.16	0.033 528 92	$-5.523\ 964\ 05$	0.051 786 598
309.02	8.33	0.06	0.003 236 037	$-4.787\ 891\ 82$	1.605 103 917

TABLE **6-2.** Least-Squares Analysis of Kinetic Data in Table 6-1.

	Analysis	
Quantity	*Weighted*	*Unweighted*
a	17.92	16.79
b	-7016	-6670
σ_a^2	4.45	4.79
σ_b^2	0.401×10^6	0.431×10^6
$A/M^{-1}\ min^{-1}$	6.1×10^7	2.0×10^7
$E_a/kcal\ mol^{-1}$	13.9	13.3
$\sigma_A/M^{-1}\ min^{-1}$	12.8×10^7	4.3×10^7
$\sigma_{E_a}/kcal\ mol^{-1}$	1.3	1.3

This weighting procedure for the linearized Arrhenius equation depends upon the validity of Eq. (6-7) for estimating the variance of $y = \ln k$. It will be recalled that this equation is an approximation, achieved by truncating a Taylor's series expansion at the linear term. With poor precision in the data this approximation may not be acceptable. A better estimate may be obtained by truncating after the quadratic term; the result is

$$\sigma_y^2 = \frac{\sigma_k^2}{k^2}\left(1 - \frac{\sigma_k}{k} + \frac{\sigma_k^2}{4k^2}\right) \tag{6-13}$$

An alternative method for estimating σ_y^2 is described by Deming[4] and illustrated by Cvetanovic and Singleton.[5]

With computers we are able to use the Arrhenius equation in the form of Eq. (6-1) and to carry out nonlinear regression of k as a function of T. The principle of nonlinear least-squares regression was developed in Section 2.3. If the absolute error (σ_k) in k is constant, unweighted nonlinear regression is appropriate. [In this case, of course, weighting would be required in fitting data to Eq. (6-2). On the other hand, if the relative error σ_k/k is constant, weighting of the nonlinear regression is required, but unweighted linear regression is appropriate. See Problem 7.]

Kinetic studies at several temperatures followed by application of the Arrhenius equation as described constitutes the usual procedure for the measurement of activation parameters, but other methods have been described. Bunce et al.[6] eliminate the rate constant between the Arrhenius equation and the integrated rate equation, obtaining an equation relating concentration to time and temperature. This is analyzed by nonlinear regression to extract the activation energy. Another approach is to program temperature as a function of time and to analyze the concentration–time data for the activation energy. This "nonisothermal" method is attractive because it is efficient, but its use is not widespread.[7,8]

Curved Arrhenius Plots

Usually the Arrhenius plot of ln k vs. $1/T$ is linear, or at any rate there is usually no sound basis for concluding that it is not linear. This behavior is consistent with the conclusion that the activation parameters are constants, independent of temperature, over the experimental temperature range. For some reactions, however, definite curvature is detectable in Arrhenius plots. There seem to be three possible reasons for this curvature.

One possibility is that the curvature is an artifact introduced by a systematic error in the measurements. This is not unlikely, because rate constants may vary by orders of magnitude over a wide temperature range, necessitating different analytical methods or data treatments in different temperature regions. Careful experimental work should be able to identify such problems.

A more interesting possibility, one that has attracted much attention, is that the activation parameters may be temperature dependent. In Chapter 5 we saw that theory predicts that the preexponential factor contains the quantity T^n, where $n = \frac{1}{2}$ according to collision theory, and $n = 1$ according to the transition state theory. In view of the uncertainty associated with estimation of the preexponential factor, it is not possible to distinguish between these theories on the basis of the observed temperature dependence, yet we have the possibility of a source of curvature. Nevertheless, the exponential term in the Arrhenius equation dominates the temperature behavior. From Eq. (6-4), we may examine this in terms either of E_a or $\Delta H^{\ddagger}$. By analogy with equilibrium thermodynamics, we write

$$\Delta C_p^{\ddagger} = \frac{d\Delta H^{\ddagger}}{dT} \tag{6-14}$$

Thus curvature in an Arrhenius plot is sometimes ascribed to a nonzero value of $\Delta C_p^{\ddagger}$, the heat capacity of activation. As can be imagined, the experimental problem is very difficult, requiring rate constant measurements of high accuracy and precision. Figure 6-2 shows a curved Arrhenius plot for the neutral hydrolysis of methyl trifluoroacetate in aqueous dimethysulfoxide.[9] The rate constants were measured by conductometry, their relative standard deviations being 0.014 to 0.076%. The value of $\Delta C_p^{\ddagger}$ was estimated to be about -200 J mol^{-1} K^{-1}, with an uncertainty of less than 10 J mol^{-1} K^{-1}.

$\Delta C_p^{\ddagger}$ is interpreted as the difference in heat capacities between the transition state and the reactants, and it may be a valuable mechanistic tool. Most reported $\Delta C_p^{\ddagger}$ values are for reactions of neutral reactants to products, as in solvolysis reactions of neutral esters or aliphatic halides.[10,11] Because of the slight curvature seen in the Arrhenius plots, as exemplified by Fig. 6-2, the interpretation, and even the existence, of $\Delta C_p^{\ddagger}$ is a matter of debate. The subject is rather specialized, so we will not explore it deeply, but will outline methods for the estimation of $\Delta C_p^{\ddagger}$.

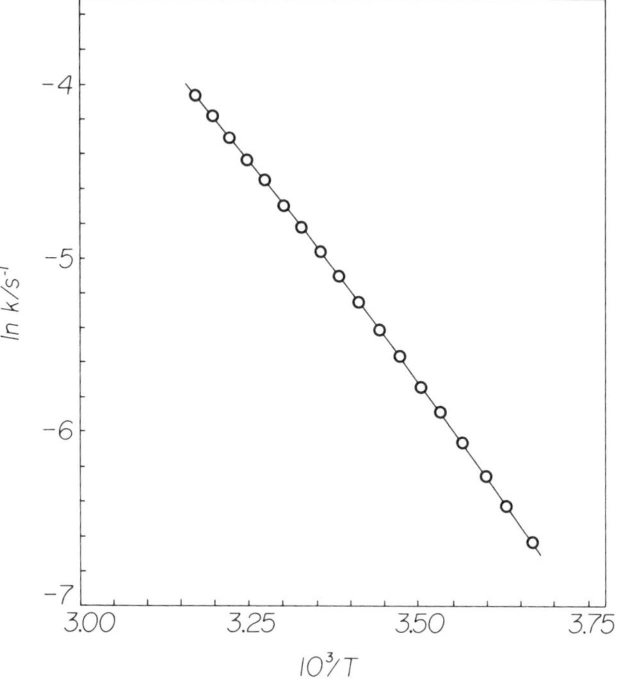

Figure 6-2. Curved Arrhenius plot for the hydrolysis of methyl trifluoroacetate in dimethylsulfoxide–water (mole fraction water = 0.973).[9]

These methods are based on extensions of the Arrhenius equation, which we can write

$$\ln k = A + B/T \tag{6-15}$$

where A and B are constants that are related to the activation parameters of the Arrhenius equation. [Note that A in Eq. (6-15) is *not* the Arrhenius preexponential factor.] By assuming that $\Delta C_p^{\ddagger}$ is a linear function of T, Eq. (6-16) can be derived.[11]

$$\ln k = A + B/T + C \ln T + DT \tag{6-16}$$

If $\Delta C_p^{\ddagger}$ is independent of temperature, the final term in Eq. (6-16) can be neglected. Clarke and Glew[12] expanded $\Delta H^{\ddagger}$ in a Taylor's series, truncating at the third derivative of $\Delta C_p^{\ddagger}$, and obtained Eq. (6-17).

$$\ln k = A + B/T + C \ln T + DT + ET^2 + FT^3 \tag{6-17}$$

They point out that it is invalid to retain a higher-order (in T) term while omitting a lower-order term, for this is inconsistent with the Taylor's series expansion. Least-squares regression is used to evaluate the parameters. Kanerva et al.[9] interpreted the data in Fig. 6-2 by means of Eq. (6-17).

Wold[13] proposed Eq. (6-18), finding that it yielded curve fits as good as Eq. (6-16), with $D = 0$.

$$\ln k = A + B/T + C/T^2 \tag{6-18}$$

Adams et al.[14] introduced an empirical expression, Eq. (6-19).

$$\ln k = A + BT^C \tag{6-19}$$

Equation (6-19) was said to provide a fit as good as or better than those with other equations. The parameters were evaluated by fixing C and carrying out a linear least-squares regression of $\ln k$ on T^C; C was then altered and the procedure was repeated. The residual sum of squares was taken as a criterion of best fit.

In each case the parameters A, B, C, . . . can be related to the activation quantities $\Delta H^{\ddagger}$, $\Delta S^{\ddagger}$, $\Delta C_p^{\ddagger}$ via the relationship $d \ln k/dT = - E_a/RT^2$, Eqs. (6-4) and (6-14), and the functional relationship assumed between $\Delta C_p^{\ddagger}$ and T.

Reported $\Delta C_p^{\ddagger}$ values for solvolytic reactions are in the range $- 10$ to $- 100$ cal mol^{-1} K^{-1}.[11] A negative value of $\Delta C_p^{\ddagger}$ means that $\Delta H^{\ddagger}$ and $\Delta S^{\ddagger}$ decrease as temperature increases. Several authors[9-11,15] have concluded that $\Delta C_p^{\ddagger}$ values are real quantities and that they constitute valuable mechanistic information, but this view is not unanimous. The difficulty, as seen by skeptics,[16] is that a curved Arrhenius plot may be a consequence of a complex reaction. This is the third of the possibilities, mentioned earlier, that can produce curvature in an Arrhenius plot. If the measured rate constant is a function of several elementary reactions, each with its own (temperature-independent) activation energy, the observed Arrhenius plot may be linear or curved depending upon the functional relationship of the observed constant to the elementary constants and the magnitudes of the activation energies.[7,17] The apparent activation energies of multistep enzyme-catalyzed reactions can be related to the rate constants and activation energies of the individual steps.[18] Evidently an interpretation of curvature in terms of a $\Delta C_p^{\ddagger}$ value, when a complex reaction is responsible, will produce a $\Delta C_p^{\ddagger}$ that is an artifact.[16] Some gas phase reactions show highly curved Arrhenius plots, and these have been interpreted in terms of a complex reaction with a stable intermediate.[19,20]

Standard States

A first-order rate constant has the dimension time^{-1}, but all other rate constants include a concentration unit. It follows that a change of concentration scale results in a change in the magnitude of such a rate constant. From the equilibrium assumption of transition state theory we developed these equations in Chapter 5:

$$k = \frac{kT}{h} K^{\ddagger} \tag{6-20}$$

$$\Delta G^{\ddagger} = -RT \ln K^{\ddagger} \tag{6-21}$$

$$\Delta G^{\ddagger} = \Delta H^{\ddagger} - T\Delta S^{\ddagger} \tag{6-22}$$

These apply to a bimolecular reaction in which two reactant molecules become a single particle in the transition state. It is evident from Eqs. (6-20) and (6-21) that a change in concentration scale will result in a change in the magnitude of $\Delta G^{\ddagger}$. An Arrhenius plot is, in effect, a plot of $\Delta G^{\ddagger}$ against $1/T$. Because a change in concentration scale alters the intercept but not the slope of an Arrhenius plot, we conclude that the values of $\Delta G^{\ddagger}$ and $\Delta S^{\ddagger}$, but not of $\Delta H^{\ddagger}$, depend upon the concentration scale employed for the expression of reactant concentrations. We, therefore, wish to know which concentration scale is the preferred one in the context of mechanistic interpretation, particularly of $\Delta S^{\ddagger}$ values.

The mole fraction (x), molality (m), and molarity (c) are the concentration scales in common use. These are related by Eqs. (6-23) and (6-24),

$$x_i = \frac{c_i}{1000\rho} \left(\frac{\Sigma n_j M_j}{\Sigma n_j} \right) \tag{6-23}$$

where ρ is the solution density, Σn_j is the sum of the number of moles, and M_j is the molecular weight of constituent j.

$$x_i = \frac{m_i}{1000} \left(\frac{n_1 M_1}{\Sigma n_j} \right) \tag{6-24}$$

In Eq. (6-24) M_1 is the molecular weight of the solvent. The quantity $1000/m_1 = M^*$ is the number of moles of solvent in 1 kg of solvent; this can be calculated for mixed solvents as well as for pure solvents.

In very dilute solutions of solute i, the several concentration scales become proportional to each other as expressed in Eqs. (6-25) and (6-26).

$$x_i = \frac{c_i M_1}{1000\rho_1} \tag{6-25}$$

$$x_i = \frac{m_i M_1}{1000} \tag{6-26}$$

The transition state theory allows us to apply results of equilibrium thermodynamics, and we, therefore, write, for a reactant or transition state species i,

$$\mu_i = \mu_i^0 + RT \ln a_i \tag{6-27}$$

where μ_i is the chemical potential (partial molar free energy) of i and a_i is its activity. The quantity μ_i^0 is the standard chemical potential, which is the value of μ_i when $a_i = 1$. The state of a system in which the activity of substance i is unity is the *standard state* for that substance.

The activity is related to concentration by

$$a_i = \gamma_i c_i \tag{6-28}$$

where γ_i is called the *activity coefficient*. The *reference state* is that state of the system for which the activity coefficient is unity. An equation like Eq. (6-28) can be written for each concentration scale, giving, in combination with Eq. (6-27),

$$\mu = \mu_x^0 + RT \ln \gamma_x x$$

$$\mu = \mu_c^0 + RT \ln \gamma_c c$$

$$\mu = \mu_m^0 + RT \ln \gamma_m m$$

For a substance in a given system the chemical potential μ has a definite value; however, the standard potentials and activity coefficients have different values in these three equations. Therefore, the selection of a concentration scale in effect determines the standard state.

There is no compelling thermodynamic reason to prefer one concentration scale to another. (The mole fraction and molality scales are temperature independent, which may be a practical reason to prefer them to the molarity scale.) However, there is a significant difference in that the standard state on the mole fraction scale is the pure substance, whereas the molality and molarity scale standard states approximate to the $1m$ and $1M$ solutions; that is, they are mixtures. Gurney[21] has analyzed the problem of the choice of standard state and shows that the mole fraction scale leads to values for the free energy and entropy changes that do not include the contribution from mixing (which is of no interest in terms of the chemical reaction being studied). This point is particularly important when comparing results in different solvents, for then the mixing contribution varies with the solvent, being a function of the quantity M^*. The arguments have been given in detail elsewhere in the context of equilibrium studies.[22,23] Gurney[21] calls the quantities evaluated on the mole fraction scale *unitary* thermodynamic quantities.

For ease in converting from the commonly used molarity scale to the mole fraction (unitary) scale, let us develop expressions suitable to the bimolecular reaction case,

$$A + B \rightarrow M^\ddagger$$

Define $K_x^\ddagger = x_M/x_A x_B$ and $K_c^\ddagger = c_M/c_A c_B$, convert molar concentrations to mole fractions using the dilute solution limit, Eq. (6-25), thus obtaining $K_x^\ddagger = \rho_1 M^* K_c^\ddagger$. Using Eqs. (6-21) and (6-22) yields

$$\Delta G_x^\ddagger = \Delta G_c^\ddagger - RT \ln \rho_1 M^* \tag{6-29}$$

$$\Delta S_x^\ddagger = \Delta S_c^\ddagger + R \ln \rho_1 M^* \tag{6-30}$$

For example, in aqueous solution $\Delta S_x^\ddagger = \Delta S_c^\ddagger + 8.0$ cal mol^{-1} K^{-1}; thus, it is advisable to be cautious in making mechanistic inferences based on the magnitude or sign of the entropy of activation. In reporting $\Delta S\ddagger$ values, the standard state (concentration scale) should be specified.

Complications in Buffered Solutions

Reactions catalyzed by hydrogen ion or hydroxide ion, when studied at controlled pH, are often described by pseudo-first-order rate constants that include the catalyst concentration or activity. Activation energies determined from Arrhenius plots using the pseudo-first-order rate constants may include contributions other than the activation energy intrinsic to the reaction of interest. This problem was analyzed for a special case by Higuchi et al.;[24] the following treatment is drawn from a more general analysis.[25]

We suppose that the molar concentration scale is used; brackets signify concentration, and parentheses signify conventional activity as determined by pH measurements. First, consider an acid-catalyzed reaction having the rate equation

$$v = k_H[\text{reactant}][H^+]$$

giving the pseudo-first-order rate constant

$$k_{obs} = k_H(H^+)$$

where we suppose that (H^+) has been determined from the measured pH. Taking logarithms and differentiating gives

$$E_{obs} = E_H - R\frac{d \ln (H^+)}{d(1/T)} \tag{6-31}$$

where Eq. (6-3) has been used; E_{obs} is the observed activation energy, and E_H is the activation energy of the acid-catalyzed reaction. Because the activity is related to concentration by $(H^+) = \gamma_H[H^+]$, Eq. (6-31) can be expressed

$$E_{obs} = E_H - R\frac{d \ln \gamma_H}{d(1/T)} - R\frac{d \ln [H^+]}{d(1/T)} \tag{6-32}$$

The last term in Eq. (6-32) describes the temperature dependence of the molar concentration; in water, this contributes only about -45 cal mol^{-1} to E_{obs} at room temperature.[25] In a strong mineral acid solution, the temperature dependence of the activity coefficient term[26] contributes about -90 cal mol^{-1}. These are small quantities relative to the uncertainty in E_{obs}.

The situation is more interesting when the solution is buffered with a weak acid, say HA. Then, if $K'_a = (H^+)[A^-]/[HA]$ is the apparent acid dissociation constant, we have

$$\frac{d \ln (H^+)}{d(1/T)} = \frac{d \ln K'_a}{d(1/T)} - \frac{d \ln r}{d(1/T)} \tag{6-33}$$

where $r = [A^-]/[HA]$, the buffer ratio. This ratio is invariant with temperature. The van't Hoff equation for the temperature dependence of the dissociation constant is

$$\frac{d \ln K_a'}{d(1/T)} = - \frac{\Delta H_a}{R} \tag{6-34}$$

Combining Eqs. (6-31), (6-33), and (6-34),

$$E_{\text{obs}} = E_H + \Delta H_a \tag{6-35}$$

In Eq. (6-35), ΔH_a is the molar heat of ionization of the buffer acid at the conditions (temperature, solvent composition) of the kinetic studies. It happens that for many commonly used acidic buffers this quantity is small. Harned and Owen[27] give $\Delta H_a^0 = -0.09$ kcal/mol for acetic acid at 25°C, for example. The very important buffer of dihydrogen phosphate–monohydrogen phosphate is controlled by pK_2 of phosphoric acid; at 25°C its heat of ionization is -0.82 kcal/mol.

Turning to base-catalyzed reactions, the most common formulation is

$$k_{\text{obs}} = k_{\text{OH}}(\text{OH}^-)$$

in which (OH^-) is derived from a pH measurement via the autoprotolysis constant, $K_w = (\text{H}^+)(\text{OH}^-)$. Combining these equations, differentiating, and making substitutions as before gives

$$E_{\text{obs}} = E_{\text{OH}} + \Delta H_w + R \frac{d \ln (\text{H}^+)}{d(1/T)} \tag{6-36}$$

where ΔH_w is the heat of ionization of water. At very high pH, where the alkali hydroxides may be used to hold pH constant, the analysis is similar to the treatment for strong acids. The activity coefficient and thermal expansion terms will usually be negligible.

Suppose now that the pH is controlled by a weak base buffer, the equilibrium being written $\text{BH}^+ \rightleftharpoons \text{B} + \text{H}^+$, where B signifies a neutral base. The apparent dissociation constant is $K_a' = (\text{H}^+)[\text{B}]/[\text{BH}^+]$. Following the earlier argument, we obtain

$$E_{\text{obs}} = E_{\text{OH}} + \Delta H_w - \Delta H_a \tag{6-37}$$

The quantities ΔH_w and ΔH_a require consideration. These are molar heats of ionization at the conditions of the kinetic measurements. The thermodynamic heat of ionization of water in pure water, ΔH_w^0, is a function of temperature; Harned and Owen[27, pp. 644–7] give for this quantity

$$\Delta H_w^0 = 23984.15 - 23.497T - 0.039025T^2 \text{ cal mol}^{-1}$$

For example, at 25°C, $\Delta H_w^0 = 13.51$ kcal mol^{-1}; at 50°C, $\Delta H_w^0 = 12.32$ kcal mol^{-1}. (This variation could conceivably lead to nonlinearity in Arrhenius plots of k_{obs}.) It seems appropriate to use, in Eq. (6-37), a heat of ionization of water in the middle of the temperature range of the kinetic studies.

Aside from its temperature dependence, ΔH_w depends upon solvent composition, this dependence being expressible by Eq. (6-38),

$$\Delta H_w = \Delta H_w^0 + a\mu^{1/2} \tag{6-38}$$

where μ is ionic strength and a is a constant. Harned and Owen[27, pp. 644-647] found that Eq. (6-38) is followed up to an ionic strength of unity; the value of a depends upon the identity of the salt. At an ionic strength of 0.1 M, the a term can contribute 50–150 cal mol^{-1} to ΔH_w.

These same dependencies will, in general, apply to the heat of ionization of the buffer acid, ΔH_a. Thermodynamic quantities, namely, ΔH_a^0, have been reported for some buffer substances, and it is found that ΔH_a^0 is temperature dependent. Bates and Hetzer[28] studied the temperature dependence of K_a for the important buffer tris(hydroxymethyl)aminomethane (TRIS), finding

$$\Delta H_a^0 = 57080 - 0.1067T^2 \text{ J mol}^{-1}$$

These authors also list thermodynamic quantities for many other amines. ΔH_a^0, at 25°C, ranges from 3.37 kcal mol^{-1} for the very weakly basic 2,2′-bipyridinium to 13.88 kcal mol^{-1} for n-butylammonium; for TRIS at 25°C, $\Delta H_a^0 = 11.38$ kcal mol^{-1}.

It may happen that ΔH_a is not available for the buffer substance used in the kinetic studies; moreover the thermodynamic quantity ΔH_a^0 is not precisely the correct quantity to use in Eq. (6-37) because it does not apply to the experimental solvent composition. Then the experimentalist can determine ΔH_a. The most direct method is to measure ΔH_a calorimetrically; however, few laboratories are equipped for this measurement. An alternative approach is to measure K_a', under the kinetic conditions of temperature and solvent; this can be done potentiometrically or by potentiometry combined with spectrophotometry. Then, from the slope of the plot of log K_a' against $1/T$, ΔH_a is calculated. Although this value is not thermodynamically defined (since it is based on the assumption that ΔH_a is temperature independent), it will be valid for the present purpose over the temperature range studied.

An alternative calculational method is based directly on Eq. (6-36), which can be put into the equivalent form, Eq. (6-39).

$$E_{obs} = E_{OH} + \Delta H_w + 2.3RT^2 \frac{dpH}{dT} \tag{6-39}$$

Bates and Bower[29] have reported values of dpH/dT for several important basic buffers.

On the basis of this analysis, there are three different experimental designs available to the kineticist for the evaluation of the intrinsic activation energy. The following discussion is directed to the case of a reaction in the presence of a weak base buffer, but the same considerations, appropriately modified, apply to the other cases.

1. Prepare the solutions and measure the pH at one temperature of the kinetic study. Of course, the pH meter and electrodes must be properly calibrated against standard buffers, all solutions being thermostated at the single temperature of measurement. Carry out the rate constant determinations at three or more tempertures; do not measure the pH or change the solution composition at the additional temperatures. Determine E_{obs} from an Arrhenius plot of log k_{obs} against $1/T$. Then calculate E_{OH} using Eq. (6-37) or (6-39) and the appropriate values of ΔH_w and ΔH_a as discussed above.

Notice that the pH is allowed to vary with temperature in this method, Eqs. (6-37) and (6-39) taking this variation into account. Notice also that the measured value of pH is nowhere used in finding E_{OH}; it is only needed to calculate k_{OH}, when the value of K_w appropriate to the experimental temperature must be used in calculating (OH$^-$).

2. Prepare the solutions, thermostat them at the temperatures to be used in the rate study, and then adjust them all to the same pH value by the addition of small volumes of concentrated strong acid or base. The pH meter must be correctly calibrated at each temperature. Now carry out the kinetic study and calculate E_{obs}. Because this procedure has set $d \ln (H^+)/d(1/T) = 0$ experimentally, use Eq. (6-36) in the form $E_{obs} = E_{OH} + \Delta H_w$.

3. Prepare the solutions, thermostat them at the rate study temperatures, and measure the pH at each temperature, taking the correct precautions concerning calibration of the pH meter. Now convert each pH to (OH$^-$), using (OH$^-$) = $K_w/(H^+)$, where K_w must be appropriate to each temperature. Then calculate $k_{OH} = k_{obs}/(OH^-)$ and make the Arrhenius plot with k_{OH}, obtaining E_{OH} directly.

In reporting the results of such studies, the experimental design should be carefully described so readers are able to assess the precise significance of the reported values.

Uses of Activation Parameters

We can make two different uses of the activation parameters $\Delta H^\ddagger$ and $\Delta S^\ddagger$ (or, equivalently, E_a and A). One of these uses is a very practical one, namely, the use of the Arrhenius equation as a guide for interpolation or extrapolation of rate constants. For this purpose, rate data are sometimes "stored" in the form of the Arrhenius equation. For example, the data of Table 6-1 may be represented (see Table 6-2) as

$$\ln k = 17.92 - 7016/T$$

or as

$$k = 6.061 \times 10^7 \, e^{-13941/RT}$$

More digits are retained in such presentations than are required to express the experimental precision in order that rounding errors be minimized.

In industrial applications the Arrhenius equation is often used to extrapolate outside the experimental temperature range. *Accelerated stability studies* are exemplified by kinetic studies on drug dosage forms; very commonly rate constants are measured at several temperatures in the range 40–70°C, and extrapolation to room temperature (25°C) is made with the Arrhenius equation. All extrapolations are potentially misleading, and the possible weaknesses in the present example are well recognized. Important among these is the possibility that more than one reaction is responsible for the observed chemical change and that the proportions of the reactions contributing to the overall process change with the temperature. Eriksen and Stelmach[7] have discussed this situation.

It is interesting to consider the values of activation energies likely to be encountered in solution reactions. In surveying literature reports we must be aware that most reactions are complex and that complications like those described earlier for buffered solutions may be present. Nevertheless, practical information is available. Kennon[30] found the average activation energy (E_a) of 38 pharmaceuticals to be 19.8 kcal mol^{-1}; this value includes multiple determinations for some drugs. Figure 6-3 shows a frequency distribution for 147 activation energies of drug decomposition reactions drawn from a compendium of literature sources.[31] The mean value of E_a is 21.0 kcal mol^{-1}, and 88% of the values fall in the range 10–30 kcal mol^{-1}. Most of these reactions are hydrolyses, but some oxidations and other reaction types are represented. Table 5-1 lists 15 more reactions; the mean value of E_a in this list is 21.3 kcal mol^{-1}. The rule of thumb that reaction rates increase by a factor of 2 to 3 for each 10°C rise in temperature is based on

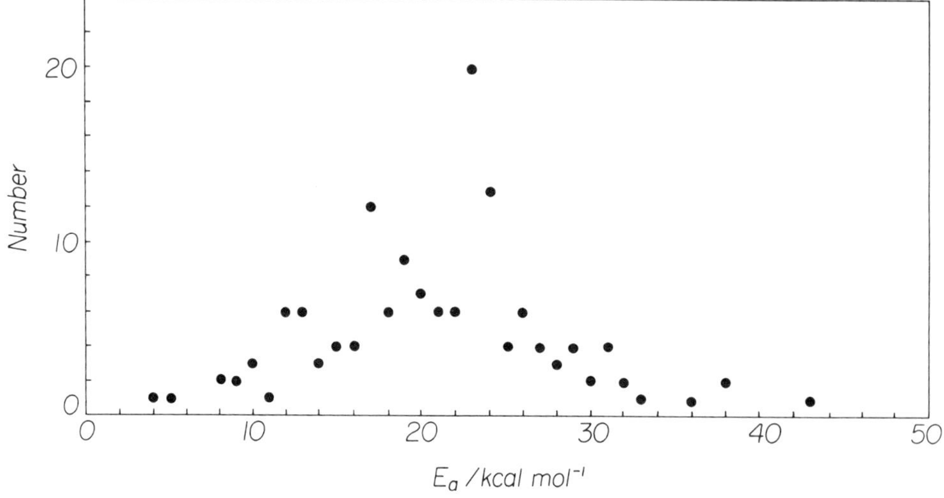

Figure 6-3. Distribution of Arrhenius activation energies for 147 drug decomposition reactions.

observations of typical E_a values like these. A convenient calculational method is provided by Eq. (6-40), which is easily obtained from the Arrhenius equation.

$$\ln \frac{k_2}{k_1} = \frac{E_a}{R} \left(\frac{T_2 - T_1}{T_1 T_2} \right)$$
(6-40)

Thus, it is found, for $T_1 = 300$ K and $T_2 = 310$ K, that if $E_a = 10$ kcal mol^{-1}, $k_2/k_1 = 1.72$; if $E_a = 15$, $k_2/k_1 = 2.24$; if $E_a = 20$, $k_2/k_1 = 2.95$; and if $E_a = 25$, $k_2/k_1 = 3.86$. Incidentally, Eq. (6-40) shows why in temperature studies of rates the activation energy is always reported, whereas the preexponential factor may not be.

The second use of activation parameters is as criteria for mechanistic interpretation. In this application the activation parameters of a single reaction are, by themselves, of little use; such quantities acquire meaning primarily by comparison with other values. Thus, the trend of activation parameters in a reaction series may be suggestive. For example, many linear correlations have been reported between $\Delta H^\ddagger$ and $\Delta S^\ddagger$ within a reaction series; such behavior is called an *isokinetic relationship*, and its significance is discussed in Chapter 7. In Section 5.3 we commented on the use of $\Delta S^\ddagger$ to determine the molecularity of a reaction. Carpenter[32] has described examples of mechanistic deductions from activation parameters of organic reactions.

6.2 THE EFFECT OF PRESSURE

The dependence of the rate constant on pressure provides another activation parameter of mechanistic utility. From thermodynamics we have $(\partial G/\partial P)_T = V$, where V is the molar volume (partial molar volume in solutions). We define the free energy of activation by $\Delta G^\ddagger = G^\ddagger - \Sigma G_R$, where ΣG_R is the sum of the molar free energies of the reactants. Thus, we obtain

$$\left(\frac{\partial \Delta G^\ddagger}{\partial P} \right)_T = V^\ddagger - \Sigma V_R$$

$$= \Delta V^\ddagger$$
(6-41)

where $\Delta V^\ddagger$ is called the *activation volume*. From transition state theory (Section 5.2) we can write $\Delta G^\ddagger = -RT \ln k + $ constant; therefore,

$$\left(\frac{\partial \ln k}{\partial P} \right)_T = - \frac{\Delta V^\ddagger}{RT}$$
(6-42)

The activation volume can be positive or negative, and its sign is a valuable piece of mechanistic information. Although the measurement of $\Delta V^\ddagger$ is not commonly carried out in solution kinetic studies, the necessary equipment is commercially available, and many $\Delta V^{\ddagger}$ values have been reported. Several reviews are available.[33-37]

$\Delta V^{\ddagger}$ is itself pressure dependent, which complicates its measurement. There is no theory for the pressure dependence of $\Delta V^{\ddagger}$, so an empirical approach is used. Usually ln k is represented as a quadratic in the pressure,

$$\ln k = a + bP + cP^2$$

It is conventional to take as "the" activation volume the value of $\Delta V^{\ddagger}$ when $P = 0$, namely $-bRT$. (This is essentially equal to the value at atmospheric pressure.) Pressure has usually been measured in kilobars (kbar), or 10^9 dyn cm^{-2}; 1 kbar $= 986.92$ atm. The currently preferred unit is the pascal (Pa), which is 1 N m^{-2}; 1 kbar $= 0.1$ GPa. Measurements of $\Delta V^{\ddagger}$ usually require pressures in the range 0–10 kbar. The units of $\Delta V^{\ddagger}$ are cubic centimeters per mole; most $\Delta V^{\ddagger}$ values are in the range -30 to $+30$ cm^3 mol^{-1}, and the typical uncertainty is 1 cm^3 mol^{-1}. Rate constant measurements should be in pressure-independent units (mole fraction or molality), not molarity.[38]

If $\Delta V^{\ddagger}$ is negative, an increase in pressure will increase the reaction rate, and this effect can be exploited to optimize yields in synthetic procedures.[39] However, the goal in most pressure studies has been the measurement of $\Delta V^{\ddagger}$ in order to learn something about the reaction mechanism. Glasstone et al.[40] showed how a simple model of the transition state could provide a basis for calculating an estimate of $\Delta V^{\ddagger}$, which could be compared with an experimental result. Many reactions of known mechanism have been studied, and these have yielded estimates of the effects to be expected from characteristic phenomena.[33-35] For example, the formation of one bond contributes about -10 cm^3 mol^{-1} to $\Delta V^{\ddagger}$, whereas the cleavage of one bond adds $+10$ cm^3 mol^{-1} to $\Delta V^{\ddagger}$. Ionization contributes about -20 cm^3 mol^{-1}, which might seem inconsistent with the bond cleavage contribution, but is a result of strong ion-solvent interactions, which increase the local density relative to the bulk density; this phenomenon is called *electrostriction*. Usually we anticipate that $\Delta V^{\ddagger}$, the activation volume, is a smaller (absolute) number than ΔV, the reaction volume for the overall reaction, although exceptions are known.[37]

$\Delta V^{\ddagger}$ measurements can be helpful in distinguishing between a concerted and a two-step process[37] or between a unimolecular and a bimolecular pathway. Thus, $\Delta V^{\ddagger}$ appears to be a useful criterion to distinguish between the unusual E1cb (unimolecular elimination from the conjugate base) ester hydrolysis and the common $B_{Ac}2$ route.[41] A distinction between S_N1 and S_N2 mechanisms is not clear-cut at low to moderate pressures because of the important contribution of the electrostriction effect to $\Delta V^{\ddagger}$ in the S_N1 ionization; but at high pressures the solvent compressibility is small, the electrostriction becomes less important, and the sign of $\Delta V^{\ddagger}$ becomes diagnostic, being positive for S_N1 and negative for S_N2 reactions. In this way it was shown that the solvolysis of isopropyl bromide is an S_N1 reaction.[42]

6.3 CATALYSIS

Definitions and Examples

Definitions of catalysis take several forms, not all of them equivalent.

1. A catalyst is a substance that increases the rate of a reaction without affecting the position of equilibrium. It follows that the rate in the reverse direction must be increased by the same factor as that in the forward direction. This is a consequence of the principle of microscopic reversibility (Section 3.3), which applies at equilibrium, and rates are often studied far from equilibrium.
2. A catalyst is a substance whose concentration appears in the rate equation to a higher power than it does in the stoichiometric equation.[43]
3. A catalyst is a substance that increases the rate of a reaction, other than by a medium effect, regardless of the ultimate fate of this substance. For example, in hydroxide-catalyzed ester hydrolysis the catalyst OH⁻ is consumed by reaction with the product acid; some writers, therefore, call this a *hydroxide-promoted reaction,* because the catalyst is not regenerated, although the essential chemical event is a catalysis.
4. A catalyst is a substance that makes available a reaction path with a lower free energy of activation than is available in its absence. (The catalyst does not "lower $\Delta G^{\ddagger}$"; the uncatalyzed reaction path remains available.)

Catalysis occurs because the catalyst in some manner increases the probability of reaction. This may result from the reactants being brought closer together [*catalysis by approximation,*[44] or the *propinquity effect*[45]], or somehow assisted to achieve the necessary relative orientation for reaction. Noncovalent interactions may be responsible for the effect. Covalent bond changes may also take place in catalysis. In a formal way, the manner in which catalysis occurs can be described by schemes such as Schemes I and II.

$$A + X \rightleftharpoons C$$

$$C + B \rightarrow M + N + X$$

Scheme I

$$A + X \rightleftharpoons C' + M$$

$$C' + B \rightarrow N + X$$

Scheme II

In each case the overall reaction is $A + B \rightarrow M + N$, the catalyst X being regenerated. The rate constant for the reaction between C and B (Scheme I) or C' and B (Scheme II) must be larger than that between A and B; this is the essence of catalysis.

Several types of catalysis are described by terms denoting the catalyst structure or function.

1. *Specific acid catalysis* is catalysis by the hydronium ion (in water) or the lyonium ion in general. Acid-catalyzed ester hydrolysis is an example.

$$
\begin{array}{c}
\text{O} \qquad\qquad\qquad\qquad\qquad\; ^{+}\text{OH} \\
\parallel \qquad\qquad \text{fast} \qquad \parallel \\
\text{R}-\text{C}-\text{OR}' + \text{H}^{+} \;\rightleftharpoons\; \text{R}-\text{C}-\text{OR}'
\end{array}
$$

$$
\begin{array}{c}
^{+}\text{OH} \qquad\qquad\qquad \text{OH} \qquad\qquad ^{+}\text{OH} \\
\parallel \qquad\quad \text{slow} \qquad | \qquad\qquad\quad \parallel \\
\text{R}-\text{C}-\text{OR}' + \text{H}_2\text{O} \;\rightleftharpoons\; \text{R}-\text{C}-\text{OR}' \;\rightleftharpoons\; \text{R}-\text{C}-\text{OH} + \text{R}'\text{OH} \\
| \\
^{+}\text{OH}_2
\end{array}
$$

$$
\begin{array}{c}
^{+}\text{OH} \qquad\qquad\qquad \text{O} \\
\parallel \qquad\qquad\qquad \parallel \\
\text{R}-\text{C}-\text{OH} \;\rightleftharpoons\; \text{R}-\text{C}-\text{OH} + \text{H}^{+}
\end{array}
$$

(Note that the reverse reaction is acid-catalyzed ester formation.) The rate equation is

$$
v = k_{\text{H}}[\text{S}][\text{H}^{+}] = k_{\text{H}}K_a[\text{SH}^{+}]
$$

where S represents the unprotonated ester. Experimentally it is found that k_{H} is about $10^{-4}\ \text{M}^{-1}\text{s}^{-1}$, and because $\text{p}K_a$ is about -6 to -8 for esters, we calculate the rate constant $k_{\text{H}}' = k_{\text{H}}K_a$ for the kinetically equivalent form to be 10^2–$10^4\ \text{s}^{-1}$.

2. *Specific base catalysis* is catalysis by the hydroxide ion (in water) or the lyate ion in general. Consider hydroxide-catalyzed ester hydrolysis:

$$
\begin{array}{c}
\text{O} \qquad\qquad\qquad\qquad \text{O}^{-} \qquad\qquad\qquad \text{O} \\
\parallel \qquad\qquad\qquad\quad | \qquad\qquad\qquad\; \parallel \\
\text{R}-\text{C}-\text{OR}' + \text{OH}^{-} \;\rightleftharpoons\; \text{R}-\text{C}-\text{OR}' \;\rightarrow\; \text{R}-\text{C}-\text{O}^{-} + \text{R}'\text{OH} \\
| \\
\text{OH}
\end{array}
$$

The second step is essentially irreversible because of the $\text{p}K_a$ of the product acid; thus, the catalyst is consumed. If the substrate is ionizable (say HS), a rate term $k[\text{HS}][\text{OH}^{-}]$ is kinetically indistinguishable from the term $k'[\text{S}^{-}]$. Thus, the mechanistic involvement of a catalytic hydroxide ion is not proved, in such a case, by the rate equation.

3. *General acid catalysis* is a catalysis by a Brønsted acid (other than the lyonium ion) acting by donating a proton. The addition of thiols to the carbonyl group is general acid catalyzed.[46]

$$
\begin{array}{ccc}
\overset{\displaystyle O}{\underset{\displaystyle R'-C-R'}{\|}} \; H-A & \rightleftharpoons & \overset{\displaystyle OH}{\underset{\displaystyle R'-C-R'}{|}} + A^- \\
\overset{\displaystyle S}{\diagup \diagdown} & & + S H \\
R \qquad H & & R \\
& & OH \\
& & R'-C-R' + H^+ \\
& & SR
\end{array}
$$

The product is a hemithioacetal. In the rate-determining step, the general acid HA donates a proton to the carbonyl oxygen, thus assisting the nucleophilic attack of the thiol on the carbonyl carbon.

4. *General base catalysis* is catalysis by a Brønsted base acting by accepting a proton. The halogenation of ketones, a reaction discussed in Section 5.3, is general base catalyzed.

$$
\overset{O}{\overset{\|}{CH_3\,C\,CH_3}} + B \rightleftharpoons \overset{O}{\overset{\|}{CH_3\,C\,CH_2^-}} + BH^+
$$

$$
\downarrow X_2
$$

$$
\overset{O}{\overset{\|}{CH_3\,C\,CH_2X}} + X^-
$$

The Brønsted base B abstracts a proton from the ketone in the rate-determining step.

5. *Electrophilic catalysis* is catalysis by an electrophile (Lewis acid) acting as an electron-pair acceptor. For example, metal ions catalyze the decarboxylation of dimethyloxaloacetic acid.[47]

$$
\begin{array}{l}
\overset{\displaystyle Cu^{2+}}{} \\
\overset{-O \quad O \quad CO_2^-}{\underset{O{=}C-C-C-Me}{|\quad\|\quad|}} \;\; \rightarrow \;\; ^-O_2C-C-\bar{C}-Me + CO_2 + Cu^{2+} \\
\qquad\qquad | \qquad\qquad\qquad\qquad | \\
\qquad\qquad Me \qquad\qquad\qquad\qquad Me
\end{array}
$$

6. *Nucleophilic catalysis* is catalysis by a general base (electron-pair donor) acting by donating its electron pair to an atom (usually carbon) other than hydrogen. Nucleophilic catalysis is exemplified by the imidazole-catalyzed hydrolysis of a phenyl acetate. (The tetrahedral intermediates are not shown.)

$$
CH_3C(=O){-}OAr \;\longrightarrow\; CH_3C(=O){-}N\text{(imidazole)} + HOAr
$$

$$
\xrightarrow{\;H_2O\;}
$$

$$
CH_3COOH \;+\; \text{imidazole}
$$

This catalysis follows Scheme II; that catalysis occurs means that N-acetylimidazole is more susceptible to hydrolysis than is the ester. (This does not explain why the N-acetylimidazole is formed, however.)

We should distinguish between the phrases *nucleophilic attack* and *nucleophilic catalysis*. Nucleophilic attack means the bond-forming approach by an electron pair of the nucleophile to an electron-deficient site on the substrate. In nucleophilic catalysis this results in an increase in the rate of reaction relative to the rate in the absence of the catalyst. However, nucleophilic attack may not result in catalysis. Thus, if methylamine is reacted with a phenyl acetate, the reaction observed is amide formation, not hydrolysis, because the product of the nucleophilic attack is more stable than is the ester to hydrolysis.

$$
CH_3C(=O){-}OAr + CH_3NH_2 \;\rightarrow\; CH_3C(=O){-}NHCH_3 + HOAr
$$

The rate equation for this reaction is expected to be $v = k[\text{RCOOR}'][\text{MeNH}_2]$, but it is possible for terms like $k'[\text{RCOOR}'][\text{MeNH}_2]^2$ or $k''[\text{RCOOR}'][\text{MeNH}_2][\text{MeNH}_3{}^+]$ to be present. The k' term could be a general base catalysis or nucleophilic attack, the k'' term, general acid catalysis of nucleophilic attack.

Yet another distinction is between *intermolecular catalysis,* in which the catalytic function and the reaction site are on different molecules, and *intramolecular catalysis,* in which the catalytic function and the reaction site are within the same molecule. All of the above examples constitute intermolecular catalyses. The following reaction, the hydrolysis of a monomaleate ester, is an intramolecular nucleophilic catalysis.[48]

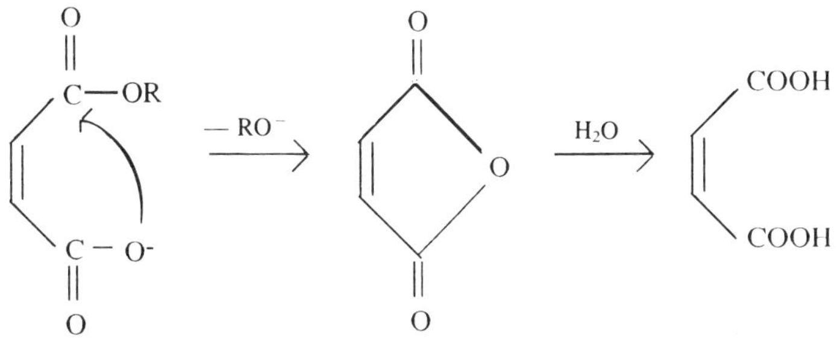

As with intermolecular catalysis, the form of the rate equation may not decisively indicate the mechanism of the catalysis because of kinetic equivalences. Consider a substrate containing the acyl function –COX and an ionizable catalytic function –YH.

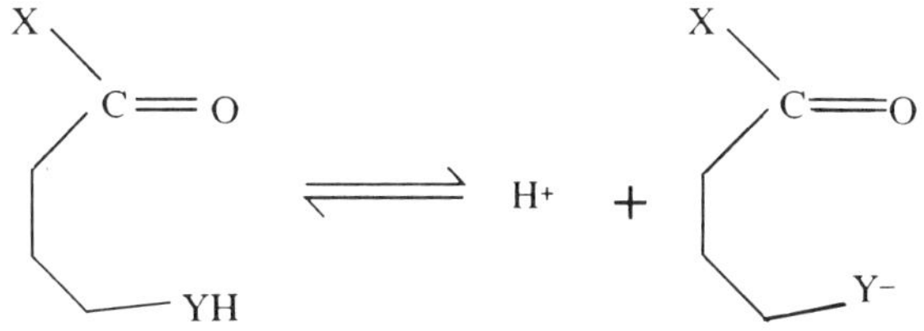

Three kinetically equivalent rate terms involving intramolecular participation are shown in Table 6-3 with representations of appropriate transition states (mechanisms). Differentiation among these possibilities can be difficult.

TABLE 6-3. Kinetically Equivalent Intramolecular Catalysis Mechanisms

Transition state	Mechanism	Rate term form
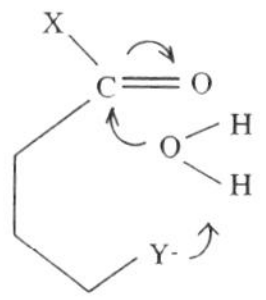	Intramolecular general base	$[\text{Anion}][\text{H}_2\text{O}]$
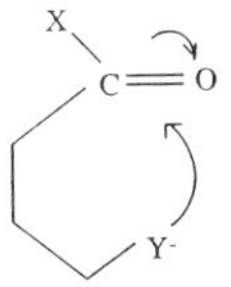	Intramolecular nucleophilic	$[\text{Anion}]$
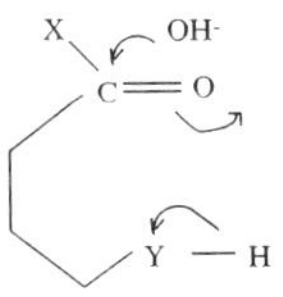	Intramolecular general acid–intermolecular nucleophilic	$[\text{Acid}][\text{OH}^-]$

Books by Bruice and Benkovic,[45] Jencks,[44] and Bender[49] are rich sources of examples of catalytic processes and their mechanisms.

Determination of Catalytic Rate Constants

Much of the study of kinetics constitutes a study of catalysis. The first goal is the determination of the rate equation, and examples have been given in Chapters 2 and 3, particularly Section 3.3, "Model Building." The subsection following this one describes the dependence of rates on pH, and most of this dependence can be ascribed to acid–base catalysis. Here we treat a very simple but widely applicable method for the detection and measurement of general acid–base or nucleophilic catalysis. We consider aqueous solutions where the pH and pK_a concepts are well understood, but similar methods can be applied in nonaqueous media.

An effective experimental design is to measure the pseudo-first-order rate constant k at constant pH and ionic strength as a function of total buffer concentration B_t. Very often the buffer substance is the catalyst. Let B represent the conjugate base form of the buffer. Because pH is constant, the ratio $[B]/[BH^+]$ is constant, and the concentrations of both species increase directly with B_t, where $B_t = [B] + [BH^+]$.

The simplest rate equation (model) to be tested is

$$k = k_1[H^+] + k_2 + k_3[OH^-] + k_b[B] + k_a[BH^+] \tag{6-43}$$

where the rate terms have these probable meanings:

k_1 Specific acid catalysis.

k_2 Uncatalyzed ("water," "spontaneous") reaction.

k_3 Specific base catalysis.

k_b General base or nucleophilic catalysis.

k_a General acid catalysis.

At any fixed pH the quantity k_0, given by Eq. (6-44), is a constant, which can be determined in the absence of B.

$$k_0 = k_1[H^+] + k_2 + k_3[OH^-] \tag{6-44}$$

Let us define fractions by

$$F_B = \frac{[B]}{B_t} = \frac{K_a}{K_a + [H^+]} \tag{6-45a}$$

$$F_{BH} = \frac{[BH^+]}{B_t} = \frac{[H^+]}{K_a + [H^+]} \tag{6-45b}$$

where K_a is the acid dissociation constant of BH^+. Combining these relationships gives

$$k = k_0 + (k_b F_B + k_a F_{BH})B_t \qquad (6\text{-}46)$$

showing that k is a linear function of total buffer concentration at constant pH if Eq. (6-43) is valid. Since $F_B + F_{BH} = 1$, we also get

$$\frac{k - k_0}{B_t} = k_a + (k_b - k_a)F_B \qquad (6\text{-}47)$$

Thus, a plot of $(k - k_0)/B_t$ against F_B should be linear; the intercept at $F_B = 0$ is k_a, and the intercept at $F_B = 1$ is k_b. Note that the quantity $(k - k_0)/B_t$ is an apparent second-order rate constant; some authors use the symbol k_2' or k_{cat} for this quantity. Observe also that k_0 has a different value at each pH.

Figure 6-4 is a plot according to Eq. (6-46) for the hydrolysis of *trans*-cinnamic anhydride in the presence of carbonate buffers.[50] The nonzero slopes indicate the existence of *buffer catalysis,* and the increasing slope value with increasing pH shows that k_b must be larger than k_a. From the intercepts k_0 is obtained at each pH, and the plot according to Eq. (6-47) is shown in Fig. 6-5. This plot shows that k_a is negligible. With this information, Eq. (6-46) can be simplified to

$$k = k_0 + k_b[B] \qquad (6\text{-}48)$$

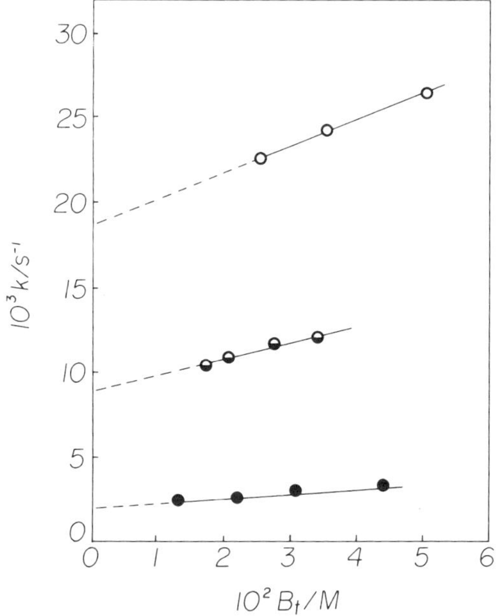

Figure 6-4. Plot of Eq. (6-46) for the hydrolysis of cinnamic anhydride at 25°C and ionic strength 0.1 M in carbonate buffers. B_t represents the total buffer concentration. From top to bottom, pH = 10.06, 9.76, and 9.14.

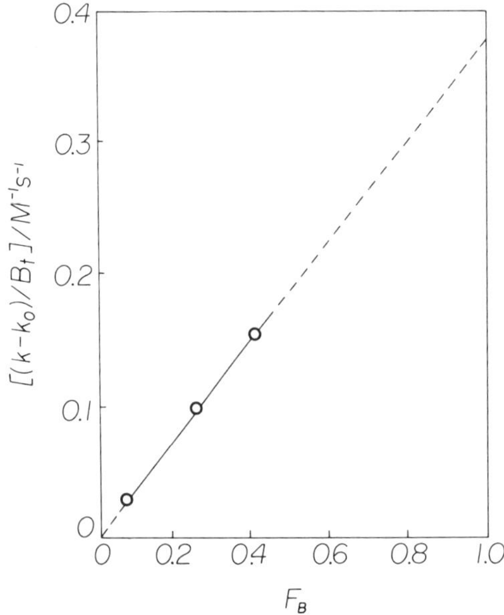

Figure 6-5. Plot according to Eq. (6-47) of the results in Fig. 6-4. The points, from top to bottom, correspond to pH 10.06, 9.76, and 9.14.

so that a plot of k against the conjugate base concentration yields k_b as the slope; this plot is shown in Fig. 6-6, where it is seen that the slopes are essentially the same at the three pH values.

Positive curvature in a plot of k against B_t (at constant pH) indicates the presence of a rate term with an order higher than unity with respect to B_t. Negative curvature can be caused by complex formation between reactants.[51]

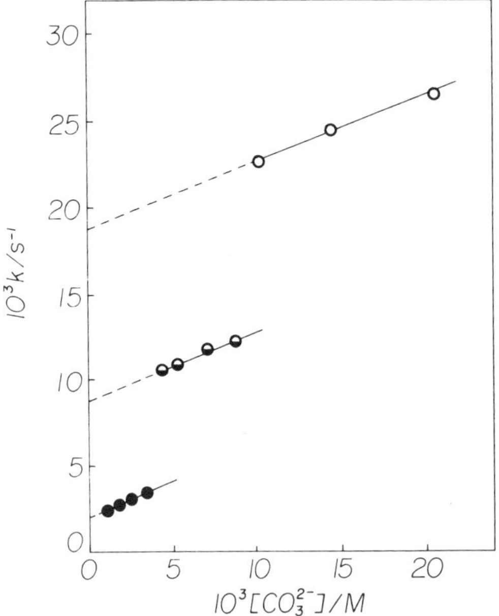

Figure 6-6. Plot of Eq. (6-48) for the data of Fig. 6-4.

Distinction between General Base and Nucleophilic Catalysis

The rate constant k_b, determined by means of Eq. (6-47) or (6-48), may describe either general base or nucleophilic catalysis. To distinguish between these possibilities requires additional information. For example, in Section 3.3, we described a kinetic model for the N-methylimidazole-catalyzed acetylation of alcohols and experimental designs for the measurement of catalytic rate constants. These are summarized in Scheme XVIII of Section 3.3, which we present here in slightly different form.

$$\begin{array}{c} k_A\,[\text{ROH}] \\ k_{AN}\,[\text{N}]\,[\text{ROH}] \\ \text{AX} \xrightarrow{\hspace{3cm}} \text{AOR} + \text{HX} \end{array}$$

$$\text{AX} + \text{N} \;\rightleftharpoons\; \text{I}^+ + \text{X}^-$$

$$\begin{array}{c} k_I\,[\text{ROH}] \\ k_{IN}\,[\text{N}]\,[\text{ROH}] \\ \text{I}^+ \xrightarrow{\hspace{3cm}} \text{AOR} + \text{NH}^+ \end{array}$$

Scheme III

In Scheme III, AX is the acetylating agent, N is N-methylimidazole, and I^+ is an intermediate formed by the reaction of N with AX; the extra piece of information in this case was the spectrophotometric detection of I^+.[52] We can describe the rate constants in this way:

k_A Uncatalyzed reaction.

k_{AN} General base catalysis.

k_I Nucleophilic catalysis.

k_{IN} General base catalysis of nucleophilic catalysis.

The detection of an intermediate is not the only kind of information that can assist in determining the mechanism of the reaction. Other methods for discriminating between general base and nucleophilic reactions are kinetic in nature. The ratio of catalytic rate constants for a substrate with imidazole and with monohydrogen phosphate is an indicator of mechanism, because these two reagents have about the same basicity. If they both operate by a general base mechanism, the rate ratio should therefore be about unity. If, on the other hand, they both operate by nucleophilic attack (or if imidazole does), the rate of imidazole catalysis will be much larger than that of phosphate catalysis, because imidazole is a much better nucleophile. The hydroxide/imidazole ratio tells if the imidazole reaction is general

TABLE 6-4. Methods for Differentiating between General Base and Nucleophilic Catalysis

Method	Criterion	Example	Reference
Product analysis	Product identity or distribution attributable only to reaction via intermediate formed by nucleophilic attack	[^{18}O]Acetate-catalyzed hydrolysis of 2,4-dinitrophenyl benzoate gives [^{18}O]-benzoic acid	86
Trapping of intermediate	Product identity attributable only to diversion of intermediate by reaction with added trapping nucleophile	Acetanilid formation in acetate-catalyzed hydrolysis of aryl acetates in presence of aniline	87
Physical detection of intermediate	Change of physical property during reaction attributable to intermediate formed by nucleophilic attack	Spectrophotometric observation of acetylpyridinium ion in pyridine-catalyzed hydrolysis of acetic anhydride	88
Demonstration of reversible nucleophilic reaction	Inhibition of catalysis by added product of reaction	Inhibition of nucleophilic component of acetate-catalyzed acetylimidazole hydrolysis by added imidazole	89
Solvent D$_2$O kinetic isotopic effect	$k_H/k_D > 2$ for general base, 0.8–1.9 for nucleophilic catalysis (usually)	$k_H/k_D = 3.1$ for imidazole-catalyzed hydrolysis of ethyl trifluorothiolacetate	90
Relative reactivity of hindered and unhindered bases	k(hindered)/k(unhindered) same order for bases of same pK_a if general base catalysis, but this ratio very small for nucleophilic catalysis	2,6-Lutidine is much less effective than pyridine in catalysis of acetic anhydride hydrolysis	91
Relative reactivity of imidazole and monohydrogen phosphate	k(imidazole)/k(HPO$_4^{2-}$) is order of unity for general base catalysis, but about 10^3 for nucleophilic catalysis	This ratio is 0.25 for ethyl acetate hydrolysis and 4.7×10^3 for p-nitrophenyl acetate hydrolysis	92
Relative reactivity of hydroxide and imidazole	k(hydroxide)/k(imidazole) is 10^5–10^6 for general base, but 10–10^3 for nucleophilic catalysis	This ratio is 9.1×10^5 for ethyl acetate hydrolysis and 160 for p-nitrophenyl acetate hydrolysis	92

base or nucleophilic, because hydroxide attacks as a nucleophile. Because imidazole is a better nucleophile than it is a base, the ratio will be high for a general base and low for a nucleophilic reaction. In Table 6-4 several criteria are described, with examples drawn from acyl-transfer reactions. Johnson[53] has given many more examples.

6.4 pH EFFECTS

Preliminaries

That the rates of many reactions are markedly dependent upon the acidity or alkalinity of the reaction medium has been known for many decades. In this section, the kinetic analysis of reactions in dilute aqueous solution in which pH is the accessible measure of acidity is presented in sufficient detail to allow the experimentalist to interpret data for most of the systems likely to be encountered and to extend the treatment to cases not covered here. This section is based on an earlier discussion.[22, pp. 76–97] The problem has also been analyzed by Van der Houwen et al.[54]

Throughout this section the hydronium ion and hydroxide ion concentrations appear in rate equations. For convenience these are written $[H^+]$ and $[OH^-]$. Usually, of course, these quantities have been estimated from a measured pH, so they are conventional activities rather than concentrations. However, our present concern is with the formal analysis of rate equations, and we can conveniently assume that activity coefficients are unity or are at least constant. The basic experimental information is k, the pseudo-first-order rate constant, as a function of pH. Within a series of such measurements the ionic strength should be held constant. If the pH is maintained constant with a buffer, k should be measured at more than one buffer concentration (but at constant pH) to see if the buffer affects the rate. If such a dependence is observed, the rate constant should be measured at several buffer concentrations and extrapolated to zero buffer to give the correct k for that pH.

Except for those reactions whose characteristic rate constants vary linearly with the hydronium or hydroxide ion concentration, the most effective presentation of pH–rate data is a graphical one. Two kinds of plot (*pH–rate profiles*) are commonly seen:

1. The observed rate constant is plotted on the vertical axis against pH on the horizontal axis. This plot is particularly suited for the estimation of dissociation constants.
2. The logarithm of the rate constant is plotted against pH. This plot has two advantages—the order with respect to H^+ or OH^- is readily apparent from linear portions of the curve, and a great range of values of the rate constant can be exhibited on one graph.

The initial goal of the kinetic analysis is to express k as a function of $[H^+]$, pH-independent rate constants, and appropriate acid–base dissociation constants. Then numerical estimates of these constants are obtained. The theoretical pH–rate profile can now be calculated and compared with the experimental curve. A quantitative agreement indicates that the proposed rate equation is consistent with experiment. It is advisable to use other information (such as independently measured dissociation constants) to support the kinetic analysis.

The manner of the analysis is to locate characteristic features of the pH–rate curve—linear segments, maxima, minima, and inflection points—and to relate the value of pH at which they occur to the parameters of the (assumed) rate equation. Because many rate equations are quite complicated, the only feasible differentiation is with respect to $[H^+]$, yielding the derivative $dk/d[H^+]$. This, however, does not correspond to the slope of either of the plots employed in kinetic graphing and must be converted. The appropriate conversion equations, which are obtained from the basic relations $d \ln = du/u$ and $\ln u = 2.303 \log u$, are

$$\frac{dk}{d\,\mathrm{pH}} = -2.3[H^+]\frac{dk}{d[H^+]} \tag{6-49}$$

$$\frac{d\log k}{d\,\mathrm{pH}} = -\frac{[H^+]}{k}\frac{dk}{d[H^+]} \tag{6-50}$$

$$\frac{d^2k}{d\,\mathrm{pH}^2} = -2.3[H^+]\frac{d^2k}{d\,\mathrm{pH}\,d[H^+]} \tag{6-51}$$

$$\frac{d^2\log k}{d\,\mathrm{pH}^2} = -2.3[H^+]\frac{d^2\log k}{d\,\mathrm{pH}\,d[H^+]} \tag{6-52}$$

One useful fact demonstrated by these formulas is that the location of maxima and minima is unaffected by the type of plot, so the derivative $dk/d[H^+]$ can be used in locating these features.

Throughout this section attention is restricted to rate equations that include concentrations of only the substrate, H^+, and OH^-. The observed first-order rate constant, therefore, contains concentrations of only H^+ and OH^- (the quantity $[OH^-]$ is often replaced by $K_w/[H^+]$). The substrate (reactant) may be ionizable. Rate equations containing the concentration of additional solutes (especially catalytic additives) can be developed as shown in Section 6.3.

Curves without Inflection Points

If a pH–rate curve does not exhibit an inflection, then very probably the substrate does not undergo an ionization in this pH range. The kinds of substrates that often lead to such simple curves are nonionizable compounds subject to hydrolysis, such as esters and amides. Reactions other than hydrolysis may be characterized by similar behavior if catalyzed by H^+ or OH^-. The general rate equation is

$$v = k_1[H^+]^n[S] + k_2[S] + k_3[OH^-]^m[S] \tag{6-53}$$

although sometimes a rate equation may contain terms in, say, both the first order and second order in some catalytic species. S represents the substrate. The experimental first-order rate constant, determined at essentially constant pH, is defined by $v = k[S]$, therefore,

$$k = k_1[H^+]^n + k_2 + k_3[OH^-]^m \qquad (6\text{-}54)$$

Plots of k or log k against pH for this equation yield a so-called U graph or a V graph. The analysis of this curve will provide estimates of k_1, k_2, k_3, n, and m.

We introduce an approximation that is subsequently used many times, and that is indispensable. This is to consider only a portion of the curve and to neglect those terms describing the rest of the curve. It is necessary to exercise some chemical discretion in applying such approximations. The relative values of the rate constants and concentrations determine the approximations that can be safely made, and the level of uncertainty that one may be willing to introduce in this way is gauged by a consideration of the experimental error in the raw data. Consider, in the present case, the very acid region ($[H^+]$ is large, pH is low). Then in most cases Eq. (6-54) reduces to (6-55) since $[OH^-] = K_w/[H^+]$.

$$k = k_1[H^+]^n \qquad (6\text{-}55)$$

Taking logarithms,

$$\log k = \log k_1 - n\text{pH} \qquad (6\text{-}56)$$

Thus in the low pH region a plot of log k against pH should give a straight line with slope equal to $-n$, yielding the order with respect to hydronium ion. Usually $n = 1$.

Similar reasoning shows that in regions of high basicity a plot of log k against pH will be linear with slope $+m$. Usually in a log k–pH plot a positive constant slope indicates dependence upon hydroxide ion, and a negative constant slope shows dependence upon the hydronium ion. This is one of the conveniences of the log k–pH plot. If k_2 is not negligible relative to the other terms, evidently a straight line will not be obtained; however, at very low or very high pH this would be a most unusual circumstance. An example of the kind of graph described is shown in Fig. 6-7, in which log k is plotted against pH for the alkaline hydrolysis of ethyl p-nitrobenzoate.[55] The slope of the line is $+1.00$; hence the reaction is first-order in hydroxide ion.

The pH at which the minimum occurs in a plot of Eq. (6-54) is calculated for the most common case, $n = m = 1$. The derivative of $dk/d[H^+] = k_1 - k_3 K_w/[H^+]^2$ is set equal to zero, giving

$$\text{pH}_{\min} = \tfrac{1}{2}\,\text{p}K_w + \tfrac{1}{2}\log\frac{k_1}{k_3} \qquad (6\text{-}57)$$

If $k_1 = k_3$, $\text{pH}_{\min} = 1/2\,\text{p}K_w$. This is an unusual condition; it has been observed[56] in the hydrolysis of acetamide at 100°C. Since $\text{p}K_w = 12.32$ at 100°C, the minimum rate in this reaction occurs at pH 6.16. For ester hydrolyses, k_3 is usually greater than k_1, and the minimum is observed near pH 5–6 (at room temperature). Equation (6-57) is used in the construction of a calculated pH–rate profile, when it allows

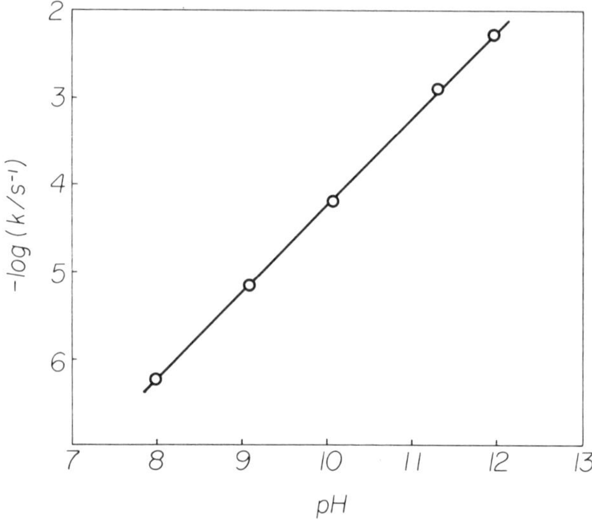

Figure 6-7.   Log k–pH plot for the alkaline hydrolysis of ethyl p-nitrobenzoate.

the position of the minimum to be easily located. This equation evidently yields the pH at which the reaction rate is minimal, and so it is useful in selecting the pH at which the substrate is most stable.

The constant k_1 is readily calculated from data at low pH, where Eq. (6-55) holds, because n is now known. Similarly k_3 can be calculated from data at high pH.

If k_2 is sufficiently large that it cannot be neglected (which will be revealed by the inconstancy of the "constant" calculated as above), then Eq. (6-54) may be put into the forms

$$k = k_2 + k_1[\text{H}^+]^n \qquad \text{(at low pH)} \tag{6-58}$$

$$k = k_2 + k_3[\text{OH}^-]^m \qquad \text{(at high pH)} \tag{6-59}$$

A plot of k against $[\text{H}^+]^n$ at low pH should give a straight line with intercept k_2 and slope k_1; at high pH the plot of k against $[\text{OH}^-]^m$ similarly permits the evaluation of k_2 and k_3.

Figure 6-8 is a pH–rate profile for the hydrolysis of p-nitrophenyl acetate. The slopes of the straight-line portions are -1, 0, and $+1$, reading in the acid to base direction, and this system can be described by

$$k = k_1[\text{H}^+] + k_2 + k_3[\text{OH}^-] \tag{6-60}$$

The constants k_1 and k_2 were evaluated from the measurements at low and high pH as described earlier, and k_2 was calculated from the measurement at pH 4.65 after taking into account any significant contribution from the k_1 and k_3 terms. The smooth

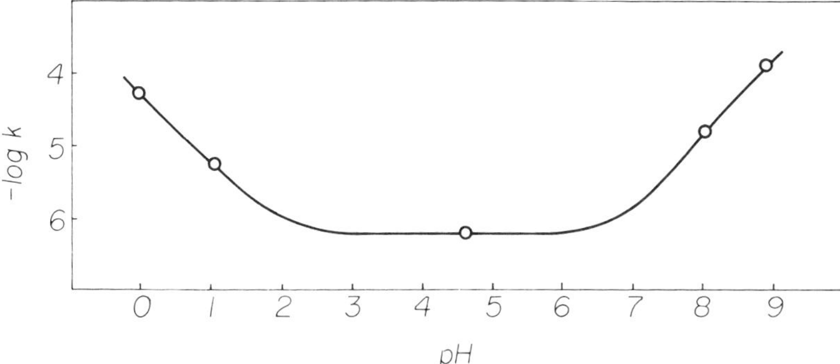

Figure 6-8. pH–rate profile for the hydrolysis of *p*-nitrophenyl acetate at 25°C in aqueous solution containing 1% acetonitrile. Ionic strength = 0.1 M except at pH 0.

curve in Fig. 6-8 was calculated with Eq. (6-60) and the estimated rate constants, which have these values: $k_1 = 5.34 \times 10^{-5} \text{ M}^{-1} \text{ s}^{-1}$; $k_2 = 5.5 \times 10^{-7} \text{ s}^{-1}$; $k_3 = 12.6 \text{ M}^{-1} \text{ s}^{-1}$. (The reaction described by the k_2 term is variously called the *water, uncatalyzed,* or *spontaneous* reaction, and some authors write $k_2 = k_w [\text{H}_2\text{O}]$, expressing [$\text{H}_2\text{O}$] as the molar concentration in calculating k_2.)

If a substrate contains an ionizable group that dissociates in the pH range being kinetically studied, one may expect a change in direction in the pH–rate curve in that pH region corresponding to the ionization of the substrate; this is because the conjugate acid and base forms are unlikely to undergo reaction at the same velocity. Such an identity of rates has apparently been observed, however, in the hydrolysis of nicotinamide.[57] The pH–rate curve is essentially V shaped with no inflection corresponding to the ionization of the substrate.

Sigmoid Curves

An inflection point in a pH–rate profile suggests a change in the nature of the reaction caused by a change in the pH of the medium. The usual reason for this behavior is an acid–base equilibrium of a reactant. Here we consider the simplest such system, in which the substrate is a monobasic acid (or monoacidic base). It is pertinent to consider the mathematical nature of the acid–base equilibrium. Let HS represent a weak acid. (The charge type is irrelevant.) The acid dissociation constant, $K_a = [\text{H}^+][\text{S}^-]/[\text{HS}]$, is taken to be appropriate to the conditions (temperature, ionic strength, solvent) of the kinetic experiments. The fractions of solute in the conjugate acid and base forms are given by

$$F_{\text{HS}} = \frac{[\text{HS}]}{S_t}$$

$$F_{\text{S}} = \frac{[\text{S}^-]}{S_t}$$

where S_t, the total molar concentration of solute, is

$$S_t = [HS] + [S^-]$$

Combining these equations,

$$F_{HS} = \frac{[H^+]}{[H^+] + K_a} \tag{6-61}$$

$$F_S = \frac{K_a}{[H^+] + K_a} \tag{6-62}$$

Figure 6-9 shows F_{HS} and F_S plotted against pH, according to Eqs. (6-61) and (6-62), for a weak acid of $pk_a = 4.0$. Because of their appearance such curves are called S-shaped or sigmoid curves.

These curves have some interesting properties. At any given pH, evidently $F_{HS} + F_S = 1$. At the point where the two curves cross, $F_{HS} = F_S = 0.5$, and from Eqs. (6-61) and (6-62), at this point $[H^+] = K_a$, or pH $= pK_a$. That this point corresponds to the inflection point can be shown by taking the second derivative d^2F/dpH^2 and setting this equal to zero; one finds $pH_{infl} = pK_a$. In the limit as $[H^+]$ becomes much greater than K_a, F_{HS} approaches unity and F_S approaches zero

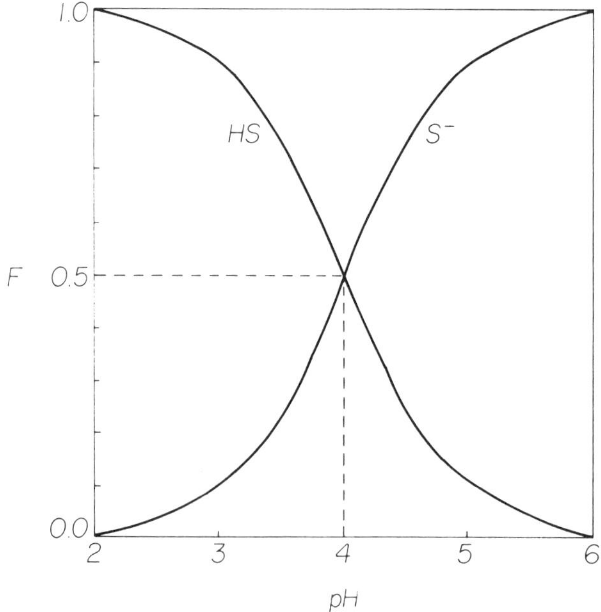

Figure 6-9. Variation with pH of the fractions F_{HS} (conjugate acid) and F_S (conjugate base) for an acid with $pK_a = 4.0$.

(although theoretically they never attain these limiting values). Similarly, as $[H^+]$ becomes much less than K_a, F_{HS} approaches zero and F_S approaches unity.

Now suppose that the substrate of a chemical reaction is a weak acid, with both the conjugate acid HS and conjugate base S^- being capable of undergoing reaction. Usually these two species will react at different rates because of the considerable difference in their electronic configurations. The rate equation for this system is

$$v = k'[HS] + k''[S^-] \qquad (6\text{-}63)$$

Combining this with the preceding equations,

$$v = \left(\frac{k'[H^+] + k''K_a}{[H^+] + K_a} \right) S_t$$

Because the experimental rate equation is $v = kS_t$, the first-order rate constant k becomes

$$k = \frac{k'[H^+] + k''K_a}{[H^+] + K_a} \qquad (6\text{-}64)$$

which relates k to the substrate dissociation constant, the hydronium ion concentration, and the pH-independent rate constants characteristic of the reactions of the two forms of the substrate.

If k' is much larger than k'', Eq. (6-64) takes the form of Eq. (6-61) for the fraction F_{HS}; thus we may expect the experimental rate constant to be a sigmoid function of pH. If k'' is larger than k', the k–pH plot should resemble the F_S–pH plot. Equation (6-64) is a very important relationship for the description of pH effects on reaction rates. Most sigmoid pH–rate profiles can be quantitatively accounted for with its use. Relatively minor modifications [such as the addition of rate terms first-order in H^+ or OH^- to Eq. (6-63)] can often extend the description over the entire pH range.

We consider first the k–pH plot corresponding to Eq. (6-64). Figure 6-10 is such a plot calculated with Eq. (6-64) and the typical values $pK_a = 4.0$, $k' = 1.10^{-6}\ s^{-1}$, and $k'' = 1 \times 10^{-3}\ s^{-1}$. Note that this plot is similar to the F_S–pH plot in Fig. 6-9. Differentiating Eq. (6-64) to find $dk/d[H^+]$ and using the identity equation (6-49) gives the first derivative of the k–pH plot. The second derivative is found with the aid of Eq. (6-51). Imposing the condition $k' \neq k''$ and setting the second derivative equal to zero leads to a simple expression for the pH at the inflection point.

$$pH_{infl} = pK_a \qquad (6\text{-}65)$$

This property of the sigmoid curve permits K_a to be easily estimated. This is an advantage of the k–pH plot. If the inflection point cannot be accurately located, the dissociation constant may still be estimated. Let $[H^+] = K_a$ in Eq. (6-64); then Eq. (6-66) results.

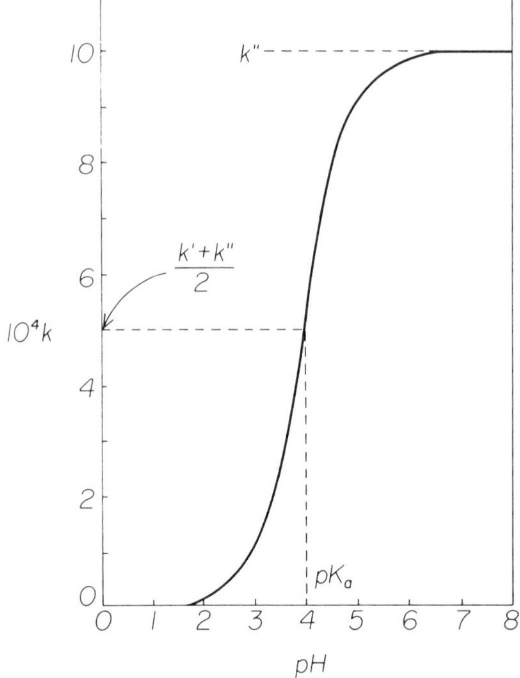

Figure 6-10. Plot of Eq. (6-64) with $pK_a = 4.0$, $k' = 1 \times 10^{-6}$ s^{-1}, $k'' = 1 \times 10^{-3}$ s^{-1}.

$$k = \frac{k' + k''}{2} \tag{6-66}$$

That is, pK_a = pH at the point where Eq. (6-66) holds. Because the larger of the two constants is usually much greater than the smaller one, this often may be interpreted that pK_a = pH when $k = k_{max}/2$ (see Fig. 6-10). Graphical methods for estimating K_a by using all of the kinetic data are considered later.

The log k–pH plot can display a large range in values on the vertical axis. The properties of this curve are different from those of the k–pH plot. Figure 6-11 is a plot of log k against pH for the same system graphed in Fig. 6-10. The first derivative is found with the help of Eq. (6-50) and the second derivative is obtained from this and Eq. (6-52). If the special case $k' = k''$ is excluded, this leads to

$$\mathrm{pH}_{infl} = pK_a - \tfrac{1}{2} \log \frac{k''}{k'} \tag{6-67}$$

This dependence of pH_{infl} on the rate constants as well as the dissociation constant has sometimes been overlooked by authors evaluating log k–pH curves, and the literature contains examples of such plots that have been erroneously used to estimate pK_a.

It was shown above that, for the k–pH plot, pK_a = pH when $k = k_{max}/2$. In terms of the log k–pH plot, this means that pK_a = pH when log k = log $k_{max} - 0.30$.

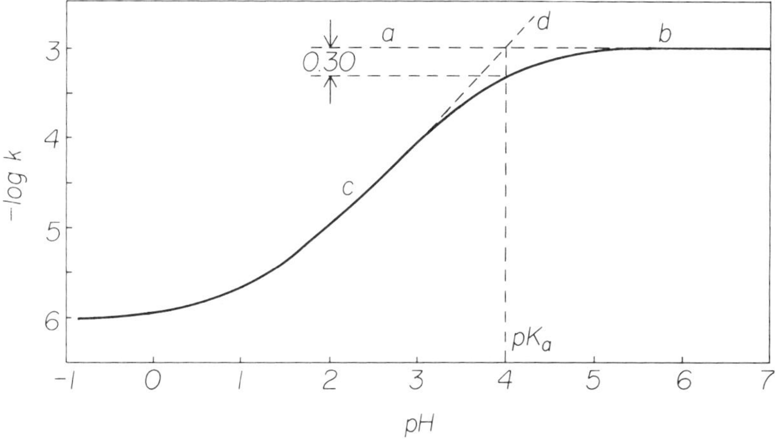

Figure 6-11. Log k–pH plot of Eq. (6-64) with $pK_a = 4.0$, $k' = 1 \times 10^{-6}$ s^{-1}, $k'' = 1 \times 10^{-3}$ s^{-1}.

Figure 6-11 shows the relationship expressed by this statement. An alternate method is developed from Eq. (6-64); in the special case that $k'' \gg k'$, this becomes $k = k''K_a/([H^+] + K_a)$. Taking logarithms and imposing the condition $[H^+] \gg K_a$,

$$\log k = pH + \log k'' - pK_a$$

Thus, under these conditions, the log k–pH plot is linear with slope $+ 1$. The extension of this segment of the plot is shown as c–d in Fig. 6-11. When $\log k = \log k''$, $pH = pK_a$; that is, the pH at which c–d and a–b intersect is numerically equal to the pK_a. A similar treatment can be given for the case $k' \gg k''$.

The dissociation constant is most accurately estimated from kinetic data when all of the data points are used in the evaluation. There are several ways to do this. The Henderson–Hasselbalch equation

$$pK_a = pH - \log \frac{[S^-]}{[HS]}$$

can be combined with the equations $k = k'F_{HS} + k''F_S$ and $F_{HS} + F_S = 1$ to give

$$pK_a = pH - \log \frac{k - k'}{k'' - k} \tag{6-68}$$

pK_a can be calculated with Eq. (6-68) at each of the experimental pH's in the rising portion of the sigmoid curve; values of k' and k'' can be estimated from the extreme low and high pH regions. Alternatively the third term can be plotted against pH; $pK_a = pH$ at the point where the logarithmic term equals zero.[44, p. 583] The slope of this plot should be unity.

In suitable circumstances the parameters of Eq. (6-64) may be evaluated by means of linear graphical methods. Consider the case in which $k'[H^+] \gg k''K_a$. Then Eq. (6-64) becomes

$$k = \frac{k'[H^+]}{[H^+] + K_a} \tag{6-69}$$

Equation (6-69) is of the form $x = y/(a + by)$, with $x = k$, $y = [H^+]$, $a = K_a/k'$, and $b = 1/k'$. Three corresponding linear equations are

$$\frac{1}{k} = \frac{K_a}{k'[H^+]} + \frac{1}{k'}$$

A plot of $1/k$ against $1/[H^+]$ should be linear; from the slope and intercept K_a and k' can be evaluated.

$$\frac{k}{[H^+]} = -\frac{k}{K_a} + \frac{k'}{K_a}$$

$k/[H^+]$ is plotted against k.

$$\frac{[H^+]}{k} = \frac{[H^+]}{k'} + \frac{K_a}{k'}$$

$[H^+]/k$ is plotted against $[H^+]$.

When $k''K_a \gg k'[H^+]$, three linear equations, suitable for evaluating k'' and K_a, can be written as shown for the first case.

The kinetic analysis of the sigmoid pH–rate profile will yield numerical estimates of the pH-independent parameters K_a, k', and k''. With these estimates the apparent constant k is calculated using the theoretical equation over the pH range that was explored experimentally. Quantitative agreement between the calculated line and the experimental points indicates that the model is a good one. A further easy, and very pertinent, test is a comparison of the kinetically determined K_a value with the value obtained by conventional methods under the same conditions.

Many sigmoid rate curves have been reported. A typical example is provided by the hydrolysis of phthalamic acid.[58]

This compound undergoes hydrolysis of the amide group intramolecularly catalyzed by the neighboring carboxylic acid group. The rate equation, in the pH range 1–

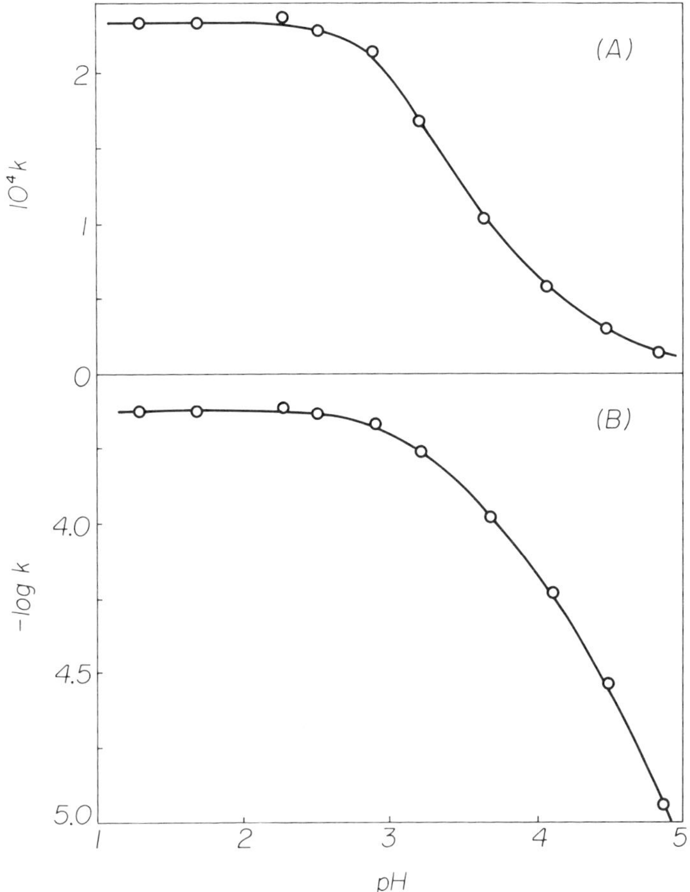

Figure 6-12. (A) pH–rate profiles plotted at k–pH and (B) log k–pH for the hydrolysis of phthalamic acid.[58]

5, is given by Eq. (6-63) with k'' essentially equal to zero; thus Eq. (6-69) describes the pH–rate curve. The kinetic results are shown as the k–pH plot (Fig. 6-12A) and as the log k–pH plot (Fig. 6-12B). Application of the simple methods described earlier gives $pK_a = 3.6$ and $k' = 2.35 \times 10^{-4}$ s^{-1}.

$$\text{(6-70)}$$

The hydrolysis of aspirin [Eq. (6-70), R = CH$_3$] is a classic example demonstrating a sigmoid pH–rate effect.[59] Figure 6-13 shows this curve for trimethyl-

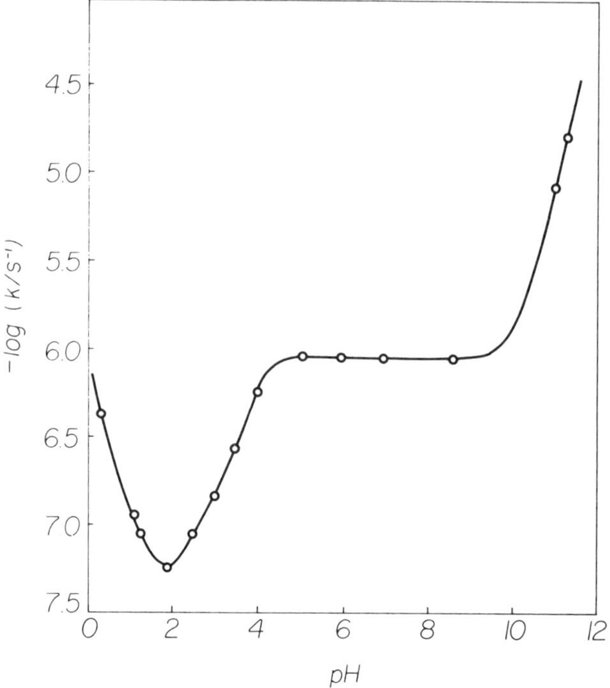

Figure 6-13. pH–rate profile for the hydrolysis of trimethylacetylsalicylic acid at 25°C (aqueous solution containing 0.5% ethanol) (60).

acetylsalicylic acid [Eq. (6-70), R = C(CH$_3$)$_3$].[60] The extreme left-hand segment of the curve has a slope of -1 and the extreme right-hand slope is $+1$, suggesting first-order rate terms in [H$^+$] and [OH$^-$], respectively. The intermediate portion is sigmoid. Writing the substrate as HS, the rate equation immediately suggested by Fig. 6-13 is, therefore,

$$v = k^0[\text{HS}][\text{H}^+] + k'[\text{HS}] + k''[\text{S}^-] + k'''[\text{S}^-][\text{OH}^-] \tag{6-71}$$

The pH-independent "plateau" from about pH 5 to 9 represents reaction of the acylsalicylate anion. It is obvious from the pH–rate profile that k'' is much larger than k'. The theoretical equation for k, the observed first-order rate constant, is derived in the usual way from Eq. (6-71).

$$k = \frac{k^0[\text{H}]^2 + k'[\text{H}^+] + k''K_a + k'''K_a[\text{OH}^+]}{[\text{H}^+] + K_a} \tag{6-72}$$

k^0 is evaluated from k values at very low pH, when the other terms become negligible, and k''', from the data at very high pH. From the plateau region, k'' can be estimated, because in this range only the k'' term is important. The value of k' can then be estimated from data in the pH 2–3 range, where the k^0, k', and k'' terms may all contribute to k, by making appropriate corrections with the known constants k^0, k'', and K_a. The same type of pH–rate profile is shown in the iodination of o-isobutyrylbenzoic acid.[61]

Bell-Shaped Curves

A frequently encountered pH–rate profile exhibits a bell-like shape or "hump," with two inflection points. This graphical feature is essentially two sigmoid curves back-to-back. By analogy with the earlier analysis of the sigmoid pH–rate curve, where the shape was ascribed to an acid–base equilibrium of the substrate, we find that the bell-shaped curve can usually be accounted for in terms of two acid–base dissociations of the substrate. The substrate can be regarded, for this analysis, as a dibasic acid H_2S, where the charge type is irrelevant; we take the neutral molecule as an example. The acid dissociation constants are

$$K_1 = \frac{[H^+][HS^-]}{[H_2S]}$$

$$K_2 = \frac{[H^+][S^{2-}]}{[HS^-]}$$

The fractions of solute in each form are given by $F_{H_2S} = [H_2S]/S_t$, $F_{HS} = [HS^-]/S_t$, and $F_S = [S^{2-}]/S_t$, where S_t, the total molar concentration of substrate is $[H_2S] + [HS^-] + [S^{2-}]$. Combining these leads to expressions for the fractions of solute as functions of the dissociation constants and the hydronium ion concentration.

$$F_{H_2S} = \frac{[H^+]^2}{[H^+]^2 + K_1[H^+] + K_1K_2} \tag{6-73}$$

$$F_{HS} = \frac{K_1[H^+]}{[H^+]^2 + K_1[H^+] + K_1K_2} \tag{6-74}$$

$$F_S = \frac{K_1K_2}{[H^+]^2 + K_1[H^+] + K_1K_2} \tag{6-75}$$

Figure 6-14 shows F_{H_2S}, F_{HS}, and F_S plotted against pH for an acid with $pK_1 = 5.0$ and $pK_2 = 10.0$. Evidently with such widely spaced dissociation constants the solution contains, at any one pH, significant fractions of only two species. The fraction of monoanion rises essentially to unity at one point. The pH at which the monoanion fraction achieves its maximum value is calculated by differentiating Eq. (6-74) and setting the result equal to zero; this gives

$$pH_{max} = \tfrac{1}{2}(pK_1 + pK_2) \tag{6-76}$$

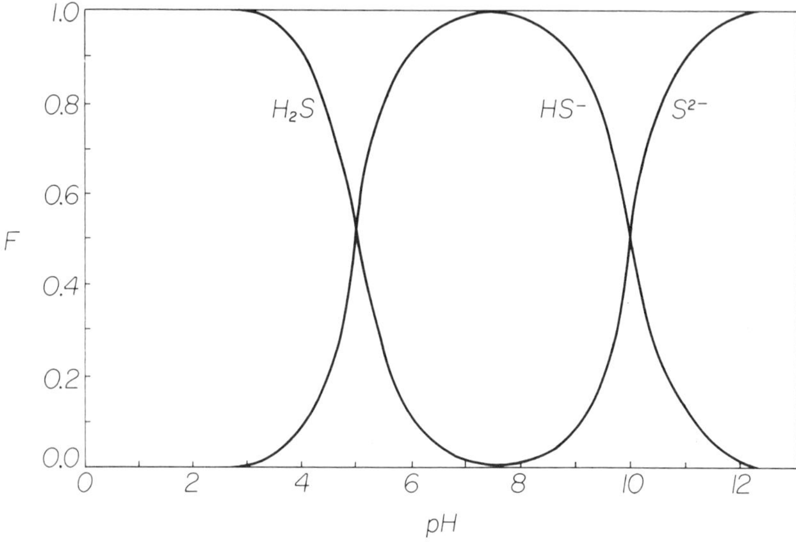

Figure 6-14.   Distribution of acid–base species for an acid with $pK_1 = 5.0$ and $pK_2 = 10.0$.

The intersection of the curves F_{H_2S} and F_{HS} occurs at $pH = pK_1$, as can be found by setting Eqs. (6-73) and (6-74) equal. Also, when $F_{HS} = F_S$, $pH = pK_2$. These are general relationships. We note, at this point, that the function F_{HS} has the previously mentioned bell shape, and it is this function that will be of later kinetic interest.

In order to find the inflection points in a plot of F_{HS} against pH, the second derivative d^2F_{HS}/dpH^2 is set to zero. The result is a quartic in $[H^+]$, which is not reproduced. When K_1 is much larger than K_2 (by at least three orders of magnitude), the location of the inflection points becomes particularly simple. Then at low pH, in the region of the left inflection point, $[H^+]$ is much greater than the quantity $(K_1K_2)^{1/2}$, and

$$pH_{infl}^{left} = pK_1$$

Similarly, when $(K_1K_2)^{1/2} \gg [H^+]$, the location of the right inflection point is given by

$$pH_{infl}^{right} = pK_2$$

These simple results will not be applicable if pK_1 and pK_2 are fairly close. Figure 6-15 is a plot of the species distribution for a hypothetical dibasic acid with $pK_1 = 7.0$ and $pK_2 = 8.0$. The inflection points on the F_{HS} curve do not coincide with the pK's of the acid. It is also important to notice that F_{HS} never reaches unity and that within the approximate pH range 6 to 8, appreciable fractions of all three forms coexist.

Now suppose that only the monoanionic form of the dibasic acid H_2S undergoes reaction and that neither the hydronium nor the hydroxide ion is directly involved. The kinetic scheme is, therefore,

$$HS^- \rightarrow \text{products} \tag{6-77}$$

and the rate equation is $v = k'[HS^-]$. The concentration of monoanion is given by $F_{HS}S_t$, and the experimental rate equation, at constant pH, is $v = kS_t$. These lead to

$$k = \frac{k'K_1[H^+]}{[H^+]^2 + K_1[H^+] + K_1K_2} \tag{6-78}$$

Equation (6-78) has the same form as Eq. (6-74) for F_{HS}, so the simple scheme embodied in Eq. (6-77) can account for a bell-shaped curve when k is plotted against pH.

The observed kinetics are seldom as simple as this. Usually one or both of the other forms of the substrate also undergo reaction, with the usual reactions being

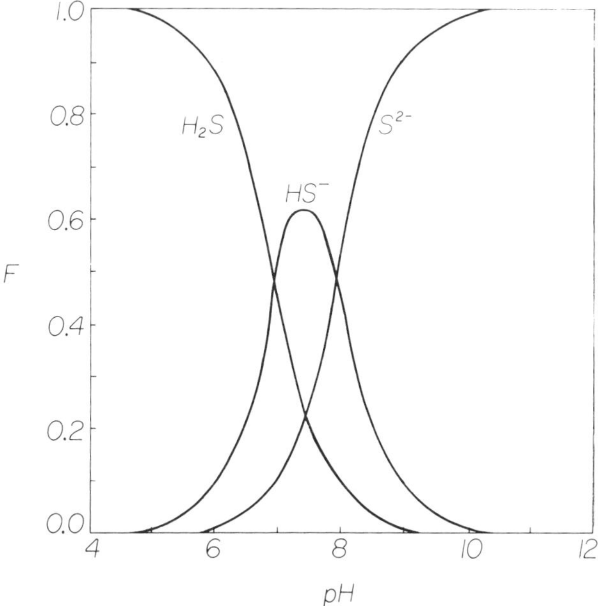

Figure 6-15. Distribution of acid–base species for an acid with $pK_1 = 7.0$ and $pK_2 = 8.0$.

hydronium ion catalysis of H_2S and hydroxide ion catalysis of S^{2-}. The kinetic scheme is then

$$H_2S + H^+ \xrightarrow{k^0} \text{products}$$

$$HS^- \xrightarrow{k'} \text{products}$$

$$S^{2-} + OH^- \xrightarrow{k''} \text{products}$$

Scheme IV

and the rate equation is

$$v = k^0[H_2S][H^+] + k'[HS^-] + k''[S^{2-}][OH^-] \tag{6-79}$$

This leads, after the usual development, to Eq. (6-80) for the pseudo-first-order rate constant k.

$$k = \frac{k^0[H^+]^3 + k'K_1[H^+] + k'K_1K_2K_w/[H^+]}{[H^+]^2 + K_1[H^+] + K_1K_2} \tag{6-80}$$

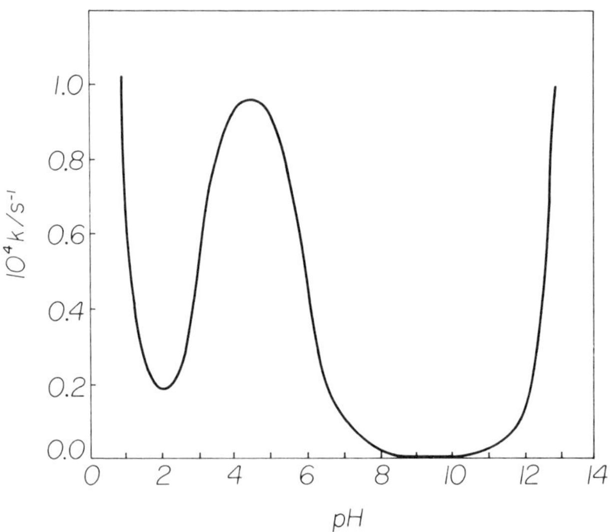

Figure 6-16. Theoretical k–pH plot according to Eq. (6-80) for a substrate with $pK_1 = 3.0$, $pK_2 = 6.0$, $k^0 = 1 \times 10^{-3}$ M^{-1} s^{-1}, $k' = 1 \times 10^{-4}$ s^{-1}, $k'' = 1 \times 10^{-3}$ M^{-1} s^{-1}.

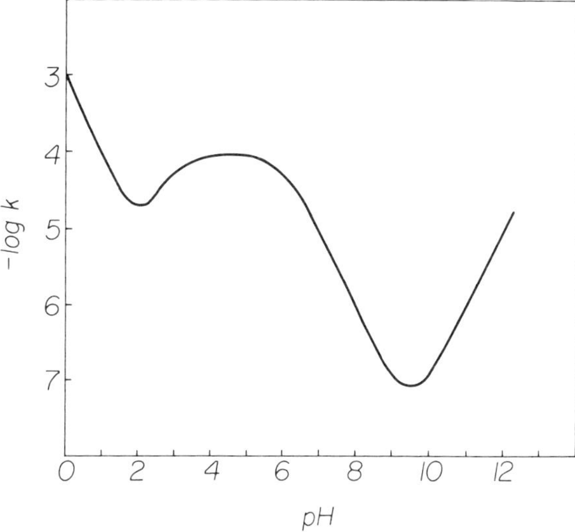

Figure 6-17. Theoretical log k–pH plot for the substrate described in Fig. 6-16.

The analysis of a k–pH curve in terms of Eq. (6-80) is treated by making approximations that are equivalent to ignoring some of the rate terms in certain pH regions. A common type of system is analyzed as an example; the approach can be modified to suit a particular demand. The evaluation of k^0 and k'' can nearly always be accomplished from rate data at very low and high pH, respectively. We are concerned with k', K_1, and K_2. Figures 6-16 and 6-17 are k–pH and log k–pH plots for a hypothetical system described by Eq. (6-80). The analysis assumes this equation and these types of parameters.

Often k'' is smaller than k^0 because attack on the dianion by hydroxide is disfavored. Moreover, the maximum often occurs on the acid side of neutrality. In the pH region near the maximum it will, therefore, often be permissible to set $k'' = 0$ in Eq. (6-80). The result is differentiated and the derivative is set equal to zero, giving, for the hydronium ion concentration at the maximum in the bell,

$$[\text{H}^+]^2_{\text{max}} \simeq \frac{k'K_1K_2}{k' - 3k^0K_2}$$

where quartic and cubic terms in [H$^+$] have been neglected.

If $k' \gg 3k^0K_2$, this becomes

$$\text{pH}_{\text{max}} \simeq \tfrac{1}{2}(\text{p}K_1 + \text{p}K_2) \tag{6-81}$$

The value of the observed rate constant at the maximum is found by substituting $[\text{H}^+]_{\text{max}} \simeq (K_1K_2)^{1/2}$ into Eq. (6-78), giving the approximate result[62]

$$k_{\max} \simeq \frac{k'}{1 + 2(K_2/K_1)^{1/2}} \tag{6-82}$$

which may be useful in estimating k'.

The placement of the minimum that occurs to the left of the maximum will not be affected by the k'' term, and it will usually be permissible to neglect the second dissociation of the substrate. Thus, Eq. (6-80) simplifies to

$$k = \frac{k^0[\mathrm{H}^+]^2 + k'K_1}{[\mathrm{H}^+] + K_1}$$

Setting the first derivative to zero gives Eq. (6-83) for the hydronium ion concentration at the left-hand minimum.

$$[\mathrm{H}^+]_{\min}^{\mathrm{left}} = K_1\left[\left(1 + \frac{k'}{k^0K_1}\right)^{1/2} - 1\right] \tag{6-83}$$

The left-hand inflection point is obtained by neglecting K_2 terms. The derivative $d^2k/d\mathrm{pH}^2$ is set equal to zero. As a very approximate solution, higher-order terms are neglected,

$$[\mathrm{H}^+]_{\mathrm{infl}}^{\mathrm{left}} \simeq \frac{k'K_1}{4k^0K_1 + k'}$$

or, in even more approximate form, $\mathrm{pH}_{\mathrm{infl}}^{\mathrm{left}} \simeq pK_1$. In many cases, however, the location of the left-hand inflection point in the k–pH plot will not be given by these relationships.

In the region of the right-hand inflection point both the k^0 and k'' terms can often be neglected. The second derivative $d^2k/d\mathrm{pH}^2$ is then set to zero. As a first approximation all terms higher than the linear one are neglected:

$$\mathrm{pH}_{\mathrm{infl}}^{\mathrm{right}} \simeq pK_2$$

Very often this will give a good estimate of K_2; K_1 can then be found with Eq. (6-81).

When estimates of k^0, k', k'', K_1, and K_2 have been obtained, a calculated pH–rate curve is developed with Eq. (6-80). If the experimental points follow closely the calculated curve, it may be concluded that the data are consistent with the assumed rate equation. The constants may be considered adjustable parameters that are modified to achieve the best possible fit, and one approach is to use these initial parameter estimates in an iterative nonlinear regression program. The dissociation constants K_1 and K_2 derived from kinetic data should be in reasonable agreement with the dissociation constants obtained (under the same experimental conditions) by other means.

The rate equation (6-79), upon which this analysis has been based, may be found

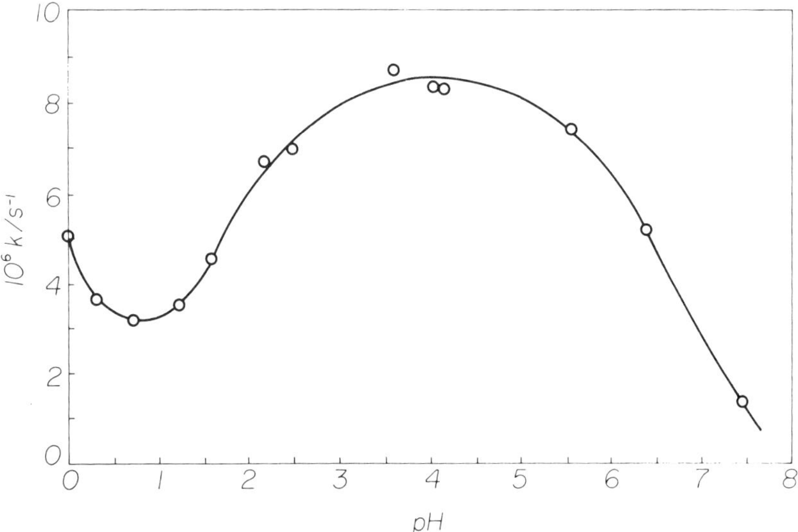

Figure 6-18. pH–rate profile for the hydrolysis of monoethyl dihydrogen phosphate in aqueous solution at 100.1°C.

inadequate to account in full for the data. Obvious alterations are the addition of the rate terms $k_1[H_2S]$ and $k_2[S^{2-}]$. These terms added to those of Eq. (6-79) include all of the possibilities (other than terms containing concentrations to orders other than unity); any other term involving these species would be kinetically equivalent to one of these five terms, as we saw in Section 3.3.

Figure 6-18 shows a bell-shaped pH–rate profile for the hydrolysis of monomethyl dihydrogen phosphate.[63] Other examples are the hydrolysis of *o*-carboxyphenyl hydrogen succinate[64] and the hydration of fumaric acid.[65,66]

Kinetic schemes other than that embodied in Eq. (6-77) can give rise to a bell-shaped curve. As in Eq. (6-77), however most of these involve two ionizations. Thus Scheme V, where HS is a monoprotic acid and B is a base (or the kinetic equivalent of $S^- + BH^+$) yields a bell-shaped curve.

$$HS + B \rightarrow products$$

Scheme V

A nonionizable substrate reacting with two ionizable species or groups as in Scheme VI also give this type of behavior.

$$S + HA + B \rightarrow products$$

Many examples are known in the field of enzyme catalysis, the groups HA and B both being situated in the active site of the enzyme.

A less obvious scheme that can lead to a bell-shaped curve has been recognized by Zerner and Bender.[67] The substrate is a weak acid or base, but possesses only one ionizable group, and no other ionizable reactant is involved (other than water). The second inflection in the curve is ascribed to an ionizable group created in an *intermediate*. An example has been discovered in the hydrolysis of *o*-carboxyphthalimide:

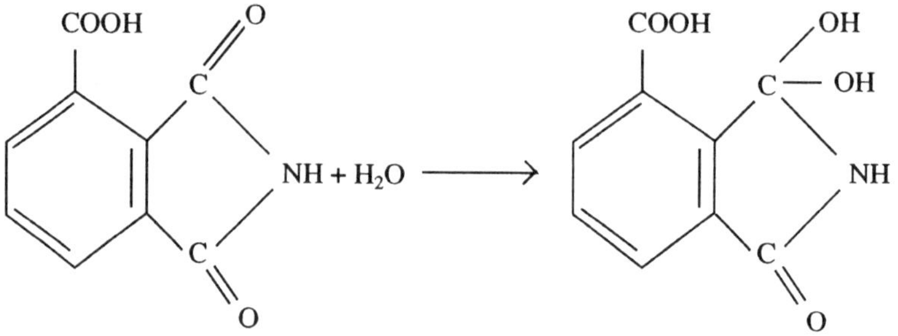

$$\text{(o-carboxyphthalimide)} \quad NH + H_2O \longrightarrow \text{(hydrated intermediate)}$$

This can be represented by Scheme VII.

$$HS + H_2O \rightleftharpoons IH$$

$$IH \rightleftharpoons I^- + H^+$$

$$I^- \longrightarrow products$$

Scheme VII

A bell-shaped pH–rate profile can also be produced in a two-step reaction involving a single ionizable group if the rate-determining step changes when the pH is altered. An example, the oximation of acetone,[68] is shown in Fig. 5-12.

A collection of pH–rate profiles for drug decomposition reactions has been published.[31]

6.5 KINETIC ISOTOPE EFFECTS

Consider a reactant molecule in which one atom is replaced by its isotope, for example, protium (H) by deuterium (D) or tritium (T), ^{12}C by ^{13}C, etc. The only change that has been made is in the mass of the nucleus, so that to a very good approximation the electronic structures of the two molecules are the same. This means that reaction will take place on the same potential energy surface for both molecules. Nevertheless, isotopic substitution can result in a rate change as a consequence of quantum effects. A rate change resulting from an isotopic substitution is called a *kinetic isotope effect*. Such effects can provide valuable insights into reaction mechanism.

This treatment is not very mathematical; it is intended to provide a physical picture for the origin of isotope effects and to show some of their uses. More detailed discussions are available in reviews by Bell,[69] Saunders,[70] Ritchie,[71] Carpenter,[32, Chap. 5] and Drenth and Kwart.[72]

Primary Isotope Effects

The most profound isotope effects are those resulting from the replacement of H by D or T, primarily because the percentage mass change is greatest for these isotopic substances. (Another factor is that quantum mechanical tunneling is important because of the low mass of H.) We will, therefore, use the isotopic pair of H and D in illustration of the isotope effect. Consider the isotopic molecules R–H and R–D, each subject to an elementary reaction in which the R–H or R–D bond is broken. (The symbol R represents an atom or group that is massive compared with H or D.) This reaction can be represented by a bond dissociation potential energy function, Fig. 6-19, in which the abscissa is the bond distance between R and H or D. As noted above, the same potential energy curve applies to both reactants.

It is a fundamental result of quantum mechanics that the vibrational energy levels of the bond are given by

$$E_n = (n + \tfrac{1}{2})h\nu \tag{6-84}$$

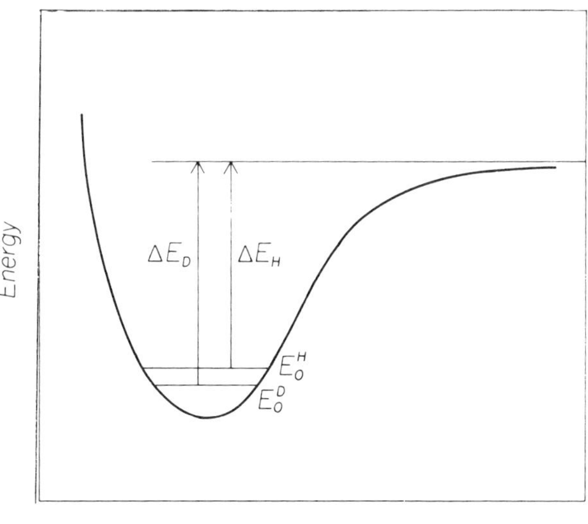

Figure 6-19. Bond dissociation curve showing the different zero-point vibrational energies of isotopic species R–H and R–D.

where v is the frequency of the stretching vibration and n is the vibrational quantum number.

At low temperatures nearly all bonds will be in their lowest vibrational level, $n = 0$, and will, therefore, possess the *zero-point vibrational energy*, $E_0 = hv/2$. Presuming the molecule behaves as a simple harmonic oscillator, the vibrational frequency is given by

$$v = \frac{1}{2\pi} \left(\frac{k'}{\mu}\right)^{1/2} \tag{6-85}$$

where k' is a force constant and μ is the reduced mass,

$$\mu = \frac{m_R m_H}{m_R + m_H} \tag{6-86}$$

where m represents mass; thus if $m_R \gg m_H$, $\mu \approx m_H$.

If H is replaced by D, μ will increase, whereas k' will not change, because it is determined by the electronic distribution. Therefore $v_D < v_H$, and $E_0^H > E_0^D$. The implication of this result can be seen graphically in Fig. 6-19; because of the difference in zero-point energies, the bond dissociation energies of R–H and R–D are different, the energy required to break the R–D bond being the greater.

We now carry the argument over to transition state theory. Suppose that in the transition state the bond has been completely broken; then the foregoing argument applies. No real transition state will exist with the bond completely broken—this does not occur until the product state—so we are considering a limiting case. With this realization of the very approximate nature of the argument, we make estimates of the maximum kinetic isotope effect. We write the Arrhenius equation for the R–H and R–D reactions

$$k_H = A_H \exp(-E_a^H/RT); \qquad k_D = A_D \exp(-E_a^D/RT)$$

To an approximation justified by this crude theory we set $A_H = A_D$, thus,

$$\frac{k_H}{k_D} = e^{-(E_a^H - E_a^D)/RT}$$

Figure 6-19 shows that we can write $E_a^D = E_a^H + (E_0^H - E_0^D)$, or $(E_a^H - E_a^D) = (E_0^D - E_0^H)$. Because $E_0^H = hhv_H/2$ and $E_0^D = hv_D/2$, we obtain

$$\frac{k_H}{k_D} = e^{h(v_H - v_D)/2RT} \tag{6-87}$$

Assuming that $\mu_H = m_H$ and $\mu_D = m_D$, we use Eq. (6-85) to get

$$\frac{v_H}{v_D} = \left(\frac{m_D}{m_H}\right)^{1/2} = 2^{1/2}$$

so $(\nu_H - \nu_D) = \nu_H(1 - 2^{-1/2}) = 0.293\ \nu_H$. This is used in Eq. (6-87):

$$\frac{k_H}{k_D} = e^{0.146h\nu_H/RT} \tag{6-88}$$

A kinetic isotope effect that is a result of the breaking of the bond to the isotopic atom is called a *primary kinetic isotope effect*. Equation (6-88) is, therefore, a very simple and approximate relationship for the maximum primary kinetic isotope effect in a reaction in which only bond cleavage occurs. Table 6-5 shows the results obtained when typical vibrational frequencies are used in Eq. (6-88). Evidently the maximum isotope effect is predicted to be very substantial.

Because of the serious approximations that were made in developing Eq. (6-88), these theoretical results cannot be safely used to make inferences of a quantitative nature, for example, by comparing an observed isotope effect with the calculated maximum effect as a measure of the location of the transition state on the reaction coordinate. We can consider some of the limitations of the theory. A very important one was the assumption that bond breaking is complete in the transition state. This cannot be, so in the transition state there will exist some residual zero-point energy difference between the hydrogen and deuterium species. Thus, the kinetic isotope effect should be smaller, from the operation of this cause, than that calculated with Eq. (6-88). This phenomenon is discussed below. Another limitation is that we considered only a bond cleavage. If bond formation also takes place, the transition state energy levels will be affected. Related to this factor is the neglect of bending vibrations. Suppose a transfer reaction A–H + B → A + H–B is studied. In the transition state bond cleavage and bond formation have both partially occurred to give a transition state structure A · · · H · · · B. This species is capable of two bending vibrations (one in the plane of the paper changing the AHB angle, one normal to this) that do not exist for the reactants. These, therefore, affect the zero-point energies of the H and D transition states. We also neglected any isotopic effect on the preexponential factors. A further factor, often important for hydrogen, is quantum mechanical tunneling, which makes a greater contribution to the rate for H than for any other atom. Bell discusses the tunneling effect for proton transfers.[69]

TABLE 6-5. Calculated Hydrogen/Deuterium Primary Kinetic Isotope Effects[a,b]

| Bond | ν_H | | $\Delta E_a/kJ\ mol^{-1}$ | k_H/k_D |
	cm^{-1}	Hz		
C–H	2800	8.4×10^{13}	4.90	7.2
N–H	3100	9.3×10^{13}	5.31	8.5
O–H	3300	9.9×10^{13}	5.86	10.6

Source: Reference 71, with minor alterations.

[a]Calculated at 25°C with Eq. (6-88).

[b]$\Delta E_a = 0.146\nu_H$.

A more rigorous theory of kinetic isotope effects begins with the transition state equation $k = (kT/h)K^{\ddagger}$. Writing this for k_H and k_D leads to

$$\frac{k_H}{k_D} = \frac{K_H^{\ddagger}}{K_D^{\ddagger}} \qquad (6\text{-}89)$$

Next, $K_H^{\ddagger}$ and $K_D^{\ddagger}$ are written in terms of partition functions (see Section 5.2), which are in principle calculable from quantum mechanical results together with experimental vibrational frequencies. The application of this approach to mechanistic problems involves postulating alternative models of the transition state, estimating the appropriate molecular properties of the hypothetical transition state species, and calculating the corresponding k_H/k_D values for comparison with experiment.[32, Chap. 5; 71, 72]

We will now consider in qualitative terms the important class of proton-transfer reactions

$$AH + B \rightleftharpoons A + HB$$

where charge types are not specified. Let A and B be massive relative to H and D, so $\mu_{AH} = m_H$, etc. A and B may be polyatomic, but vibrations within these groups are ignored.

For this reaction the transition state is the linear species $A \cdots H \cdots B$, which possesses $3n - 5 = 4$ vibrational modes.[70]

$$\overleftarrow{A} \cdots \overrightarrow{H} \cdots \overleftarrow{B} \qquad\qquad \overleftarrow{A} \cdots H \cdots \overrightarrow{B}$$

Asymmetric stretch   Symmetric stretch

$$\overset{\uparrow}{A} \cdots \underset{\downarrow}{H} \cdots \overset{\uparrow}{B}$$

Bending
vibration
(doubly degenerate)

We will neglect the bending mode, which may not always be acceptable.

In the asymmetric stretching vibration the hydrogen is increasing its distance from A and decreasing its distance from B. This is, therefore, motion along the reaction coordinate; it is the vibrational mode that is factored out of the expression for $K^{\ddagger}$ and, therefore, does not contribute to the zero-point energy of the transition state. The symmetric stretch is a true vibrational mode of the transition state and, therefore, contributes to the transition state zero-point energy. In comparing AH and AD as reactants, we therefore anticipate, in general, a transition state zero-point energy difference.

Now consider the position of the proton in the transition state, that is, the extent to which the proton has been transferred from A to B. First suppose H is equidistant from A and B in the transition state. Then the symmetric stretch consists of A and

B synchronously moving away from and toward H; the H atom does not move (if A and B are of equal mass). If H does not move in a vibration, its replacement with D will not alter the vibrational frequency. Therefore, there will be no zero-point energy difference between the H and D transition states, so the difference in activation energies is equal to the difference in initial state zero-point energies, just as calculated with Eq. (6-88). The kinetic isotope effect will therefore have its maximal value for this location of the proton in the transition state.

If the proton is not equidistant between A and B, it will undergo some movement in the symmetric stretching vibration. Isotopic substitution will, therefore, result in a change in transition state vibrational frequency, with the result that there will be a zero-point energy difference in the transition state. This will reduce the kinetic isotope effect below its maximal possible value. For this type of reaction, therefore, k_H/k_D should be a maximum when the proton is midway between A and B in the transition state and should decrease as H lies closer to A or to B.

It appears that the bending vibrations, which are ignored in the above discussion, may act to increase k_H/k_D, as can the tunneling effect, so it is possible for the primary kinetic isotope effect to be larger than the estimates calculated wih Eq. (6-88).

Study of the primary isotope effect is useful in establishing whether or not a bond to hydrogen is broken in the rate-determining step. (Interpretation can be complicated by acid–base preequilibria or isotopic exchange with the solvent.) A good example is Bell's work confirming the rate-determining step in the halogenation of ketones, which, on the basis of kinetic evidence (see Section 5.3, "Composition of the Transition State"), appears to undergo rate-determining proton removal assisted by specific or general bases:

$$\begin{array}{ccccc} O & & O & & O \\ \| & slow & \| & X_2 & \| \\ CH_3CCH_3 \;+\; OH^- &\rightleftharpoons& CH_3\,C\,CH_2^- &\rightarrow& CH_3\,C\,CH_2Br \\ & & + & & + \\ & & H_2O & & Br^- \end{array}$$

In confirmation, when acetone-d_6 was studied, a very substantial kinetic isotope effect was observed, k_H/k_D being about 7 to 12 depending upon the base catalyst.

The E2 β-elimination reaction is a base-catalyzed concerted mechanism:

$$\begin{array}{c} Br \\ | \\ CH_3CHCH_3 \;+\; OEt^- \;\rightarrow\; CH_3CH{=}CH_2 \;+\; EtOH \;+\; Br^- \end{array}$$

The S_N2 substitution competes with the elimination, but a bond to hydrogen is not cleaved in the substitution.

$$\begin{array}{cc} Br & OEt \\ | & | \\ CH_3CHCH_3 \;+\; OEt^- \;\rightarrow\; CH_3CHCH_3 \;+\; Br^- \end{array}$$

As expected, k_H/k_D is essentially unity for the substitution, but is 6.7 for the elimination.

Much evidence suggests that electrophilic aromatic substitution takes place in a two-step reaction with the formation of a benzenonium ion intermediate.

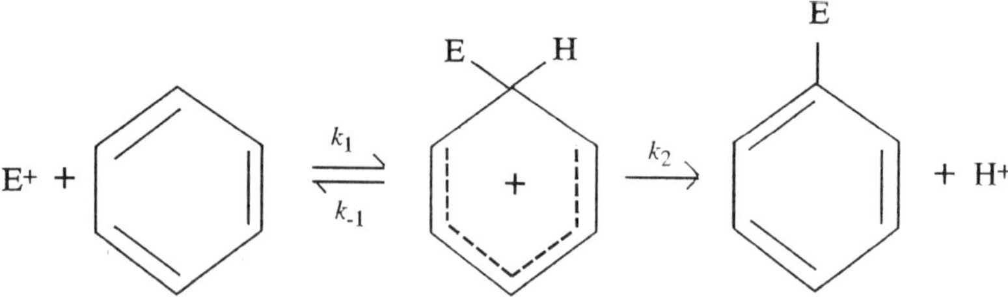

Either step could be rate determining. Study of many reactions has shown that most occur with a significant isotopic effect, but for some reactions the isotope effect is absent.[73,74] If we apply the steady state approximation to the intermediate, this reaction scheme leads to

$$k = \frac{\alpha k_1}{1 + \alpha}$$

where k is the observed second-order rate constant and $\alpha = k_2/k_{-1}$. We expect that neither k_1 nor k_{-1} should show an appreciable kinetic isotope effect, for the C–H bond is not broken in this first step of the reaction; k_2, on the other hand, should have a primary isotope effect. The net effect observed, therefore, depends upon the magnitude of α. If $\alpha \gg 1$, then $k \approx k_1$ and no isotope effect will be observed; this appears to be the case for nitration ($E^+ = {}^+NO_2$). If $a \ll 1$, then $k \approx k_1k_2/k_{-1}$, and an isotope effect is observed. Thus, either step can be rate determining, depending upon the identity of the electrophile.

Secondary Isotope Effects

A kinetic isotope effect that results when the bond to the isotopic atom is *not* broken is called a *secondary isotope effect*. Here are two examples;

$$\underset{CH_3}{\overset{D}{C_6H_5-\underset{|}{\overset{|}{C}}-H}} + B^- \quad \rightarrow \quad \underset{CH_3}{\overset{}{C_6H_5-\overline{C}-D}} + BH$$

$$\underset{CD_3}{\overset{H}{C_6H_5-\underset{|}{\overset{|}{C}}-H}} + B^- \quad \rightarrow \quad \underset{CD_3}{\overset{}{C_6H_5-\overline{C}-H}} + BH$$

In these examples B^- is a base. The first example is called a *secondary isotope effect of the first kind*; the next one is a *secondary isotope effect of the second kind*. The distinction between these is that in the first kind bonds to the isotopic atom have undergone spatial (i.e., structural) change. Halevi[75] has reviewed secondary isotope effects on equilibria and rates.

Secondary kinetic isotope effects are small, values of k_H/k_D being less than 2. (Sometimes k_H/k_D is less than unity, this being called an *inverse isotope effect*.) Nevertheless, the occurrence of a secondary isotope effect means, in a general way, that a vibrational mode of a hydrogen has been changed by isotopic substitution with deuterium and that this change is transmitted to the bond undergoing cleavage with kinetic consequences. One way to interpret such changes is to view them as substituent effects. That is, replacement of H by D outside the site of reaction is analogous to replacement of H by, say, methyl. From this point of view, we are led to inquire how H and D differ as substituents in molecules.

In our discussion of the primary isotope effect, and of the bond dissociation curve of Fig. 6-19, we saw that the compounds R–H and R–D have identical potential energy functions but different zero-point energies. The potential well of Fig. 6-19 approximates to a parabola, corresponding to treating the vibrational modes as those of a harmonic oscillator. In reality the potential well is anharmonic (asymmetric) in the sense shown in Fig. 6-20, because it is easier to stretch a bond than to compress it. As indicated in Fig. 6-20, this means that both the maximum length and the average length of the R–H bond are greater than those of the R–D bond. That is, hydrogen has a greater steric requirement than does deuterium; H has the greater van der Waals radius.

Now, since the R–D bond is shorter than the R–H bond, but the electronic natures of H and D are identical, there must be a greater electron density at R in R–D than in R–H. Thus, solely as a consequence of the quantum mechanical zero-point energy

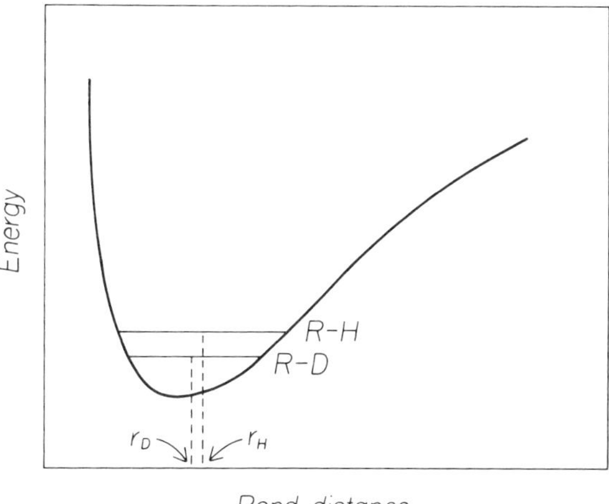

Figure 6-20. Bond dissociation curve showing that the average R–H bond distance is greater than the R–D bond distance because of their zero-point energy difference.

difference, we have arrived at two equivalent chemical bases for the interpretation of secondary isotope effects: They may be viewed as the result of different steric requirements of H and D, or they may be seen as the result of different electronic effects. Note that this second interpretation is equivalent to concluding that D has a greater inductive effect of electron release than does H. This is the basis for the interpretation of secondary isotope effects in the context of substituent effects.

These are usually very small effects, and, although real, their mechanistic interpretation is seldom clear-cut.

Solvent Isotope Effects

If an isotopic substitution is made in the solvent molecule, a resultant kinetic effect is called a *solvent isotope effect*. The most common comparison is between rates in H_2O and in D_2O. These solvents have significantly different properties; for example, their respective melting points are $0°C$ and $3.82°C$, their boiling points $100°C$ and $101.42°C$, and their autoprotolysis constants (at $25°C$) are 1.00×10^{14} and 1.54×10^{-15}. The melting and boiling point differences indicate significant structural differences in these solvents, possibly resulting from the steric or electronic difference described in preceding paragraphs. The difference in pK_w means that D_2O is a weaker acid than is H_2O.

The solvent isotope effect may be difficult to interpret because it can be a resultant of several contributions. One of these is a general medium effect, which is probably small because H_2O and D_2O are the comparable polarity. If an acid or base is involved in the reaction, its pK_a will be different in H_2O and in D_2O, so the concentration of the reactive species will be different; this can be a very important effect. If an O–H or O–D bond of the solvent is broken in the rate-determining step, a primary kinetic isotope effect contributes to the solvent isotope effect, and this effect, which is the one often being investigated, has been described earlier. Finally, the reactant species may undergo isotopic exchange with the solvent, so that a mixture of species may exist. All O–H, S–H, and N–H protons undergo exchange with H or D of the solvent, resulting in rapid equilibration. Most C–H bonds do not equilibrate rapidly, but if they are activated, as in keto-enol tautomeric systems, they too may exchange. Thus, we can study CH_3COOH in H_2O, or CH_3COOD in D_2O, or CD_3COOH in H_2O, but we cannot study CH_3COOH in D_2O.

Values of k_{H_2O}/k_{D_2O} tend to fall in the range 0.5 to 6. The direction of the effect, whether normal or inverse, can often be accounted for by combining a model of the transition state with vibrational frequencies,[76] although quantitative calculation is not reliable. Because of the difficulty in applying rigorous theory to the solvent isotope effect, a phenomenological approach has been developed. We define ϕ_i to be the ratio of D to H in site i of a reactant relative to the ratio of D to H in a solvent site. That is,

$$\phi_i = \frac{[D]_i/[H]_i}{[D]_s/[H]_s} \tag{6-90}$$

ϕ_i is called the *fractionation factor*. Let x be the mole fraction of D in the solvent, so $x = [D]_s/([D]_s + [H]_s)$, or $[D]_s/[H]_s = x/(1 - x)$, giving

$$\phi_i = \frac{[D]_i}{[H]_i} \cdot \frac{1 - x}{x} \tag{6-91}$$

Let us first treat an equilibrium, which in H_2O we write, for reactant RH and product PH,

$$RH \rightleftharpoons PH; \qquad K_H = [PH]/[RH]$$

and in D_2O

$$RD \rightleftharpoons PD; \qquad K_D = [PD]/[RD]$$

The equilibrium solvent isotope effect is defined to be

$$\frac{K_H}{K_D} = \frac{[PH][RD]}{[PD][RH]} \tag{6-92}$$

We use Eq. (6-91) to define fractionation factors for the positions R and P, thus,

$$\phi_R = \frac{[RD]}{[RH]} \cdot \frac{1 - x}{x} \tag{6-93a}$$

$$\phi_P = \frac{[PD]}{[PH]} \cdot \frac{1 - x}{x} \tag{6-93b}$$

Combining Eqs. (6-92) and (6-93) gives the solvent isotope effect:

$$\frac{K_H}{K_D} = \frac{\phi_R}{\phi_P} \tag{6-94}$$

We now wish to generalize this to include the isotope effect in H_2O/D_2O mixtures. The equilibrium constant K_x is defined

$$K_x = \frac{[PH] + [PD]}{[RH] + [RD]} \tag{6-95}$$

Into Eq. (6-95) we substitute the definitions of K_H and K_D and use Eqs. (6-93) and (6-94) to obtain

$$K_x = K_H \frac{(1 - x + x\phi_P)}{(1 - x + x\phi_R)} \tag{6-96}$$

The extension to rates draws on the equilibrium assumption of transition state theory to yield the analogous result, with rate constants replacing the equilibrium constants of Eq. (6-96). Kresge[77] has generalized this argument, the result being

$$\frac{k_x}{k_H} = \frac{\overset{P}{\Pi}(1 - x + x\phi_p)}{\underset{R}{\Pi}(1 - x + x\phi_R)} \tag{6-97}$$

That is, terms of the form $(1 - x + x\phi_i)$ appear in the denominator for all reactant sites having exchangeable protons and similarly in the numerator for all transition state sites. If there is no change in the fractionation factor for a site, its contribution cancels. If the solvent is a reactant, its term disappears because the solvent fractionation factor is unity by definition.

Fractionation factors can be measured for stable species,[77] but they must be postulated (on the basis of a model) for transition states. Rather than using Eq. (6-97) as a means for predicting the solvent isotope effect, it is more effective to measure the isotope effect and, by fitting the data (consisting of k_x as a function of x) to Eq. (6-97), to infer the number and mechanistic roles of protons involved in the rate-determining step.[78,79] This approach has been called the *proton inventory technique*. It has been used to study the mechanisms of hydrolyses, especially of the "uncatalyzed" route of reaction.[80] For example, Gandour et al.[81] examined the hydrolysis of phthalic anhydride, finding that the data were best fit by the equation $10^6 k_x = 1058 (1 - x + 0.69x)^3(1 - x + 1.33x)\,\mathrm{s}^{-1}$. This implicates four protons in the transition state, three of them being identical, with $\phi = 0.69$, the other having $\phi = 1.33$. It was inferred that the transition state includes an essentially fully formed hydronium ion. One possibility is that tetrahedral intermediate formation is rate determining, with a very late transition state, as in **1**.

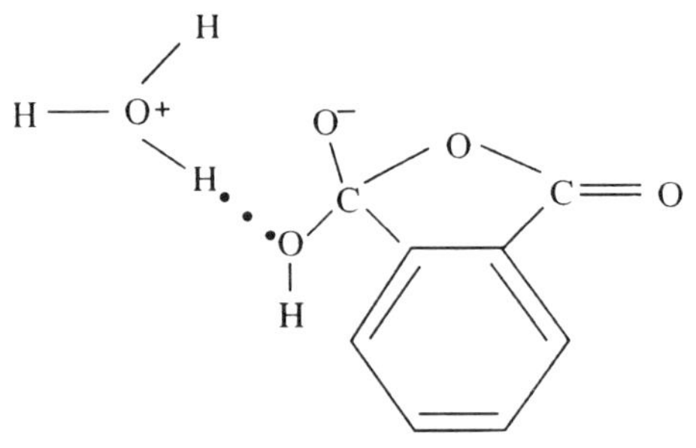

1

Another possibility is that breakdown of the tetrahedral intermediate is rate determining, the transition state being very early, as in **2**.

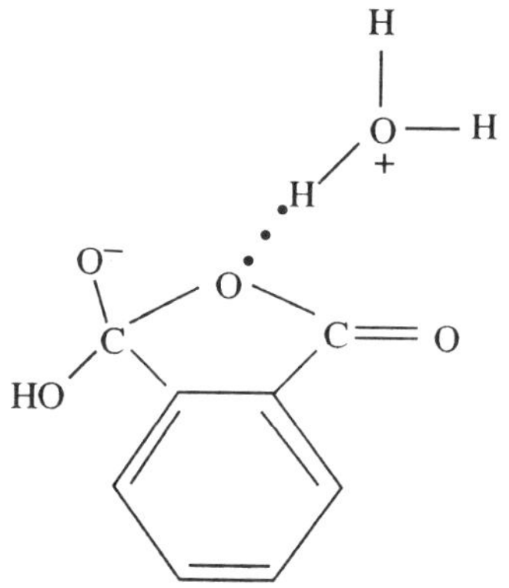

2

The technique has also been applied to the distinction between general base and nucleophilic catalysis.[82]

It must be appreciated that the selection of the best model—that is, the best equation having the form of Eq. (6-97)—may be a difficult problem, because the number of parameters is a priori unknown, and different models may yield comparable curve fits. A combination of statistical testing and chemical knowledge must be used, and it may be that the proton inventory technique is most valuable as an independent source capable of strengthening a mechanistic argument built on other grounds.

REFERENCES

1. Anet, F.A.L.; Bourn, A.J.R. *J. Am. Chem. Soc.* **1967,** *89,* 760.
2. Scheiner, P.; Schomaker, J.H.; Deming, S.; Libbey, W.J.; Nowack, G.P. *J. Am. Chem. Soc.* **1965,** *87,* 306.
3. Heberger, K.; Kemeny, S.; Vidoczy, T. *Int. J. Chem. Kinet.* **1977,** *19,* 171.
4. Deming, W.E. "Statistical Adjustment of Data"; Wiley: New York, 1943; p 202. (Dover Publications reprint, 1964.)
5. Cvetanovic, R.J.; and Singleton, D.L. *Int. J. Chem. Kinet.* **1977,** *9,* 481, 1007.
6. Bunce, N.J.; Forber, C.L.; McInnes, C.; Hutson, J.M. *J. Chem. Soc., Perkin Trans. 2* **1988,** 383.
7. Eriksen, S.P.; Stelmach, H. *J. Pharm. Sci.* **1965,** *54,* 1029.
8. Sestak, J. *J. Thermal Anal.* **1988,** *33,* 1263.
9. Kanerva, L.T.; Euranto, E.K.; Cleve, N.J. *Acta Chem. Scand.* **1983,** *B37,* 85.
10. Robertson, R.E. *Prog. Phys. Org. Chem.* **1967,** *4,* 213.
11. Kohnstam, G. *Adv. Phys. Org. Chem.* **1967,** *5,* 121.
12. Clarke, E.C.W.; Glew, D.N. *Trans. Faraday Soc.* **1966,** *62,* 539.
13. Wold, S. *Acta Chem. Scand.* **1970,** *24,* 2321.
14. Adams, P.A.; Sheppard, J.G.; Ridler, G.M.; Ridler, P.F. *J. Chem. Soc., Faraday Trans. 1* **1978,** *72,* 1500.

15. Euranto, E.K.; Kanerva, L.T. *J. Chem. Soc., Faraday Trans. 1* **1983**, *79*, 1483.
16. Blandamer, M.J.; Scott, J.W.M. *J. Phys. Chem.* **1988**, *92*, 6264.
17. Laidler, K.J. *J. Chem. Educ.* **1972**, *49*, 343.
18. Adams, P.A.; Swart, E.R.; Vernon C.A. *J. Chem. Soc., Faraday Trans. 1* **1976**, *72*, 397.
19. Mozurkewich, M.; Benson, S.W. *J. Phys. Chem.* **1984**, *88*, 6429.
20. Mozurkewich, M.; Lamb, J.J.; Benson, S.W. *J. Phys. Chem.* **1984**, *88*, 6435.
21. Gurney, R.W. "Ionic Processes in Solution"; McGraw-Hill: New York, 1953; Chapters 5 and 6. (Dover Publications reprint, 1962.)
22. Connors, K.A. "Reaction Mechanisms in Organic Analytical Chemistry"; Wiley-Interscience: New York, 1973; pp 15–21.
23. Connors, K.A. "Binding Constants: The Measurement of Molecular Complex Stability"; Wiley-Interscience: New York, 1987; pp 28–42.
24. Higuchi, T.; Havinga, A.; Busse, L.W. *J. Am. Pharm. Assoc., Sci Ed.* **1950** *39*, 405.
25. Connors, K.A. *J. Parenteral. Sci. Technol.* **1982** *36*, 205.
26. Bates, R.G. "Determination of pH", 2nd ed.; Wiley-Interscience: New York, 1973; p 124.
27. Harned, H.S.; Owen, B.B. "The Physical Chemistry of Electrolytic Solutions", 3rd ed.; Reinhold: New York, 1958; p 667.
28. Bates, R.G.; Hetzer, H.B. *J. Phys. Chem.* **1961**, *65*, 667.
29. Bates, R.G.; Bower, V.E. *Anal. Chem.* **1956**, *28*, 1322.
30. Kennon, L. *J. Pharm. Sci.* **1964**, *53*, 815.
31. Connors, K.A.; Amidon, G.L.; Stella, V.J. "Chemical Stability of Pharmaceuticals", 2nd ed.; Wiley-Interscience: New York, 1986.
32. Carpenter, B.K. "Determination of Organic Reaction Mechanisms"; Wiley-Interscience: New York, 1984; pp 123–37.
33. le Noble, W.J. *Prog. Phys. Org. Chem.* **1967**, *5*, 207.
34. Asano, T.; le Noble, W.J. *Chem. Revs.* **1978**, *78*, 407.
35. Van Eldik, R.; Asano, T.; le Noble, W.J. *Chem. Revs.* **1989**, *89*, 549.
36. Whalley, E. *Adv. Phys. Org. Chem.* **1974**, *2*, 93.
37. McCabe, J.R.; Eckert, C.A. *Acc. Chem. Res.* **1974**, *7*, 251.
38. Hamann, S.D.; le Noble, W.J. *J. Chem. Educ.* **1984**, *61*, 658.
39. Dauben, W.G.; Gerdes, J.M.; Look, G.C. *Synthesis* **1986**, 532.
40. Glasstone, S.; Laidler, K.J.; Eyring, H. "The Theory of Rate Processes"; McGraw-Hill: New York, 1948; pp 472–3.
41. Isaacs, N.S.; Najem, T.S. *J. Chem. Soc., Perkin Trans. 2* **1988**, 557.
42. Cameron, C.; Saluja, P.P.S.; Floriano, A.; Whalley, E. *J. Phys. Chem.* **1988**, *92*, 3417.
43. Bell, R.P. "Acid-Base Catalysis"; Oxford University Press: London, 1941; p 3.
44. Jencks, W.P. "Catalysis in Chemistry and Enzymology"; McGraw-Hill: New York, 1969; Chapter 1.
45. Bruice, T.C.; Benkovic, S.J. "Bioorganic Mechanisms"; Benjamin: New York, 1966; Vol. I, p 119.
46. Lienhard, G.E.; Jencks, W.P. *J. Am. Chem. Soc.* **1966**, *88*, 3982.
47. Steinberger, R.; Westheimer, F.H. *J. Am. Chem. Soc.* **1951**, *73*, 429.
48. Bruice, T.C.; Pandit, U.K. *J. Am. Chem. Soc.* **1960**, *82*, 5858.
49. Bender, M.L. "Mechanisms of Homogeneous Catalysis from Protons to Proteins"; Wiley-Interscience: New York, 1971.
50. Lin, S.-F.; Connors, K.A. *J. Pharm. Sci.* **1981**, *70*, 235.
51. Hand, E.S.; Jencks, W.P. *J. Am. Chem. Soc.* **1975**, *97*, 6221.
52. Pandit, N.K.; Connors, K.A. *J. Pharm. Sci.* **1982**, *71*, 485.
53. Johnson, S.L. *Adv. Phys. Org. Chem.* **1967**, *5*, 237.
54. van der Houwen, O.A.; Beijnen, J.H.; Bult, A.; Underberg, W.J. *Int. J. Pharmaceut.* **1988**, *45*, 181.
55. Connors, K.A.; Bender, M.L. *J. Org. Chem.* **1961**, *26*, 2498.
56. Bruice, T.C.; Marquardt, F.-H. *J. Am. Chem. Soc.* **1962**, *84*, 365.
57. Finholt, P.; Higuchi, T. *J. Pharm. Sci.* **1962**, *51*, 655.
58. Bender, M.L.; Chow, Y.-L.; Chloupek, F. *J. Am. Chem Soc.* **1958**, *80*, 5380.

59. Edwards, L.J. *Trans. Faraday Soc.* **1950,** *46,* 723; **1952,** *48,* 696.
60. Garrett, E.R. *J. Am. Chem. Soc.* **1957,** *79,* 3401.
61. Harper, E.T.; Bender, M.L. *J. Am. Chem. Soc.* **1965,** *87,* 5625.
62. Alberty, R.A.; Massey, V. *Biochim. Biophys. Acta* **1954,** *13,* 347.
63. Bunton, C.A., Llewellyn, D.R.; Oldham, K.G.; Vernon, C.A. *J. Chem. Soc.* **1958,** 3574.
64. Morawetz, H.; Oreskes, I. *J. Am. Chem. Soc.* **1958,** *80,* 2591.
65. Bender, M.L.; Connors, K.A. *J. Am. Chem. Soc.* **1962,** *84,* 1980.
66. Bada, J.L.; Miller, S.L. *J. Am. Chem. Soc.* **1969,** *91,* 3948.
67. Zerner, B.; Bender, M.L. *J. Am. Chem Soc.* **1961,** *83,* 2267.
68. Jencks, W.P. *J. Am. Chem. Soc.* **1959,** *81,* 475.
69. Bell, R.P. "The Proton in Chemistry"; Cornell University Press; Ithaca, N.Y., 1959; Chapter XI.
70. Saunders, W.H., Jr. In "Investigation of Rates and Mechanisms of Reactions", 3rd ed.; Lewis, E.S., Ed.; Wiley-Interscience: New York, 1974; Part I, Chapter V.
71. Ritchie, C.D. "Physical Organic Chemistry: The Fundamental Concepts"; Marcel Dekker: New York, 1975; Chapters 8 and 9.
72. Drenth, W.; Kwart, H. "Kinetics Applied to Organic Reactions"; Marcel Dekker: New York, 1980; Chapter 5.
73. Berliner, E. *Prog. Phys. Org. Chem.* **1964,** *2,* 253.
74. Zollinger, H. *Adv. Phys. Org. Chem.* **1964,** *2,* 163.
75. Halevi, E.A. *Prog. Phys. Org. Chem.* **1963,** *1,* 109.
76. Bunton, C.A.; Shiner, V.J., Jr. *J. Am. Chem. Soc.* **1961,** *83,* 3207, 3214.
77. Kresge, A.J. *Pure Appl Chem.* **1964,** *8,* 243.
78. Batts, B.D.; Gold, V. *J. Chem. Soc. A* **1969,** 984.
79. Schowen, R.L. *Prog. Phys. Org. Chem.* **1972,** *9,* 275.
80. Hegazi, M.; Mata-Segreda, J.F.; Schowen, R.L. *J. Org. Chem.* **1980,** *45,* 307.
81. Gandour, R.D.; Coyne, M.; Stella, V.J.; Schowen, R.L. *J. Org. Chem.* **1980,** *45,* 1733.
82. Gopalakrishnan, G.; Hogg, J.L. *J. Org. Chem.* **1981,** *46,* 4959.
83. Oesterling, T.O. *J Pharm. Sci.* **1970,** *59,* 63.
84. Gassman, P.G.; Macmillan, J.G. *J. Am. Chem. Soc.* **1969,** *91,* 5527.
85. Frost, D.V.; McIntire, F.C. *J. Am. Chem. Soc.* **1944,** *66,* 425.
86. Bender, M.L.; Neveu, M.C. *J. Am. Chem. Soc.* **1958,** *80,* 5388.
87. Oakenfull, D.G.; Riley, T.; Gold, V. *Chem. Commun.* **1966,** 385.
88. Fersht, A.R.; Jencks, W.P. *J. Am. Chem. Soc.* **1969,** *91,* 2125.
89. Jencks, W.P.; Barley, F.; Barnett, R.; Gilchrist, M. *J. Am. Chem. Soc.* **1964,** *88,* 4464.
90. Fedor, L.R.; Bruice, T.C. *J. Am. Chem. Soc.* **1965,** *87,* 4138.
91. Butler, A.R.; Gold, V. *J. Chem. Soc.* **1961,** 4362.
92. Kirsch, J.F.; Jencks, W.P. *J. Am. Chem. Soc.* **1964,** *86,* 837.

PROBLEMS

1. These rate constants are for the degradation of clindamycin at pH 4.0 in 0.2 M citrate buffers.[83]

$t/°C$	$10^6 k/s^{-1}$
47	0.0124
53	0.0279
59	0.0831
70	0.249
80	0.969
92	3.62

Make the Arrhenius plot and estimate E_a.

2. These rate constants are for the reaction

$$CH_3I + C_6H_5ON_a \rightarrow CH_3OC_6H_5 + NaI$$

in ethanol. Estimate E_a and A. What is the rate constant at 36°C?

$t/°C$	$10^5 k/M^{-1}s^{-1}$
0	5.60
6	11.8
12	24.5
18	48.8
24	100
30	208

3. This first-order decarboxylation reaction

$$CO(CH_2COOH)_2 \rightarrow CO(CH_3)_2 + 2CO_2$$

has an activation energy of 23.2 kcal mol^{-1} in aqueous solution. At 10°C the half-life is 107 min. What is the rate constant for this reaction at 50°C?

4. These rate constants are for the acetolysis of exo-2-tosyloxybicyclo[2.2.1]-heptan-7-one ethylene glycol ketal.[84]

$t/°C$	$10^4 k/s^{-1}$
80	2.71
90	8.05
100	25.6

Estimate $\Delta H^{\ddagger}$ and $\Delta S^{\ddagger}$.

5. The rate constant for the acid-catalyzed hydrolysis of ethyl acetate is

$$\ln k = 7.532 - 16200/RT \ (R \text{ in cal})$$

Calculate $\Delta H^{\ddagger}$ and $\Delta S^{\ddagger}$.

6. Analyze these data for the hydrolysis of pantothenate.[85]

$t/°C$	k/day^{-1}
10	0.000 554
23	0.003 10
39	0.013 7
55	0.089 4
75	0.42
100	2.28

7. Show that if the relative error σ_k/k is constant, then an unweighted linear Arrhenius least-squares analysis is correct.

8. le Noble[33] gives these data for the dependence of the rate of chloroform hydrolysis on pressure.

$P/kbar$	$10^5 k/M^{-1}s^{-1}$
0.00	7.39
1.07	3.53
2.13	2.15
3.17	1.62
4.21	1.08
5.24	0.66
6.45	0.70

Calculate $\Delta V^{\ddagger}$.

9. These are rate constants for the hydrolysis of cinnamic anhydride in bicarbonate-carbonate buffers. The pK_a' of bicarbonate is 10.22. Find the rate constant for hydrolysis, at each pH, at zero buffer concentration. Analyze the data to determine if the acid or base component of the buffer, or both, are responsible for catalysis, and give the catalytic rate constant(s).

pH	Total buffer concentration/M	$10^3 k_{obs}/s^{-1}$
10.06	0.025 0	22.6
	0.035 0	24.4
	0.050 0	26.5
9.76	0.017 0	10.5
	0.020 4	11.0
	0.027 2	11.8
	0.034 0	12.2
9.14	0.013 2	2.53
	0.022 0	2.73
	0.030 8	3.10
	0.044 0	3.40

10. The following data are for the hydrolysis of cinnamic anhydride in (2-amino-2-hydroxymethyl-1,3-propane diol buffers. Extrapolate them to zero buffer concentration, and, together with data from Problem 9, plot the pH–rate profile. Determine the order with respect to hydroxide, and calculate the rate constant for hydrolysis.

pH	Total buffer concentration/M	$10^3 k_{obs}/s^{-1}$
8.80	0.010 5	3.95
	0.012 6	4.49
	0.016 8	5.77
	0.021 0	6.72
8.48	0.012 0	3.15
	0.014 4	3.68
	0.019 2	4.64
	0.024 0	5.74
8.17	0.009 8	1.86
	0.011 8	2.10
	0.019 6	3.49

11. These rate constants are for the hydrolysis of cinnamic anhydride in carbonate buffer, pH 8.45, total buffer concentration 0.024 M, in the presence of the catalysts pyridine, *N*-methylimidazole (NMIM), or 4-dimethylaminopyridine (DMAP). In the absence of added catalyst, but the presence of buffer, the rate constant was 0.005 24 s^{-1}. You may assume that only the conjugate base form of each catalyst is catalytically effective. Calculate the catalytic rate constant for the three catalysts. What is the catalytic power of NMIM and of DMAP relative to pyridine?

Catalyst	Total catalyst concentration/M	$10^3 k_{obs}/s^{-1}$
Pyridine	0.001 21	8.60
($pK_a' = 5.29$)	0.001 94	10.28
	0.002 42	11.85
NMIM	2.94×10^{-4}	9.06
($pK_a' = 7.19$)	4.90×10^{-4}	11.60
	6.86×10^{-4}	14.09
DMAP	1.17×10^{-5}	6.48
($pK_a' = 9.68$)	1.95×10^{-5}	7.42
	2.74×10^{-5}	8.27

12. These are rate data for the hydrolysis of succinimide.

pH	$10^4 k_{obs}/s^{-1}$
7.90	0.08
9.11	0.31
9.65	0.53
10.24	0.81
11.10	0.95
11.90	1.02
13.00	0.95

Analyze these data; that is, find the rate equation and evaluate the constants.

13. Suppose the rate equation for the aminolysis of an ester is

$$v = k_1[E][OH^-] + k_2[E][RNH_2] + k_3[E][RNH_2]^2$$
$$+ k_4[E][RNH_2][OH^-] + k_5[E][RNH_2][RNH_3^+]$$

Develop an experimental design that will permit all of the rate constants to be determined.

14. Apply the steady-state approximation to the general base-catalyzed halogenation of ketones:

$$RCOR + B \underset{k_{-1}}{\overset{k_1}{\rightleftharpoons}} I^- + BH^+$$

$$I^- + X_2 \overset{k_2}{\rightarrow} products$$

(a) Show that the observed rate constant is proportional to the general base concentration.

(b) Show how the observed rate constant is independent of the halogen concentration when the proton abstraction is the rate-determining step.

15. Sketch the kinetic scheme that is kinetically equivalent to Scheme VII.

16. Calculate the difference in activation energies corresponding to a primary kinetic isotope effect of $k_H/k_D = 7$ at 25°C.

CHAPTER 7

Structure–Reactivity Relationships

Very early in the development of organic chemistry, it was recognized that similar structural changes tend to produce similar chemical effects. This orderly behavior permitted generalizations to be made in describing the relationships between chemical structure and reactivity. In this chapter we describe many of these generalizations. A useful approach is to seek patterns in the form of correlations of rate data with other quantities such as equilibrium constants. Such patterns define "normal" or "expected" structure–reactivity behavior and, therefore, provide a basis for detecting and exploring the often more interesting deviations from typical behavior.

7.1 EXTRATHERMODYNAMIC RELATIONSHIPS

The magnitude of a change in a thermodynamic property of a system is determined only by the values of the property in the initial and final states of the system. An infinite number of routes may exist between these two states, and each route necessarily results in the same net change in the property. Thus, knowledge of the overall change cannot distinguish between alternate paths. The concepts and methods of chemical kinetics provide considerable insight into the actual paths (mechanisms) accessible to the process, because kinetics is concerned with changes in properties between an initial state and some intermediate state, namely, the transition state.

The great generality of thermodynamics is a consequence of its minimal use of specific and detailed models; on the other hand, it is the absence of such models that prevents thermodynamics from providing insight into molecular mechanisms. The combination of detailed models with the concepts of thermodynamics is called the *extrathermodynamic approach*. Because it involves model building, the technique lacks the rigor of thermodynamics, but it can provide information not otherwise accessible. Extrathermodynamic relationships often take the form of correlations among rates and equilibria, and the models used to account for these include hypotheses of chemical bonding, transmission of electronic effects, and interactions among electronic effects.

The most common manifestation of extrathermodynamic relationships is a linear correlation between the logarithms of rate or equilibrium constants for one reaction series and the logarithms of rate or equilibrium constants of a second reaction series, both sets being subjected to the same variation, usually of structure. For illustration, suppose the logarithm of the rate constants for a reaction series B is linearly correlated with the logarithm of the equilibrium constants for a reaction series A, with substituent changes being made in both series. The empirical correlation is

$$\log k_B = m \log K_A + b \tag{7-1}$$

As seen earlier, these logarithmic terms are linearly related to free energy changes:

$$\log k_B = \frac{-\Delta G_B^{\ddagger}}{2.3RT} + \log \frac{kT}{h} \tag{7-2}$$

$$\log K_A = \frac{-\Delta G_A^0}{2.3RT} \tag{7-3}$$

Combining Eqs. (7-1)–(7-3),

$$\Delta G_B^{\ddagger} = m \Delta G_A^0 + b' \tag{7-4}$$

Such correlations are therefore called *linear free energy relationships* (LFERs). Often it is convenient to express the correlation in terms of ratios of constants by referring all members of the series to a reference member of the series; thus the correlation in Eq. (7-1) can be expressed

$$\log \frac{k_B}{k_B^0} = m \log \frac{K_A}{K_A^0} \tag{7-5}$$

where k_B^0 and K_A^0 are the constants for the reference substituent.

It is useful at this point to introduce a new symbol. We are accustomed to use the operator Δ to signify a change in a quantity associated with a chemical reaction, e.g., ΔG, ΔH, or ΔS. It is convenient to define different symbols to describe changes caused by alterations in structure and in the medium. We will employ the δ operators of Leffler and Grunwald[1] for this purpose. A change in a quantity caused by a structural alteration is indicated by the symbol δ_R, and a change caused by a change in the medium by δ_M. These operators can be combined to describe changes in changes; thus, $\delta_R \Delta G^0 = \Delta G_R^0 - \Delta G_{R_0}^0$ is a change in standard free energy change caused by a structural variation from R_0 to R. (Some authors write $\Delta \Delta G^0$ instead of $\delta_R \Delta G^0$). It can be demonstrated that these operators commute, so $\delta_R \Delta G = \Delta \delta_R G$, and so on.[1] With this symbolism, Eq. (7-5) can be written

$$\delta_R \Delta G_B^{\ddagger} = m \delta_R \Delta G_A^0 \tag{7-6}$$

Though LFERs are not a necessary consequence of thermodynamics, their occurrence suggests the presence of a real connection between the correlated quantities, and the nature of this connection can be explored. This treatment follows Leffler and Grunwald.[1, pp. 140–143] Standard free energy changes ΔG^0 will pertain to either equilibrium or rate processes. Consider the reaction of a molecule that, for convenience, we divide into two zones, R and X. Zone R includes the variable substituent, and X is the reaction site. The standard free energy of the substance is considered to be an additive function of the free energies of these two zones plus a term describing the interaction between the two zones.

$$G^0_{RX} = G^0_R + G^0_X + I_{R.X} \tag{7-7}$$

Equation (7-7) represents the model, which will be applied to a reaction series A, in which a set of compounds varying only in the substituent R undergoes the common reaction

$$RA \rightleftharpoons RA' \tag{7-8}$$

Writing $\Delta G^0_{RA} = G^0_{RA'} - G^0_{RA}$ and applying Eq. (7-7),

$$\Delta G^0_{RA} = (G^0_{A'} - G^0_A) + (I_{R.A'} - I_{R.A}) \tag{7-9}$$

in which the G^0_R term has disappeared because it does not undergo any change in the chemical reaction. A second reaction series B, with the same variations in substituents but a different reaction site, is similarly described;

$$RB \rightleftharpoons RB' \tag{7-10}$$

$$\Delta G^0_{RB} = (G^0_{B'} - G^0_B) + (I_{R.B'} - I_{R.B}) \tag{7-11}$$

Variations in the substituent R within the reaction series are described by the terms $\delta_R \Delta G^0_A$ and $\delta_R \Delta G^0_B$,

$$\delta_R \Delta G^0_A = \Delta G^0_{RA} - \Delta G^0_{R_0A} \tag{7-12}$$

$$\delta_R \Delta G^0_B = \Delta G^0_{RB} - \Delta G^0_{R_0B} \tag{7-13}$$

where R_0 is the reference substituent. Equations (7-12) and (7-13) are combined with (7-9) and (7-11):

$$\delta_R \Delta G^0_A = (I_{R.A'} - I_{R.A}) - (I_{R_0.A'} - I_{R_0.A}) \tag{7-14}$$

$$\delta_R \Delta G^0_B = (I_{R.B'} - I_{R.B}) - (I_{R_0.B'} - I_{R_0.B}) \tag{7-15}$$

It is necessary to introduce one further assumption to make this model lead to an LFER; this is a separability postulate,

$$I_{R,X} = I_R \cdot I_X \tag{7-16}$$

Then, from Eqs. (7-14) and (7-15),

$$\delta_R \Delta G_A^0 = (I_R - I_{R_0})(I_{A'} - I_A) \tag{7-17}$$

$$\delta_R \Delta G_B^0 = (I_R - I_{R_0})(I_{B'} - I_A) \tag{7-18}$$

Combining these equations,

$$\delta_R \Delta G_B^0 = \left(\frac{I_{B'} - I_B}{I_{A'} - I_A}\right) \delta_R \Delta G_A^0 \tag{7-19}$$

Equation (7-19) has the form of an LFER [compare with Eq. (7-6)]. The quantity in parentheses is independent of the nature of the substituent, depending only upon the reaction types; it is called the *reaction parameter*. Now suppose that reaction series A is selected as a standard reaction; then $\delta_R \Delta G_A^0$ becomes dependent only on the substituent and is called the *substituent parameter*. (For the standard reaction, the reaction parameter is arbitrarily set equal to unity.) Wells[2,3] has given an equivalent treatment.

Later sections describe several empirical relationships with the form of Eq. (7-19). Some free energy relationships require expression as four-parameter equations of the form

$$\delta_R \Delta G_B^0 = m\delta_R \Delta G_1^0 + n\delta_R \Delta G_2^0 \tag{7-20}$$

Such an equation implies two interaction mechanisms in the model.[1, pp. 140–3]

The preceding treatment applies to LFERs between rates or equilibria of two different reactions. A special case of this situation is the relationships between the rate constants and the equilibrium constants of the *same* reaction series. Equation (7-19) could apply to this case, with $\delta_R \Delta G_B^0$ proportional to $\log(k/k_0)$ and $\delta_R \Delta G_A^0$ proportional to $\log(K/K_0)$ for the same reaction. Thus, we anticipate an LFER. This case happens to be susceptible to theoretical analysis because the rates and equilibria apply to movement over the same potential energy surface; we have reviewed this problem in Section 5.3, "Position and Height of the Energy Barrier," where we obtained the Marcus equation, Eq. (7-21), as the relationship between the free energy of activation $\Delta G^{\ddagger}$ and the standard free energy change ΔG^0 of the same reaction.

$$\Delta G^{\ddagger} = \Delta G_0^{\ddagger} + \frac{\Delta G^0}{2} + \frac{(\Delta G^0)^2}{16\Delta G_0^{\ddagger}} \tag{7-21}$$

In Eq. (7-21), $\Delta G_0^{\ddagger}$ is the intrinsic barrier, the free energy of activation of the (hypothetical) member of the reaction series having $\Delta G^0 = 0$. It is evident that the Marcus equation predicts a nonlinear free energy relationship, although if a limited range is explored experimentally the nonlinearity may not be evident. We also saw in Section 5.3 that the Marcus equation applies to a relatively simple system and that more complicated behavior is possible. These considerations should caution us not to expect great generality in empirical LFERs, but rather to view these relationships as fortunate, although limited, simple forms of more complicated behavior.

The literature in this area is extensive and some of the concepts and symbolism may be transitory. This chapter reviews the field at a level and with a coverage adequate for the experimentalist to use the standard relationships and to follow their use in the mechanistic literature. Research on the meaning of the extrathermodynamic relationships themselves is beyond our needs; the interested reader can explore these ideas further in the references cited. Grunwald[4] has reviewed the early history of LFERs.

7.2 SUBSTITUENT EFFECTS IN AROMATIC COMPOUNDS

The designation of one reactant as the *substrate* and another as the *reagent* is arbitrary but useful in discussing chemical reactivity. The substrate always undergoes some change in the reaction. A catalyst is always considered to be a reagent. When a reaction is studied under the usual pseudo-first-order conditions, the concentration of the substrate does not appear in the apparent first-order rate constant, whereas the reagent concentration does. In the reaction between an acid anhydride and an amine, the physical organic chemist might regard the amine as the reagent, whereas the analytical chemist may consider the anhydride as the reagent. This dual use of the term should not cause any confusion here. The present section concerns LFERs describing substituent effects in aromatic substrates; very generally, a series of substrates R–X varying in the R substituent is subjected to a common reaction at site X, and the quantity $\delta_R \Delta G^{\ddagger}$ is correlated with a standard reaction series (the "model" reaction).

The Hammett Equation

The classic example, and still the most useful one, of a LFER is the Hammett equation,[5,6] which correlates rates and equilibria of many side-chain reactions of *meta-* and *para*-substituted aromatic compounds. The standard reaction is the aqueous ionization equilibrium at 25°C of *meta-* and *para*-substituted benzoic acids.

The reference compound is benzoic acid. The substituent parameter, which is called the *Hammett substituent constant* σ, is defined by

$$\sigma = \log \frac{K_a}{K_a^0} = pK_a^0 - pK_a \qquad (7\text{-}22)$$

σ is related to $\delta_R \Delta G^0$ by

$$\delta_R \Delta G^0 = -2.3RT\sigma \qquad (7\text{-}23)$$

The Hammett equation is written as

$$\log \frac{k}{k^0} = \rho\sigma \qquad (7\text{-}24)$$

where k is a rate constant for m- or p-substituted substrates, and ρ is the Hammett reaction constant. Equation (7-24) obviously has the same form as Eqs. (7-5) and (7-19).

The form of Eq. (7-24) suggests, by comparison with Eq. (7-19), that a single interaction mechanism is operative between the substituent R and the reaction site X. That this is not necessarily true is shown as follows. Suppose two interaction mechanisms control the substituent effect. Then a Hammett relationship may be written as in

$$\log \frac{k}{k^0} = \rho_1\sigma_1 + \rho_2\sigma_2 \qquad (7\text{-}25)$$

which can be rearranged to

$$\log \frac{k}{k^0} = \rho_1 \left(\sigma_1 + \frac{\rho_2}{\rho_1}\sigma_2 \right) \qquad (7\text{-}26)$$

If the two interaction mechanisms maintain the same relative importance throughout the reaction series, that is, if the ratio ρ_2/ρ_1 is constant, Eq. (7-26) becomes identical in form with Eq. (7-24). This in fact appears to be the usual situation in Hammett correlations,[1, pp. 192–4] and the two interaction mechanisms are commonly discussed in terms of the inductive and resonance effects of electronic displacement.

Two points of view have developed concerning the "best" values of the Hammett substituent constants. One of these, exemplified by the remarks of Jaffé[7] and Ehrenson,[8] treats σ as an adjustable parameter whose value should be chosen to best fit the entire body of experimental data. The assignment of σ values in this context becomes a statistical problem of curve fitting, best handled by computer methods. The advantage of the method is that it provides substituent constants capable of generating ρ values, and of regenerating rate constants, with reasonable accuracy over the entire range of reactions used in establishing the values. One of its disadvantages is that these σ values are subject to periodic change as new data become available.

The other point of view[9,10] is that the σ values should be based on the original defining relation of Hammett, Eq. (7-22). Besides the advantage of providing

TABLE 7-1. Hammett Substituent Constants Based on Ionization of Benzoic Acids[a]

Group	σ_m	σ_p
—NHMe	—	-0.84
—NMe$_2$	—	-0.83
—NH$_2$	-0.21	-0.63
—OH	$+0.13$	-0.38
—OMe	$+0.11$	-0.28
—OEt	$+0.10$	-0.24
—OC$_6$H$_5$	$+0.25$	-0.32
tert-Bu	0.00	-0.19
iso-Pr	—	-0.15
Et	-0.07	-0.15
Me	-0.06	-0.17
—CH$_2$CH=CH$_2$	-0.06	-0.12
—C$_6$H$_5$	$+0.06$	-0.01
—H	(0.00)	(0.00)
—SiMe$_3$	$+0.11$	$+0.01$
—CO$_2^-$	-0.10	0.00
—NHCOMe	$+0.21$	0.00
—SH	$+0.25$	$+0.15$
—SMe	$+0.15$	$+0.00$
—CH$_2$OH	—	$+0.04$
F	$+0.34$	$+0.06$
Cl	$+0.37$	$+0.22$
Br	$+0.34$	$+0.22$
I	$+0.35$	$+0.24$
—OCOMe	$+0.39$	$+0.31$
—SCOMe	$+0.39$	$+0.44$
—CHO	$+0.25$	$+0.45$
—COMe	$+0.38$	$+0.50$
—COEt	—	$+0.48$
—S(O)Me	$+0.52$	$+0.54$
—SO$_3^-$	$+0.05$	$+0.48$
—CF$_3$	$+0.43$	$+0.54$
—CN	$+0.61$	$+0.65$
—NO$_2$	$+0.74$	$+0.78$
—NH$_3^+$	$+0.90$	$+0.71$
—NMe$_3^+$	$+1.04$	$+0.97$
—SMe$_2^+$	$+1.11$	$+1.14$
—N$_2^{+}$ [b]	$+1.76$	$+1.91$

Source: References 9 and 31.

[a] $pK_a^0 = 4.203$ at 25°C for benzoic acid.

[b] From Reference 199.

essentially permanent σ values, this approach takes a fundamentally different, and more optimistic, view of deviations from Hammett plots. In this context such deviations represent different mechanistic effects in the correlated reaction series and the standard reaction series, and the deviations can guide mechanistic interpretations. The statistical method of evaluating substituent constants tends to minimize deviations. The dispute really concerns the utility of LFERs. If their principal use is for data storage and retrieval, then the statistically evaluated σ values will, on the average, provide more accurate estimates of rate and equilibrium constants. If they are helpful primarily as tools for mechanistic interpretation, the definition of σ in terms of a standard reaction is to be preferred, because it yields substituent constants of definite value and unambiguous meaning (even though the meaning may not be understood). The practice of estimating "secondary" σ values by measuring ρ for a given reaction series and then calculating σ for a group subjected to this reaction, without having available its corresponding benzoic acid dissociation constant, has contributed to the range of published σ values for a single substituent.

We take the view of McDaniel and Brown[9] that the Hammett substituent constants should be defined by Eq. (7-22). Table 7-1 lists many of these constants based on the ionization of *meta*- and *para*-substituted benzoic acids.

The Hammett equation is said to be followed when a plot of log k against σ is linear. Most workers take as the criterion of linearity the correlation coefficient r, which is required to be at least 0.95 and preferably above 0.98. A weakness of r as a statistical measure of goodness of fit is that r is a function of the slope ρ; if the slope is zero, the correlation coefficient is zero. A slope of zero in an LFER is a chemically informative result, for it demonstrates an absence of a substituent

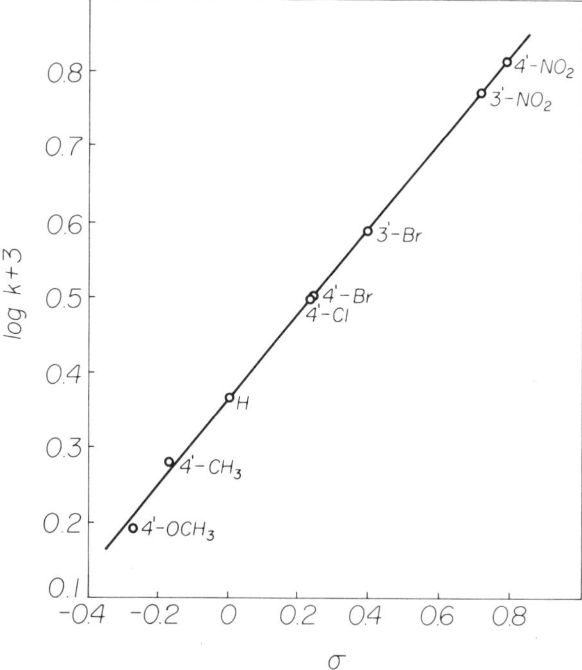

Figure 7-1. Hammett plot for the alkaline hydrolysis of substituted ethyl *p*-biphenylcarboxylates at 40°C in 88.7% (w/w) ethanol.[12]

TABLE 7-2. Hammett ρ Values for Some Rate Processes[a]

No.	Reaction	Solvent	$t/°C$	ρ	$-log\ k^{0}$[b]
1	$ArCOOH + CH_3OH + H^+$ $\rightarrow ArCOOCH_3$	CH_3OH	25	-0.229	3.841
2	$ArCOOH + HN_3 \rightarrow ArNH_2 + CO_2$	$CHCl{=}CHCl$	40	-1.415	5.289
3	$ArCOOCH_3 + OH^- \rightarrow ArCOO^-$	$60\%\ Me_2CO$	25	2.229	2.075
4	$ArCOOC_2H_5 + OH^- \rightarrow ArCOO^-$	$60\%\ Me_2CO$	25	2.265	2.557
5	$ArCOOC_2H_5 + H^+ \rightarrow ArCOOH$	$60\%\ Me_2CO$	100	0.106	4.086
6	$p\text{-}ArC_6H_4COOC_2H_5 + OH^-$ $\rightarrow p\text{-}ArC_6H_4COO^-$	$88\%\ C_2H_5OH$	40	0.583	2.636
7	$ArCH_2COOC_2H_5 + OH^-$ $\rightarrow ArCH_2COO^-$	$87.83\%\ C_2H_5OH$	30	0.824	1.813
8	$ArCH_2CH_2COOC_2H_5 + OH^-$ $\rightarrow ArCH_2CH_2COO^-$	$87.83\%\ C_2H_5OH$	30	0.489	2.198
9	$ArCH{=}CHCOOC_2H_5 + OH^-$ $\rightarrow ArCH{=}CHCOO^-$	$87.83\%\ C_2H_5OH$	30	1.329	2.752
10	$ArCOCl + H_2O \rightarrow ArCOOH$	$95\%\ Me_2CO$	25	1.782	4.200
11	$ArCONH_2 + OH^- \rightarrow ArCOO^-$	$60\%\ C_2H_5OH$	100.1	1.100	3.523
12	$ArCONH_2 + H^+ \rightarrow ArCOOH$	$60\%\ C_2H_5OH$	99.6	-0.222	3.806
13	$(ArCO)_2O + H_2O \rightarrow ArCOOH$	75% dioxane	58.25	1.568	5.113
14a	$ArOCOC_6H_5 + OH^- \rightarrow ArOH$	$60\%\ Me_2CO$	-20	1.051	1.801
14b	$ArOCOC_6H_5 + OH^- \rightarrow ArOH$	$60\%\ Me_2CO$	-10	1.034	1.408
14c	$ArOCOC_6H_5 + OH^- \rightarrow ArOH$	$60\%\ Me_2CO$	0	0.976	1.015
14d	$ArOCOC_6H_5 + OH^- \rightarrow ArOH$	$60\%\ Me_2CO$	15	0.930	0.490
15	$ArNH_2 + C_6H_5COCl$ $\rightarrow ArNHCOC_6H_5$	C_6H_6	25	-2.781	2.888

[a]Abstracted from the compilation by Jaffé,[7] where original references may be found.
[b]Value of log k on the least-squares regression line where $\sigma = 0$; the time unit is seconds.

effect; yet the correlation coefficient criterion would reject this result. Davis and Pryor[11] have analyzed the relationship among r, the slope, and the scatter of points about the line and urge that r be abandoned as a criterion of linearity in LFERs. They propose instead that the standard deviation of the slope, σ_ρ, or the confidence limits of the slope at a stated level of significance, $\pm t\sigma_\rho$, be reported. Figure 7-1 is a Hammett plot for the hydrolysis of substituted ethyl p-biphenylcarboxylates, **1**.[12] This is a very good example; statistical analysis is not required to conclude that the Hammett equation is followed. The slope of the line, and therefore ρ, is $+0.583$.

Y — CO_2C_2H_5

1

Table 7-2 lists 15 reactions whose rates are correlated by the Hammett equation.[7] Besides the reaction constant ρ, the table gives a value for k^0 (from the fitted line), which provides all the information needed to estimate k for any member of the series, if the corresponding σ is available, by means of Eq. (7-24). Note that k^0 in Table 7-2 is generally not identical to the experimental value of k for the $\sigma = 0$ member of the series, because this experimental point may deviate from the regression line.

Deviations and Variations

The discussion in Section 7.1 should prepare us to expect deviations from such a simple relationship as the Hammett equation if the reaction being correlated differs greatly from the standard reaction. When this happens we have two choices (within this extrathermodynamic approach): We can select a different standard reaction, or we can increase the number of parameters.

A Hammett plot of the pK_a values of p-substituted phenols against the σ_ρ values shows serious deviations for the members of the series at the extremes of the σ scale, that is, for substituents that are strongly electron donating or electron withdrawing.[13] It was recognized very early[14] that such deviations could be rectified by choosing an appropriate σ value for such substituents; in effect, this means a different model reaction was adopted. The chemical basis of the procedure can be illustrated with the p-nitro substituent. The p-nitrophenolate ion is stabilized by *through resonance* as shown in **2.**

O⁻   O

NO₂   NO₂⁻

2

This phenomenon is not possible in p-nitrobenzoic acid; hence, p-nitrophenol is a stronger acid with respect to p-nitrobenzoic acid than is expected on the basis of a comparison of substituents in which this resonance delocalization is not an important factor. It was, therefore, recommended[14] that $\sigma_\rho = 1.27$ be used for p-nitro derivatives of phenols and anilines, rather than the $\sigma_\rho = 0.78$ given in Table 7-10. These "enhanced" sigma constants, symbolized σ^-, apply primarily to electron-withdrawing groups in reactions aided by low electron density at the reaction site,

TABLE 7-3. Enhanced *para*-Substituent Constants, σ^-

Substituent	σ^-	Substituent	σ^-
—CO$_2$H	0.728	—CN	1.000[a]
—CO$_2$CH$_3$	0.636	—NO$_2$	1.27[b]
—CO$_2$C$_2$H$_5$	0.678	—SO$_2$CH$_3$	1.049
—CO$_2$C$_4$H$_9$	0.674	—C$_6$H$_4$N═NC$_6$H$_5$	1.088
—CO$_2$CH$_2$C$_6$H$_5$	0.667	—CH═CH—C$_6$H$_5$	0.619
—CONH$_2$	0.627	—N$_2^+$	3.2[c]
—CHO	1.126		

Source: From Jaffé's[7] collection except as indicated.
[a]Biggs and Robinson[200] give 0.983 and 0.874 based on ionization of anilines and phenols, respectively.
[b]Biggs and Robinson[200] give 1.24.
[c]Reference 199.

that is, reactions in which a negative charge is developed at the reaction site. For groups in the *meta* position, which cannot undergo extensive resonance delocalization with the reactive site, the Hammett σ_m values are used for correlating these reactions; likewise for *para*-substituted groups incapable of extensive resonance interaction, the σ_p constants are adequate. Table 7-3 gives σ^- constants for those *para*-substituted groups exhibiting these enhanced effects. σ^- has yielded improved correlations for aromatic nucleophilic substitutions.[15]

Reactions that occur with the development of an electron deficiency, such as aromatic electrophilic substitutions, are best correlated by substituent constants based on a more appropriate defining reaction than the ionization of benzoic acids. Brown and Okamoto[16] adopted the rates of solvolysis of substituted phenyldimethylcarbinyl chlorides (*t*-cumyl chlorides) in 90% aqueous acetone at 25°C to define electrophilic substituent constants symbolized σ^+. Their procedure was to establish a conventional Hammett plot of log (k/k^0) against σ_m for 16 *meta*-substituted *t*-cumyl chlorides, because *meta* substituents cannot undergo significant direct resonance interaction with the reaction site. The resulting ρ value of -4.54 was then used in a modified Hammett equation,

$$\log \frac{k}{k^0} = \rho\sigma^+ \tag{7-27}$$

to calculate σ^+ constants for a total of 41 substituents. This procedure has the result of placing the σ^+ and σ constants on essentially the same scale. The σ^+ constants are given in Table 7-4. Some of the reactions correlated by Eq. (7-27) are listed in Table 7-5.

The σ^- and σ^+ scales of substituent effects result from changes in the standard reaction that defines the σ scale. An alternative approach to dealing with substituents that possess more than one mechanism of electronic interaction with the reaction site is to make use of more than one substituent constant. Yukawa and Tsuno[17]

TABLE 7-4. Electrophilic Substitutent Constants, σ^+

Substitutient	σ_m^+	σ_p^+
—OCH$_3$	0.047	-0.778
—SCH$_3$	0.158	-0.604
—CH$_3$	-0.066	-0.311
—C$_2$H$_5$	-0.064	-0.295
—CH(CH$_3$)$_2$	-0.060	-0.280
—C(CH$_3$)$_3$	-0.059	-0.256
—C$_6$H$_5$	0.109	-0.179
—(2-Naphthyl)	—	-0.135
—H	0	0
—Si(CH$_3$)$_3$	0.011	0.021
—F	0.352	-0.073
—Cl	0.399	0.114
—Br	0.405	0.150
—I	0.359	0.135
—COOH	0.322	0.421
—COOCH$_3$	0.368	0.489
—COOC$_2$H$_5$	0.366	0.482
—CF$_3$	0.520	0.612
—CN	0.562	0.659
—NO$_2$	0.674	0.790
—COO$^-$(K$^+$)	-0.028	-0.023
—N(CH$_3$)$_3^+$(Cl$^-$)	0.359	0.408

Source: Reference 16.

TABLE 7-5. Reaction Constants for σ^+ Correlations

No.	Reaction	t/°C	ρ
	Electrophilic nuclear substitutions		
1	Bromination of monosubstituted benzenes by Br$_2$ in HOAc	25	-12.14
2	Chlorination of monosubstituted benzenes by Cl$_2$ in HOAc	25	-8.06
3	Nitration of monosubstituted benzenes by HNO$_3$ in CH$_3$NO$_2$ or Ac$_2$O	0 or 25	-6.22
4	Bromination of monosubstituted benzenes by HOBr and HClO$_4$ in 50% dioxane	25	-5.78
5	Protonolysis of aryltrimethylsilanes by H$_2$SO$_4$ in HOAc		-4.60
6	Brominolysis of aryltrimethylsilanes in HOAc	25	-6.04
7	Brominolysis of benzeneboronic acids in 20% HOAc and 0.40 M NaBr	25	-4.44
8	Bromination of polymethylbenzenes by Br$_2$ in CH$_3$NO$_2$	25	-8.10
	Electrophilic side-chain reactions		
9	Solvolysis of *t*-cumyl chlorides in 90% acetone[a]	25	-4.54
10	Solvolysis of *t*-cumyl chlorides in ethanol	25	-4.67
11	Acid-catalyzed rearrangement of α-phenylethyl chlorocarbonates in dioxane	80	-3.01
12	Equilibrium constants for carbonium ion formation from triphenylcarbinols in aq H$_2$SO$_4$	25	-3.64
13	Equilibrium constants for ionization of triphenylcarbinyl chlorides in SO$_2$	0	-3.73

Source: Reference 16.
[a]Model reaction.

proposed the difference $\sigma^+ - \sigma$ as a measure of the through-resonance effect and wrote Eq. (7-28).

$$\log \frac{k}{k_0} = \rho[\sigma + r(\sigma^+ - \sigma)] \qquad (7\text{-}28)$$

Equation (7-28) is equivalent to Eq. (7-29), with $\rho' = \rho(1 - r)$ and $\rho'' = \rho r$.

$$\log \frac{k}{k_0} = \rho'\sigma + \rho''\sigma^+ \qquad (7\text{-}29)$$

Equation (7-29) is linear in the parameters and can be regarded as a multiple LFER. Many other equations of this form have been proposed.

The Substituent Constant

In the early decades of this century a successful qualitative theory of substituent effects was developed with the concept of the electronic nature of the covalent bond as applied to rates and equilibria of organic reactions. The electronic effect of a substituent at a reaction site can be described as a combination of inductive and resonance effects. *Inductive* (polar) effects include through-bond displacements of electron density as a consequence of bonding between two unlike atoms and the electrostatic field effect operating through space (or through the solvent) between the substituent and the reaction center. *Resonance* effects are a consequence of electron delocalization or conjugative interaction, primarily of π electrons, but including hyperconjugation. (The English school uses the word *mesomerism* for resonance.) Symbols have been assigned to these effects, I representing the inductive effect and M, T, or R, the resonance effect. The direction of the electronic effect is represented by a sign, but it unfortunately happens that there are two sign conventions. In Ingold's convention,[18] an electron-attracting substituent is associated with a negative sign, an electron-repelling substituent with a positive sign; thus, the $p\text{-}NO_2$ group is said to exert a $-R$ effect, whereas the $p\text{-}OMe$ group has a $+R$ effect. This convention is sympathetic to the terminology in which an electronegative group attracts electrons (relative to hydrogen), whereas an electropositive groups repels electrons. This sign convention has been widely used.[6,18–21] The opposite sign convention is attributed to Robinson and has been used by Wells,[2,3] Taft and Lewis,[13] and Dewar.[22] This convention has the advantage that the sign of the effect ($\pm I$, $\pm R$) agrees with the sign of σ; thus $p\text{-}NO_2$ has the $+R$ effect, and $\sigma_p = +0.78$.

A further complication arises with Ingold's suggestion[18] that both the inductive and resonance effects are composed of initial state equilibrium displacements that reveal themselves in equilibrium properties like dipole moments and equilibrium constants and of time-dependent displacements produced during reaction by the approach of an attacking reagent, observed rate effects being resultants of both types of electronic effects. Hammett,[5,6] however, claims that it is not necessary or possible to make this distinction.

The Hammett equation and LFER in general added no new concepts to the qualitative picture that had been built up of electronic effects in organic reactions, but they did provide a quantitative measure that had been lacking and that has been found very useful. Here we will describe the further development of ideas concerning the substituent constant.

We have already defined σ, σ^-, and σ^+. It is apparent that by choosing additional standard reactions we could generate further sigma values, and van Bekkum et al.[23] have indeed shown that σ for a substituent appears to have a range of values (of which σ^- and σ^+ may mark approximate extremes). We are thus faced with the problem of choosing the "best" value in making the LFER plot. The source of the variability in the substituent constant is the electronic requirements of the particular reaction and the capability of the substituent to respond to this requirement. Direct resonance interaction seems to be the principal mechanism contributing to variability in σ values. For this reason a set of σ values has been proposed in which direct resonance interaction is absent; these substituent constants may therefore be quantitative measures of the inductive transmission of electronic effects. These *normal* or *primary* σ values were taken to be the Hammett σ_m values for "well-behaved" substituents that are expected to undergo no resonance delocalization interactions. van Bekkum et al.[23] symbolized these by σ^n, and Taft,[24] by σ^0. The σ^n and σ^0 scales are nearly identical with each other and with the Hammett σ_m valves except for *para* substituents with resonance capabilities. Table 7-6 gives valves of σ^n and σ^0.

The reasonable extension of these ideas is to express σ as a sum of contributions from the inductive and resonance effects. Branch and Calvin[25] suggested this, and much of the research on LFER of the past three decades has been concerned with

TABLE 7-6. The Normal (σ^n) and Primary (σ^0)
Substituent Constants

	σ^n		σ^0	
Group	*meta*	*para*	*meta*	*para*
Me	-0.07	-0.13	-0.07	-0.15
NMe_2	-0.05	-0.17	-0.15	-0.44
NH_2	-0.04	-0.17	-0.14	-0.38
OMe	$+0.08$	-0.11	$+0.13$	-0.12
OH	$+0.10$	-0.18	—	—
COOMe	$+0.32$	$+0.39$	$+0.36$	$+0.46$
COOH	—	$+0.41$	—	—
F	$+0.34$	$+0.06$	$+0.35$	$+0.17$
Cl	$+0.37$	$+0.24$	$+0.37$	$+0.27$
Br	$+0.39$	$+0.27$	$+0.38$	$+0.26$
I	$+0.35$	$+0.30$	$+0.35$	$+0.27$
COMe	$+0.38$	$+0.50$	$+0.34$	$+0.46$
CF_3	$+0.47$	$+0.53$	—	—
CN	$+0.61$	$+0.67$	$+0.62$	$+0.69$
NO_2	$+0.71$	$+0.78$	$+0.70$	$+0.82$

Source: References 23 and 24.

the separation of inductive and resonance electronic effects. The goal is to express the observed substituent effect in the form of Eq. (7-30),

$$\log \frac{k}{k_0} = \rho_I \sigma_I + \rho_R \sigma_R \qquad (7\text{-}30)$$

a multiple LFER. The problem is to evaluate σ_I (the inductive substituent constant) and σ_R (the resonance substituent constant). Because a $\log k$ quantity is proportional to an energy, it is acceptable to represent it as a sum for a specific case, but the intent of Eq. (7-30) is to express substituent effects as sums for many reactions, σ_I and σ_R being dependent on substituent but not on the reaction.

Taft and Lewis[13] began with Eq. (7-30), adding the assumptions that the inductive effect is identical from the *meta* and *para* positions and that the resonance effect from the *meta* position is some fraction α of its effect from the *para* position. These assumptions give Eq. (7-31):

$$\log \left(\frac{k}{k_0} \right)_p = \rho_I \, \sigma_I + \rho_R \sigma_R \qquad (7\text{-}31\text{a})$$

$$\log \left(\frac{k}{k_0} \right)_m = \rho_I \sigma_I + \alpha \rho_R \sigma_R \qquad (7\text{-}31\text{b})$$

Eliminating the resonance effect $R = \rho_R \sigma_R$ gives

$$\log \left(\frac{k}{k_0} \right)_m - \alpha \log \left(\frac{k}{k_0} \right)_p = (1 - \alpha)\rho_I \sigma_I \qquad (7\text{-}32)$$

The constants σ_I were taken equal to a scaled value of the aliphatic polar substituent constants σ^* (which are defined in Section 7.3), and α was set at $\frac{1}{3}$ (or $\alpha = \frac{1}{10}$ for substituents capable of through resonance). The resulting plots of Eq. (7-32) gave good LFER, which was interpreted to justify the approach. Refinements,[26,27] of this treatment showed that α depends upon the reaction, although most values fell in the range $\alpha = 0.2$–0.6.

Subsequent work has attempted to refine the estimates of σ_I. We can apply the Hammett equation to Eq. (7-32), obtaining

$$\sigma_I = \frac{\rho_m \sigma_m - \alpha \rho_p \sigma_p}{\rho_I(1 - \alpha)}$$

or, with the assumption $\rho_m = \rho_p = \rho_I$,

$$\sigma_I = \frac{\sigma_m - \alpha \sigma_p}{1 - \alpha} \qquad (7\text{-}33)$$

and

$$\sigma_R = \sigma_p - \sigma_I \qquad (7\text{-}34)$$

TABLE 7-7. Inductive and Resonance Substituent Constants

| Group | Exner[a] | | Ehrenson et al.[b] | | Charton[c] |
	σ_I	σ_R	σ_I	σ_R	σ_I
H	0.00	0.00	0.00	0.00	0.00
Me	-0.06	-0.10	-0.04	-0.11	-0.01
Et	-0.06	-0.09	—	—	-0.01
iso-Pr	-0.07	-0.09	—	—	-0.01
tert-Bu	-0.08	-0.08	—	—	-0.01
—C_6H_5	$+0.10$	-0.10	$+0.10$	-0.11	$+0.12$
—NH_2	$+0.11$	-0.48	$+0.12$	-0.48	$+0.17$
—NMe_2	$+0.11$	-0.54	$+0.06$	-0.52	$+0.17$
—NHCOMe	$+0.29$	-0.25	$+0.26$	-0.25	$+0.28$
—OH	$+0.28$	-0.40	—	—	$+0.24$
—OMe	$+0.31$	-0.41	$+0.27$	-0.45	$+0.30$
—OC_6H_5	—	—	$+0.38$	-0.34	$+0.40$
—OCOMe	$+0.40$	-0.21	—	—	$+0.38$
—SH	$+0.28$	-0.15	—	—	$+0.27$
—SMe	$+0.22$	-0.24	$+0.23$	-0.20	$+0.30$
—SOMe	—	—	$+0.50$	0.00	—
—CF_3	$+0.46$	$+0.01$	$+0.45$	$+0.08$	$+0.40$
—CN	$+0.61$	$+0.07$	$+0.56$	$+0.13$	$+0.57$
—CHO	$+0.35$	—	—	—	—
—COMe	$+0.34$	$+0.18$	$+0.28$	$+0.16$	$+0.30$
—COOH	$+0.34$	$+0.16$	—	—	$+0.30$
—COOR	$+0.35$	$+0.16$	$+0.30$	$+0.14$	$+0.31$
—NO_2	$+0.70$	$+0.10$	$+0.65$	$+0.15$	$+0.67$
—NMe_3^+	$+0.99$	0.00	—	—	$+1.07$

[a]Reference 28.
[b]Reference 29.
[c]Reference 31.

Statistical methods can be used on a large body of data to generate σ_I and σ_R values by means of Eqs. (7-33) and (7-34). The results of Exner[28] and Ehrenson et al.[29] are listed in Table 7-7.

Another approach to evaluating σ_I was taken by Roberts and Moreland,[30] who defined inductive substituent constants in terms of the acid dissociation constants of 4-substituted bicyclo[2.2.2]octane-1-carboxylic acids, **3**.

X——COOH

3

In these acids the geometric relationship of the substituent X to the reactive site COOH approximates that in 4-substituted benzoic acids, but in **3** the resonance effect is absent because of the insulating effect of the saturated system; therefore, only the inductive effect remains. Charton[31] scaled σ_I values based on the ionization of **3** so that they are on the same scale as those based on the ionization of benzoic acids, and these σ_I values are also given in Table 7-7. NMR chemical shifts have also been used as measures of the inductive effect.[32,33]

Although σ_I estimates by different methods or from different data sets may disagree, it is generally held that the inductive effect of a substituent is essentially independent of the nature of the reaction. It is otherwise with the resonance effect, and Ehrenson et al.[29] have defined four different σ_R values for a substituent, depending upon the electronic nature of the reaction site. An alternative approach is to add a third term, sometimes interpreted as a polarizability factor, and to estimate the inductive and resonance contribution statistically; with the added parameter the resonance effect appears to be substantially independent of reaction site.[34-37]

How is the experimental kineticist to cope with this overabundance of substituent constants? Let us first examine some implications of the separation of substituent effects into inductive and resonance effects. Consider the p-NO$_2$ group, which, according to Table 7-6, is electron-withdrawing by both the inductive and the resonance effects, just as expected on the basis of classical arguments. The p-OMe group, on the other hand, is electron-withdrawing by the inductive effect (because oxygen is electronegative relative to carbon), but electron-donating by the resonance effect as shown in **4**; this too is consistent with chemically based conclusions.

OMe $\longleftrightarrow$ +OMe

4

The net effect is, therefore, a balance of these two components. Some of the results of the separation may be surprising, such as the relatively small contribution that resonance makes to the overall substituent effect of p-NO$_2$. Another interesting result is the predominant resonance contribution in the alkyl series; this (if real) must be attributed to hyperconjugation.

The nonspecialist reading Table 7-7 will probably be impressed by the substantial consistency among σ_I values evaluated by different methods, but the specialist tends to concentrate on the differences. There is one very interesting difference in Table 7-7, that for σ_I of alkyl groups based on Eq. (7-33) compared with σ_I based on the ionization of **3**, the latter values showing practically no effect of inductive electron release and certainly no trend with increased branching. (The uncertainties associated with these substituent constants can be found in the original literature.) Swain et al.,[34,35] using a statistical method, reached the same conclusion concerning alkyl groups.

The general chemical reasonableness of the results in Table 7-7 is gratifying, but this does not constitute a demonstration that the separation of substituent effects into inductive effects and resonance effects is quantitatively possible, for these effects may interact so as to be nonadditive. Ritchie and Sager[38] express reservations about the approach in general, and other authors[33-36] have criticized results based on Eqs. (7-33) and (7-34).

The chemical information available through LFER is primarily the reaction constant ρ, but this value depends upon the substituent constants selected for the construction of the LFER. The σ values available are σ, σ^-, σ^+, σ^n or σ^0, and σ_I; these quantities are listed, for many substituents, in Tables 7-1, 7-3, 7-4, 7-6, and 7-7. A reasonable approach is to plot $\log k$ against the substituent constant defined by a standard reaction that is expected to be most like the reaction under study. It is also reasonable to plot $\log k$ against several of the σ quantities, seeking the best correlation. [In choosing between two types of substituent constants, it is necessary to make use of substituents for which the two scales (say σ and σ^+, for example) are not themselves correlated, for otherwise both LFERs will be acceptable.[39]] The σ^n or σ^0 constants should be applicable to reactions that do not combine reaction sites and *para* substituents of the $+$ and $-$ type (*push–pull* systems capable of through resonance); for example, one would not expect σ^n or σ^0 to provide good correlations for reactions of phenols or anilines substituted with nitro or cyano or for reactions of benzoic acids substituted with amino or methoxy.

If the reaction series cannot be correlated with one of these univariate LFER, it may be possible to fit the data to Eq. (7-30), a multivariate LFER. Examples of this approach are given by Ehrenson et al.[29]

The Reaction Constant

The reaction constant ρ is a quantitative measure of the sensitivity of the reaction to the influence of substituents. Three factors combine to determine the value of ρ:

1. The transmission of electronic effects from the substituent to the reaction site.
2. The reaction conditions of temperature and solvent.
3. The susceptibility of the reaction to electronic effects.

The transmission effect is illustrated by the ionizations of $ArCOOH$, $ArCH_2COOH$, and $ArCH_2CH_2COOH$, which have respective ρ values of 1.000, 0.489, and 0.212. The insulating effect of the methylene group interposed between the aromatic ring and the reaction site produces a progressive decrease in ρ, amounting to a transmission factor of about 0.4–0.5 per CH_2 group. Similarly the transmission factor of the $CH{=}CH$ group is 0.4–0.5, making this unsaturated pair of carbons about as insulating as a methylene group.

The effect of reaction conditions on ρ can be pronounced, and this factor must be remembered when comparing ρ values. For example, ρ for the ionization of benzoic acids is 1.000, 1.537, and 1.957 in water, methanol, and ethanol, respectively. This phenomenon may be related to the ease of transmission of electronic effects by a field effect through the solvent, and limited correlations of ρ with the reciprocal of the dielectric constant have been observed (see Section 8.3). The data for Reaction 14 in Table 7-2 show the effect of temperature on ρ. The expected effect can be found by writing the Hammett equation in this form:

$$\delta_R \Delta G^{\ddagger} = -2.3RT \log\frac{k}{k_0} = -2.3RT\rho\sigma = \delta_R \Delta H^{\ddagger} - T\delta_R \Delta S^{\ddagger} \quad (7\text{-}35)$$

$$\rho = -\frac{\delta_R \Delta H^{\ddagger}}{2.3RT\sigma} + \frac{\delta_R \Delta S^{\ddagger}}{2.3R\sigma} \quad (7\text{-}36)$$

The substituent constant σ is defined at a given temperature. If $\delta_R \Delta H^{\ddagger}$ and $\delta_R \Delta S^{\ddagger}$ are independent of temperature, ρ should be linearly related to $1/T$ if all other conditions are held constant. (The constancy of other conditions is not strictly possible, however, because the dielectric constant is a function of temperature.) Data are in reasonable agreement with this prediction.[27,38]

The third factor controlling ρ, the susceptibility of the reaction to electronic effects, is of greatest chemical interest. First, consider the sign of ρ. This merely expresses whether a reaction is facilitated by electron-withdrawing groups (ρ is positive) or by electron-donating groups (ρ is negative). As a mechanistic inference, it is often stated that a positive ρ means that a negative charge is developed at the reaction site in the transition state (or positive charge is lost); correspondingly, a negative ρ means that a positive charge develops at the reaction site (or a negative charge disappears). Consider, for example, the saponification (alkaline hydrolysis) of ethyl benzoates, Reaction 4 in Table 7-2, for which $\rho = +2.27$. This ρ value is consistent with nucleophilic attack by hydroxide on the carbonyl carbon, the electron density of which is reduced by electron-withdrawing substituents. A negative charge is developed at the reaction site in the transition state, **5.**

$$(\rho = +2.27)$$

5

Reaction 12, the acid-catalyzed hydrolysis of substituted benzamides, has $\rho = -0.22$, so the reaction is modestly facilitated by electron-donating groups. This suggests that protonation of the amide is kinetically important:

$$H_2O \cdots \overset{\overset{+OH}{\|}}{C} - NH_2 \quad \longrightarrow \quad \left[H_2O \cdots \overset{\overset{OH}{|}}{C} - NH_2 \right] \qquad (\rho = -2.22)$$

6

The electrophilic reactions of Table 7-5 are clear-cut examples of positive charge development and negative ρ values.

Some care is necessary, however, in making generalizations based on the sign of ρ. Let us compare Reactions 10 and 15, both of these being nucleophilic attacks on acid chlorides. Yet they have ρ values opposite in sign:

$$\text{(C}_6\text{H}_4\text{X)COCl} + H_2O \longrightarrow \text{(C}_6\text{H}_4\text{X)COOH} + HCl \qquad (\rho = +1.78)$$

$$\text{(C}_6\text{H}_5)COCl + \text{(C}_6\text{H}_4\text{X)NH}_2 \longrightarrow \text{C}_6\text{H}_5\text{C(=O)}-\text{NHC}_6\text{H}_4\text{X} + HCl \quad (\rho = -2.78)$$

When written in this way it is clear what is happening. The mechanisms of these reactions are probably similar, despite the different ρ values. The distinction is that in Reaction 10 the substituent X is on the *substrate*, its usual location; but in Reaction 15 the substituent changes have been made on the *reagent*. Thus, electron-withdrawing substituents on the benzoyl chloride render the carbonyl carbon more positive and more susceptible to nucleophilic attack, whereas electron-donating substituents on the aniline increase the electron density on nitrogen, also facilitating nucleophilic attack. The mechanism may be an addition–elimination via a tetrahedral intermediate:

$$\left[H_2O_0^{\delta+} \cdots \overset{\overset{\displaystyle O^{\delta-}}{|}}{\underset{\underset{\displaystyle Ph}{|}}{C}} - Cl \right] \longrightarrow H_2\overset{+}{O} - \overset{\overset{\displaystyle O^-}{|}}{\underset{\underset{\displaystyle Ph}{|}}{C}} - Cl \longrightarrow H_2\overset{+}{O} - \overset{\overset{\displaystyle O^-}{\|}}{\underset{\underset{\displaystyle Ph}{|}}{C}} - Cl \quad + \; Cl^-$$

or it may proceed by a direct nucleophilic displacement:

$$\left[H_0^{\delta+} \cdots \overset{\overset{\displaystyle O}{\|}}{\underset{\underset{\displaystyle Ph}{|}}{C}} \cdots Cl_0^{\delta-} \right] \qquad H_2\overset{+}{O} - \overset{\overset{\displaystyle O}{\|}}{\underset{\underset{\displaystyle Ph}{|}}{C}} \qquad + \; Cl^-$$

In any case, for both these reactions, 10 and 15, the same charge displacements are occurring, despite the opposite signs of ρ. The phrase *development of positive (negative) charge at the reaction site in the transition state* can be ambiguous. In Reactions 10 and 15 all reactants are neutral, and in the transition state some negative charge has been transferred from the nucleophile to the substrate, within which a redistribution of negative charge takes place. Charge separation is certainly taking place, but it is not clear whether there is any meaning in the statement that positive or negative charge is developing at the reaction site.

Next we turn to the magnitudes of the ρ constants. Evidently if $\rho = 0$, there is no substituent effect on reactivity. Moreover because $\rho = +1.000$ by definition for the aqueous ionization of benzoic acids, we have a scale calibration of sorts. Wiberg[40] gives examples of ρ as a measure of the extent of charge development in the transition state. McLennan[41] has pointed out that ρ values must first be adjusted for the transmission factor before they can be taken as measures of charge development. Poh[42,43] takes the ratio of ρ for the reaction rate to ρ for a calibrating equilibrium as a measure of fractional charge development, and Williams[44] has developed a similar system.

More elaborate approaches make use of multiple substitutions to generate additional information. Consider a reaction in which substitutions can be made at two sites, x and y, so that the observed response is a function of both substituents, $f(x,y)$. Expanding this in a Taylor's series gives

$$
\begin{aligned}
f(x,y) = f(a,b) &+ (x - a)\frac{\partial f}{\partial x} + (y - b)\frac{\partial f}{\partial y} \\
&+ \tfrac{1}{2}\left[(x - a)^2\frac{\partial^2 f}{\partial x^2} + 2(x - a)(y - b)\frac{\partial^2 f}{\partial x\partial y} + (y - b)^2\frac{\partial^2 f}{\partial y^2} \right] \\
&+ \dots
\end{aligned}
\tag{7-37}
$$

If y is held constant, Eq. (7-37) becomes, for small variations in x,

$$
f(x) = f(a) + (x - a)\frac{df}{dx} \tag{7-38}
$$

which may be seen as a Hammett equation for variation in the substituent x, i.e.,

$$
\log k_x = \log k_{0y} + \rho_x\sigma_x \tag{7-39}
$$

where we equate ρ_x with df/dx and σ_x with $(x - a)$. Similarly, if x is held constant, we get

$$
\log k_y = \log k_{x0} + \rho_y\sigma_y \tag{7-40}
$$

Returning to Eq. (7-37), if the Hammett equations (7-39) and (7-40) are obeyed for separate variations in x and y, we may take as a good approximation that $\partial^2 f/\partial x^2 = d\rho_x/dx = 0$ and $\partial^2 f/\partial y^2 = d\rho_y/dy = 0$, and Eq. (7-37) becomes, in LFER form,

$$
\log \frac{k_{xy}}{k_{00}} = \rho_x\sigma_x + \rho_y\sigma_y + \rho_{xy}\sigma_x\sigma_y \tag{7-41}
$$

The quantity ρ_{xy} is called the *cross-interaction constant*. Only if ρ_{xy} is negligible does Eq. (7-41) become the additive relationship

$$
\log \frac{k_{xy}}{k_{00}} = \rho_x\sigma_x + \rho_y\sigma_y \tag{7-42}
$$

Miller[45] first used Eq. (7-41) to correlate multiple variations, and this approach has more recently been subjected to considerable development.[46] Many cross-interaction constants have been evaluated; multiple regression analysis is one technique, but Miller[45] and Dubois et al.[47] discuss other methods. Lee et al.[48,49] consider ρ_{xy} to be a measure of the distance between groups x and y in the transition state and have developed empirical relationships between the cross-interaction constant and intramolecular distance. We can expect further application of Eq. (7-41) to mechanistic problems.

Not all free energy relationships (Hammett plots) are linear. Aside from occasional scatter or unexplained deviations from an otherwise linear plot, it sometimes happens that the plot is smooth but nonlinear. Two causes have been identified, and many examples are known.[1, pp. 187–91; 28, 51] One cause is a change in a rate-determining step of a complex reaction. Each step (rate or equilibrium) will have a characteristic susceptibility to electronic effects and, therefore, its own ρ value. If the rate-determining step changes at some position along the σ axis, the observed ρ will change, leading to curvature in the Hammett plot. Because any step can a priori have any ρ value, the net result could be curvature either concave upward or concave downward. Jencks[52] has described the analysis of such plots.

The second possible cause of nonlinearity is a change in mechanism. Within a reaction series any change in mechanism must be such as to provide a smaller free energy of activation for the reaction (otherwise the mechanism would not change). If a substituent effect can produce a change in mechanism, the result must therefore be curvature that is concave upward. Figure 7-2 is a $\rho\sigma^+$ plot for the S_N1 solvolyses

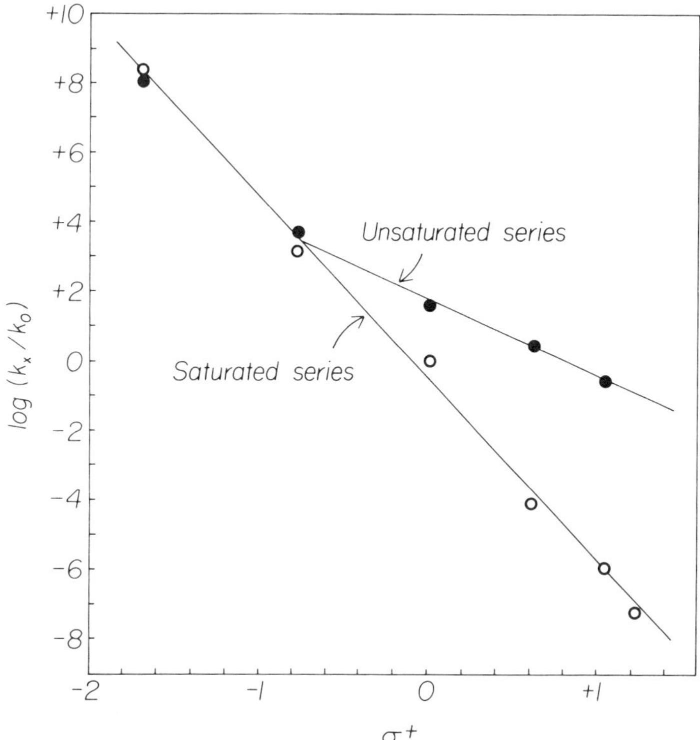

Figure 7-2. Hammett plot for solvolysis of substituted norbornyl esters (**7**) and norbornenyl esters (**8**) in 70:30 dioxane:water at 25°C.[53]

of two related reaction series,[53] the *syn*-7-aryl-*anti*-7-norbornyl *p*-nitrobenzoates
(**7**) and the *syn*-7-aryl-*anti*-7-norbornenyl *p*-nitrobenzoates (**8**).

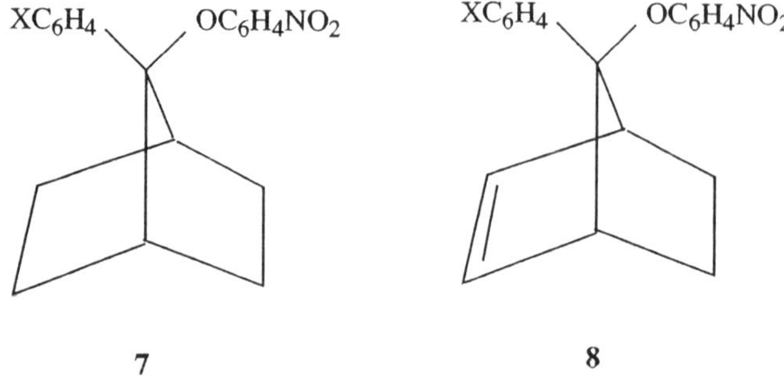

Series **7** gives a linear plot with $\rho = -5.27$, consistent with the S_N1 mechanism.
Series **8,** however, shows a discontinuity, which Gassmann and Fentiman[53] interpreted as a change in mechanism. Compounds in series **8** are capable of intramolecular assistance (*neighboring group participation*) by electron donation from the
double bond to stabilize the cation, as in **9.**

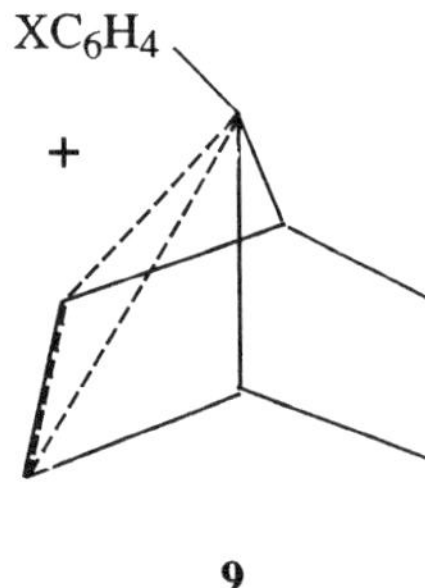

Figure 7-2 indicates that this intramolecular assistance takes place with those compounds having substituents to the right of X = OMe ($\sigma^+ = -0.78$), for in this
portion of the σ scale compounds **8** solvolyze more rapidly than do **7**. At the X =
OMe member, however, the two series have essentially identical reactivities, and
this behavior continues at more negative σ^+. It, therefore, appears that intramolecular participation by the double bond occurs when it is needed; when the X
substituent is sufficiently electron-donating to stabilize the cation, intramolecular
assistance is not needed, so it does not occur, and the saturated and unsaturated
series show the same reactivity.

The Ortho Effect

It will be obvious that the treatment thus far includes only substituents that are *meta*
or *para* to the reaction site. When Hammett plots are made with data for *ortho*-
substituted reactants, scatter diagrams usually result. This failure might be attributed
to "steric effects," but this is not very helpful, and many attempts have been made

to fit *ortho*-substituted reactants into the LFER framework. Much has been learned from this work, and some success has been achieved, but it is a measure of our ignorance that when we carry out a structure–reactivity study of reaction mechanism we usually exclude *ortho* substituents.

Taft[54] began the LFER attack on steric effects as part of his separation of electronic and steric effects in aliphatic compounds, which is discussed in Section 7.3. For our present purposes we abstract from that treatment the portion relevant to aromatic substrates. Hammett ρ values for alkaline ester hydrolysis are in the range $+2.2$ to $+2.8$, whereas for acid ester hydrolysis ρ is close to zero (see Table 7-2). Taft, therefore, concluded that electronic effects of substituents are much greater in the alkaline than in the acid series and, in fact, that they are negligible in the acid series. This left the steric effect alone controlling relative reactivity in the acid series. A steric substituent constant E_S^o was defined [by analogy with the definition of σ in Eq. (7-22)] by Eq. (7-43), where k is the rate constant for acid-catalyzed hydrolysis of an *ortho*-substituted benzoate ester and k_o is the corresponding rate constant for the *ortho*-methyl ester; note that CH_3, not H, is the reference substituent.[51]

$$E_S^o = \log \left(\frac{k}{k_o}\right)_A \tag{7-43}$$

Table 7-8 lists these *ortho* E_S^o constants. An LFER can now be written by analogy with the Hammett equation,

$$\log \frac{k}{k^o} = \delta E_S^o \tag{7-44}$$

where k_o is the rate constant for the *o*-methyl-substituted member of a series. Taft[56] has had some success in fitting data to Eq. (7-44). More commonly, of course, both electronic and steric components are significant contributors to the observed rate effect, and Eq. (7-45), a multiple LFER, can be written.

$$\log \frac{k}{k_o} = \rho_I \sigma_I + \rho_R \sigma_R + \delta E_S^o \tag{7-45}$$

TABLE 7-8. *Ortho* Steric Constants, E_s^o

Group	E_s^o
OMe	$+0.99$
OEt	$+0.90$
F	$+0.49$
Cl	$+0.18$
Br	0.00
CH_3	(0.00)
I	-0.20
NO_2	-0.75
C_6H_5	-0.90

The *ortho* effect may consist of several components. The "normal" electronic effect may receive contributions from inductive and resonance factors, just as with *meta* and *para* substituents. There may also be a "proximity" or field electronic effect that operates directly between the substituent and the reaction site. In addition there may exist a true steric effect, as a result of the space-filling nature of the substituent (itself ultimately an electronic effect). Finally it is possible that non-covalent interactions, such as hydrogen bonding or charge transfer, may take place. The role of the solvent in both the initial state and the transition state may be different in the presence of *ortho* substitution. Many attempts have been made to separate these several effects. For example, Farthing and Nam[57] defined an *ortho* substituent constant in the usual way by $\sigma_o = \log (K/K_o)$ for the ionization of benzoic acids, postulating that σ_o includes both electronic and steric components. They assumed that the electronic portion of the *ortho* effect is identical to the *para* effect, writing $\sigma_E = \sigma_p$, and that the steric component is equal to the difference between the total effect and the electronic effect, or $\sigma_S = \sigma_o - \sigma_E$. They then used a multiple LFER to correlate data for *ortho*-substituted reactants.

$$\log \frac{k}{k_o} = \rho_S \sigma_S + \rho_E \sigma_E \tag{7-46}$$

Many other definitions of an *ortho* substituent constant σ_o have been made; Shorter[58] has reviewed these. Charton[59] analyzed σ_o in terms of σ_I and σ_R, i.e., $\sigma_o = a\sigma_I + b\sigma_R$, finding that the distribution of inductive and resonance effects (the ratio a/b) varies widely with the substituent and, therefore, that no general σ_o scale is possible. Charton also subjected E_S^o to analysis according to Eq. (7-47),

$$E_S^o = \alpha\sigma_I + \beta\sigma_R + \psi r_v \tag{7-47}$$

where r_v is a group van der Waals radius. He found, for aromatic *ortho* substituents, that ψ was negligible and, therefore, concluded that the *ortho* effect is purely electronic, having no steric component. This conclusion is not universally accepted, however. In later work, on the reaction of acids with diphenyldiazomethane, Charton[60] detected a significant steric contribution to the *ortho* effect; this steric factor could be either acceleratory or inhibitory.

Fujita and Nishioka[61] have attempted to place *ortho* effects on the same numerical scale as *meta* and *para* effects. They assume that the normal *ortho* electronic effect can be represented by the standard substituent constant appropriate to the reaction $(\sigma, \sigma^+, \sigma^-, \sigma^0)$, that the steric effect is given by E_S^o, and that the proximity effect is measured by the Swain-Lupton F.[35] Then a multiple LFER is written

$$\log \frac{k_{o,m,p}}{k_o} = \rho\sigma_{o,m,p} + \delta E_S^o + fF \tag{7-48}$$

where data for *ortho, meta,* and *para* substitution are combined. This is an interesting approach, although the particular choices of quantitative measures may be improved.

Thus, Segura[61] replaced σ with a scaled energy change, obtained by a molecular orbital calculation, for this proton transfer:

Berg et al.[62] defined a different *ortho* steric constant. The model reaction is the quaternization of substituted pyridines with methyl iodide in acetonitrile solution.

A plot of log (k/k_H) against pK_a (this is called a *Brønsted-type plot*, as we will see in Section 7.4) for *meta-* and *para*-substituted pyridines gave an LFER describing "normal" behavior. On this same plot the *ortho*-substituted pyridines all fell below the *m* and *p* reference line. The steric constant was defined as the vertical distance between the reference line [whose equation was log $(k_{m,p}/k_H) = 0.35\ pK_a - 1.73$] and the result for the *ortho* compound, or

$$S^0 = \log \frac{k_o}{k_H} - (0.35\ pK_a - 1.73) \tag{7-49}$$

Because S^0 is not well correlated with E_S^0, these two steric constants contain different information. Berg et al.[62] conclude that S^0 is a measure of the pure steric effect.

Conclusions

Research on the nature of substituent constants continues, with results that can bewilder the nonspecialist. The dominant approach is a statistical one, and the main goal is to dissect substituent effects into separate electronic causes. This has led to a proliferation of terms, symbols, and conclusions. A central issue is (here we change terminology somewhat from our earlier usage) to determine the balance of field and inductive effects contributing to the observed polar electronic effect. In this termi·· ·lo· y *inductive* refers to through-bond displacements as a consequence of differei ··s · ·electronegativities of the bonded atoms; the *field* component is the direct electrostatic interaction through space (i.e., through the solvent or through molecular space), which is dependent upon group dipole moment and the effective dielectric constant. The inductive substituent constant σ_I actually includes both the inductive and field effects, and opinion varies as to the balance of these effects represented in σ_I; Reynolds[63,64] concludes that the field effect predominates. Taft and Topson[65] write

$$\log \frac{k}{k_0} \; \rho_F \sigma_F \; + \; \rho_\chi \sigma_\chi \; + \; \rho_\alpha \alpha \; + \; \rho_R \sigma_R$$

where subscript F denotes the field effect; χ, the electronegativity contribution; α, the polarizability; and R, the resonance effect. The electronegativity term is negligible for distant substituents. Levitt and Widing[66] find a simple relationship between σ_I and the number of carbons in an alkyl group; these authors disagree with Charton's view[31] that σ_I does not vary with the size of the alkyl group. (We will return to this controversy over the inductive effects of alkyl groups in Section 7.3.) Charton[67] has attempted to resolve the resonance effect into components as follows:

$$\log \frac{k}{k_0} \; = \; L\sigma_I \; + \; D\sigma_d \; + \; R\sigma_e$$

Here the σ_I term includes the field and inductive effects, σ_d describes the *intrinsic delocalized resonance* when the electron demand by the active site is negligible, and σ_e describes the sensitivity of the substrate in responding to electron demand by the active site.

Because the entire subject of LFER is empirical in nature, these attempted extensions of the field may be justified, provided they meet reasonable criteria of chemical and statistical significance. Whether or not they will successfully extend our ability to correlate data or will lead to improved physical insight must be established by further effort. For the present, however, the experimentalist probably should base interpretations on the quantities and concepts outlined earlier in this section, for the essential worth of these simpler ideas has been established by example and practice.

7.3 SUBSTITUENT EFFECTS IN ALIPHATIC COMPOUNDS

Electronic Effects

The Hammett type of correlation fails with aliphatic substrates just as it does with *ortho*-substituted aromatics. Taft[54] provided the first successful correlations of aliphatic reactions by developing quantitatively an early suggestion of Ingold's[68] that polar and steric effects might be separated by measuring the relative rates of alkaline and acid hydrolysis of esters. Taft's treatment requires these assumptions:

1. $\delta_R\Delta G^{\ddagger}$ terms are additive functions of polar, resonance, and steric effects.
2. In the acidic and alkaline hydrolysis rates of the same ester, the steric and resonance effects are the same.
3. The polar effects are much greater in alkaline hydrolysis than in acid hydrolysis.

The difference $\delta_R\Delta G^{+}_{\text{alkaline}} - \delta_R\Delta G^{\ddagger}_{\text{acid}}$ for an ester hydrolysis should, if these assumptions are valid, be a measure of the polar effect in the substituent. Taft defined a polar substituent constant σ^* based on these ideas:

$$\sigma^* = \frac{1}{2.48}\left[\log\left(\frac{k}{k^0}\right)_B - \log\left(\frac{k}{k^0}\right)_A\right] \qquad (7\text{-}50)$$

$(k/k^0)_B$ is the relative rate of alkaline hydrolysis of ester RCOOR$'$, with k^0 being the rate constant for R $=$ CH$_3$; $(k/k^0)_A$ is the relative rate of acid hydrolysis of the same ester with the same solvent and temperature; and 2.48 is a factor that places σ^* on about the same scale as the Hammett σ values. Table 7-9 lists σ^* values for aliphatic substituents.[54,69]

An equation can now be written by analogy with the Hammett equation.

$$\log\frac{k}{k^0} = \rho^*\sigma^* \qquad (7\text{-}51)$$

TABLE 7-9. Polar Substituent Constants, σ^*, for Aliphatic Substrates X-R

R	σ^*	R	σ^*	R	σ^*
t-Bu	-0.30	—(CH$_2$)CF$_3$	$+0.32$	—CH$_2$Cl	$+1.05$
—CH(Me)(CMe$_3$)	-0.28	—CH=CHMe	$+0.36$	—CH$_2$CO$_2$H	$+1.05$
—CH$_2$SiMe$_3$	-0.26	—(CH$_2$)$_2$Br	$+0.37^a$	—CH$_2$F	$+1.10$
—CHEt$_2$	-0.23	—(CH$_2$)$_2$Cl	$+0.39$	—CH(Me)Cl	$+1.11^a$
s-Bu	-0.21	—CHPh$_2$	$+0.41$	—CH$_2$CN	$+1.30$
$cyclo$-C$_5$H$_9$	-0.20	—CH=CHPh	$+0.41$	—CH$_2$SO$_2$Me	$+1.32$
i-Pr	-0.19	H	$+0.49$	—C=CPh	$+1.35$
—(CH$_2$)$_4$Me	-0.17^a	—(CH$_2$)$_2$NO$_2$	$+0.50$	—COMe	$+1.65$
$cyclo$-C$_6$H$_{11}$	-0.15	–CH$_2$OEt	$+0.50^a$	—C$\equiv$CH	$+1.66^a$
n-Bu	-0.13	–CH$_2$OMe	$+0.52$	—CH$_2$NMe$_3^{+}$	$+1.90$
i-Bu	-0.13	—CH$_2$OH	$+0.56$	–CHCl$_2$	$+1.94$
n-Pr	-0.12	—Ph	$+0.60$	–CHBr$_2$	$+1.99^a$
Et	-0.10	–CH$_2$COMe	$+0.60$	—CO$_2$Me	$+2.00$
Me	(0.00)	—CH$_2$SH	$+0.63^a$	—CHF$_2$	$+2.05$
—(CH$_2$)$_3$Ph	$+0.02$	—CH(OH)Ph	$+0.77$	—CF$_2$CHF$_2$	$+2.22^a$
—CH(Et)Ph	$+0.04$	—CH$_2$OPh	$+0.85$	—CBr$_3$	$+2.48^a$
—(CH$_2$)$_2$Ph	$+0.08$	—CH$_2$I	$+0.85$	—CF$_3$	$+2.60^a$
—CH$_2$Ph	$+0.22$	–CH$_2$CF$_3$	$+0.92$	—CClF$_2$	$+2.64^a$
—CH=CH$_2$	$+0.28^a$	—CH$_2$Br	$+1.00$	—CCl$_3$	$+2.65$

Source: Reference 54, except as noted.
[a]From Reference 69.

Equation (7-51), the Taft equation, provides good correlations for many aliphatic reactions. The scope of this relationship is illustrated by Table 7-10. The reaction constant ρ^* has been interpreted along the lines described for the Hammett ρ values.

Correlations with σ^* in carboxylic acid derivative reactions have been most successful for variations in the acyl portion, R in RCOX. Variation in the alkyl portion of esters, R$'$ in RCOOR$'$, has not led to many good correlations, although use of relative rates of alkaline and acidic reactions, as in the defining relation, can generate linear correlations. The failure to achieve satisfactory correlations with σ^* for such substrates may be a consequence of the different steric effects of substituents in the acyl and alkyl locations.[70] It has been shown[71,72] that solvolysis rates of some acetates are related to the pK_a of the *leaving group,* that is, of the parent alcohol. The pK_a of alcohols has been correlated with σ^*,[54,73] but this relationship requires σ^* to refer to R in RCH_2OH. When pK_a is plotted against σ^* for R in ROH, the linear correlation is lost, presumably because of a steric effect. This steric influence is paralleled in ester reactions, and Robinson and Matheson[74] have demonstrated good LFERs between many ester solvolysis rates and pK_a of the parent alcohols. Because of the dearth of accurate pK_a values for alcohols, they then used this correlation to justify adoption of a secondary standard, the rate of alkaline hydrolysis of esters of 3,5-dinitrobenzoic acid at 25°C in 40% acetonitrile. $\delta_R \Delta G^{\ddagger}$ for this reaction, with the methyl ester as the reference compound, was used as the substituent parameter to give LFERs for many ester reactions. Interestingly, some electrophilic aromatic substitution reactions of substrates with aliphatic substituents also gave LFERs with the substituent parameter.

Fundamental to the interpretations of σ^* as a measure of electronic effects is the validity of the assumption that the steric effect is identical in acid- and base-catalyzed

TABLE 7-10. Some Reactions Correlated by σ^*

No.	Reaction	Solvent	$t/°C$	ρ^*
1	Hydrolysis of $RCO_2C_2H_5$			$(2.48)^a$
2	Ionization of RCO_2H	Water	25	1.72
3	Acid hydrolysis of $RCH(OC_2H_5)_2$	50% dioxane	25	-3.65
4	Ionization of RCH_2OH	Isopropyl alcohol	27	1.36
5	Sulfation of alcohols, ROH	H_2SO_4/H_2O	25	4.60
6	Acetone iodination catalyzed by RCOOH	Water	25	1.14
7	Nitramide decompn. catalyzed by $RCOO^-$	Water	15	-1.43
8	Acid hydrolysis of $H_2C(OR)_2$	Water	25	-4.17
9	log (k_B/k_A) for hydrolysis of CH_3COSR	43% acetone	30	1.49
10	Ionization of $RNH_3^{+\,b}$	Water	25	3.14
11	Ionization of $RC(NO_2)_2H^c$	Water	25	3.60

Source: Reference 54, except as noted.
[a]Defining reaction.
[b]Reference 201. pK_a plotted against $\Sigma\sigma^*$ for R plus the H atoms.
[c]Reference 202.

ester hydrolysis. These reactions are believed to proceed via a tetrahedral intermediate,[75] formed as shown in Scheme I:

$$
\text{Acid} \qquad
\begin{array}{c}
{}^+OH \\
\parallel \\
R{-}C{-}OR' \\
| \\
O{-}H \\
| \\
H
\end{array}
\quad\rightleftharpoons\quad
\begin{array}{c}
OH \\
| \\
R{-}C{-}OR' \\
| \\
{}^+OH_2
\end{array}
$$

$$
\text{Base} \qquad
\begin{array}{c}
O \\
\parallel \\
R{-}C{-}OR' \\
| \\
O^- \\
| \\
H
\end{array}
\quad\rightleftharpoons\quad
\begin{array}{c}
O^- \\
| \\
R{-}C{-}OR' \\
| \\
OH
\end{array}
$$

Scheme I

The assumption, therefore, consists of the belief that the steric effect of R on the attack by H_2O is identical to that by OH^-, which is reasonable on the basis of the sizes of these species (although solvation effects may be different in the two reactions). It appears likely that this assumption of steric identity is a very good one when the electronic substituent effect is large (so that a small error introduced by incomplete steric cancellation is relatively unimportant), whereas it may not be a sound assumption if the electronic effect is small or comparable to the steric effect. This latter circumstance may be the cause of a continuing controversy concerning the meaning of σ^*. On one side are those workers who believe that σ^* is a quantitative measure of polar effects, as originally proposed by Taft.[54] Kanerva and Euranto[69] make mechanistic use of σ^*, although they caution against basing correlations only on alkyl substituents. Levitt and Widing[66] have reevaluated σ^* for alkyl groups and conclude that it measures the inductive effect. It has become common practice to write the Taft equation in terms of the inductive constants σ_I, thus $\log (k/k_0) = \rho_I\sigma_I$, and some authors give equations for interconverting σ_I and σ^*.[66,76] Discussions of the validity of σ^* are therefore sometimes phrased in terms of σ_I. Recall from Table 7-7 that Charton found σ_I to be invariant for alkyl groups; thus Levitt and Widing[66] and Reynolds[63] disagree with Charton[31] on the meaning of σ^*. Charton[31,77] and Detar[76,78] have concluded that σ^* of alkyl groups is simply an artifact, probably measuring some residual steric effect. It is interesting that quantum mechanical calculations of the net charge on methyl groups in alkanes $R{-}CH_3$ correlate linearily with σ^* of the R group; this result has been taken as substantiating the calculations,[79] so it would be circular reasoning to use it as evidence for the validity of σ^* as a measure of the electronic inductive effect.

At this time it seems reasonable to accept σ^* or σ_I as measures of the polar effect (consisting of the inductive and field effects) for groups that are quite polar relative to alkyl groups. Whether or not σ^* for alkyl groups is an artifact or a measure of electron-releasing ability is unresolved, and both points of view can be found in the literature.

Steric Effects

If the assumptions underlying the Taft treatment of the separation of electronic and steric effects are valid, then the relative rates of acid-catalyzed reactions of esters should be a measure of the steric effect. Taft[54] accordingly defined a steric constant E_S by Eq. (7-52).

$$E_S = \log \left(\frac{k}{k_0} \right)_A \tag{7-52}$$

In this definition k_0 is the rate constant for CH_3COOR' and k is the constant for $RCOOR'$; thus $E_S = 0$ for $R = CH_3$. Table 7-11 lists some E_S values. Taft's E_S steric constants are in some instances based on averages of several different reactions, so MacPhee et al.[60] have defined a steric constant E_S' by Eq. (7-52) for a single reaction, namely, the acid-catalyzed esterification of carboxylic acids in methanol at 40°C. E_S' values are also given in Table 7-11. Additional E_S and E_S' values are available.[54,80,81]

TABLE 7-11. Steric Constants for R in RCOOR'

R	E_s^a	$E_s'^b$	R	E_s^a	$E_s'^b$
H	+1.24	+1.12	Ph_2CH—	−1.76	−1.50
Me	(0.00)	(0.00)	Ph_2MeC—	—	−3.73
Et	−0.07	−0.08	Ph_2EtC—	—	−4.55
n-Pr	−0.36	−0.31	Ph_3C—	—	−4.91
i-Pr	−0.47	−0.48	Ph_2CHCH_2—	−0.38	−0.35
n-Bu	−0.39	−0.31	FCH_2—	−0.24	−0.20
s-Bu	−1.13	−1.00	F_2CH—	−0.67	−0.32
i-Bu	−0.93	−0.93	F_3C—	−1.16	−0.78
t-Bu	−1.54	−1.43	$ClCH_2$—	−0.24	−0.18
c-C_3H_5		−1.09	Cl_2CH—	−1.54	−0.58
c-C_4H_7	−0.06	−0.03	Cl_3C—	−2.06	−1.75
c-C_5H_9	−0.51	−0.41	$BrCH_2$—	−0.27	−0.24
c-C_6H_{11}	−0.79	−0.69	Br_2CH—	−1.86	−0.76
c-C_7H_{13}	−1.10	−0.92	Br_3C—	−2.43	−2.14
CH_2=CH—	—	−2.07	I_3C—	—	−2.62
CH_2=CHCH_2—	—	−0.31	F	—	+0.57
Ph—		−2.31	Cl	—	−0.02
$PhCH_2$—	−0.38	−0.39	Br	—	−0.22
PhC(Me)H—	−1.19	−0.90	I	—	−0.50
PhC(Et)H—	−1.50	−1.30	$PhOCH_2$—	−0.33	−0.32

[a]From Reference 54.
[b]From Reference 80.

An LFER for steric effects is now written in the familiar form:

$$\log \left(\frac{k}{k_0}\right) = \delta E_S \qquad (7\text{-}53)$$

If electronic effects are not constant in the reaction series, a multiple LFER is used:

$$\log \left(\frac{k}{k_0}\right) = \rho^*\sigma^* + \delta E_S \qquad (7\text{-}54)$$

(Some authors use σ_I instead of σ^* as the substituent constant in such correlations.) An example is provided by the aminolysis of phenyl esters in dioxane[82]; the substrates RCOOPh were reacted with n-butylamine, and the observed first-order rate constants were related to amine concentration by $k_{obs} = k_2$ [amine] $+ k_3$ [amine]^2. The rate constants k_2 and k_3 could be correlated by means of Eq. (7-54), the reaction constants being $\rho^* = +2.14$, $\delta = +1.03$ (for k_2) and $\rho^* = +3.03$, $\delta = +1.08$ (for k_3). Thus, the two reactions are about equally sensitive to steric effects, whereas the amine-catalyzed reaction is more susceptible to electronic effects than is the "uncatalyzed" reaction.

It is obvious, from some of the series of substituents in Table 7-11, that the steric constant reflects intuitive notions of group size, and Taft[54] pointed out close parallels with van der Waals radii. For asymmetrical groups a single van der Waals radius cannot be defined, and the situation is complicated by conformational flexibility. Nevertheless, empirical relationships can be established, and Kutter and Hansch[83] gave Eq. (7-55),

$$E_S = -1.839\bar{r}_v + 3.484 \qquad (7\text{-}55)$$

where $\bar{r}_v$ is the mean of the minimum and maximum van der Waals radii of the group. Equation (7-55) has been used to obtain E_S estimates of groups for which reactivity data are unavailable. Dubois et al.[84] have analyzed their E_S' constants in terms of conformational preferences of substituted alkyl groups. Charton[60] defined a steric parameter in terms of van der Waals radii, but Gallo[85] has shown that Charton's parameter is linearly related to E_S, so it contains the same information. DeTar[76] concludes that E_S' corrects some errors in the E_S values, but that E_S' is on a slightly different scale than is E_S (that is, if $\delta = 1.000$ for E_S, then $\delta \neq 1.000$ for E_S').

Within some limited series of substituents it appears that E_S is correlated with σ^*, which is not unreasonable, because the electronic effect of a group is in part related to the size of the group. DeTar[76] has discussed this matter. Kramer[86] has demonstrated well-defined familial relationships between E_S' and σ^* and concludes that E_S' possesses some polar character.

The steric constant E_S and related quantities do not constitute the only approach to the study of steric effects on reactivity. Steric strain energy calculations and topological indices are more recent approaches.[85,87] Qualitative concepts have been important in accounting for steric effects in organic chemistry; a 1956 book edited by Newman[88] remains a classic resource. Brown,[89] who made important contributions to steric effect studies, has called attention to the role of steric strain in the initial state and its relief in the transition state as a contributor to steric enhancement of reactivity.

A very interesting steric effect is shown by the data in Table 7-12 on the rate of acid-catalyzed esterification of aliphatic carboxylic acids. The dissociation constants of these acids are all of the order 10^{-5}, the small variations presumably being caused by minor differences in polar effects. The variations in esterification rates for these acids are quite large, however, so that polar effects are not responsible. Steric effects are, therefore, implicated; indeed, this argument and these data were used to obtain the E_S steric constants. Newman[88] has drawn attention to the conformational role of the acyl group in limiting access to the carboxyl carbon. He represents maximum steric hindrance to attack as arising from a coiled conformation, shown for *n*-butyric acid in **5**.

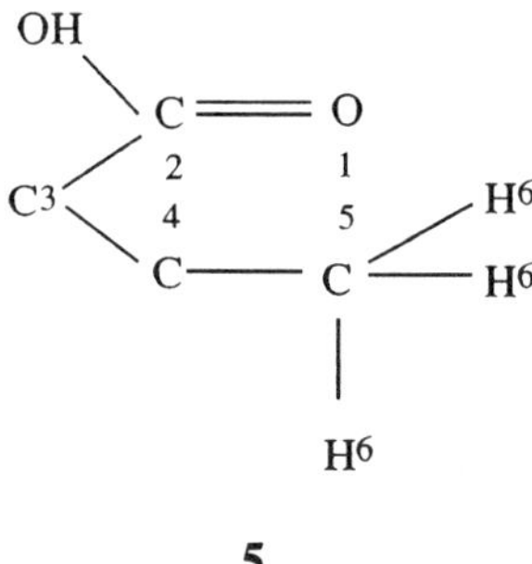

5

Maximum interference arises from atoms in the 6 position, numbering from the carbonyl oxygen. Thus, *n*-butyric acid, which has three atoms in the 6 position, is said to have a *6 number* of 3. The higher the 6 number, the greater the possible steric resistance to ester formation.[88, Chap. 4] This is called *Newman's rule of 6*. The rates in Table 7-12 show that this rule is severely limited as a quantitative tool, but a plot of relative rates k_{RCOOH}/k_{CH_3COOH} against 6 numbers does reveal that a smooth envelope of maximum rates is developed; individual acids may react much slower than this, but the rule gives a rough guide for estimated maximum esterification rates. Brownstein et al.[90] have presented ^{13}C NMR data showing nonbonded interactions consistent with the rule of 6.

7.4 SUBSTITUENT EFFECTS IN REAGENTS

General Acid–Base Catalysis

The rate constant for a general acid- or general base-catalyzed reaction increases as the acid or base strength of the catalyst is increased. For many such systems

TABLE 7-12. Kinetics of Acid-Catalyzed Esterification of Aliphatic Acids

Acid	10^3 k/M^{-1} s^{-1a}	$\dfrac{k_{RCOOH}}{k_{CH_3COOH}}$	6 number
CH_3COOH	132	1.00	0
CH_3CH_2COOH	111	0.84	0
$CH_3CH_2CH_2COOH$	65.2	0.495	3
$(CH_3)_2CHCOOH$	44.0	0.333	0
$(CH_3)_3CCOOH$	4.93	0.037	0
$(CH_3)_2CHCH_2COOH$	15.4	0.117	6
$(CH_3)_3CCH_2COOH$	3.09	0.023	9
$(CH_3)_2CHCH_2CH_2COOH$	63.4	0.48	3
$(CH_3)_3CCH_2CH_2COOH$	61.1	0.46	3
$CH_3CH_2CH(CH_3)COOH$	13.1	0.099	3
$(CH_3)_3CCH(CH_3)COOH$	0.0817	0.000 62	9
$(CH_3)_3CC(CH_3)_2COOH$	0.0170	0.000 13	9
$(CH_3)_3CCH_2CH(CH_3)COOH$	2.03	0.001 5	3
$(CH_3)_3CHCH(C_2H_5)COOH$	0.0780	0.000 59	9
$(CH_3CH_2)_3CCOOH$	0.0214	0.000 16	9
$[(CH_3)_2CH]_2CHCOOH$	Nil		12
$(n\text{-}C_4H_9)_2CHCOOH$	1.10	0.008 3	6
$(CH_3)_3CCH(C_2H_5)COOH$	Nil		12

Source: Reference 88 (Chap. 4).
[a]At 40°C in 0.005 M HCl in methanol.

good LFER are observed; indeed, it was in these systems that the concept of LFER was first recognized. The relationships are often written as Eq. (7-56) for general acid catalysis and Eq. (7-57) for general base catalysis,

$$k_A = G_A K_a{}^{\alpha} \tag{7-56}$$

$$k_B = G_B(1/K_a)^{\beta} \tag{7-57}$$

where K_a is the acid dissociation constant of the conjugate acid. These can also be written in linearized form:

$$\log k_A = -\alpha pK_a + \log G_A \tag{7-58}$$

$$\log k_B = \beta pK_a + \log G_B \tag{7-59}$$

These equations are called *Brønsted relationships* and the parameters α and β are *Brønsted coefficients*.[91,92]

We can gain some insight into the meaning of the Brønsted relationships by means of the following development. Let us write a Hammett LFER for the acid dissociation constants of a series of acid catalysts, namely, $\log (K_a/K_a^0) = \rho_{equil}\sigma$;

and another Hammett equation for the rates of acid catalysis by the same set of acids, log (k_A/k_A^0) = $\rho_{rate}\sigma$. Dividing one of these by the other gives

$$\log k_A = \alpha \log K_A + \log [k_A^0/(K_a^0)^\alpha]$$

where we have written $(\rho_{rate}/\rho_{equil})$ = α. This result is identical in form with Eq. (7-58), and it leads to an interpretation of the coefficient α. If the catalytic reaction is an elementary reaction, the catalytic process is a proton transfer that is partially accomplished in the transition state. The acid dissociation equilibrium is a process in which the proton is fully transferred in the product state. Thus, the ratio ρ_{rate}/ρ_{equil} should be a fraction denoting the extent of proton transfer in the transition state of the catalytic process. A similar argument applies to the coefficient β. In qualitative terms these limits of 0 and 1 for α and β have this meaning: Taking acid catalysis for illustration, a negative value of α would mean that increasing acid strength results in decreased catalytic activity, and this is inconsistent with the notion of acid catalysis as a proton-transfer phenomenon. An α value greater than unity, on the other hand, would mean that log k_A is increasing faster than is log K_a within the reaction series, and this too is not consistent with the normal view of acid catalysis. All known Brønsted plots for nitrogen and oxygen acids and bases appear to yield α and β values between 0 and 1; examples are given by Bell.[93,94] Figure 7-3 is a Brønsted plot for the decomposition of nitramide catalyzed by substituted anilines; this was the first example of a general base-catalyzed reaction.[91,92]

$$H_2N_2O_2 \rightarrow H_2O + N_2O$$

The slope of the line, and, therefore, β, is 0.75. It is very commonly observed that a Brønsted relationship holds well only for a series of acid or base catalysts that are structurally very similar.

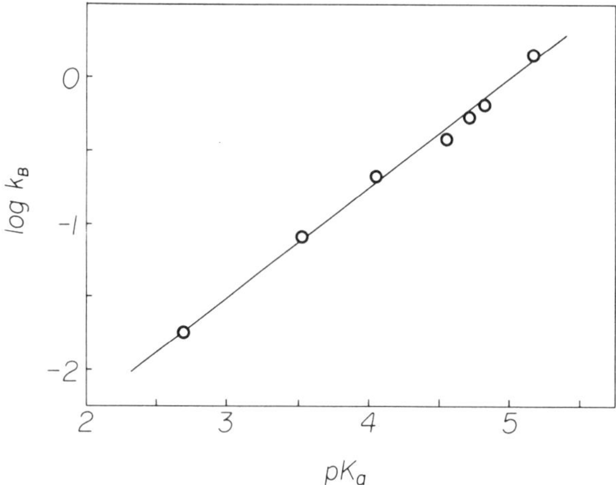

Figure 7-3. Brønsted plot for the decomposition of nitramide catalyzed by substituted anilines.

When Brønsted plots apply to the rates and equilibria of the *same* reaction, a further restriction can be written for the coefficients. Let the generalized reaction scheme be

$$A \underset{k_B}{\overset{k_A}{\rightleftharpoons}} B$$

and assume that Brønsted relationships apply to both the forward (acid-catalyzed) and reverse (base-catalyzed) steps, the rate constant k_A correlating with the equilibrium constant $K_a = k_A/k_B$, and k_B with $1/K_a$. Then, applying Eqs. (7-56) and (7-57) in the equivalent form common to all linear free energy relationships:

$$\log \frac{k_A}{k_A^0} = \alpha \log \frac{K_a}{K_a^0}$$

$$\log \frac{k_B}{k_B^0} = \beta \log \frac{(1/K_a)}{(1/K_a^0)}$$

where the superscripted quantities refer, as usual, to a reference compound. Equation (7-60) follows.

$$\frac{(k_A/k_B)}{(k_A^0/k_B^0)} = \frac{K_a^{(\alpha + \beta)}}{K_a^{0(\alpha + \beta)}} \tag{7-60}$$

Therefore $\alpha + \beta = 1$.

The reader may now recall the discussion in Section 5.3, "Position and Height of the Energy Barrier," on correlations of rates and equilibria of the same reactions and the interpretation of the slope α from such LFER as a measure of the position of the transition state along the reaction coordinate. It will now be apparent why the term *Brønsted coefficient* is applied both to this quantity and also to the slope of LFER according to Eqs. (7-58) and (7-59). The interpretation of α and β from Eqs. (7-58) and (7-59) as measures of fractional progress along a reaction coordinate may be misleading when the reaction is complex, and caution is appropriate.[1, pp. 238–41]

The existence of Brønsted relationships affects the experimental problem of detecting general acid or base catalysis. This is clearly shown by an example given by Bell.[93] Consider the reaction under study as carried out in an aqueous solution containing 0.10 M acetic acid and 0.10 M sodium acetate, and suppose that the Brønsted equation applies. Three catalytic species are present; these are H_3O^+, with $pK_a = -1.74$; H_2O, pK_a 15.74; and HOAc, pK_a 4.76.[52, pp. 171–3; 93, pp. 91–5; 95] The concentrations of these acids are 1.76×10^{-5} M, 55.5 M, and 0.10 M, respectively. Then the fraction of the total catalytic effect produced by any one of the acids is given by

$$\text{fraction of catalysis by acid } i = \frac{(K_a^i)^\alpha [i]}{\sum\limits_i (K_a^i)^\alpha [i]}$$

Table 7-13 gives these fractions for the three acids assuming hypothetical Brønsted coefficients of 0.0, 0.5, and 1.0. α clearly measures the susceptibility of the reaction to catalysis by acids of corresponding strength. When $\alpha = 0$, the reaction is very sensitive to the weakest acid, which is the solvent; if $\alpha = 1$, the lyonium ion, which is the strongest possible acid, is the most effective catalyst. It, therefore, appears that general acid (or general base) catalysis can only be detected if the Brønsted coefficient is in the intermediate range. This is correct if the Brønsted equation is accurately followed. If, however, the slope is unity but the point corresponding to H_3O^+ should show a negative deviation from the line, then general acid catalysis can be detected. This fortuitous circumstance occurs in the general acid catalysis of the hemiacetal formed by the addition of hydrogen peroxide to p-chlorobenzaldehyde.[96] The detection of general base catalysis in the presence of specific base catalysis has been analyzed.[97]

When a Brønsted plot includes acids or bases with different numbers of acidic or basic sites, "statistical" corrections are sometimes applied; in effect, the rate and equilibrium constants are corrected to a "per functional group" basis. If an acid has p equivalent dissociable protons and its conjugate base has q equivalent sites for proton addition, the statistically corrected forms of the Brønsted relationships are

$$\frac{k_A}{p} = G_A \left(\frac{qK_a}{p} \right)^\alpha \tag{7-61}$$

$$\frac{k_B}{q} = G_B \left(\frac{p}{qK_a} \right)^\beta \tag{7-62}$$

The equivalency of sites required for the application of these equations is seldom found in practice, although many authors apply these corrections. Benson[98] has described an alternative procedure in which the rate and equilibrium constants are

TABLE 7-13. Acid Catalysis in 0.1 M Acetic Acid–0.1 M
Sodium Acetate Buffer

	Percentage of catalysis by		
Brønsted α	H_3O^+	*HOAc*	H_2O
0.00	0.000 3	0.2	99.8
0.5	23.7	76.1	0.14
1.0	99.8	0.2	1.0×10^{-11}

corrected by factors consisting of statistical mechanical symmetry numbers. In designing catalytic studies it seems advisable to choose acids and bases that require minimal adjustments of this type.

It may be necessary and possible to achieve a good Brønsted relationship by adding another term to the equation, as Toney and Kirsch[99] did in correlating the effects of various amines on the catalytic activity of a mutant enzyme. A simple Brønsted plot failed, but a multiple linear regression on the variables pK_a and molecular volume (of the amines) was successful.

Finally we should note that the demonstration of a Brønsted relationship does not constitute proof that general acid or general base catalysis is occurring. Because of the problem of kinetic equivalence of rate terms, we may not be able unequivocally to distinguish between these possibilities:

1. General acid catalysis (HA).
   or
2. General base-specific acid catalysis $(A^-)(H^+)$.
   and
3. General base catalysis (B).
   or
4. General acid-specific base catalysis $(BH^+)(OH^-)$.

There is also the possibility of mistaking nucleophilic catalysis for general base catalysis. Table 6-4 outlines some experimental techniques for distinguishing between these possibilities.

Nucleophilicity in Acyl Transfers

When a Brønsted base functions catalytically by sharing an electron pair with a proton, it is acting as a general base catalyst, but when it shares the electron with an atom other than the proton it is (by definition) acting as a nucleophile. This other atom (electrophilic site) is usually carbon, but in organic chemistry it might also be, for example, phosphorus or silicon, whereas in inorganic chemistry it could be the central metal ion in a coordination complex. Here we consider nucleophilic reactions at unsaturated carbon, primarily at carbonyl carbon. Nucleophilic reactions of carboxylic acid derivatives have been well studied. These acyl transfer reactions can be represented by

$$\overset{-}{N}: + R-\overset{\overset{\displaystyle O}{\|}}{C}-X \;\rightarrow\; R-\overset{\overset{\displaystyle O}{\|}}{C}-N + \overset{-}{X}:$$

where $\overset{-}{N}:$ is the nucleophile and $\overset{-}{X}:$ is called the *leaving group*.

The most common manifestation of a structure–reactivity correlation in a reaction series of this type is a plot of log k for the reaction against pK_a of the conjugate acid of the nucleophile. Of course, this is identical with the graphical presentation we called a Brønsted plot in studies of general acid–base catalysis. In the literature the term *Brønsted plot* is often applied to nucleophilic reactions, although some authors qualify such applications by speaking of Brønsted-type plots, as we will. If the substrate RCOX is held constant and the structure of the nucleophile is altered, the slope of the Brønsted-type plot is labeled β_{nuc}; if the nucleophile is held constant and the leaving group is altered, the slope of the plot of log k against pK_a of the conjugate acid of the leaving group is β_{lg}.

These Brønsted-type plots often seem to be scatter diagrams until the points are collated into groups related by specific structural features. Thus, *p*-nitrophenyl acetate gives four separate, but parallel, lines for reactions with pyridines, anilines, imidazoles, and oxygen nucleophiles.[100] Figure 7-4 shows such a plot for the reaction of *trans*-cinnamic anhydride with primary and secondary aliphatic amines to give substituted cinnamamides.[101] All of the primary amines without substituents on the α carbon (R–CH$_2$–NH$_2$) fall on a line of slope 0.62; cyclopentylamine also lies on this line. If this line is characteristic of "normal" behavior, most of the deviations become qualitatively explicable. The line drawn through the secondary amines (slope 1.98) connects amines with the structure R–CH$_2$–NH–CH$_2$–R. The different steric requirements in the acylation reaction and in the model process

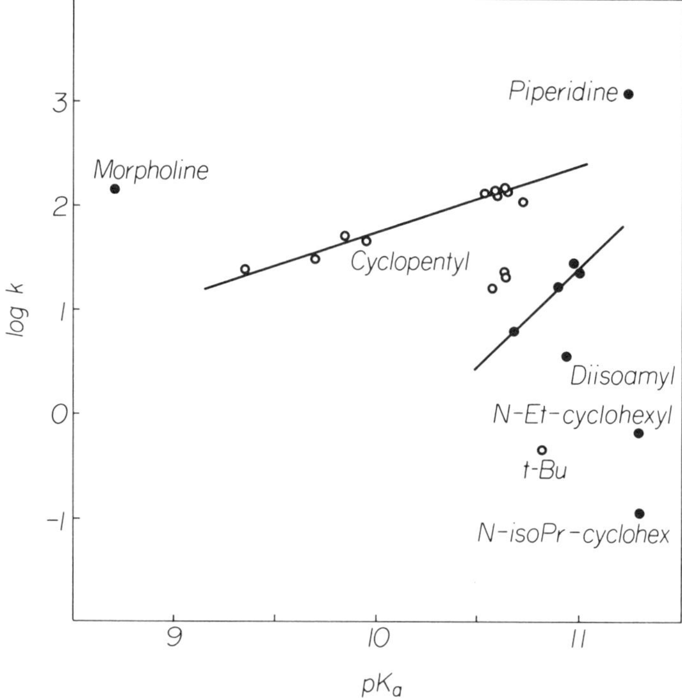

Figure 7-4. Brønsted-type plot for reaction of aliphatic amines with cinnamic anhydride at 25°C in acetonitrile;[101] the pK_a values are for the conjugate acids in water. Open circles: primary amines; closed circles: secondary amines.

(protonation) can account for the negative deviations, nearly all of which occur with amines carrying α substituents. The nucleophilicity of the alicyclic amines is greater than their basicity would suggest; this enhanced reactivity might be a consequence of C–N–C bond angle contraints.[102] Some of the behavior in Fig. 7-4 may follow from the use of a model process in water to correlate a reaction in acetonitrile.

In a true Brønsted plot of general base catalysis the slope β may be a reasonable measure of the fractional bond formation between the catalyst and the proton in the transition state, because the standard equilibrium process closely models the rate process being correlated. It is in this sense that LFERs are said to be *probes of transition state structure*. Whether or not β_{nuc} and β_{lg} are measures of fractional bond formation and bond cleavage, respectively, in nucleophilic reactions is debatable. Certainly these quantities carry information about the transition state, but they may not be simple fractional bond orders, although this interpretation is often made. One problem, however, is that in these Brønsted-type plots the standard reaction (equilibrium basicity toward the proton) is substantially different from the rate process being correlated (nucleophilic attack on sp^2 carbon). This difference contributes an ambiguity to the meaning of β_{nuc} and β_{lg}; in fact, the slopes of these Brønsted-type plots may even exceed unity. Some authors have suggested that ratios of the type $\beta_{nuc}/\beta_{equil}$ are the appropriate normalized measure of fractional bond formation, where β_{equil} is the slope of a plot of log K_{obs} against pK_a, K_{obs} being the equilibrium constant for the reaction.[42,103] Another problem is that few acyl transfer reactions are elementary reactions; this is discussed below. Yet another issue is the possible interaction between bond formation and bond cleavage. This can be investigated by examining the dependence of β_{nuc} on the pK_a of the leaving group, and the dependence of β_{lg} on the pK_a of the nucleophile. The existence of such dependencies means that interpretation of only one bond change (say, by means of β_{nuc}) when another is also occurring may be misleading. We earlier discussed this matter in the context of Hammett plots, obtaining Eq. (7-41) and the cross-interaction constant ρ_{xy}, which describes the interactions between substituents and bond changes at two locations in the transition state. Ba-Saif et al.[104,105] found linear dependencies of β_{nuc} on pK_{lg} and of β_{lg} on pK_{nuc} for these acetyl-transfer reactions:

$$ArO^- + CH_3COOAr' \rightarrow CH_3COOAr + {}^-OAr'$$

They were able to infer β for the identity reaction in which Ar $\equiv$ Ar', and interpreted the results in terms of a More O'Ferrall–Jencks diagram of the type described in Section 5.3.

Sometimes Brønsted-type plots are curved. Many possible causes of such curvature have been discussed.

1. The curvature may be an artifact of a selection of nucleophiles of mixed structural types chosen to display a wide range in pK_a.[106] Buncel et al.[103] varied pK_a by changing the solvent composition over a limited range rather than by changing the structure. They studied the reaction between X–C_6H_4–O^- and *p*-nitrophenyl acetate in 40–90 mol% dimethylsulfoxide–water mixtures; with just three X substituents

combined with the solvent variation a pK_a range of 9 to 18 was achieved. This kind of experiment possesses its own difficulties, but the approach is interesting.
2. Diffusion-limited rate control at high basicity may set in. This is more commonly seen in a true Brønsted plot. If the rate-determining step is a proton transfer, and if this is diffusion controlled, then variation in base strength will not affect the rate of reaction. Thus, β may be zero at high basicity, whereas at low basicity a dependence on pK_a may be seen.[107] Yang and Jencks[108] show an example in the nucleophilic attack of aniline on methyl formate catalyzed by oxygen bases.
3. If the reaction is complex, a change in the rate-determining step may take place along the pK_a scale. Most acyl transfers are thought to take place via a tetrahedral intermediate, as in Scheme II.

$$
\text{N: + RCOX} \underset{k_{-1}}{\overset{k_1}{\rightleftharpoons}} \overset{\displaystyle O^-}{\underset{\displaystyle N^+}{R-C-X}} \overset{k_2}{\rightarrow} \text{RCON + X:}
$$

Scheme II

Applying the steady-state approximation to the intermediate gives $k_{obs} = k_1/[(k_{-1}/k_2) + 1]$, where k_{obs} is the observed second-order rate constant. The ratio k_{-1}/k_2 describes the "partitioning" of the intermediate between the reactant and product states, or the relative leaving abilities (nucleofugalities) of N: and X:. Each of the rate constants in Scheme II has its characteristic sensitivity to changes in N: and X:, and the observed coefficients β_{nuc} and β_{lg}, therefore, depend upon the composition of k_{obs}. If $k_{-1} \ll k_2$, then $k_{obs} = k_1$, the first step is rate limiting, β_{nuc} should be a fairly straightforward measure of the sensitivity of nucleophilic attack to basicity, and k_{obs} should be correlate with pK_a of the leaving group. [These observations have been taken as evidence for the tetrahedral intermediate mechanism.[103]] On the other hand, if $k_{-1} \gg k_2$, then $k_{obs} = k_1 k_2/k_{-1}$, the second step is rate limiting, and k_{obs} is now a composite quantity.

For these two limiting cases let us write Brønsted-type relationships for variation in the nucleophile, namely,

$$k_1 = G_1(1/K_a)^{\beta_1} \tag{7-63}$$

$$\frac{k_1 k_2}{k_{-1}} = G_2(1/K_a)^{\beta_2} \tag{7-64}$$

The steady-state expression can be written

$$\frac{1}{k_{obs}} = \frac{k_{-1}}{k_1 k_2} + \frac{1}{k_1} \tag{7-65}$$

Combining Eqs. (7-63)–(7.65) yields

$$k_{obs} = \frac{G_1G_2(1/K_a)^{(\beta_1+\beta_2)}}{G_1(1/K_a)^{\beta_1} + G_2(1/K_a)^{\beta_2}} \qquad (7\text{-}66)$$

When the first step is rate determining, $\beta_{nuc} = \beta_1$; when the second step is rate determining, $\beta_{nuc} = \beta_2$. If a change in the rate-determining step occurs over the pK_a range studied, a curved Brønsted-type plot will be observed (unless $\beta_1 = \beta_2$), Eq. (7-66) describing the curve quantitatively. In the special case $k_{-1} = k_2$, we have $k_{obs} = k_1/2$; letting $k_{obs} = k_{obs}^0$ and $K_a = K_a^0$ for this case, we get

$$k_{obs}^0 = G_1(1/K_a)^{\beta_1}/2 = G_2(1/K_a^0)^{\beta_2}/2$$

Solving for K_a^0 gives

$$pK_a^0 = \frac{1}{(\beta_2-\beta_1)} \log (G_1/G_2) \qquad (7\text{-}67)$$

Equations (7-66) and (7-67), or related versions, have been used by Hupes and Jencks[109] and by Castro and co-workers[110,111] to account for curvature. The quantity pK_a^0 defines the *center of curvature* of the plot and is expected to occur when the pK_a of the nucleophile is equal to the pK_a of the leaving group.[109] For weaker nucleophiles ($pK_a < pK_a^0$), breakdown of the tetrahedral intermediate will be rate determining, because the leaving group X: is a stronger nucleophile than is N:, so $k_2 < k_{-1}$; if, however, $pK_a > pK_a^0$, the nucleophilic attack is rate determining.
4. The reaction mechanism may change in the reaction series, and this should reveal itself in the Brønsted-type plot as a change in slope. As noted above, most acyl transfers seem to take place via the two-step process shown in Scheme II. A direct displacement is also possible, as in Scheme III.

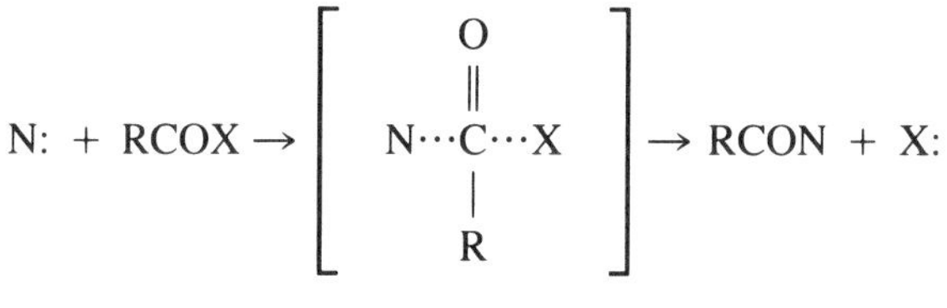

$$N\text{: } + RCOX \rightarrow \left[\begin{array}{c} O \\ \parallel \\ N\cdots C\cdots X \\ | \\ R \end{array} \right] \rightarrow RCON + X\text{:}$$

Scheme III

There is disagreement about the importance of the direct displacement mechanism in these reactions.[103–5] Yet another mechanistic possibility is for reaction via the acylium ion, Scheme IV.

$$RCOX \rightleftharpoons RCO^+ + X\text{:}^-$$

$$RCO^+ + N\text{:}^- \rightarrow RCON$$

Scheme IV

Acid chlorides, which have a very good leaving group (low basicity and nucleophilicity), may react in this manner.

Another type of mechanistic change is a switch from general base catalysis to nucleophilic catalysis. A discussion of this subject will not be given here; many reviews are available.[52, Chaps. 2, 3; 112–4] An example of this mechanistic switch is seen in the imidazole-catalyzed hydrolysis of acetate esters.[115,116] Esters with very good leaving groups (e.g., *p*-nitrophenyl acetate) undergo simple nucleophilic catalysis. As the leaving group is made poorer (as in phenyl acetate), general base catalysis of the nucleophilic catalysis can be detected. With still poorer leaving groups (alkyl acetates), only general base catalysis is seen.

5. Curvature in a Brønsted-type plot is sometimes attributed to a change in transition state structure. This is not a change in mechanism; rather it is interpreted as a shift in extent of bond cleavage and bond formation within the same mechanistic pattern. Thus, Ba-Saif et al.[104] found curvature in the Brønsted-type plot for the identity reactions in acetyl transfer between substituted phenolates; this reaction was shown earlier. They concluded that a change in transition state structure occurs in the series. Jencks et al.[117] caution against this type of conclusion solely on the evidence of curvature, because of the other possible causes.

6. A qualitative difference in the type of solvation (not simply in the strength of solvation) in a series of nucleophiles may contribute to curvature. Jencks has examined this possibility.[109,117,118] An example is the reaction of phenoxide, alkoxide, and hydroxide ions with *p*-nitrophenyl thiolacetate, the Brønsted-type plot showing $\beta_{nuc} = 0.68$ for phenoxide ions (the weaker nucleophiles) and $\beta_{nuc} = 0.17$ for alkoxide ions. It is suggested that the need for desolvation of the alkoxide ions prior to nucleophilic attack results in their decreased nucleophilicity relative to the phenoxide ions, which do not require this desolvation step.

This discussion of sources of curvature in Brønsted-type plots should suggest caution in the interpretation of observed curvature. There is a related matter, concerning particularly item 5 in this list, namely, the effect of a change in transition state structure. Brønsted-type plots are sometimes linear over quite remarkable ranges, of the order 10 pK_a units, and this linearity has evoked interest because it seems to be incompatible with Marcus theory, which we reviewed in Section 5.3. The Marcus equation (Eq. 5-69) for the plot of $\log k$ against $\log K$ of the *same* reaction series requires curvature, the slope of the plot being the coefficient α, given by Eq. (5-67). A Brønsted plot, however, is not a Marcus plot, because it correlates rates and equilibria of *different* reactions. The slope β of a Brønsted plot is defined $\beta = d \log k_{obs}/d\, pK_a$, which we can expand as

$$\beta = \frac{d \log k_{obs}}{d \log K_{obs}} \cdot \frac{d \log K_{obs}}{d\, pK_a}$$

where K_{obs} is the equilibrium constant of the reaction whose rate constant is k_{obs}. Therefore $d \log k_{obs}/d \log K_{obs}$ is the slope of a Marcus plot, and if the Marcus equation is valid for the reaction series, this should contribute curvature according to Eq. (5-67). If this effect is substantially compensated by the term $d \log K_{obs}/d$ pK_a, a straight Brønsted-type plot will be observed. This possibility indicates that a straight Brønsted-type plot is not necessarily evidence that the transition state structure is constant in a reaction series.

The α Effect

Figure 7-5 is a Brønsted-type plot for nucleophilic reactions of p-nitrophenyl acetate (a favorite substrate for studies of this type); some of the reactions are nucleophilic catalyses and some are nucleophilic attacks. This figure illustrates several interesting features. It shows dispersion into different lines for different families of nucleophiles, it includes curvature resulting from the reduced sensitivity of nucleophilicity to basicity for highly basic oxyanions, and it reveals that the reactivity of hydroxide is significantly less than its basicity would predict, a common observation. Most interesting, for our present topic, are the anomalously high nucleophilicities, relative

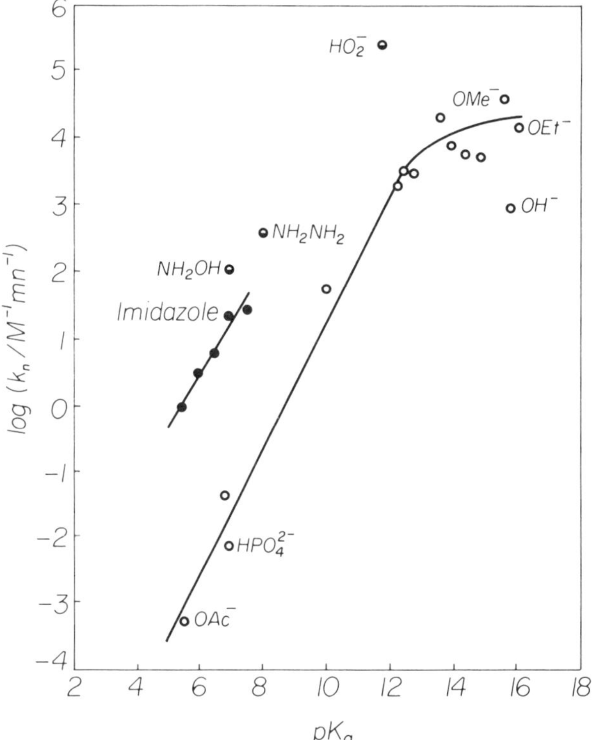

Figure 7-5. Brønsted-type plot for nucleophilic reactions of p-nitrophenyl acetate. Key: ●, simple imidazoles in 28.5% ethanol at 30°C, β = 0.80 [data from Ref. 197]; ○, oxygen anions, in water at 25°, β = 0.95 for linear portion [data from Ref. 119, 198]; ◐, α effect nucleophiles. Several of the nucleophiles are identified.

to their basicities, of several nucleophiles (hydroxylamine, hydrazine, and hydrogen peroxide anion), all of which possess an atom with an unshared electron pair adjacent to the attacking atom. This phenomenon of enhanced nucleophilicity by such nucleophiles is called the α *effect*.

The α effect has been extensively discussed.[119–125] It is probable that the term collects nucleophiles that owe their high reactivity to different factors into a single class, but it has some utility in calling attention to the common structural feature. The role of the α-electron pair in α-effect nucleophiles is sometimes ascribed to relief of the electron deficiency at the attacking atom in the transition state. For the presumed transition state in the attack of the α-effect nucleophile hydrazine on a carboxylic acid derivative, to give a tetrahedral intermediate, **6** shows the developing positive charge on the attacking atom.

$$H_2N = N^{\delta+} - - - - C = O^{\delta-}$$

6

The hypothesis that electron-pair donation from the α atom will stabilize this transition state leads to the difficulty that the attacking atom must carry more bonds than conventional valence bond symbolism admits. Despite this problem, the general idea is expressed by **7** and its relationship to **6** by resonance. It is possible that transition state stabilization can be obtained in this way by rehybridization of the entire molecule.[52, pp. 107–11] Klopman et al.[122] suggest that the α effect arises from orbital splitting that raises the

$$H_2N^{\delta+} = N - - - - C = O^{\delta-}$$

7

energy of the highest occupied orbital of the nucleophile, thus destabilizing the initial state relative to the transition state; on this basis the α effect will be observed mainly in those reactions in which nucleophilic attack is the rate-determining step. Hoz[124], on the other hand, making use of a model in which an attacking nucleophile acquires some diradical character, attributes the α effect to transition state stabilization of the radicallike attacking atom by the adjacent electron pair. There is some support for the idea of nucleophilic attack as a single electron transfer.[126]

Hudson[125] has noted that any explanation of the α effect must account both for the enhanced nucleophilicity and the lack of effect on the pK_a of the nucleophile; he attributes the α effect to a balance (which is different for nucleophile–carbon and nucleophile–proton interactions) between an orbital splitting contribution and an electrostatic bond polarity factor.

Some examples of α-effect nucleophilicity may be the result of concurrent general acid (or Lewis acid) assistance to the nucleophilic attack. The high nucleophilic reactivity of hypochlorite may be rationalized in this way, as suggested by **8.**

8

Hydroxylamine, NH_2OH, is a remarkable α effect nucleophile that may function with internal general acid assistance. Upon reaction with activated acyl groups (anhydrides, esters) a substantial fraction of the acylated hydroxylamine is found to be O-acylhydroxylamine.[127] Jencks proposes that the nucleophilic agent in the attack leading to O-acylhydroxylamine is the zwitterionic form of hydroxylamine, $^+NH_3O^-$. This species combines in one molecule a good nucleophile and a general acid. A concerted mechanism is favored, with the oxygen anion attacking the carboxyl carbon while the ammonium function polarizes the carboxyl group by general acid interaction with the carboxyl oxygen, as in **9.**

9

Nucleophilicity at Saturated Carbon

The S_N2 direct displacement reaction at saturated carbon can be represented as

$$Y\bar{:} + R_3CX \rightarrow R_3CY + X\bar{:}$$

where $Y\bar{:}$ is the attacking nucleophile and $X\bar{:}$ is the leaving group. This reaction can also be described as an alkyl group transfer from X to Y. As a group transfer, the reaction possesses a symmetry that allows any combination of X and Y to be considered in terms of the identity reactions

$$X{:}^- + R_3CX \rightarrow R_3CX + X{:}^-$$

$$Y{:}^- + R_3CY \rightarrow R_3CY + Y{:}^-$$

As a consequence, it appears to be valid to apply Marcus theory (Section 5.3) to S_N2 reactions. Note that we may expect structure–reactivity relationships in S_N2 reactions to be functions of both the bond formation and bond cleavage processes, just as in acyl transfers.

There is another type of reaction to be considered, namely,

$$Y{:}^- + R_3C^+ \rightleftharpoons R_3CY$$

This is the reverse of the first step in the S_N1 mechanism. As written here, this reaction is called *cation–anion recombination*, or an *electrophile–nucleophile reaction*. This type of reaction lacks the symmetry of a group transfer reaction, and we should therefore not expect Marcus theory to be applicable, as Ritchie et al.[128] have emphasized. Nevertheless, the electrophile–nucleophile reaction possesses the simplifying feature that bond formation occurs in the absence of bond cleavage.

We consider first the S_N2 type of process. (In some important S_N2 reactions the solvent may function as the nucleophile. We will treat solvent nucleophilicity as a separate topic in Chapter 8.) Basicity toward the proton, that is, the pK_a of the conjugate acid of the nucleophile, has been found to be less successful as a model property for reactions at saturated carbon than for nucleophilic acyl transfers, although basicity must have some relationship to nucleophilicity. Bordwell et al.[106,129] have demonstrated very satisfactory Brønsted-type plots for nucleophilic displacements at saturated carbon when the basicities and reactivities are measured in polar aprotic solvents like dimethylsulfoxide. The problem of establishing such simple correlations in hydroxylic solvents lies in the varying solvation stabilization within a reaction series in H-bond donor solvents.

Another approach is therefore to adopt a model process that is very similar to the reactions of interest. Swain and Scott[30] selected as a standard reaction the nucleophilic substitution reaction of methyl bromide in water at 25°C.

$$CH_3Br + Y^- \rightarrow CH_3Y + Br^-$$

A nucleophilic parameter n_{MeBr} is defined

$$n_{MeBr} = \log \frac{k_{MeBr}}{k^0_{MeBr}} \tag{7-68}$$

where k_{MeBr} is the rate constant for the reaction with nucleophile Y^- and k^0_{MeBr} is the rate constant with the standard nucleophile, H_2O. Then an LFER is written in the usual form,

$$\log \frac{k}{k_0} = sn_{\text{MeBr}} \tag{7-69}$$

where s is a substrate constant, which for the standard reaction is defined to be unity. Table 7-14 gives values of the Swain-Scott nucleophilic parameter.

It was soon learned that the nucleophilic parameter n_{MeBr} is not a universal measure of reagent nucleophilicity, and Edwards[131,132] took a more fundamental approach, writing a two-term LFER, Eq. (7-70),

$$\log \frac{k}{k^0} = aE_N + bH_N \tag{7-70}$$

where

$$H_N = pK_a + 1.74$$

$$E_N = E^0 + 2.60$$

E^0 is the standard potential of the reaction $2Y^- = Y_2 + 2e^-$. The additive constants set H_N and E_N equal to zero for H_2O. The quantities a and b are parameters measuring susceptibility to the corresponding nucleophilic parameters. Evidently H_N describes

TABLE 7-14. Nucleophile Parameters for Correlation of Nucleophilic Substitutions

Nucleophile	n_{MeBr}	n_{Pt}	E_N	H_N
H_2O	(0.00)	—	(0.00)	(0.00)
NO_3^-	1.03	—	0.29	0.4
F^-	2.0	<2.2	-0.27	4.9
SO_4^{2-}	2.5	—	0.59	3.74
OAc^-	2.72	<2	0.95	6.49
Cl^-	3.04	3.04	1.24	-3.0
C_5H_5N	3.6	3.19	1.20	7.04
Br^-	3.89	4.18	1.51	-6.0
N_3^-	4.00	3.58	1.58	6.46
OH^-	4.20	—	1.65	17.48
$C_6H_5NH_2$	4.49	3.16	1.78	6.32
SCN^-	4.77	5.75	1.83	1.0
I^-	5.04	5.46	2.06	-9.0
SH^-	5.1	—	2.10	9.54
CN^-	5.1	7.14	2.79	10.88
$S_2O_3^{2-}$	6.36	7.34	2.52	3.64

Sources: References 134 and 135.

proton basicity, and E_N describes the ability of the nucleophile to donate an electron. The basis of the oxidation reaction as a model for nucleophilicity is the analogy between oxidation (loss of electrons) and nucleophilic displacement (donation of an electron pair). The Edwards equation provides the *oxibase* scale of nucleophilicities.[133] Table 7-14 lists E_N and H_N values drawn from the collections of Ibne-Rasa[134] and Pearson et al.[135] This table also gives values of the nucleophilic parameter n_{Pt} defined by means of the standard reaction

$$Y^- + Pt(py)_2Cl_2 \rightarrow Pt(py)_2ClY + Cl^-$$

where py is pyridine.[135] Comparisons of the quantities in Table 7-14 reveal some interesting features. For example, n_{MeBr} and n_{Pt} are not well correlated; nucleophilicity toward carbon and toward platinum evidently depends differently on the contributing factors. Note also the greater nucleophilicity of iodide than of hydroxide as expressed in n_{MeBr}; clearly basicity alone is not a good measure of S_N2 reactivity.

Other measures of nucleophilicity have been proposed. Brauman et al.[136] studied S_N2 reactions in the gas phase and applied Marcus theory to obtain the intrinsic barriers of identity reactions. These quantities were interpreted as intrinsic nucleophilicities. Streitwieser[137] has shown that the reactivity of anionic nucleophiles toward methyl iodide in dimethylformamide (DMF) is correlated with the overall heat of reaction in the gas phase; he concludes that bond strength and electron affinity are the important factors controlling nucleophilicity. The dominant role of the solvent in controlling nucleophilicity was shown by Parker,[138] who found solvent effects on nucleophilic reactivity of many orders of magnitude. For example, most anions are more nucleophilic in DMF than in methanol by factors as large as 10^5, because they are less effectively shielded by solvation in the aprotic solvent. Liotta et al.[139] have measured rates of substitution by anionic nucleophiles in acetonitrile solution containing a crown ether, which forms an inclusion complex with the cation (K^+) of the nucleophile. These rates correlate with gas phase rates of the same nucleophiles, which, in this crown ether–acetonitrile system, are considered to be "naked" anions. The solvation of anionic nucleophiles is treated in Section 8.3.

Pearson's hard–soft acid–base (HSAB) concept has been useful as a qualitative guide to reactivity in nucleophilic reactions.[144] A hard acid is one in which the electron-pair acceptor atom is small in size, with high positive charge density and low polarizability. A soft acid is large and polarizable. A hard base has high electronegativity and low polarizability; a soft base is easily polarizable. Table 7-15 shows examples of hard and soft bases. The HSAB principle is that hard acids (electrophiles) prefer hard bases (nucleophiles), and soft acids prefer soft bases. Alkyl groups are soft electrophiles, and attachment of electron-withdrawing groups makes a carbon increasingly hard. Thus, the carbonyl group is a hard electrophile. On this basis the success of Brønsted-type plots in correlating nucleophilic reactions at acyl carbon results because pK_a measures reaction at a hard electrophile (H^+), and acyl transfer is also a reaction at a hard electrophile. On the other hand, S_N2 reactions at saturated carbon, a soft electrophile, should not be correlated with pK_a; rather nucleophilic reactivity at saturated carbon should be greatest for soft nucleophiles. Pearson states that an extra stabilization of the transition state (called the *symbiotic effect*) takes place when both the entering and leaving groups are

TABLE 7-15. Examples of Hard–Soft Acid–Bases

	Acids	Bases
Hard	H^+, Li^+, Na^+, Mg^{2+}, Ca^{2+}, Al^{3+}, RCOR	H_2O, OH^-, F^-, Cl^-, NO_3^-, NH_3
Soft	Cu^+, Ag^+, Hg^{2+}, Pt^{2+}, R_4C	I^-, SCN^-, CN^-

soft.[141] The HSAB theory has been placed on a quantitative basis[142] with the definition of hardness η and softness σ by

$$\eta = \frac{(I - A)}{2} = \frac{1}{\sigma}$$

where I is ionization potential and A is electron affinity. In molecular orbital theory, the difference $(I - A)$ is equal to the energy difference between the highest occupied molecular orbital (HOMO) and the lowest unoccupied molecular orbital (LUMO).

Both the Edwards equation and Pearson's HSAB concept take as primary determinants of nucleophilicity the polarizability and basicity. A two-term equation of Bartoli and Todesco[143] uses these ideas also, but as a measure of polarizability the

TABLE 7-16. Nucleophile Parameters for Cation–Anion Recombinations[a]

Nucleophile	Solvent	N_+
H_2O	H_2O	(0.0)
MeOH	MeOH	1.18
$CF_3CH_2NH_2$	H_2O	2.89
CN^-	H_2O	3.67
$C_6H_5NH_2$	H_2O	4.10
OH^-	H_2O	4.75
$C_6H_5O^-$	H_2O	5.6
CN^-	MeOH	5.94
$CF_3CH_2O^-$	H_2O	6.42
OCl^-	H_2O	7.13
MeO^-	H_2O	7.28
MeO^-	MeOH	7.68
SO_3^{2-}	H_2O	7.90
HOO^-	H_2O	8.08
CN^-	Me_2SO	8.60
N_3^-	MeOH	8.85
CN^-	DMF	9.33
N_3^-	Me_2SO	10.07
$C_6H_5S^-$	MeOH	10.51
$C_6H_5S^-$	Me_2SO	12.83

Source: Reference 147.

[a]Some neutral nucleophiles are included.

bond refraction is taken. Their approach is to attempt to separate the roles of basicity and polarizability by comparing the reactivities of nucleophiles of low and high polarizability. Pytela and Zima[144] have applied a statistical procedure to a large body of nucleophilic rate data, obtaining a correlation equation whose substituent parameters were subsequently identified with softness or polarizability and with hardness or basicity.

Turning to cation–anion recombination reactions we find that most of the quantitative studies have been by Ritchie,[145–7] who defined a nucleophilic constant by Eq. (7-71),

$$\log \frac{k_N}{k_{H_2O}} = N_+ \tag{7-71}$$

where k_{H_2O} is the rate constant for the reaction of the cation with water (the standard nucleophile) in water as the solvent, and k_N is the rate constant for the anionic nucleophile, not necessarily in water. Some of the cations studied were triarylmethyl cations (**10**), tropylium ions (**11**), and diazonium ions (**12**).

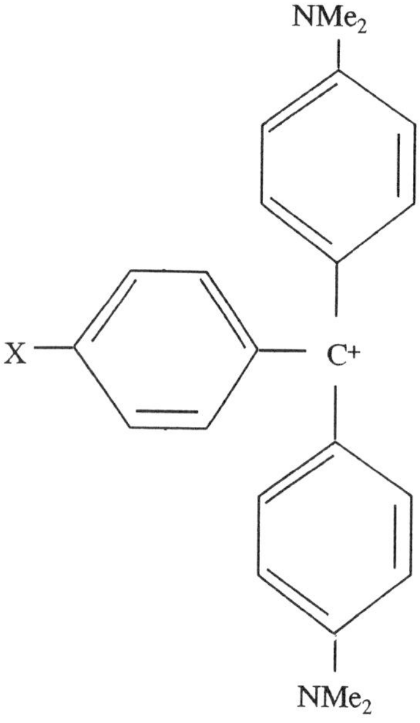

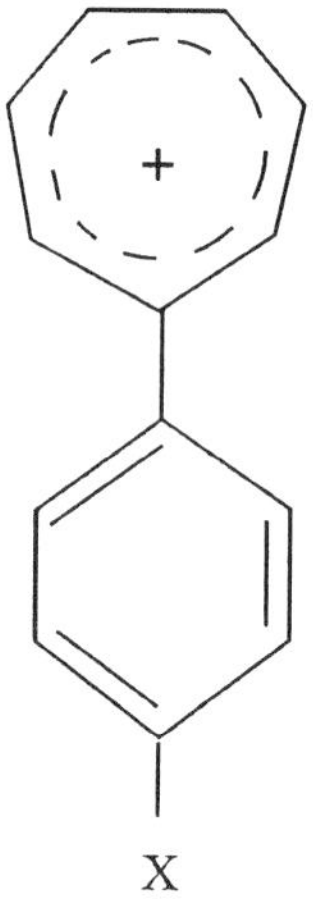

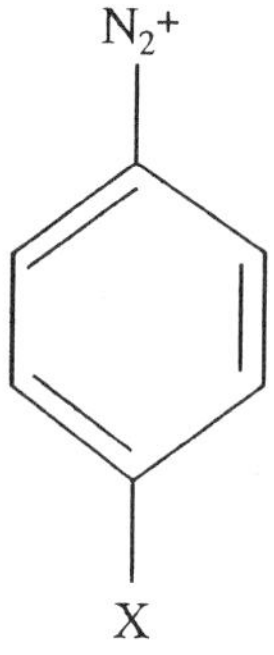

Quite remarkably it was found that all cations yielded the same N_+ values, or, to put this another way, the LFER of $\log k_N$ against N_+ had slope $= 1.00$ for all cations; only the intercepts ($\log k_{H_2O}$) differed.[145–7] Table 7-16 gives values of N_+. [Additions of nucleophiles to some activated olefins also have been found to correlate with N_+, although slopes were different from unity.[124]]

Rates and equilibria within these cation–anion recombination reactions are not correlated.[128,146,147] Ritchie considers that extensive desolvation of the reactant ions occurs before appreciable bond formation takes place. It is very interesting that the same anion in different solvents constitutes different nucleophiles that fall on the same LFER. Table 7-16 shows the magnitudes of some of these solvent effects.

Intramolecular Reactivity

An intramolecular reaction requires the presence of two suitably positioned functional groups, the substrate function and the reagent function, in the same molecule. The demonstration of an intramolecular process is obvious when the product is cyclic, as in this lactonization:

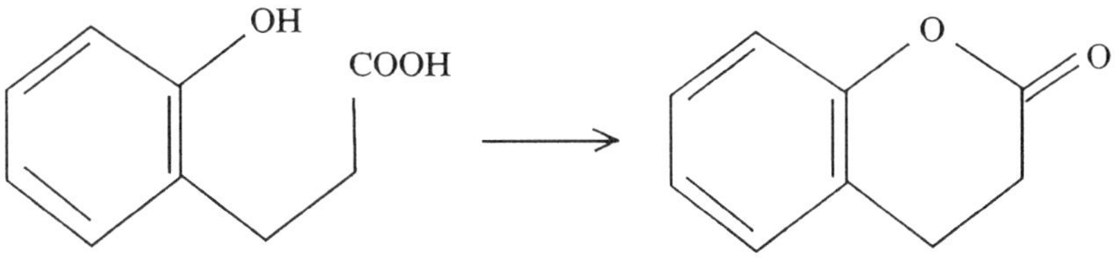

but if the reaction might occur intermolecularly, as in the hydrolysis of aspirin anion, below, evidence of its intramolecularity must be sought.

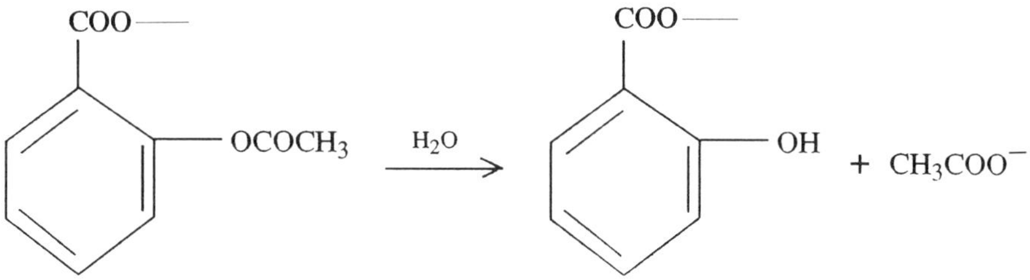

A favorite strategem is to compare the reactivities of *ortho-* and *para*-substituted aromatic substrates, such as **13** and **14**; the electronic (inductive and resonance) effects of the substituent should be about the same in the two cases, but only the *ortho*-substituted compound can undergo intramolecular catalysis.

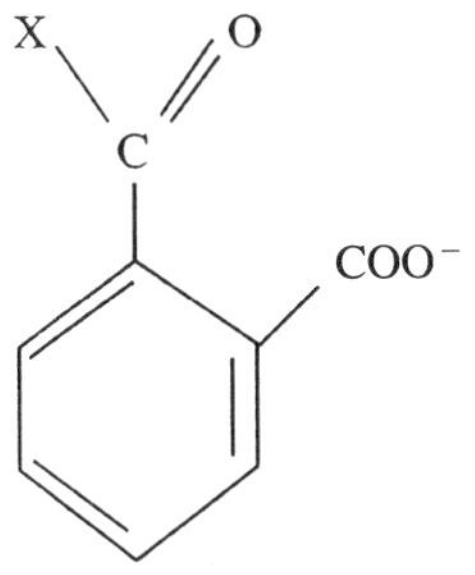

13    **14**

Another comparison is between o-hydroxy and o-methoxy compounds, as in **15** and **16**; the electronic and steric effects will be, very roughly, similar, but only the hydroxy group is capable of intramolecular catalysis.

15

16

The mechanisms available to intramolecular reactions are the same as those of intermolecular reactions. The same problems of kinetic equivalence of rate terms may arise, and Table 6-3 shows some kinetically equivalent mechanisms for intramolecular reactions of the acyl function. The efficiency of intramolecular reactivity may be difficult to assess. One technique, described above as a method for the detection of an intramolecular reaction, is to make a comparison with an analog incapable of the intramolecular process. Thus p-nitrophenyl 5-nitrosalicylate, **17**, hydrolyzes about 2500 times faster than p-nitrophenyl 2-methoxy-5-nitrobenzoate, **18**.[148]

17

18

The currently popular approach is to compare the rate of the intramolecular reaction with the rate of an intermolecular reaction in which the reacting groups are closely similar. The intermolecular reaction will usually be overall second-order, in accordance with the rate equation

$$v = k_{\text{inter}} [RX][Y] \qquad (7\text{-}72)$$

whereas the analogous intramolecular reaction will be first-order,

$$v = k_{\text{intra}} [\text{XRY}] \tag{7-73}$$

First-order and second-order rate constants have different dimensions and cannot be directly compared, so the following interpretation is made.[149] The ratio $k_{\text{intra}}/k_{\text{inter}}$ has the units mole per liter and is the molar concentration of reagent Y in Eq. (7-72) that would be required for the intermolecular reaction to proceed (under pseudo-first-order conditions) as fast as the intramolecular reaction. This ratio is called the *effective molarity* (*EM*); thus $EM = k_{\text{intra}}/k_{\text{inter}}$. An example is the nucleophilic catalysis of phenyl acetate hydrolysis by tertiary amines, which has been studied as both an intermolecular and an intramolecular process.[150]

$$CH_3COOPh + NMe_3 \xrightarrow{-PhO^-} CH_3\overset{O}{\overset{\|}{C}}\text{-}NMe_3^+ \xrightarrow{OH^-} CH_3COOH + NMe_3$$

For this example $EM = 1260$ M. In calculating EM it is necessary that the intermolecular reaction selected for the comparison possess the same mechanism as the intramolecular reaction.

Kirby[151] has assembled a valuable collection of EM values. The intramolecular rate enhancements can be remarkable, with EM values in the range 10^4 to 10^8 being quite common. Even larger enhancements are known, and these are comparable with the rate enhancements produced by enzymes. Mandolini[152] has discussed the cyclization reactions of chain molecules in terms of EM values. EM does not have to be greater than unity, and very small values are known, but obviously the very large rate enhancements have been of greatest interest. We will briefly summarize, with some leading references, the extensive literature on the origin of rate enhancements in intramolecular reactions. There is considerable disagreement in this field. Essentially four ideas have been proposed to account for large EM values.

1. *The proximity effect.* This is the simple idea that in an intramolecular reaction the substrate function may be exposed to a larger "local concentration" of the reagent than in an intermolecular reaction, because the two functions are covalently constrained to occupy adjacent space. This effect has been called the *approximation* or *propinquity effect*. The proximity effect certainly seems physically reasonable and is likely to make some contribution to intramolecular reactivity, but it cannot be a major contributor when EM is large, because EM is itself a measure of a presumed local concentration, and the observed large EM values are physically impossible concentrations. The magnitude of rate enhancement achievable by prox-

imity is shown in a study by Menger and Venkataram[153] in which an intermolecular S_N2 reaction was carried out with the solvent (pyridine) as the nucleophile. Thus, in pure pyridine the substrate was fully surrounded by reagent. On reducing the pyridine concentration by incorporating another aprotic solvent of similar polarity, a smooth decrease in rate was seen, but the effect was not dramatic, and very roughly varied with the pyridine concentration. Proximity of the reagent to the substrate is obviously necessary for reaction, but by itself it cannot account for large rate enhancements. It is a contributing factor.

2. *Orientation effects.* Because few reactant species are spherical, some directional constraints to effective reaction must exist. For example, there is strong evidence that the S_N2 displacement requires "backside" attack of the nucleophile with respect to the leaving group. Koshland and co-workers[154,155] have developed this idea by trying to define, experimentally, the magnitude of angle of approach of reagent to substrate site that leads to reaction. The concept is that large intramolecular rate enhancements are the result of favorable juxtaposition of the reactant species in the relatively rigid intramolecular framework; in the comparable intermolecular situation the mutual orientation is nearly random. This concept of the angle dependence for successful reaction was called *orbital steering.* It was concluded that a change in angle of approach of about $10°$ might result in a rate change of 10^4. This theory has been harshly criticized on the basis that reaction *cones,* or *windows,* are not this narrow, and Menger[156] has designed experimental studies that demonstrate the existence of considerable flexibility in the approach of the reagent to the reaction site. The orbital steering hypothesis has been attacked not because of the concept that a preferred direction of approach is required, for this is agreed to by all, but because of Koshland's claim that the angle of acceptance is very small. No doubt this angle depends somewhat upon the type of reaction.

TABLE 7-17. Relative Rates of Intramolecular Catalysis of Ester Hydrolysis by Carboxylate Groups

Ester[a]	*Relative rate*[b]
COOR / COO⁻	1.0
Me, Me — COOR / COO⁻	19.3
COOR / COO⁻	232
O, COOR / COO⁻	10,300
COOR / COO⁻	53,100

Source: Reference 157.

[a] R = *p*-bromophenyl or *p*-methoxyphenyl.

[b] In 50% dioxane at 30°C.

Structural features that render the initial state conformation more like the cyclic transition state will enhance intramolecular reactivity. The conformational mobility is such a feature. Table 7-17 shows relative rates of some intramolecularly catalyzed ester hydrolyses.[157,158] The glutarate possesses two methylene–methylene single bonds, rotation about which can oppose the juxtaposition of ester and carboxyl groups required for intramolecular catalysis. In the β,β-dimethylglutarate, *steric compression* by the *gem*-dimethyl groups reduces the conformational freedom, and this is reflected in a higher rate. The succinate possesses one bond fewer than the glutarate, leading to restricted rotamer distribution and a rate increase. Further "freezing" of the two groups in opposition, as with the maleate and the 3,6-endoxo-Δ^4-tetrahydrophthalate, gives large rate enhancements. Even larger rate accelerations have been achieved by means of this *stereopopulation control,* which is a narrowing of the distribution of conformational populations, with the fortuitous elimination of unproductive conformers. A dramatic illustration is the relative rate of intramolecular acid-catalyzed lactonization of **19** and **20**.[159]

19   **20**

Compound **20** reacts 10^{11} times faster than **19**. This level of enhancement is comparable with that achieved by enzymes.

3. *The entropy argument.* By making use of Eq. (7-2) we can write

$$2.3\ RT \log EM = -\delta_R \Delta G^{\ddagger} = -\delta_R \Delta H^{\ddagger} + T\delta_R \Delta S^{\ddagger} \tag{7-72}$$

The foregoing discussion suggests that entropic factors are important in determining intramolecular reactivity, and if we suppose that $\delta_R \Delta H^{\ddagger} = 0$ we can relate EM to $\delta_R \Delta S^{\ddagger}$, the difference in entropy of activation between the intramolecular reaction and its intermolecular analog. Thus, if $\delta_R \Delta S^{\ddagger} = 8$ eu, $EM = 56$ M (which, incidentally, is the molar concentration of pure water and therefore, in one view, a maximum measure of the proximity effect). A $\delta_R \Delta S^{\ddagger}$ of 20 eu corresponds to an EM of 2.4×10^4, and $\delta_R \Delta S^{\ddagger} = 40$ eu is equivalent to $EM = 5.7 \times 10^8$. These numbers indicate that it is reasonable to seek the source of intramolecular reactivity in entropic effects. Page and Jencks[160,161] have pursued this notion in great detail and conclude that translational and rotational motions provide most of the driving force for intramolecular reactions. Their point of view is that the entropy argument is a different concept from the orientation concept; it is not simply a quantitative analysis of orientation effects. Menger[153,162] has cited some weaknesses of the entropy argument, one of these being the lack of correlation of *EM* values with entropies of activation. DeTar and Luthra[163] have found that enthalpic effects can be more important than entropic effects in some S_N2 cyclizations.

4. *The spatiotemporal hypothesis.* Menger[162,164] has postulated that "the rate of reaction between functionalities A and B is proportional to the time that A and B reside within a critical distance." This critical distance is considered to be the distance at which bond formation begins between a nucleophile and an electrophile. If a rigid covalent framework holds the substrate and reagent functions within the critical distance (time not being a factor because of the molecular rigidity), very high *EM* values are possible. The function of an enzyme is to hold the substrate within the critical distance of the catalytic group for a sufficient length of time. This time/distance concept bypasses the entropic description and rephrases the proximity/orientation arguments in different terms and variables. Theoretical and experimental studies indicate that reaction rate is a very sensitive function of distance, but is not highly angle dependent. Menger estimates that the critical distance for nucleophilic attack on a carbonyl is about 2.8 Å. Because this is less than the diameter of a water molecule, desolvation is implicated as an essential step in determining reaction rate.

7.5 RELATED TOPICS

The Isokinetic Relationship

It can be shown[1, p. 55] that if an LFER is observed over a range of temperatures, and if the enthalpy and entropy changes are temperature independent, then the enthalpy changes must be directly proportional to the entropy changes for the reaction series. Let us start with the proposition that a real effect of this type has been demonstrated for a reaction series; we write this as

$$\delta\Delta H^{\ddagger} = \beta\delta\Delta S^{\ddagger} \tag{7-73}$$

This can be combined with the general relationship

$$\delta\Delta G^{\ddagger} = \delta\Delta H^{\ddagger} - T\delta\Delta S^{\ddagger}$$

to give

$$\delta\Delta G^{\ddagger} = (\beta - T)\delta\Delta S^{\ddagger} \tag{7-74}$$

$$\delta\Delta G^{\ddagger} = \left(1 - \frac{T}{\beta}\right)\delta\Delta H^{\ddagger} \tag{7-75}$$

The proportionality constant β has the dimension of absolute temperature, and it is called the *isokinetic temperature*.[165] It has the significance that when $T = \beta$,

$\delta\Delta G^{\ddagger} = 0$; that is, all substituent (or medium) effects on the free energy change vanish at the isokinetic temperature. At this temperature the ΔH and $T\Delta S$ terms exactly offset each other, giving rise to the term *compensation effect* for isokinetic behavior.

Two extreme situations should be noted. If $\beta = 0$, then $\delta\Delta G^{\ddagger} = -T\delta\Delta S^{\ddagger}$, and the reaction series is entirely entropy controlled; it is said to be isoenthalpic. If $1/\beta = 0$, then $\delta\Delta G^{\ddagger} = \delta\Delta H^{\ddagger}$, and the series is enthalpy controlled, or isoentropic. All of these relationships apply also to equilibria, but we will be concerned with kinetic quantities.

The demonstration of a valid isokinetic relationship has several implications. First, an isokinetic relationship is a necessary condition for the existence of a reaction series, or at any rate this is a condition that may be taken to define a reaction series. Exceptions are then indications of changes in interaction mechanisms. Second, substituent effects change their direction at the temperature β, so if the experimental temperature is near β, caution is needed in interpreting effects; the notion of a reversal in reactivity order with a change in temperature should itself be cautionary. Third, when an isokinetic relationship holds, $\Delta G^{\ddagger}$, $\Delta H^{\ddagger}$, and $\Delta S^{\ddagger}$ all change in a parallel fashion, so discussions of substituent effects on $\Delta H^{\ddagger}$ and $\Delta S^{\ddagger}$ provide no information or insight that is not already available in effects on reactivity, i.e., $\Delta G^{\ddagger}$. Fourth, related to the preceding item, the problem of whether $\Delta G^{\ddagger}$ or $\Delta H^{\ddagger}$ is the better approximation to the potential energy is solved, because they are equivalent.[166]

Most of the isokinetic relationships in the literature have been established from plots of $\Delta H^{\ddagger}$ against $\Delta S^{\ddagger}$; collections of these have been published[1, Chap. 9; 165, 167] and several authors have discussed the mechanistic implications.[1, Chap. 9; 6, Chap. 12; 168] There is a well-recognized difficulty with $\Delta H^{\ddagger}$ vs. $\Delta S^{\ddagger}$ plots, namely, that both quantities are evaluated from the same set of data, so their errors are correlated. In effect, $\Delta H^{\ddagger}$ is found from the slope of an Arrhenius or Eyring plot, and $\Delta S^{\ddagger}$ from its intercept. Not only are the errors correlated, but there is an a priori interdependence of the two quantities even in the absence of error. Of course, error is always present; Petersen et al.[169] analyzed the errors in $\Delta H^{\ddagger}$ and $\Delta S^{\ddagger}$, showing that the error in $\Delta S^{\ddagger}$ is directly proportional to the error in $\Delta H^{\ddagger}$. A reaction series can generate an apparent isokinetic relationship solely through the operation of this error effect, which predicts that a plot of $\Delta H^{\ddagger}$ against $\Delta S^{\ddagger}$ will be linear with slope about equal to the experimental temperature. The observation that many apparent isokinetic temperatures are in the experimental range may be related to this error effect. The error contour in the ΔH, ΔS plane is a very elongated elipse,[170–2] with the slope of the major axis being close to the experimental temperature. It has been suggested that to distinguish between a true isokinetic relationship and a spurious error correlation the spread of enthalpy and entropy values should be several times the maximum error. Another criterion is that the points on the line fall in a rational rather than a random order, as established by correlation with other properties. Petersen[173] believes that a linear enthalpy–entropy relationship can never be, by itself, a sufficient demonstration of an isokinetic relationship; he concludes that the only unambiguous evidence for an isokinetic relationship is the intersection of all $\log (k/T)$ vs. $1/T$ plots for the series at a common point (which will correspond to $1/\beta$).

It was not until the 1970s that the statistics of the isokinetic relationship was satisfactorily worked out.[166,171,174] Exner[175,176] first took this approach: Let k_1 and k_2 be the rate constants for a member of a reaction series at temperatures T_1 and T_2, with $T_2 > T_1$, and let k_1^0 and k_2^0 be the corresponding values for the reference member of the series. Then Eqs. (7-76) and (7-77) are easily derived for the reaction series.

$$\delta_R \Delta H^{\ddagger} = R \left(\frac{T_1 T_2}{T_2 - T_1} \right) \left(\ln \frac{k_2}{k_2^0} - \ln \frac{k_1}{k_1^0} \right) \tag{7-76}$$

$$\delta_R \Delta S^{\ddagger} = R \left(\frac{1}{T_2 - T_1} \right) \left(T_2 \ln \frac{k_2}{k_2^0} - T_1 \ln \frac{k_1}{k_1^0} \right) \tag{7-77}$$

Equation (7-76) shows that, in general, a plot of log k_2 against log k_1 is not expected to be linear. Linearity in such a plot can be assured, however, if an isokinetic relationship, $\delta_R \Delta H^{\ddagger} = \beta \delta_R \Delta S^{\ddagger}$, is followed by the system. By incorporating this relationship into Eqs. (7-76) and (7-77), we obtain

$$\ln \frac{k_2}{k_2^0} = \frac{T_1(T_2 - \beta)}{T_2(T_1 - \beta)} \ln \frac{k_1}{k_1^0} \tag{7-78}$$

Thus, a linear plot of log k_2 against log k_1 for a reaction series implies an isokinetic relationship for the series. The reason that this plot is a reliable test for such a relationship is that the errors in k_1 and k_2 are independent (unlike the errors in $\Delta H^{\ddagger}$ and $\Delta S^{\ddagger}$). From the slope b of the straight line the isokinetic temperature β can be found:

$$\beta = \frac{T_1 T_2(b - 1)}{b T_2 - T_1} \tag{7-79}$$

Obviously for this method to work the ratio T_1/T_2 must be appreciably smaller than unity. Provided this condition is met, this method is a simple and reliable way to test for an isokinetic relationship or to detect deviations from such a relationship. Exner[166] shows examples of systems plotted both as log k_2 vs. log k_1 and as $\Delta H^{\ddagger}$ vs. $\Delta S^{\ddagger}$, demonstrating the inadequacy of the latter plot. Exner[166] has also developed a statistical analysis of the Petersen method;[173] this analysis yields β and an uncertainty estimate of β. Exner[177] has applied his statistical methods to 100 reaction series, finding that 78 of them follow approximately valid isokinetic relationships.

Krug et al.[171] have shown that a plot of $\Delta G^{\ddagger}$ against $\Delta H^{\ddagger}$ [see Eq. (7-75)] is not subject to the error correlation problem of the $\Delta H^{\ddagger}$ vs. $\Delta S^{\ddagger}$ plot provided $\Delta G^{\ddagger}$ is evaluated at the harmonic mean temperature T_{hm} of the experimental range. (The harmonic mean is the reciprocal of the mean of the reciprocals, i.e., $T_{hm} = <1/T>^{-1}$.) They write Eq. (7-75) in the form

$$\delta_R \Delta H^{\ddagger} = \gamma \delta_R \Delta G^{\ddagger}_{hm} \tag{7-80}$$

where

$$\gamma = 1/(1 - T_{hm}/\beta) \tag{7-81}$$

Thus, γ is the slope of the plot of $\Delta H^{\ddagger}$ against $\Delta G^{\ddagger}$ at the harmonic mean temperature, and from γ the isokinetic temperature β is calculated. Tomlinson[178] has shown many examples of this type of analysis.

Earlier analyses making use of $\Delta H^{\ddagger}$ vs. $\Delta S^{\ddagger}$ plots generated many β values in the experimentally accessible range, and at least some of these are probably artifacts resulting from the error correlation in this type of plot. Exner's treatment[166,177] yields β values that may be positive or negative and that are often experimentally inaccessible. Some authors have associated isokinetic relationships and β values with specific chemical phenomena, particularly solvation effects and solvent structure,[167] but skepticism seems justified in view of the treatments of Exner and Krug et al. At the present time an isokinetic relationship should not be claimed solely on the basis of a plot of $\Delta H^{\ddagger}$ vs. $\Delta S^{\ddagger}$, but should be examined by the Exner or Krug methods.

The Reactivity–Selectivity Principle

Consider a bimolecular reaction between a substrate and a reagent. Upon each encounter of these two species there is a probability P that reaction will occur. If the solution contains two substrates S_1 and S_2, each characterized by a probability of reaction P_1 and P_2 with the common reagent, evidently the ratio P_2/P_1 is a measure of the selectivity of this process for S_2 relative to S_1. If the two substrates are not markedly dissimilar, the ratio P_2/P_1 will be similar to the ratio of rate constants, k_2/k_1. Leffler and Grunwald[1, pp. 162–8] define the selectivity as

$$\text{selectivity of the reagent for } S_2$$

$$\text{relative to } S_1 = -\delta_R \Delta G^{\ddagger}$$

$$= RT \ln (k_2/k_1) \tag{7-82}$$

The reactivity of a reagent can be defined as its rate constant with a standard substrate. Equation (7-82), therefore, relates selectivity to reactivity. The significance of this statement can be developed by imagining that a substrate is progressively changed in such a way that its probability of reaction increases toward the limit of unity. Comparing two substrates, each of whose reactivity is made successively higher in this way, the ratio P_2/P_1 will tend toward unity, as will its estimator k_2/k_1. Thus, in the limit of "infinite" reactivity, the selectivity becomes zero. (Since the designation of one reactant as the substrate and the other as the reagent is arbitrary, a similar statement can be made about a pair of reagents reacting with a common substrate.) The *reactivity–selectivity principle* (RSP) states that reactivity and selectivity are inversely related.

Instead of the definition in Eq. (7-82), the selectivity is often written as log (k_2/k_1). Another way to consider a selectivity–reactivity relationship is to compare the relative effects of a series of substituents on a pair of reactions. This is what is done when Hammett plots are made for a pair of reactions and their ρ values are compared. The slope of an LFER is a function of the sensitivity of the process being correlated to structural or solvent changes. Thus, in a family of closely related LFERs, the one with the steepest slope is the most selective, and the one with the smallest slope is the least selective.[179] Moreover, the intercept (or some arbitrarily selected abscissa value, usually log k_0 for the reference substituent) should be a measure of reactivity in each reaction series. Thus, a correlation should exist between the slopes (selectivity) and intercepts (reactivity) of a family of related LFERs. It has been suggested that the slopes and intercepts should be linearly related,[45] but the conditions required for linearity are seldom met, and it is instead common to find only a rough correlation, indicative of normal selectivity–reactivity behavior.[179] The Brønsted slopes, β, for the halogenation of a series of carbonyl compounds catalyzed by carboxylate ions show a smooth but nonlinear correlation with log k_0.[1, pp. 162–8] Jencks[52, pp. 195–8] uses the selectivity–reactivity relationship between Brønsted slopes and nucleophilic reactivity to distinguish between general acid catalysis and specific acid–general base catalysis.

In contrast to the comparison of a pair of reactions, with variations in substituents, Brown and his co-workers[180] compared a pair of substituents subjected to a large number of reactions. These were all electrophilic aromatic substitutions, the most extensively studied substrates being benzene and toluene. When a monosubstituted benzene derivative undergoes substitution of hydrogen, there are three possible sites for the entering group: *ortho* (two equivalent positions), *meta* (two), and *para* (one). Thus, the observed rate constant for the substitution reaction does not give a complete picture of the course of the reaction, because it does not distinguish among the positional isomers. If $k_{C_6H_5Y}$ is the observed rate constant for reaction of the monosubstituted derivative:

$$+ \; X^+ \quad \xrightarrow{k_{C_6H_5Y}} \quad + \; H^+ \tag{7-83}$$

then $k_{C_6H_5Y} = 2k_{o-Y} + 2k_{m-H} + k_{p-Y}$. The corresponding reaction for benzene,

$$+ \; X^+ \quad \xrightarrow{k_{C_6H_6}} \quad + \; H^+$$

leads to $k_{C_6H_6} = 6k_H$, because the reagent has an equal probability of attacking each of the six carbon atoms. The isomer distributions can be related to the individual rate constants, for Eq. (7-83), by

$$\frac{\%\ ortho}{100} = \frac{2k_{0-Y}}{k_{C_6H_5Y}}$$

$$\frac{\%\ meta}{100} = \frac{2k_{m-Y}}{k_{C_6H_5Y}}$$

$$\frac{\%\ para}{100} = \frac{k_{p-Y}}{k_{C_6H_5Y}}$$

Partial rate factors are defined as rates of substitution at each position relative to benzene, on a per site basis, thus,

$$o_f^Y = \frac{k_{0-Y}}{k_H}$$

$$m_f^Y = \frac{k_{m-Y}}{k_H}$$

$$p_f^Y = \frac{k_{p-Y}}{k_H}$$

Combining these relationships gives the following equations for calculation of partial rate factors:[18, pp. 245–7; 180, 181]

$$o_f^Y = \frac{3k_{C_6H_5Y}}{k_{C_6H_6}} \times \frac{\%\ ortho}{100}$$

$$m_f^Y = \frac{3k_{C_6H_5Y}}{k_{C_6H_6}} \times \frac{\%\ meta}{100}$$

$$p_f^Y = \frac{6k_{C_6H_5Y}}{k_{C_6H_6}} \times \frac{\%\ para}{100}$$

Partial rate factors have the significance that a factor greater than unity represents activation of that site by the substituent Y relative to hydrogen, and a factor less than unity represents deactivation.

Brown[180,182] defined a selectivity factor S_f by

$$S_f = \log\frac{p_f^Y}{m_f^Y} \tag{7-84}$$

Evidently S_f is a measure of intramolecular selectivity; because it involves a ratio, the contribution of the benzene substitution rate disappears, and the selectivity factor expresses the selectivity of the reagent X^+ in Eq. (7-83) for the *para* position relative to the *meta* position. Each individual partial rate factor, on the other hand, is expressive of an intermolecular selectivity; thus p_f^Y is a measure of the selectivity of the reagent for the *para* position in C_6H_5Y relative to benzene. It was observed that Eq. (7-85), where c_{Me} is a constant, is satisfied for a large number of electrophilic substitutions of toluene.[180]

$$\log p_f^{Me} = c_{Me}S_f \tag{7-85}$$

Equation (7-85) is a selectivity–reactivity relationship, with lower values of S_f denoting lower selectivity. Lower values of p_f correspond to greater reactivity, with the limit being a partial rate factor of unity for an infinitely reactive electrophile. This selectivity–reactivity relationship is followed for the electrophilic substitution reactions of many substituted benzenes, although toluene is the best studied of these.

It is probably inappropriate that the RSP has been called a *principle,* which implies a statement of wide generality, because many examples of its failure are known. For example, Ritchies' cation–anion recombination reactions follow Eq. (7-71), so they are LFER with the same slope; this is an instance of constant selectivity. Anti-RSP behavior is also known. As a consequence, the validity of the RSP is currently a controversial matter. There are several aspects of this problem.

One difficulty is in the choice of substrates (or reagents) when defining the selectivity. It is possible to convert RSP to anti-RSP behavior by altering the definition.[183] RSP behavior may be masked or revealed by changing the graphical presentation; we noted above that Ritchie plots of log k against N_+ yield the same slopes for different substrates, implying constant selectivity; yet when the reactivities are plotted against σ^+ (now holding solvent constant, as is appropriate), different ρ values are observed. Johnson and Stratton[183] conclude that real selectivity differences are obscured by the much larger range of reactivities that can be accommodated in the N_+ plot. Ta-Shma and Rappoport[184] studied the reactivity–selectivity behavior of Scheme V:

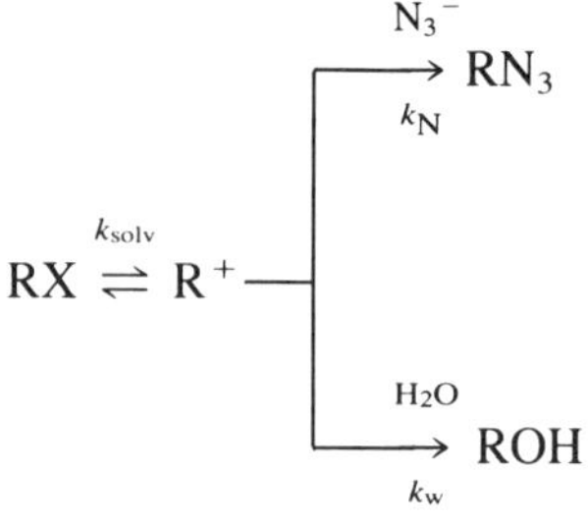

Scheme V

The rate constant k_{solv} for solvolysis is assumed to reflect the stability and reactivity of R^+ (i.e., faster solvolysis gives a more stable cation, which, therefore, reacts more slowly with nucleophiles). The ratio k_N/k_w, measured by product distribution studies, is a measure of selectivity with respect to the nucleophiles azide ion and water. The plot of log k_{solv} against log (k_N/k_w) is messy and rather confusing, containing regions of constant selectivity, RSP behavior, and possibly anti-RSP behavior. The present point is that the measures of reactivity and selectivity may affect the conclusions drawn. The role of solvent effects can be important in modifying RSP behavior.[185,186]

A more fundamental issue concerns the connection between free energy relationships (FERs) and reactivity–selectivity behavior. Recall from Chapter 5 that the Marcus equation, Eq. (5-69), an FER between rates and equilibria of the same reaction series, is predicted to be curved, which is consistent with the RSP, the selectivity decreasing as the reactivity increases. Thus, the RSP is consistent with Marcus theory and the Hammond postulate, and as a consequence it is inconsistent with *linear* FERs. Yet, as we have seen, many examples are known of LFERs. No doubt some of these LFERs extend over ranges of variables too limited to detect curvature if it exists, yet LFERs covering many orders of magnitude are known. This circumstance has elicited much discussion. One extreme point of view[187-9] is that LFERs are fundamental relationships and RSP behavior is not a general phenomenon. The other extreme is that LFERs cannot exist as general relationships and only appear to be linear because a limited range of variable is accessible.[190] Curvature in an FER may be evidence of the RSP,[191] but other sources of curvature may exist.[188]

We saw in Section 5.3 that simple Marcus theory (upon which the above discussion is based) applies only when a single reaction variable suffices to describe progress along the reaction coordinate. For more complex systems we extended the treatment by considering the changes in two reaction variables over a potential energy surface or by a configuration mixing model. We may, therefore, suspect a breakdown in the simple RSP when the reacting system is of this more complex type in which the transition state cannot be fully described in terms of the initial and final states. This appears to be the direction that current interpretations of the RSP are taking.[192,193]

REFERENCES

1. Leffler, J.E.; Grunwald, E. "Rates and Equilibria of Organic Reactions"; Wiley: New York, 1963; p 22.
2. Wells, P.R. *Chem. Revs.* **1963**, *63*, 171.
3. Wells, P.R. "Linear Free Energy Relationships"; Academic Press: New York, 1968.
4. Grunwald, E. *CHEMTECH* **1984**, *14*, 698.
5. Hammett, L.P. "Physical Organic Chemistry"; McGraw-Hill: New York, 1940; Chapter VII.
6. Hammett, L.P. "Physical Organic Chemistry", 2nd ed.; McGraw-Hill; New York, 1970; Chapter 11.
7. Jaffé, H.H. *Chem. Revs.* **1953**, *53*, 191.

8. Ehrenson, S. *Prog. Phys. Org. Chem.* **1964**, *2*, 195.

9. McDaniel, D.H.; Brown, H.C. *J. Org. Chem.* **1958**, *23*, 420.

10. Wiberg, K.B. "Physical Organic Chemistry"; Wiley: New York, 1964; pp 408–9.

11. Davis, W.H., Jr.; Pryor, W.A. *J. Chem. Educ.* **1976**, *53*, 285.

12. Berliner, E.; Liu, L.H. *J. Am. Chem. Soc.* **1953**, *75*, 2417.

13. Taft, R.W., Jr.; Lewis, I.C. *J. Am. Chem. Soc.* **1958**, *80*, 2436.

14. Hammett, L.P. *J. Am. Chem. Soc.* **1937**, *59*, 96.

15. Bunnett, J.F.; Draper, F., Jr.; Ryason, P.R., Noble, P., Jr.; Tonkyn, R.G.; Zahler, R.E. *J. Am. Chem. Soc.* **1953**, *75*, 642.

16. Brown, H.C.; Okamoto, Y. *J. Am. Chem. Soc.* **1958**, *80*, 4979.

17. Yukawa, Y.; Tsuno, Y. *Bull. Chem. Soc. Jap.* **1959**, *32*, 971.

18. Ingold, C.K. "Structure and Mechanism in Organic Chemistry"; Cornell University Press: Ithaca, N.Y., 1953; pp 64–6, 71.

19. Alexander, E.R. "Principles of Ionic Organic Reactions"; Wiley: New York, 1950; p 6.

20. Gould, E.S. "Mechanism and Structure in Organic Chemistry"; Holt, Rinehart, & Winston: New York, 1959; pp 207, 217.

21. Chapman, N.B.; Shorter, J., Eds., "Advances in LFER"; Plenum: London, 1972; pp xi–xiii.

22. Dewar, M.J.S. "The Electronic Theory of Organic Chemistry"; Oxford University Press: London, 1949; p 52.

23. van Bekkum, H.; Verkade, P.E.; Wepster, B.M. *Rec. Trav. Chim.* **1959**, *78*, 815.

24. Taft, R.W., Jr. *J. Phys. Chem.* **1960**, *64*, 1805.

25. Branch, G.E.K.; Calvin, M. "The Theory of Organic Chemistry"; Prentice-Hall: New York, 1941; p 417.

26. Roberts, J.L.; Jaffé, H.H. *J. Am. Chem. Soc.* **1959**, *81*, 1635.

27. Taft, R.W.; Lewis, I.C. *J. Am. Chem. Soc.* **1959**, *81*, 5343.

28. Exner, O. In "Advances in LFER"; Chapman, N.B.; Shorter, J., Eds.; Plenum: London, 1972; Chapter 1.

29. Ehrenson, S.; Brownlee, R.T.C.; Taft, R.W. *Prog. Phys. Org. Chem.* **1972**, *10*, 1.

30. Roberts, J.D.; Moreland, W.T., Jr. *J. Am. Chem. Soc.* **1953**, *75*, 2167.

31. Charton, M. *Prog. Phys. Org. Chem.* **1981**, *13*, 119.

32. Taft, R.W.; Ehrenson, S.; Lewis, I.C.; Glick, R.E. *J. Am. Chem. Soc.* **1959**, *81*, 5352.

33. Fadhil, G.F.; Godfrey, M. *J. Chem. Soc., Perkin Trans. 2* **1988**, 133.

34. Swain, C.G.; Unger, S.H.; Rosenquist, N.R.; Swain, M.S. *J. Am. Chem. Soc.* **1983**, *105*, 492.

35. Swain, C.G.; Lupton, E.C. *J. Am. Chem. Soc.* **1968**, *90*, 4328.

36. Afanas'ev, I.B. *J. Chem. Soc., Perkin Trans. 2* **1984**, 1589.

37. Taft, R.W.; Abboud, J.L.M.; Anvia, F.; Berthelot, M.; Fujio, M.; Gal, J.-F.; Headley, A.D.; Henderson, W.G.; Koppel, I.; Qian, J.H.; Mishima, M.; Taagepera, M.; Ueji, S. *J. Am. Chem. Soc.* **1988**, *110*, 1797.

38. Ritchie, C.D.; Sager, W.F. *Prog. Phys. Org. Chem.* **1964**, *2*, 323.

39. Gawley, R.E. *J. Org. Chem.* **1981**, *46*, 4595.

40. Wiberg, K.B. "Physical Organic Chemistry"; Wiley: New York, 1964; pp 404–8.

41. McLennan, D.J. *Tetrahedron* **1978**, *34*, 2331.

42. Poh, B.-L. *Can. J. Chem.* **1977**, *55*, 3721.

43. Poh, B.-L. *Can. J. Chem.* **1979**, *57*, 255.

44. Williams, A. *Acc. Chem. Res.* **1984**, *17*, 425.

45. Miller, S.I. *J. Am. Chem. Soc.* **1959**, *81*, 101.

46. Jencks, D.A.; Jencks, W.P. *J. Am. Chem. Soc.* **1977**, *99*, 7948.

47. Dubois, J.-E.; Ruasse, M.-F.; Argile, A. *J. Am. Chem. Soc.* **1984**, *106*, 4840.

48. Lee, I.; Sohn, S.C. *J. Chem. Soc., Chem. Commun.* **1986**, 1055.

49. Lee, I.; Kim, H.Y.; Kang, H.K. *J. Chem. Soc., Chem. Commun.* **1987**, 1216.

50. Lee, I.; Shim, C.S.; Chung, S.Y.; Kim, H.Y.; Lee, H.W. *J. Chem. Soc., Perkin Trans. 2*, **1988**, 1919.

51. Sykes, P. "A Guidebook to Mechanism in Organic Chemistry", 6th ed.; Longman: Harlow, England, 1986; pp 375–83.

52. Jencks, W.P. "Catalysis in Chemistry and Enzymology"; McGraw-Hill: New York, 1969; pp 480–3.

53. Gassman, P.G.; Fentiman, A.F., Jr. *J. Am. Chem. Soc.* **1970**, *92*, 2549.

54. Taft, R.W. "Steric Effects in Organic Chemistry"; Newman, M.S., Ed., Wiley: New York, 1956; Chapter 13.

55. Taft, R.W. *J. Am. Chem. Soc.* **1952**, *74*, 3126.

56. Taft, R.W. *J. Am. Chem. Soc.* **1953**, *75*, 4538.

57. Farthing, A.C.; Nam, B. In "Steric Effects in Conjugated Systems"; Gray, G.W., Ed.; Academic Press: New York, 1958; Chapter 11.

58. Shorter, J. In "Advances in LFER"; Chapman, N.B.; Shorter, J., Eds.; Plenum: London, 1972; Chapter 2.

59. Charton, M. *Prog. Phys. Org. Chem.* **1971**, *12*, 235.

60. Charton, M. *J. Org. Chem.* **1975**, *40*, 407.

61. Segura, P. *J. Org. Chem.* **1985**, *50*, 1045.

62. Berg, U.; Gallo, R.; Klatte, G.; Metzger, J. *J. Chem. Soc., Perkin Trans. 2* **1980**, 1350.

63. Reynolds, W.F. *Prog. Phys. Org. Chem.* **1983**, *14*, 165.

64. Reynolds, W.F. *J. Chem. Soc., Perkin Trans. 2* **1980**, 985.

65. Taft, R.W.; Topsom, R.D. *Prog. Phys. Org. Chem.* **1987**, *16*, 1.

66. Levitt, L.S.; Widing, H.F. *Prog. Phys. Org. Chem.* **1976**, *12*, 119.

67. Charton, M. *Prog. Phys. Org. Chem.* **1987**, *16*, 287.

68. Ingold, C.K. *J. Chem. Soc.* **1930**, 1032.

69. Kanerva, L.T.; Euranto, E.K. *J. Chem. Soc., Perkin Trans. 2* **1987**, 441.

70. Jones, R.W.A.; Thomas, J.D.R. *J. Chem. Soc.* **1966B**, 661.

71. Bruice, T.C.; Fife, T.H.; Bruno, J.T.; Brandon, N.E. *Biochemistry* **1962**, *1*, 7.

72. (a) Kirsch, J.F.; Jencks, W.P. *J. Am. Chem. Soc.* **1964**, *86*, 837.
    (b) Jencks, W.P.; Gilchrist, M. *J. Am. Chem. Soc.* **1968**, *90*, 2622.

73. Ballinger, P.; Long, F.A. *J. Am. Chem. Soc.* **1960**, *82*, 795.

74. Robinson, J.R.; Matheson, L.E. *J. Org. Chem.* **1969**, *34*, 3630.

75. DeTar, D.F. *J. Am. Chem. Soc.* **1982**, *104*, 7205.

76. DeTar, D.F. *J. Org. Chem.* **1980**, *45*, 5166.

77. Charton, M. *J. Am. Chem. Soc.* **1977**, *99*, 5687.

78. DeTar, D.F. *J. Am. Chem. Soc.* **1980**, *102*, 7988.

79. Fliszár, S. "Charge Distributions and Chemical Effects"; Springer-Verlag: New York, 1983; Chapter 2.

80. MacPhee, J.A.; Panaye, A.; Dubois, J.-E. *Tetrahedron* **1978**, *34*, 3553.

81. Unger, S.H.; Hansch, C. *Prog. Phys. Org. Chem.* **1976**, *12*, 91.

82. Shawali, A.S.A.S.; Biechler, S.S. *J. Am. Chem. Soc.* **1967**, *89*, 3020.

83. Kutter, E.; Hansch, C. *J. Med. Chem.* **1969**, *12*, 647.

84. Dubois, J.-E.; MacPhee, J.A.; Panaye, A. *Tetrahedron* **1980**, *36*, 919.

85. Gallo, R. *Prog. Phys. Org. Chem.* **1983**, *14*, 115.

86. Kramer, C.-R. *Z. Phys. Chem. (Leipzig)* **1989**, *270*, 257, 271.

87. Mueller, P.; Mareda, J. *Tetrahedron Lett.* **1984**, *25*, 1703.

88. Newman, M.S., Ed., "Steric Effects in Organic Chemistry"; Wiley: New York; 1956.

89. Brown, H.C. In "Nonclassical Ions"; Bartlett, P.D., Ed. Benjamin: New York, 1965; pp 438–55.

90. Brownstein, S.; Burton, G.W.; Hughes, L.; Ingold, K.U. *J. Org. Chem.* **1989**, *54*, 560.

91. Brønsted, J.N.; Pedersen, K.J. *Z. Phys. Chem.* **1924**, *108*, 185.

92. Brønsted, J.N. *Chem. Revs.* **1928**, *5*, 231.

93. Bell, R.P. "Acid-Base Catalysis"; Oxford University Press: London, 1941; Chapter V.

94. Bell, R.P. "The Proton in Chemistry"; Cornell University Press: Ithaca, N.Y., 1959; Chapter X.

95. Starkey, R.; Norman, J.; Hintze, M. *J. Chem. Educ.* **1986**, *63*, 473.

96. Sander, E.G.; Jencks, W.P. *J. Am. Chem. Soc.* **1968**, *90*, 4377.

97. Kirsch, L.E.; Notari, R.E. *J. Pharm. Sci.* **1984**, *73*, 724.

98. Benson, S.W. *J. Am. Chem. Soc.* **1958**, *80*, 5151.

99. Toney, M.D.; Kirsch, J.F. *Science* **1989**, *243*, 1485.
100. Bruice, T.C.; Lapinski, R. *J. Am. Chem. Soc.* **1958**, *80*, 2265.
101. Hong, W.-H.; Connors, K.A. *J. Pharm. Sci.* **1968**, *57*, 1789.
102. Fedor, L.R.; Bruice, T.C.; Kirk, K.L.; Meinwald, J. *J. Am. Chem. Soc.* **1966**, *88*, 108.
103. Buncel, E.; Um, I.H.; Hoz, S. *J. Am. Chem. Soc.* **1989**, *111*, 971.
104. Ba-Saif, S.; Luthra, A.K.; Williams, A. *J. Am. Chem. Soc.* **1989**, *111*, 2647.
105. Williams, A. *Acc. Chem. Res.* **1989**, *22*, 387.
106. Bordwell, F.G.; Hughes, D.L. *J. Org. Chem.* **1982**, *47*, 3224.
107. Eigen, M. *Angew. Chem. Intern. Ed.* **1964**, *3*, 1.
108. Yang, C.C.; Jencks, W.P. *J. Am. Chem. Soc.* **1988**, *110*, 2972.
109. Hupes, D.J.; Jencks, W.P. *J. Am. Chem. Soc.* **1977**, *99*, 451.
110. Castro, E.A.; Santander, C.L. *J. Org. Chem.* **1985**, *50*, 3595.
111. Castro, E.A.; Ureta, C. *J. Org. Chem.* **1989**, *54*, 2153.
112. Bruice, T.C.; Benkovic, S.J. "Bioorganic Mechanisms"; Benjamin: New York, 1966; Vol. I, pp 46–67.
113. Johnson, S.L. *Advan. Phys. Org. Chem.* **1967**, *5*, 237.
114. Bender, M.L. "Mechanisms of Homogeneous Catalysis from Protons to Proteins"; Wiley-Interscience: New York, 1971; Chapters 4–6.
115. Kirsch, J.F.; Jencks, W.P. *J. Am. Chem. Soc.* **1964**, *86*, 837.
116. Fersht, A.R.; Kirby, J. *J. Am. Chem. Soc.* **1968**, *90*, 5818.
117. Jencks, W.P.; Brant, S.R.; Gandler, J.R.; Fendrich, G.; Nakamura, C. *J. Am. Chem. Soc.* **1982**, *104*, 7045.
118. Jencks, W.P. In "Nucleophilicity"; Harris, J.M.; McManus, S.P., Eds.; Advan. Chem. Ser. No. 215; American Chemical Society: Washington, D.C., 1987; Chapter 10.
119. Jencks, W.P.; Carriuolo, J. *J. Am. Chem. Soc.* **1960**, *82*, 1778.
120. Sander, E.G.; Jencks, W.P. *J. Am. Chem. Soc.* **1968**, *90*, 6154.
121. Edwards, J.O.; Pearson, R.G. *J. Am. Chem. Soc.* **1962**, *84*, 16.
122. Klopman, G.; Tsuda, K.; Louis, J.B.; Davis, R.E. *Tetrahedron* **1970**, *26*, 4549.
123. Dixon, J.E.; Bruice, T.C. *J. Am. Chem. Soc.* **1971**, *93*, 3248.
124. Hoz, S. In "Nucleophilicity"; Harris, J.M.; McManus, S.P., Eds.; Advan. Chem. Ser. No. 215; American Chemical Society: Washington, D.C., 1987; Chapter 12.
125. Hudson, R.F. In "Nucleophilicity"; Harris, J.M.; McManus, S.P., Eds.; Advan. Chem. Ser. No. 215; American Chemical Society: Washington, D.C., 1987; pp 201–7.
126. Buncel, E.; Shaik, S.S., Um, I.-H.; Wolfe, S. *J. Am. Chem. Soc.* **1988**, *110*, 1275.
127. Jencks, W.P. *J. Am. Chem. Soc.* **1958**, *80*, 4581, 4585.
128. Ritchie, C.D.; Kubisty, C.; Ting, G.Y. *J. Am. Chem. Soc.* **1983**, *105*, 279.
129. Bordwell, F.G.; Cripe, T.A.; Hughes, D.L. In "Nucleophilicity"; Harris, J.M.; McManus, S.P., Eds.; Advan. Chem. Ser. No. 215; American Chemical Society: Washington, D.C., 1987; Chapter 9.
130. Swain, C.G.; Scott, C.B. *J. Am. Chem. Soc.* **1953**, *75*, 141.
131. Edwards, J.O. *J. Am. Chem. Soc.* **1954**, *76*, 1540.
132. Edwards, J.O. *J. Am. Chem. Soc.* **1956**, *78*, 1819.
133. Davis, R.E.; Suba, L.; Klimishin, P.; Carter, J. *J. Am. Chem. Soc.* **1969**, *91*, 104.
134. Ibne-Rasa, K.M. *J. Chem. Educ.* **1967**, *44*, 89.
135. Pearson, R.G.; Sobel, H.; Songstad, J. *J. Am. Chem. Soc.* **1968**, *90*, 319.
136. Brauman, J.I.; Dodd, J.A.; Han, C.-C. In "Nucleophilicity"; Harris, J.M.; McManus, S.P., Eds.; Advan. Chem. Ser. No. 215; American Chemical Society: Washington, D.C., 1987; Chapter 2.
137. Streitwieser, A.J. *Proc. Natl. Acad. Sci. U.S.A.* **1985**, *82*, 8288.
138. Parker, A.J. *Advan. Phys. Org. Chem.* **1967**, *5*, 173.
139. Liotta, C.L.; Grisdale, E.E.; Hopkins, H.P., Jr. *Tetrahedron Lett.* *1975*, 4205.
140. Pearson, R.G. In "Advances in LFER"; Chapman, N.B.; Shorter, J., Eds.; Plenum: London, 1972; Chapter 6.
141. Pearson, R.G. *J. Org. Chem.* **1987**, *52*, 2131.
142. Pearson, R.G. *J. Chem. Educ.* **1987**, *64*, 561.
143. Bartoli, G.; Todesco, P.E. *Acc. Chem. Res.* **1977**, *10*, 125.

144. Pytela, O.; Zima, V. *Coll. Czech. Chem. Commun.* **1989,** *54,* 117.

145. Ritchie, C.D.; Virtanen, P.O.I. *J. Am. Chem. Soc.* **1972,** *94,* 4966.

146. Ritchie, C.D. *Acc. Chem. Res.* **1972,** *5,* 348.

147. Ritchie, C.D. In "Solute-Solvent Interactions"; Coetzee, J.F.; Ritchie, C.D., Eds.; Dekker: New York, 1976; Vol. 2, Chapter 12.

148. Bender, M.L.; Kézdy, F.J.; Zerner, B. *J. Am. Chem. Soc.* **1963,** *85,* 3017.

149. Bender, M.L.; Neveu, M.C. *J. Am. Chem. Soc.* **1958,** *80,* 5388.

150. Bruice, T.C.; Benkovic, S.J. *J. Am. Chem. Soc.* **1963,** *85,* 1.

151. Kirby, A.J. *Advan. Phys. Org. Chem.* **1980,** *17,* 183.

152. Mandolini, L. *Advan. Phys. Org. Chem.* **1986,** *22,* 2.

153. Menger, F.M.; Venkataram, U.V. *J. Am. Chem. Soc.* **1985,** *107,* 4706.

154. Storm, D.R.; Koshland, D.E., Jr. *J. Am. Chem. Soc.* **1972,** *94,* 5805, 5815.

155. Dafforn, A.; Koshland, D.E., Jr. *Biochem. Biophys. Res. Commun.* **1973,** *52,* 779.

156. Menger, F.M. *Tetrahedron* **1983,** *39,* 1013.

157. Bruice, T.C.; Pandit, U.K. *Proc. Natl. Acad. Sci. U.S.A.* **1960,** *46,* 402.

158. Bruice, T.C.; Pandit, U.K. *J. Am. Chem. Soc.* **1960,** *82,* 5858.

159. Milstein, S.; Cohen, L.A. *Proc. Natl. Acad. Sci. U.S.A.* **1970,** *67,* 1143.

160. Page, M.I.; Jencks, W.P. *Proc. Natl. Acad. Sci. U.S.A.* **1971,** *68,* 1678.

161. Jencks, W.P.; Page, M.I. *Biochem. Biophys. Res. Commun.* **1974,** *57,* 887.

162. Menger, F.M. *Acc. Chem. Res.* **1985,** *18,* 128.

163. DeTar, D.F.; Luthra, N.P. *J. Am. Chem. Soc.* **1980,** *102,* 4505.

164. Menger, F.M. In "Nucleophilicity"; Harris, J.M.; McManus, S.P., Eds.; Advan. Chem. Ser. No. 215; American Chemical Society: Washington, D.C., 1987; Chapter 14.

165. Leffler, J.E. *J. Org. Chem.* **1955,** *20,* 1202.

166. Exner, O. *Prog. Phys. Org. Chem.* **1973,** *10,* 411.

167. Lumry, R.; Rajender, S. *Biopolymers* **1970,** *9,* 1125.

168. Bunnett, J.F. In "Investigation of Rates and Mechanisms of Reactions", 3rd Ed.; Lewis, E.S., Ed.; Wiley-Interscience: New York, 1974; Part I, pp 412–21.

169. Petersen, R.C.; Markgraf, J.H.; Ross, S.D. *J. Am. Chem. Soc.* **1961,** *83,* 3819.

170. Leffler, J.E. *J. Org. Chem.* **1966,** *31,* 533.

171. Krug, R.R.; Hunter, W.G.; Grieger, R.A. *J. Phys. Chem.* **1976,** *80,* 2335, 2341.

172. Heberger, K.; Kemeny, S.; Vidoczy, T. *Int. J. Chem. Kin.* **1987,** *19,* 171.

173. Petersen, R.C. *J. Org. Chem.* **1964,** *29,* 3133.

174. Exner, O. *Coll. Czech. Chem. Commun.* **1972,** *37,* 1425.

175. Exner, O. *Nature* **1964,** *201,* 488B.

176. Exner, O. *Nature* **1970,** *227,* 366.

177. Exner, O. *Coll. Czech. Chem. Commun.* **1975,** *40,* 2762.

178. Tomlinson, E. *Int. J. Pharmaceut.* **1983,** *13,* 115.

179. Bender, M.L. *Chem. Revs.* **1960,** *60,* 53.

180. Stock, L.M.; Brown, H.C. *Advan. Phys. Org. Chem.* **1963,** *1,* 35.

181. Stock, L.M. "Aromatic Substitution Reactions"; Prentice-Hall: Englewood Cliffs, N.J., 1968; p 45.

182. Brown, H.C.; Smoot, C.R. *J. Am. Chem. Soc.* **1956,** *78,* 6255.

183. Johnson, C.D.; Stratton, B. *J. Chem. Soc., Perkin Trans. 2* **1988,** 1903.

184. Ta-Shma, R.; Rappoport, Z. *J. Am. Chem. Soc.* **1983,** *105,* 6082.

185. McManus, S.P.; Zutaut, S.E. *Isr. J. Chem.* **1985,** *26,* 400.

186. Bernasconi, C.F.; Bunwell, R.D. *Isr. J. Chem.* **1985,** *26,* 420.

187. Johnson, C.D. *Chem. Revs.* **1975,** *75,* 755.

188. Johnson, C.D. *Tetrahedron* **1980,** *36,* 3461.

189. Bordwell, F.G.; Branca, J.C.; Cripe, T.A. *Isr. J. Chem.* **1985,** *26,* 357.

190. Lewis, E.S.; Shen, C.C.; More O'Ferrall, R.A. *J. Chem. Soc., Perkin Trans. 2* **1981,** 1084.

191. Hupe, D.J.; Pohl, E.R. *Isr. J. Chem.* **1985,** *26,* 395.

192. Pross, A. *Isr. J. Chem.* **1985,** *26,* 390.

193. Buncel, E.; Wilson, H. *J. Chem. Educ.* **1987,** *64,* 475.

194. Berliner, E.; Altschu, L.H. *J. Am. Chem. Soc.* **1952,** *74,* 4110.

195. Goering, H.L.; Rubin, T.; Newman, M.S. *J. Am. Chem. Soc.* **1954,** *76,* 787.
196. Fife, T.H.; Natarajan, R.; Werner, M.H. *J. Org. Chem.* **1987,** *52,* 740.
197. Bruice, T.C.; Schmir, G.L. *J. Am. Chem. Soc.* **1958,** *80,* 148.
198. Jencks, W.P.; Gilchrist, M. *J. Am. Chem. Soc.* **1962,** *84,* 2910.
199. Lewis, E.S.; Johnson, M.D. *J. Am. Chem. Soc.* **1959,** *81,* 2070.
200. Biggs, A.I.; Robinson, R.A. *J. Chem. Soc.* **1961,** 388.
201. Hall, H.K., Jr. *J. Am. Chem. Soc.* **1957,** *79,* 5441.
202. Sitzmann, M.E.; Adolf, H.G.; Kamlet, M.J. *J. Am. Chem. Soc.* **1968,** *90,* 2815.

PROBLEMS

1. These data are for the hydrolysis of substituted benzoic anhydrides in 75% dioxane.[194]

	$10^4 k/min^{-1}$			
Substituent	*58.25°C*	*79.65°C*	$E_a/kcal\ mol^{-1}$	$\Delta S^{\ddagger}/eu$
p-OMe	0.547	3.50	20.1	-27.8
p-Me	1.60	8.07	17.6	-33.1
p-*t*-Bu	1.77	8.74	17.3	-33.8
m-Me	2.84	14.7	17.8	-31.4
H	4.55	20.0	16.1	-35.6
m-OMe	5.86	25.2	15.8	-36.0
p-Cl	23.5	98.4	15.5	-34.2
m-NO$_2$	861	354	11.6	-38.8
p-NO$_2$	1360	608	10.7	-40.5

(a) Make the Hammett plots and evaluate ρ.

(b) Examine the data for possible adherence to the isokinetic relationship.

2. These rate constants are for the basic hydrolysis of methyl 4-substituted 2,6-dimethylbenzoates at 125°C in 60% dioxane.[195]

CH$_3$

X — COOCH$_3$

CH$_3$

X	$10^3\ k/M^{-1}s^{-1}$
4-NH$_2$	0.884
4-CH$_3$	1.35
4-H	1.73
4-Br	6.83
4-NO$_2$	35.2

Make the Hammett plot, neglecting the 2,6-dimethyl substituents, and calculate ρ.

3. Estimate the rate constants for these reactions.

(a)

(3-Et-phenyl)—COOMe + OH⁻ $\xrightarrow[25°C]{\substack{60\% \\ Me_2CO}}$ (3-Et-phenyl)—COO⁻ + MeOH

(b)

Br_2 + (4-SCH₃-phenyl) $\xrightarrow[25°C]{HOAc}$ (2-Br-4-SCH₃-phenyl) + HBr

(c)

$$CH_3COCH_3 + I_2 \xrightarrow[\substack{H_2O \\ 25°C}]{C_2H_5COOH} CH_3COCH_2I + HI$$

4. These are rate constants for the catalyzed hydrolysis of N-(3,3-dimethylbutyryl)-4(5)-nitroimidazole in water at 30°C.[196]

Base	pK_a	$k_B/M^{-1}s^{-1}$
H_2O	-1.74	2.25×10^{-6}
Chloroacetate	2.72	1.81×10^{-4}
Formate	3.60	8.99×10^{-4}
Acetate	4.60	1.60×10^{-3}
Pyridine	5.25	4.10×10^{-3}
2-Morpholinoethanesulfonic acid	6.10	1.10×10^{-3}
Cacodylate	6.19	4.76×10^{-2}
2,6-Lutidine	6.60	2.52×10^{-4}
HPO_4^{2-}	6.75	2.46×10^{-2}
Imidazole	7.05	2.26
N-Ethylmorpholine	7.70	1.23×10^{-2}

Make the Brønsted plot, calculate β, and discuss deviations from the line.

5. These data are for the nucleophilic catalysis of the hydrolysis of p-nitrophenyl acetate by imidazoles and benzimidazoles at pH 8.0. The apparent second-order catalytic rate constants are defined by

$$k_{obs} = k_0 + k_{cat}[IM]_T$$

where $[IM]_T$ is the total imidazole concentration. pK_1 is for dissociation of the protonated amine. pK_2 is for the production of the anion by dissociation of the neutral species. Some pK_a values are listed for additional functional groups for certain compounds. Analyze the data as thoroughly as you can.

No.	Catalyst	pK_1	$k_{cat}/M^{-1}min^{-1}$
1	2-Methylimidazole	7.75	2.7
2	4-Methylimidazole	7.45	25.1
3	*N*-Acetylhistidine	7.05	11.2
4	Imidazole	6.95	20.2
5	2-Methyl-4-hydroxy-6-aminobenzimidazole	6.65	1.5
6	4-(2′,4′-Dihydroxyphenyl)-imidazole	6.45	9.4
7	4-Hydroxymethylimidazole	6.45	5.6
8	2-Methylbenzimidazole	6.1	0.0375
9	Histamine	6.0	7.0
10	6-Aminobenzimidazole	6.0 (NH_2 3.0)	2.95
11	4-Hydroxy-6-aminobenzimidazole	5.9	6.15
12	Benzimidazole	5.4	0.96
13	4-Hydroxybenzimidazole	5.3 (OH 9.5)	2.80
14	4-Methoxybenzimidazole	5.1	0.31
15	2-Methyl-4-hydroxy-6-nitrobenzimidazole	3.9	1.1
16	4-Bromoimidazole	3.7	0.28
17	6-Nitrobenzimidazole	3.05 (pK_2 10.6)	4.8
18	4-Hydroxy-6-nitrobenzimidazole	3.05	3.75
19	4-Nitroimidazole	1.5 (pK_2 9.1)	35.5

6. Suppose it is known that an isokinetic relationship holds for a reaction series. Then give the slope of a plot of log k against E_a for the series.

7. Discuss possible reasons for the curvature in the Brønsted-type plot of Fig. 7.5 for the nucleophilic reactions of oxygen nucleophiles.

8. These rate constants are for the cinnamoylation of hydroxy compounds by cinnamic anhydride catalyzed by *N*-methylimidazole. The reaction is first-order in each reactant. The kinetics were followed spectrophotometrically in acetonitrile solution. Analyze the data; that is, attempt to account for the relationship between structure and reactivity.

Hydroxy compound	$10^2 \ k/M^{-2} \ s^{-1}$
Phenol	96.4
iso-Amyl alcohol	10.7
n-Amyl alcohol	9.22
iso-Butyl alcohol	7.88
n-Butyl alcohol	3.47
iso-Propyl alcohol	0.72
n-Propyl alcohol	3.24
sec-Butyl alcohol	0.65
tert-Butyl alcohol	0.58

9. From the last four digits of the office telephone numbers of the faculty in your department, systematically construct pairs of "rate constants" as two-digit numbers times 10^{-5} s^{-1} at temperatures 300 K and 315 K (obviously the larger rate constant of each pair to be associated with the higher temperature). Make a two-point Arrhenius plot for each faculty member, evaluating $\Delta H^{\ddagger}$ and $\Delta S^{\ddagger}$. Examine the plot of $\Delta H^{\ddagger}$ against $\Delta S^{\ddagger}$ for evidence of an isokinetic relationship. (This problem was suggested by D. Khossravi.)

CHAPTER 8

Medium Effects

The effect of the medium (solvent) on chemical reactivity is a subject of great difficulty, one that can be studied at several levels of understanding. The literature of the field is large, and research interest continues to be high. In this chapter we can only summarize much that has been learned; each topic can be pursued in detail by means of the citations to original work. Many authors have reviewed solvent effects on reaction rates.[1–5] Section 8.1 introduces a few ideas that are treated more thoroughly in the rest of the chapter.

8.1 INTRODUCTION TO MEDIUM EFFECTS

Types of Effects

Perhaps the most obvious experiment is to compare the rate of a reaction in the presence of a solvent and in the absence of the solvent (i.e., in the gas phase). This has long been possible for reactions proceeding homolytically, in which little charge separation occurs in the transition state; for such reactions the rates in the gas phase and in the solution phase are similar. Very recently it has become possible to examine polar reactions in the gas phase, and the outcome is greatly different, with the gas-phase reactivity being as much as 10^{15} greater than the reactivity in polar solvents.[5] This reduced reactivity in solvents is ascribed to inhibition by solvation; in such reactions the role of the solvent clearly overwhelms the intrinsic reactivity of the reactants. Gas-phase kinetic studies are a powerful means for interpreting the reaction coordinate at a molecular level.

Most of what we know about solvent effects is a result of studies in which the reactivity is compared in a series of solvents. There are two main types of experimental design: in one of these the reaction is carried out in different pure solvents; in the other design the reaction is studied in mixed solvents, often a binary mixture whose composition is varied across the entire range. Experimental limitations often control the kind of solvent study that can be conducted. Each of these designs has advantages and disadvantages, but all solvent studies share at least this problem: When the solvent is changed, many solvent properties change. That is, it is not possible to isolate only one factor for variation (e.g., the dielectric constant), keeping all others constant; any change in solvent generates changes in many properties, not all of which may be recognized as important factors. This problem introduces uncertainty into interpretations.

There is a third experimental design often used for studies in electrolyte solutions, particularly aqueous solutions. In this design the reaction rate is studied as a function of ionic strength, and a rate variation is called a *salt effect*. In Chapter 5 we derived this relationship between the observed rate constant k and the activity coefficients of reactants (γ_A, γ_B) and transition state ($\gamma^{\ddagger}$):

$$k = k_0 \frac{\gamma_A \gamma_B}{\gamma^{\ddagger}} \tag{8-1}$$

An effect of ionic strength on k as a consequence of effects on the activity coefficient ratio is called a *primary salt effect*. We will, in Section 8.3, consider this effect quantitatively.

If the rate equation contains the concentration of a species involved in a pre-equilibrium step (often an acid–base species), then this concentration may be a function of ionic strength via the ionic strength dependence of the equilibrium constant controlling the concentration. Therefore, the rate constant may vary with ionic strength through this dependence; this is called a *secondary salt effect*. This effect is an artifact in a sense, because its source is independent of the rate process, and it can be completely accounted for by evaluating the rate constant on the basis of the actual species concentration, calculated by means of the equilibrium constant appropriate to the ionic strength in the rate study.

The observed solvent effect can be expressed quantitatively with the aid of the Leffler–Grunwald operator δ_M introduced in Chapter 7. For rate constant k_M measured in medium M we have, from transition state theory, $k_M = (\mathbf{k}T/h)\exp(-\Delta G_M^{\ddagger}/RT)$ and similarly for rate constant k_0 measured in a reference solvent. Combining these two expressions gives

$$\log\frac{k_M}{k_0} = -\frac{(\Delta G_M^{\ddagger} - \Delta G_0^{\ddagger})}{2.3RT} \tag{8-2}$$

Defining $\delta_M \Delta G^{\ddagger} = \Delta G_M^{\ddagger} - \Delta G_0^{\ddagger}$, we get

$$\delta_M \Delta G^{\ddagger} = -2.3RT \log\frac{k_M}{k_0} \tag{8-3}$$

The quantity $\delta_M \Delta G^{\ddagger}$ is the solvent effect on changing the medium from 0 to M. If the reference solvent is held constant in a study in which solvent M is altered, the solvent effect may be expressed simply by $\log k_M$.

The solvent may serve only as the medium for the reaction, or it may in addition be a reactant, as in a solvolysis reaction. It is possible that the reaction mechanism may be changed by a change in solvent (e.g., from S_N1 to S_N2) or that the rate-determining step of a complex reaction may be altered. All of these phenomena can be studied by examining the solvent dependence. One goal of research on medium effects is to achieve a level of understanding that will allow us to make mechanistic interpretations from such data. To the present time most solvent effect studies have consisted of this preliminary phase—reaching some understanding—rather than of confident applications to mechanistic problems.

Physical Theories

To go from experimental observations of solvent effects to an understanding of them requires a conceptual basis that, in one approach, is provided by physical models such as theories of molecular structure or of the liquid state. As a very simple example consider the electrostatic potential energy of a system consisting of two ions of charges Z_A and Z_B in a medium of dielectric constant ε,

$$V = \frac{Z_A Z_B e^2}{\varepsilon r} \tag{8-4}$$

where e is the electronic charge and r is the interionic distance. This relationship leads us to expect that the dielectric constant may be an important factor in reactions of ions and (by extension) of polar molecules and further that the function $1/\varepsilon$ may be a pertinent variable. This idea is developed in Section 8.3.

A qualitative application of the transition state theory provides a useful prediction of the direction of a solvent effect upon the rate of a reaction. Hughes and Ingold[6,7] introduced this idea in the context of aliphatic nucleophilic substitutions, but it is applicable to all types of reactions. The idea is that if the transition state is more polar than the initial state, an increase in polarity of the solvent will stabilize the transition state relative to the initial state and, thus, lead to an increase in reaction rate. If the transition state is less polar than the initial state, an increase in solvent polarity will decrease the rate. These statements leave unspecified the meaning of *polarity,* although all chemists share intuitive notions of the concept of polarity, based on the expected ability of a solvent to solvate (stabilize) a charged species. Thus, we can readily agree that the solvents water, ethanol, acetonitrile, benzene, and cyclohexane are here listed in order of decreasing polarity, although we might be unable to agree on a ranking of acetic acid, *t*-butyl alcohol, diethyl ether, and acetone. The concept of polarity is a major concern in discussions of medium effects, and it will be treated at several points in this chapter, particularly in Sections 8.2 and 8.4.

Table 8-1 shows the application of the Hughes–Ingold hypothesis to aliphatic nucleophilic reactions of various charge types. These predictions are borne out by observations on many reactions. It should be noted[6,7] that the Hughes–Ingold rule

TABLE 8-1.   Predicted Solvent Effects on Rates of S_N Reactions

Mechanism	Initial state	Transition state	Relative charge in transition state	Rate effect of increase in solvent polarity
S_N2	$Y^- + RX$	$^\delta{}^-Y\text{---}R\text{---}X^{\delta-}$	Dispersed	Small decrease
	$Y + RX$	$^{\delta+}Y\text{---}R\text{---}X^{\delta-}$	Increased	Large increase
	$Y^- + RX^+$	$^\delta{}^-Y\text{---}R\text{---}X^{\delta+}$	Reduced	Large decrease
	$Y + RX^+$	$^{\delta+}Y\text{---}R\text{---}X^{\delta+}$	Dispersed	Small decrease
S_N1	RX	$^{\delta+}R\text{---}X^{\delta-}$	Increased	Large increase
	RX^+	$^{\delta+}R\text{---}X^{\delta+}$	Dispersed	Small decrease

is based on solvation stabilization and is, therefore, a statement about $\Delta H^\ddagger$; possible changes in $\Delta S^\ddagger$ may alter the results.

Ultimately physical theories should be expressed in quantitative terms for testing and use, but because of the complexity of liquid systems this can only be accomplished by making severe approximations. For example, it is often necessary to treat the solvent as a continuous homogeneous medium characterized by bulk properties such as dielectric constant and density, whereas we know that the solvent is a molecular assemblage with short-range structure. This is the basis of the current inability of physical theories to account satisfactorily for the full scope of solvent effects on rates, although they certainly can provide valuable insights and they undoubtedly capture some of the essential features and even cause–effect relationships in solution kinetics. Section 8.3 discusses physical theories in more detail.

Empirical Correlations

Another method for studying solvent effects is the extrathermodynamic approach that we described in Chapter 7 for the study of structure–reactivity relationships. For example, we might seek a correlation between $\log(k_A/k_A^0)$ for a reaction A carried out in a series of solvents and $\log(k_R/k_R^0)$ for a reference or model reaction carried out in the same series of solvents. A linear plot of $\log(k_A/k_A^0)$ against $\log(k_R/k_R^0)$ is an example of a linear free energy relationship (LFER). Such plots have in fact been made. As with structure–reactivity relationships, these solvent–reactivity relationships can be useful to us, but they have limitations.

The benefit of such LFERs is that they establish patterns of regular behavior, isolating apparent simplicity and defining normal or expected reactivity. Against such patterns it becomes possible to detect widely deviant or unexpected behavior. As we saw in Chapter 7, we cannot expect great generality from the extrathermodynamic approach, so it may be necessary to define numerous model processes so as to fit a full range of situations.

The essential weakness of the correlation approach is that it lacks a "linkage" to molecular events. A correlation is not a cause–effect relationship. Nevertheless, with sufficient weight of evidence it becomes reasonable to seek an underlying physical meaning to these solvent LFER, and most of the recent research on solvent effects has been concerned with this issue. Section 8.4 is devoted to this type of investigation.

8.2 SOLVENT PROPERTIES

Physical Properties

Table 8-2 lists several physical properties pertinent to our concern with the effects of solvents on rates for 40 common solvents. The dielectric constant ε is a measure of the ability of the solvent to separate charges; it is defined as the ratio of the electric permittivity of the solvent to the permittivity of the vacuum. (Because physicists use the symbol ε for permittivity, some authors use D for dielectric constant.) Evidently ε is dimensionless. The dielectric constant is the property most often associated with the polarity of a solvent; in Table 8-2 the solvents are listed in order of increasing dielectric constant, and it is evident that, with a few exceptions, this ranking accords fairly well with chemical intuition. The dielectric constant is a bulk property.

The dipole moment μ is a molecular property defined as the product of charge (usually just a fraction of the electronic change, of course) and distance between the centers of positive and negative charge in the molecule. The dipole moment is usually expressed in debyes (D), where $1\,D = 10^{-18}$ esu; in SI units $1\,D = 3.3356 \times 10^{-30}$ C-m, so, for example, the dipole moment of water is 1.84 D or 6.14 in units of 10^{-30} C-m.[8] Again a rough correspondence is seen between this property of a molecule and its "polarity," though ε and μ are not precisely correlated.

The molar refraction, R_M, is a measure of the size of a molecule. It is calculated with Eq. (8.5), the Lorenz–Lorentz equation, where n, d, and M are the refractive index, the density, and the molecular weight, respectively.

$$R_M = \frac{M}{d} \cdot \frac{n^2 - 1}{n^2 + 2} \tag{8-5}$$

R_M is an estimate of the volume occupied by the molecules (not the free space) per mole.

The three properties ε, μ, and R_M are related. One way to determine μ is by means of measurements of the dielectric constant of dilute solutions of the molecule in an inert solvent. Equation (8-6) was derived by Debye.[9]

$$P_M = \frac{M}{d} \cdot \frac{\varepsilon - 1}{\varepsilon + 2} = \frac{4\pi N}{3}\left(\alpha + \frac{\mu^2}{kT}\right) \tag{8-6}$$

The quantity on the left side of the equation is called the *molar polarization,* and this expression is the Clausius–Mosotti equation. On the right side the quantity α is the polarizability, which measures the ease with which an induced moment is

TABLE 8-2. Physical Properties of Some Solvents[a]

No.	Solvent	ε	μ/D	R_M/cm^3	$\gamma/dyn \cdot cm^{-1}$
1	n-Pentane	1.84	0	25.3	15.5
2	n-Hexane	1.89	0.08	29.9	17.9
3	n-Heptane	1.92	0	34.6	19.8
4	Cyclohexane	2.02	0	27.8	24.4
5	1,4-Dioxane	2.21	0	21.7	32.9
6	Carbon tetrachloride	2.23	0	26.7	26.2
7	Benzene	2.28	0	26.2	28.2
8	Toluene	2.38	0.36	31.1	27.9
9	Diethyl ether	4.34	1.15	22.5	16.5
10	Chloroform	4.70	1.87	21.5	26.5
11	Bromobenzene	5.40	1.70	34.0	35.7
12	Chlorobenzene	5.62	1.69	31.1	32.7
13	Ethyl acetate	6.02	1.78	22.1	23.2
14	Acetic acid	6.19	1.74	13.0	20.4
15	Methyl acetate	6.7	1.72	17.5	24.1
16	Tetrahydrofuran	7.32	1.63	19.9	26.9
17	Dichloromethane	8.9	1.60	16.3	27.3
18	t-Butyl alcohol	10.9	1.66	22.2	20.0
19	Pyridine	12.3	2.19	24.1	36.3
20	Benzyl alcohol	13.1	1.71	32.5	39.0
21	2-Methoxyethanol	16	2.2	19.6	30.8
22	n-Butyl alcohol	17.1	1.66	22.2	24.2
23	i-Butyl alcohol	17.7	1.64	22.4	22.5
24	i-Propyl alcohol	18.3	1.66	17.6	20.8
25	n-Propyl alcohol	20.1	1.68	17.5	23.4
26	Acetone	20.7	2.88	16.2	22.9
27	Acetic anhydride	21	2.8	22.4	31.9
28	Ethanol	24.3	1.69	14.9	21.8
29	Methanol	32.6	1.70	8.2	22.4
30	Nitrobenzene	35	4.22	32.9	42.8
31	Acetonitrile	36.2	3.92	11.2	28.5
32	N,N-Dimethylformamide	36.7	3.86	20.0	35.2
33	Ethylene glycol	37.7	2.28	14.5	48.1
34	N,N-Dimethylacetamide	37.8	3.81	24.4	33.3
35	Nitromethane	38.6	3.46	12.5	50.7
36	Glycerol	42.5	2.56	20.5	62.5
37	Dimethyl sulfoxide	49	3.96	20.1	42.8
38	Formic acid	58	1.41	8.6	37.1
39	Water	78.5	1.84	3.7	71.8
40	Formamide	110	3.7	10.7	58.5

Source: Most of the data in this table are from References 197 and 198.
[a]At 25°C or room temperature.

produced under the influence of an electric field. The permanent dipole moment μ is determined by plotting P_M against $1/T$.

The connection between the molar polarization P_M and the molar refraction R_M is through Maxwell's theory of electromagnetism, according to which $\varepsilon = n^2$ (at low-frequency fields). This is the basis for considering the molar refraction a measure of polarizability.

The surface tension γ is a measure of the work required to create unit area of surface from molecules in the bulk; it is expressed in ergs per square centimeter or dynes per centimeter. The surface tension is a "bulk" property, not a molecular property. There appears to be some trend of γ with other measures of polarity, but a lower limit of γ is reached with very nonpolar liquids; this limit (evidently about 15 dyn/cm) reflects the ever-present dispersion force between the molecules of liquid.

The bulk properties of mixed solvents, especially of binary solvent mixtures of water and organic solvents, are often needed. Many dielectric constant measurements have been made on such binary mixtures.[10,11] The surface tension of aqueous binary mixtures can be quantitatively related to composition.[12]

In later sections additional solvent properties will be introduced as needed.

Intermolecular Forces

In a solution of a solute in a solvent there can exist noncovalent intermolecular interactions of solvent–solvent, solvent–solute, and solute–solute pairs. The non-covalent attractive forces are of three types, namely, electrostatic, induction, and dispersion forces. We speak of forces, but physical theories make use of inter-molecular energies. Let $V(r)$ be the potential energy of interaction of two particles and $F(r)$ be the force of interaction, where r is the interparticle distance of sepa-ration. Then these quantities are related by

$$V(r) = \int_r^\infty F(r)dr$$

$$F(r) = -\frac{dV(r)}{dr}$$

Thus, for example, if $V(r)$ were proportional to r^{-n}, $F(r)$ would be proportional to $r^{-(n+1)}$. It is conventional to take as the zero of potential energy the state in which the particles are infinitely separated. A negative $V(r)$ is attractive, a positive value is repulsive. We are interested in the dependence of $V(r)$ on r. The following treatment is drawn from Hirschfelder et al. (13).

Consider two particles, 1 and 2, in the absence of a surrounding medium. The *electrostatic* interactions consist of interactions between moments of ions or polar molecules. These moments are charges (C), dipole moments (μ), and quadrupole moments (Q). Except for the charge–charge interactions, the electrostatic potential energies depend upon the mutual orientation of the moments; however, the average potential energy $\overline{V}$, which is the angle-dependent potential energy averaged over all angles, weighted by Boltzmann factors, is dependent only on the interparticle distance. The potential energy functions, with subscripts indicating the type of interactions, are

$$\overline{V}_{C,C} = +\frac{C_1 C_2}{r} \tag{8-7}$$

$$\overline{V}_{C,\mu} = -\frac{1}{3kT}\frac{C_1^2\mu_2^2}{r^4} \tag{8-8}$$

$$\overline{V}_{C,Q} = -\frac{1}{20kT}\frac{C_1^2Q_2^2}{r^6} \tag{8-9}$$

$$\overline{V}_{\mu,\mu} = -\frac{2}{3kT}\frac{\mu_1^2\mu_2^2}{r^6} \tag{8-10}$$

$$\overline{V}_{\mu,Q} = -\frac{1}{kT}\frac{\mu_1^2Q_2^2}{r^8} \tag{8-11}$$

$$\overline{V}_{Q,Q} = -\frac{7}{40kT}\frac{Q_1^2Q_2^2}{r^{10}} \tag{8-12}$$

In Eq. (8-7), which is Coulomb's law, the charges are to be accompanied with their signs. Because of the high-order reciprocal dependence on distance in Eqs. (8-11) and (8-12), these quadrupolar interactions are usually negligible. For uncharged polar molecules the dipole–dipole interaction of Eq. (8-10), which has the r^{-6} dependence, is the most important contributor to the electrostatic potential energy.

The *induction* (polarization) forces arise from the effect of a moment in a polar molecule inducing a charge separation in an adjacent molecule. The average potential energy functions are

$$\overline{V}_{C,\mathrm{ind}\mu} = \frac{C_1^2\alpha_2^2}{2r^4} \tag{8-13}$$

$$\overline{V}_{\mu,\mathrm{ind}\mu} = -\frac{\mu_1^2\alpha_2^2}{r^6} \tag{8-14}$$

where α is the polarizability. Notice the r^{-6} dependence in the dipole-induced dipole interaction energy.

The *dispersion* (London) force is a quantum mechanical phenomenon. At any instant the electronic distribution in molecule 1 may result in an instantaneous dipole moment, even if 1 is a spherical nonpolar molecule. This instantaneous dipole induces a moment in 2, which interacts with the moment in 1. For nonpolar spheres the *induced dipole-induced dipole* dispersion energy function is

$$V_{\mathrm{disp}} = -\frac{3}{2}\left(\frac{I_1I_2}{I_1 + I_2}\right)\frac{\alpha_1\alpha_2}{r^6} \tag{8-15}$$

where I_1, I_2 are ionization potentials. Again we see the r^{-6} distance dependence. Observe also the dependence on polarizability, which appears in any expression describing an induced charge separation.

As the distance between the two particles varies, they are subject to these long-range r^{-n} attractive forces (which some authors refer to collectively as van der Waals forces). Upon very close approach they will experience a repulsive force due to electron–electron repulsion. This repulsive interaction is not theoretically well characterized, and it is usually approximated by an empirical reciprocal power of distance of separation. The net potential energy is then a balance of the attractive and repulsive components, often described by Eq. (8-16), the Lennard–Jones 6–12 potential.

$$V(r) = 4V_{\text{min}} \left[\left(\frac{r_0}{r} \right)^{12} - \left(\frac{r_0}{r} \right)^{6} \right] \qquad (8\text{-}16)$$

In Eq. (8-16), V_{min} is the minimum in the potential well, which is where $r = r_e$, the equilibrium interparticle distance, and r_0 is the value of r when $V(r) = 0$. Figure 8-1 is a plot of Eq. (8-16). The r^{-12} term is the repulsive contribution, and the attractive term has the r^{-6} dependence that we saw in Eqs. (8-10), (8-14), and (8-15).

If we now transfer our two interacting particles from the vacuum (whose dielectric constant is unity by definition) to a hypothetical continuous isotropic medium of dielectric constant $\varepsilon > 1$, the electrostatic attractive forces will be attenuated because of the medium's capability of separating charge. Quantitative theories of this effect tend to be approximate, in part because the medium is not a structureless continuum and also because the bulk dielectric constant may be an inappropriate measure on the molecular scale. Further discussion of the influence of dielectric constant is given in Section 8.3.

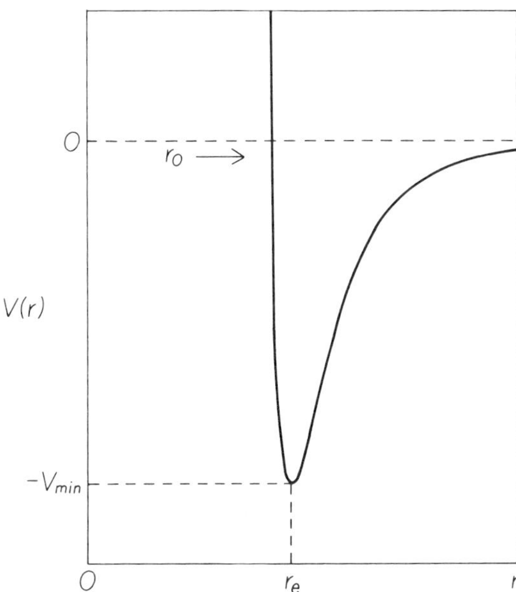

Figure 8-1. A plot of Eq. (8-16), the Lennard–Jones 6–12 potential energy function.

Chemical Interactions

Chemists often call upon certain chemical types of interaction to account for solvent–solvent, solvent–solute, or solute–solute interaction behavior, and we should consider how these chemical interactions are related to the long-range noncovalent forces discussed above. The important chemical interactions are charge transfer, hydrogen bonding, and the hydrophobic interaction.

The *charge-transfer* concept of Mulliken[14,15] was introduced to account for a type of molecular complex formation in which a new electronic absorption band, attributable to neither of the isolated interactants, is observed. The iodine (solute)–benzene (solvent) system studied by Benesi and Hildebrand[16] shows such behavior. Let D represent an interactant capable of functioning as an electron donor and A an interactant that can serve as an electron acceptor. The ground state of the 1:1 complex of D and A is described by the wave function ψ_N:

$$\psi_N = a\psi_0(D,A) + b\psi_1(D^+\!\!-\!\!-\!\!-A^-)$$

In this representation (D,A) is often called the *no-bond form*, but ψ_0 is defined to include all of the noncovalent interactions. $(D^+\!\!-\!\!-\!\!-A^-)$ is the *dative form*, and ψ_1 represents a covalent contribution to the complex; the physical concept is that an electron is transferred from D to A. The resulting product of the interaction is called a *charge-transfer (CT) complex*, or an *electron donor–acceptor (EDA) complex*. The parameters a and b are related by the normalization $a^2 + b^2 + 2abS_{01} = 1$, where S_{01} is the overlap integral between ψ_0 and ψ_1. For weak complexes S_{01} may be negligible, but in general the fractions of the no-bond (F_0) and dative (F_1) structures contributing to the complex are as follows:

$$F_0 = a^2 + abS_{01}; \qquad F_1 = b^2 + abS_{01}$$

The wave function for the excited state of the complex is

$$\psi_V = a^\star\psi_1(D^+\!\!-\!\!-\!\!-A^-) - b^\star\psi_0(D,A)$$

where $a^\star \approx a$ and $b^\star \approx b$. Normally, $a^2 \gg b^2$, so the no-bond structure makes the principal contribution to the ground state, whereas the dative structure makes the main contribution to the excited state ($a^{\star 2} \gg b^{\star 2}$).

There has been much discussion of the relative contributions of the no-bond and dative structures to the strength of the CT complex. For most CT complexes, even those exhibiting intense CT absorption bands, the dative contribution to the complex stability appears to be minor, and the interaction forces are predominantly the noncovalent ones. However, the readily observed absorption effect is an indication of the CT phenomenon. It should be noted, however, that electronic absorption shifts are possible, even likely, consequences of intermolecular interactions of any type, and their characterization as CT bands must be based on the nature of the spectrum and the structures of the interactants. This subject is dealt with in books on CT complexes.[17,18]

Turning to *hydrogen bonding,* the obvious requirements are an acid (proton donor) and base (proton acceptor), these terms not being restricted to the range of acid–base strengths exhibited by measurable pK_a values in water. Pimentel and Mc-Clellan[19] give this definition of a hydrogen bond (H bond): a hydrogen bond exists when (1) there is evidence of a bond and (2) this bond involves a hydrogen already bonded to another atom. In schematic form:

$$A—H + B \rightleftharpoons A—H \cdots B$$

where the H bond is denoted by dots. This is a partial proton transfer from A to B, but it can also be viewed as a partial electron-pair transfer from B to A, so H bonding can be interpreted as a form of CT interaction, though the strength of the H bond is largely due to the electrostatic (dipole–dipole) interaction. H bonding is an important phenomenon in solution chemistry, and the structures of hydroxylic solvents, particularly of water, are largely controlled by this mode of interaction.

Hydrophobicity ("water-hate") can dominate the behavior of nonpolar solutes in water. The key observations are (1) that very nonpolar solutes (such as saturated hydrocarbons) are nearly insoluble in water and (2) that nonpolar solutes in water tend to form molecular aggregates. Some authors refer to item 1 as the *hydrophobic effect* and to item 2 as the *hydrophobic interaction.* Two extreme points of view have been taken to account for these observations.

One of these explanations is based upon the structure of water. This has developed in large part from the concept of Frank and Evans[20] that nonpolar solutes induce in water a highly ordered local solvent structure, with the consequence that dissolution is entropically unfavorable. Kauzmann[21] on this basis developed an interpretation of the hydrophobic interaction as driven by the release of structured water as the solutes associate, the driving force being mainly a favorable entropy change. This structural interpretation of hydrophobic phenomena has been reviewed by Richards,[22] Tanford,[23] and Israelachvili.[24] Because this "classical" hydrophobic interaction is driven by a favorable entropy change, the sign and magnitude of ΔS^0 have been taken by many workers as diagnostic criteria for the importance of the hydrophobic interaction in an association equilibrium. This view is undergoing change, as will be noted below.

The other viewpoint is the cavity theory. Cavity models have a long history, especially in theories of solubility,[25] but the most extensive application to hydrophobic interactions has been by Sinanoglu.[26–29] (In general, we could speak of *solvophobic* interactions.) The essence of this model is that dissolution of a solute in a solvent requires creation of a cavity in the solvent, at the expense of energy $A\gamma$, where A is the cavity surface area and γ is the solvent surface tension; the solute molecule is then inserted in the cavity, and solute–solvent interactions follow. When two solute molecules associate, the change in free energy associated with the coalescence of two cavities into one is $\Delta A\gamma$, where ΔA, the change in cavity surface area, is negative. The hydrophobic interaction, on this view, is a consequence of the very large surface tension of water (itself related to the structure of water). Both the cavity effect and the solute–solvent interaction term contribute to the observed free energy change.

Attempts have been made to distinguish between these theories on the basis of the ΔH^0 and ΔS^0 values anticipated for the two theories,[30] but it may be illusory to think of them as independent alternatives. The cavity model has been criticized[31] on the basis that it cannot account for certain observations such as the denaturing effect of urea, but it must be noted that the cavity theory includes not only the cavity term $\Delta A\gamma$, but also a term (or terms) for the interaction of the solutes and the solvent. A more cogent objection might be to the extension of the macroscopic concepts of surface area and tension to the molecular scale. A demonstration of the validity of the cavity concept has been made with silanized glass beads, which aggregate in polar solvents and disperse in nonpolar solvents.[32]

Most of the work on the hydrophobic interaction has dealt with equilibria, and we can justify its application to rates on the basis of the equilibrium assumption of transition state theory, according to which the initial state and transition state are in quasi-equilibrium. The theory of the hydrophobic effect deals with nonpolar solutes. Let us consider the important extension to solutes of moderate polarity. The "classical" solvent-structure interpretation is that a favorable entropy change is the major driving force for hydrophobic association in water. If water structuring is the predominant feature, addition of organic solvents should result in a more negative ΔS^0 as the solvent structure becomes less important. Crothers and Ratner[30] observed such an effect for complex formation between actinomycin and deoxyguanosine. Albergo and Turner[33] found that alcohol–water mixtures gave thermodynamic quantities consistent with the solvent structure view of the hydrophobic interaction, whereas mixtures of water and dipolar aprotic solvents gave, for double helix formation, a less favorable ΔH^0 and more favorable ΔS^0. Changes consistent with structural control were seen for methylene blue dimerization,[34] whereas the complex of dopamine with ATP, although destabilized by the incorporation of organic solvent, yields values of ΔH^0 and ΔS^0 that are not consistent with classical hydrophobic interaction.[35]

Notice that these interactants are not "nonpolar" compounds. Jencks[36] has given a qualitative description of hydrophobic interaction that incorporates both the cavity and the solvent structure models. The extent to which the solvent structure is modified after insertion of the solute molecule into the cavity is determined by the polarity of the solute, with the result that the driving force for association may appear as either a favorable enthalpy change or a favorable entropy change. It is, therefore, unwise to rely on enthalpy and entropy changes as decisive criteria of mechanism for the driving force of the association.[37] The view that the entropic effect is the dominant one is being questioned. Carbon-13 relaxation time measurements suggest that motional restriction of solute molecules in aqueous solution may account for most of the anomalously large entropy loss on dissolution of a gaseous solute.[38] Cramer[39] points out that this entropy change is not very sensitive to solute size, and it, therefore, seems not to be a consequence of solvent structuring. The transfer of a nonpolar solute from water to an organic solvent may be more indicative of the solute's *lipophilicity* than of its *hydrophobicity*.[39,40] Mirejovsky and Arnett[41] have concluded that the low solubility of nonpolar solutes in water is not due to the large unfavorable entropy change, but rather to the energy required for cavity formation. Thus, the standard picture of the dominance of water structural effects in the hydrophobic interaction is being modified.

An associated feature of this problem is the possibility of enthalpy–entropy compensation. Jencks[36,37] has cautioned against confusing cause and effect, pointing out that ΔH^0 and ΔS^0 do not necessarily "cause" ΔG^0, and several authors[36,37,40,41] have noted that changes in ΔH^0 and ΔS^0 produced by structural changes in the solvent may compensate to give little or no change in ΔG^0. Ben-Naim[42,43] and Grunwald[44] have given theoretical reasons that solvent structural changes induced by a process cannot contribute to the free energy change for the process, although they may affect enthalpy, entropy, and other thermodynamic quantities. As Ben-Naim[42] concludes, ΔG^0 for the hydrophobic interaction may depend upon the *structure* of water, but not upon structural *changes* that occur as a result of the hydrophobic interaction.

We will return to the cavity model in Section 8.3, giving a description in terms of the so-called solubility parameter.

Classification of Solvents

The grouping of solvents into classes with common characteristics can be useful in focusing attention on features that may play a role in experimental solvent effects. Reichardt's[5, Chap. 3] review of classification schemes is thorough.

Structural classes are often defined, as in this scheme:

1. Aliphatic hydrocarbons.
2. Aromatic hydrocarbons.
3. Halogenated hydrocarbons.
4. Hydroxylic solvents.
5. Nitrogen compounds.
6. Oxygen compounds.
7. Sulfur compounds.

Such a list can obviously be subdivided as appropriate.

Classification according to Brønsted acid–base properties is useful.

1. Proton donors (protogenic solvents): H_2SO_4, carboxylic acids.
2. Proton acceptors (protophilic solvents): amines, ethers.
3. Proton donor/acceptors (amphoteric solvents): water, alcohols.
4. Aprotic solvents: hydrocarbons, CCl_4.

A scheme based on H-bonding properties is similar, but leads to a somewhat different identification of characteristic behavior.

1. H-bond donors: $CHCl_3$, CH_2Cl_2.
2. H-bond acceptors: carbonyls, ethers, esters, aromatic hydrocarbons, tertiary amines.
3. H-bond donor/acceptors: water, alcohols, carboxylic acids, amines, amides.
4. Non-H-bonding solvents: CCl_4, saturated hydrocarbons.

The aprotic class of solvents can be divided into the nonpolar (apolar) aprotics and the polar (dipolar) aprotics. Dipolar aprotic solvents play an important role in solvent effect studies, in part because of their solvation effects on anion nucleophilicity. Parker[45] defined dipolar aprotic solvents as aprotic solvents having dielectric constants greater than about 15. Common examples are acetone, *N,N*-dimethylformamide (DMF), *N,N*-dimethylacetamide (DMA), acetonitrile, nitromethane, and dimethyl sulfoxide (DMSO). Their dipole moments are typically greater than about 2 D. The designation aprotic for most of these solvents is an approximate description, for under the influence of a sufficiently strong base these molecules can donate a proton, because the functional group responsible for the large dipole moment is acid strengthening. Even in the pure solvent some dissociation may occur; for example, the autoprotolysis constant of acetonitrile is about 3×10^{-27}.

Dack (4) has suggested that solvents be classified on the basis of the product $\varepsilon\mu$, the *electrostatic factor*. There seems to be no theoretical basis for this function [indeed, the dipole moment appears as μ^2 in physical theory, as in Eqs. (8-10) and (8-6)], but $\varepsilon\mu$ does contain more information than either of the quantities alone,

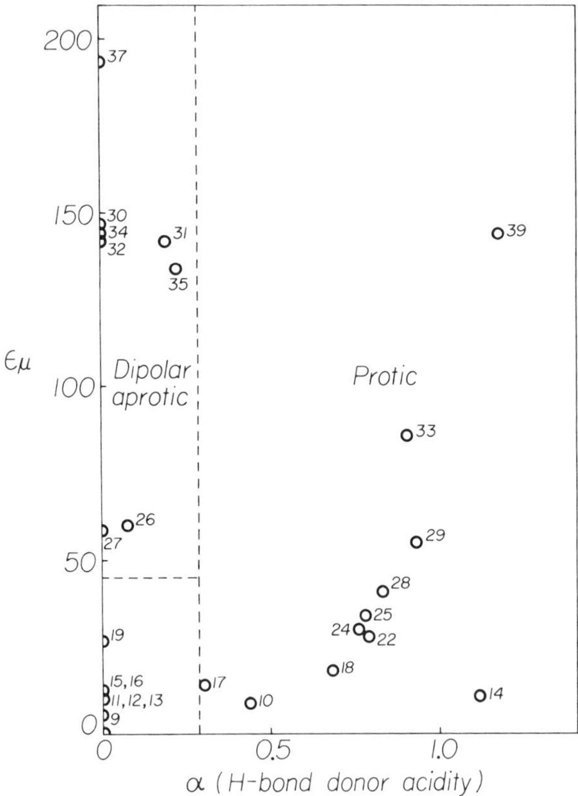

Figure 8-2. Plot of $\varepsilon\mu$ (from Table 8-2) against α (from Reference 46), a measure of H-bond donor ability. The numbers identify solvents in Table 8-2.

because ε and μ are not closely correlated (see Table 8-2). However, the scheme possesses the weakness that overlap occurs; for example, $\varepsilon\mu$ for water is similar to that for the dipolar aprotic solvents. If, however, $\varepsilon\mu$ is combined with a second dimension measuring the protic quality of the solvent, the solvent classes are dispersed in two-dimensional space. Figure 8-2 is a plot of $\varepsilon\mu$ against α, an empirical measure of H-bond donor ability[46] that will be described in Section 8.4. Now the dipolar aprotic solvents form a distinct class, and the behavior of water is seen to be a normal extension of the class of hydroxylic solvents.

Instead of using a chemical approach, it is possible to use statistics to organize solvents.[5, Chap. 3] Chastrette et al.[47] applied principal component analysis to data on 83 solvents, defining each solvent as a point in the eight-dimensional space created with the eight variables μ, R_M, n, a dielectric constant function $f(\varepsilon)$, the solubility parameter δ, boiling point (bp), and the energies of the highest occupied molecular orbital (E_H) and the lowest unoccupied molecular orbital (E_L). Principal component analysis reduced this eight-dimensional space to three dimensions with the loss of only 18% of the information. One principal component, F_1, was strongly correlated with M_R, n, and E_H; it is, therefore, associated with polarizability. F_2 was correlated with $f(\varepsilon)$, μ, and bp and was, therefore, considered to reflect polarity. The third component, F_3, is dominated by E_L and was identified with electron affinity. The manner in which the principal components were clustered led to the definition of these nine classes.

1. Aprotic dipolar (AD).
2. Aprotic highly dipolar (AHD).
3. Aprotic highly dipolar and highly polarizable (AHDP).
4. Aromatic apolar (ARA).
5. Aromatic relatively polar (ARP).
6. Electron-pair donor (EPD).
7. H bonding (HB).
8. H bonding strongly associated (HBSA).
9. Miscellaneous (MISC).

Some of the solvent assignments seem capricious, such as CCl_4 in ARP, CF_3COOH in AD, and $CHCl_3$ and $C_6H_5NH_2$ in MISC. From one point of view, it is a triumph that a purely statistical treatment, devoid of chemical experience or intuition, can generate a scheme more or less in accord with chemical concepts. However, it does not seem to be superior to the simple application of chemical ideas.

Solvent Polarity

The central role of the concept of polarity in chemistry arises from the electrical nature of matter. In the context of solution chemistry, solvent polarity is the ability of a solvent to stabilize (by solvation) charges or dipoles.[7,48] We have already seen that the physical quantities ε (dielectric constant) and μ (dipole moment) are quantitative measures of properties that must be related to the qualitative concept of polarity; ε has served as a numerical estimate of polarity more frequently than any other property. The product $\varepsilon\mu$, as used in Fig. 8-2, also conveys a quantitative sense of polarity. Nevertheless, despite very convincing overall trends of these measures with chemical intuition, serious exceptions reveal that no general quantitative measure of solvent polarity is to found among the physical properties cited earlier. For example, consider dioxane (solvent 5 in Table 8-2), which by the measures ε and μ must be as nonpolar as a hydrocarbon. Yet dioxane is miscible with water in all proportions! This apparent inconsistency is a consequence of the symmetry of the dioxane molecule, its zero molecular dipole moment being the resultant of two group moments that exactly compensate. The water solubility of dioxane must be due to its H-bond donor properties. Thus, we see that exceptions must be made on chemical bases.

The preceding example implied that water solubility is related to polarity, and indeed the equating of hydrophilic character with polarity (and of hydrophobic character with nonpolarity) is often made. Thus, we may add water solubility to the list of pertinent physical (or chemical) properties related to polarity. If a substance is infinitely soluble in water, high polarity is usually inferred.

Closely related to water solubility as a polarity measure is the partition coefficient of a substance between water and an immiscible organic solvent. Most commonly the organic solvent is selected to be n-octanol, and the symbol P is given to the octanol/water partition coefficient. Then $\log P$ is a quantitative measure of hydrophobicity and, therefore, of nonpolarity. Table 8-3 gives $\log P$ values for many of

TABLE 8-3. Octanol/Water Partition Coefficients of Solvents

No.[a]	Solvent	Log P	No.	Solvent	Log P
5	1,4-Dioxane	−0.42	22	n-Butyl alcohol	+0.88
6	Carbon tetrachloride	+2.64	23	i-Butyl alcohol	+0.74
7	Benzene	+2.14	24	i-Propyl alcohol	0.00
8	Toluene	+2.73	25	n-Propyl alcohol	+0.34
9	Diethyl ether	+0.80	26	Acetone	−0.24
10	Chloroform	+1.98	28	Ethanol	−0.32
11	Bromobenzene	+2.99	29	Methanol	−0.74
12	Chlorobenzene	+2.84	30	Nitrobenzene	+1.86
13	Ethyl acetate	+0.70	31	Acetonitrile	−0.34
14	Acetic acid	−0.24	32	N,N-Dimethylformamide	−0.73
15	Methyl acetate	+0.18	33	Ethylene glycol	−1.93
16	Tetrahydrofuran[b]	+0.22	35	Nitromethane	−0.21
18	t-Butyl alcohol	+0.37	36	Glycerol	−2.56
19	Pyridine	+0.65	37	Dimethyl sulfoxide	−2.03
20	Benzyl alcohol	+1.10	38	Formic acid	−0.54
21	2-Methoxyethanol	−0.71	39	Water	−1.15
			40	Formamide	−1.64

Source: Reference 49 except as noted.
[a]Solvent numbers as in Table 8-2.
[b]From Reference 50.

the solvents in Table 8-2.[49,50] There is a very rough inverse relationship of log P with ε, but the two properties are so poorly correlated that they must be responding in different ways to the solvent.

When the range of chemical types is restricted, regular behavior is often observed. For example, one might choose to study a series of hydroxylic solvents, thus holding approximately constant the H-bonding capabilities within the series. This is a motivation, also, for solvent studies in a series of binary mixed solvents, often an organic–aqueous mixture whose composition may be varied from pure water to pure organic. Mukerjee et al.[51] defined a quantity H for hydroxylic and mixed hydroxylic–water solvents by Eq. (8-17).

$$H = \frac{\text{molar concentration of OH groups}}{\text{molar concentration of OH in } H_2O} = \frac{[OH]}{55.5} \qquad (8\text{-}17)$$

For example, [OH] in pure n-propyl alcohol ($d^{25} = 0.8016$, MW $= 60.09$) is 13.34, so $H = 0.240$. It is found that H is a linear function of ε.

(In Section 8.4 we will encounter many empirical measures of solvent polarity. These are empirical in the sense that they are model dependent; that is, they are defined in terms of a particular standard reaction or process. Thus, these empirical measures play a role in the study of solvent effects exactly analogous to that of the substituent constants in Chapter 7.)

The widely observed lack of precise correlations among properties that chemical experience leads us to believe are somehow related to polarity has two important meanings: First, it tells us that no single property may be taken as a generally applicable measure of polarity. Second, it conveys the message that some combination of the properties may capture the essence of what we mean by polarity.

Solvation

The interaction between a solute species and solvent molecules is called *solvation,* or hydration in aqueous solution. This phenomenon stabilizes separated charges and makes possible heterolytic reactions in solution. Solvation is, therefore, an important subject in solution chemistry. The solvation of ions has been most thoroughly studied.[48,52]

First, let us consider the formation of ions from covalently bound species, i.e., the heterolytic cleavage of the covalent (or partially covalent) bond. Charge separation under the influence of the solvent generates an *ion pair* in a process called *ionization;* this ion pair may then separate into free ions in a *dissociation* step (Eq. 8-18).

$$\text{R—X} \underset{\longleftarrow}{\overset{\text{ionization}}{\rightleftharpoons}} \text{R}^+\text{X}^- \underset{\longleftarrow}{\overset{\text{dissociation}}{\rightleftharpoons}} \text{R}^+ + \text{X}^- \qquad (8\text{-}18)$$

ion pair

Ionization constant K_i, dissociation constant K_d, and overall dissociation constant K_{RX} are defined:

$$K_i = \frac{[R^+X^-]}{[RX]}; \qquad K_d = \frac{[R^+][X^-]}{[R^+X^-]}$$

$$K_{RX} = \frac{[R^+][X^-]}{c_{RX}}$$

where $c_{RX} = [RX] + [R^+X^-]$. It follows that[53]

$$K_{RX} = \frac{K_i K_d}{1 + K_i}$$

The ionization constant should be a function of the *intrinsic heterolytic ability* (e.g., intrinsic acidity if the solute is an acid HX) and the *ionizing power* of the solvents, whereas the dissociation constant should be primarily determined by the *dissociating power* of the solvent. Therefore, K_d is expected to be under the control of ε, the dielectric constant. As a consequence, ion pairs are not detectable in high-ε solvents like water, which is why the terms *ionization constant* and *dissociation constant* are often used interchangeably. In low-ε solvents, however, dissociation constants are very small and ion pairs (and higher aggregates) become important species. For example, in ethylene chloride ($\varepsilon = 10.23$), the dissociation constants of substituted phenyltrimethylammonium perchlorate salts are of the order 10^{-5}.[54] Overall dissociation constants, expressed as $pK_{RX} = -\log K_{RX}$, for some substances in acetic acid ($\varepsilon = 6.19$) are:[53] perchloric acid, 4.87; sulfuric acid, 7.24; sodium acetate, 6.68; sodium perchlorate, 5.48. Acid–base equilibria in acetic acid have been carefully studied because of the analytical importance of this solvent in titrimetry.

Ionization is obviously important in the S_N1 mechanism of nucleophilic substitution, and indeed two ion pair intermediates have been invoked.[55] These are related as in Eq. (8-19), where (s) represents the solvent.

$$R\!-\!X \rightleftharpoons R^+X^- \rightleftharpoons R^+(s)X^-(s) \rightleftharpoons R^+ + X^- \tag{8-19}$$

The species R^+X^- is called an *internal, contact,* or *intimate ion pair,* and $R^+(s),X^-(s)$, sometimes symbolized $R^+\|X^-$, is an *external* or *solvent-separated ion pair.*

The forces involved in solute–solvent interactions are those described earlier in this section, namely, the electrostatic, induction, and dispersion forces, the balance among these obviously depending upon the particular species. The CT and H-bonding interactions may be important. Now, it is evident that interaction of solvent molecules with a solute molecule or ion will result in a change in the mutual arrangement of the solvating solvent molecules relative to their arrangement in the bulk solution, distant from the solute particle. That is, some modification of the solvent structure takes place in the region immediately adjacent to the solute particle. This region was called the *cosphere* by Gurney.[56] (Gurney used the term *cosphere* more particularly to describe the solvation sphere around a spherical ion.) The cosphere is also called the *solvation shell,* and this may be subdivided (particularly for aqueous solutions) into region A, the primary solvation shell, where the solvent molecules are in intimate contact with the solute particle and are more ordered, by the solute–solvent interaction, than they are in the absence of solute; region B, the secondary shell, where the solvent molecules are next-nearest neighbors of the region A solvent molecules, and where they may be more disordered than in the bulk; and region C, the arrangement of solute molecules in the bulk, distant from any solute particle. The cosphere or solvation shell is also called the *cybotactic region.*[57, p. 262] The problem of the mutual orientation of solvent molecules in the solvation shell, particularly in aqueous solutions, touches many fields of solution chemistry; it is, for example, central to the water–structure model of the hydrophobic effect described earlier in this section. One important conclusion that may be reached, independently of any details of this issue, is that the physical properties characteristic of the bulk solvent may not be appropriate to the solvation shell. In particular, we may reasonably expect that the dielectric constant on the molecular level (often referred to as the *microscopic dielectric constant*) in the region of the solvation shell is different from the measured value in the bulk.

Solvation can be studied by thermodynamic methods, often combined with extrathermodynamic assumptions so as to express results for individual ions (rather than for neutral electrolytes). The solvation energy is the free energy change upon transferring a molecule or ion from the gas phase into a solvent at infinite dilution.[48] This sometimes can be obtained from a consideration of the following processes, written for a 1:1 electrolyte:

$$C^+A^-(\text{crystal}) \rightarrow C^+(g) + A^-(g) \qquad \Delta G^0_{\text{lattice}}$$

$$C^+(g) + A^-(g) \rightarrow C^+(\text{soln}) + A^-(\text{soln}) \qquad \Delta G^0_{\text{solv}}$$

$$C^+A^-(\text{crystal}) \rightarrow C^+(\text{soln}) + A^-(\text{soln}) \qquad \Delta G^0_{\text{soln}}$$

Evidently $\Delta G^0_{\text{soln}} = \Delta G^0_{\text{latt}} + \Delta G^0_{\text{solv}}$. The free energy of solution, ΔG^0_{soln}, is obtained from the molar solubility s, $\Delta G^0_{\text{soln}} = -RT \ln s$; the quantity ΔG^0_{latt} is the lattice energy of the crystal, obtainable as the sublimation energy. ΔG^0_{soln} is often small because the interaction energy within the solid is comparable with the energy of interaction between solute and solvent.[5, Chap. 2; 48] We will return to solvation energies in Section 8.3.

The *coordination number* is the number of solvent molecules in the primary solvation shell. This quantity can be estimated (for ions) by conductance measurements and by NMR.[48,52]

Because the key operation in studying solvent effects on rates is to vary the solvent, evidently the nature of the solvation shell will vary as the solvent is changed. A distinction is often made between general and specific solvent effects, general effects being associated (by hypothesis) with some appropriate physical property such as dielectric constant, and specific effects with particular solute–solvent interactions in the solvation shell. In this context the idea of *preferential solvation* (or selective solvation) is often invoked. If a reaction is studied in a mixed solvent, preferential solvation of a solute by one component of the solvent mixture will lead to a solvation shell composition enriched in the preferred component. (This is called *solvent sorting*.) Preferential solvation effects may be revealed by widely discordant members in an otherwise well-behaved correlation of reactivity with a solvent property or by a correlation that trends in the reverse of the expected direction. An example is the reaction of ethyl bromide and *N,N*-diethylaniline in a series of normal alcohols.[58] This is called a *Menschutkin reaction*.

$$EtBr + C_6H_5NEt_2 \rightarrow C_6H_5N^+Et_3I^-$$

The reactants are neutral, the product is charged, so some charge separation must have occurred in the transition state. According to the Hughes–Ingold hypothesis (Section 8.1), we therefore expect the reaction rate to increase as the solvent polarity is increased. Such behavior is often observed for Menschutkin reactions, as we will see in Section 8.3. In this study, however, it was found that the reactivity decreases as the polarity (measured by the dielectric constant) increases. It can be conjectured that preferential solvation of the polar transition state by the polar alcohol occurs, this selective effect overcoming the general polarity control of the rate. The greater effectiveness of long-chain alcohols indicates that these may interact with the hydrocarbon substituents in the transition state. The very large entropy decreases observed ($\Delta S^{\ddagger} \approx -70$ eu) are consistent with the restriction of solvent molecules in the activation process.

Another example of preferential solvation is provided by the inhibition by dioxane of the aryl alkylation of phenol by *t*-butyl chloride.[59]

$$t\text{-}C_4H_9Cl + C_6H_5OH \rightarrow t\text{-}C_4H_9\text{-}C_6H_4\text{-}OH + HCl$$

The decrease in rate was proportional to the concentration of dioxane in the reaction mixture. An equivalent concentration of *p*-xylene (whose dielectric constant is similar to that of dioxane) produced a smaller decrease, consistent with simple dilution of the reactants. It was, therefore, hypothesized that dioxane forms an H-bonded molecular complex with phenol, the complexed form of the phenol being unreactive. The data could be accounted for with a 2:1 stoichiometry (phenol:dioxane). This argument was supported by experiments with tetrahydrofuran, which also decreased the rate, but which required a 1:1 stoichiometry to describe the rate data.

The remarkable enhancement of anion nucleophilicity in S_N2 reactions carried out in dipolar aprotic solvents is a solvation effect.[45] Solvents like DMF and DMSO are very polar owing to the charge separation indicated in **1** and **2**.

$$O^{\delta-} \qquad\qquad O^{\delta}$$
$$\|\qquad\qquad\qquad \|$$
$$H\text{---}\underset{\delta+}{C}\text{---}NMe_2 \qquad\qquad Me\text{---}\underset{\delta+}{S}\text{---}Me$$

1   **2**

Because the positive pole is relatively "sheltered" from close approach, these solvents cannot effectively solvate anions, which, therefore, display an enhanced nucleophilicity.[60]

8.3. PHYSICAL MODELS OF MEDIUM EFFECTS

Neutral–Neutral Molecule Reactions

We have two cases to consider here. If the reactants are nonpolar and the products are nonpolar, it is a reasonable hypothesis that the transition state also is nonpolar. In this case, the Hughes–Ingold hypothesis (Section 8.1) leads us to expect that the solvent will have little effect on the rate. The decomposition of azo compounds provides an example.[61] The reaction is

$$R-N{=}N-R \rightarrow N_2 + 2R.$$

the final disposition of the free radicals depending upon the reaction conditions; dimerization to R–R may occur. Table 8-4 gives kinetic data of Peterson et al.[62] for the decomposition of 2,2'-azobis-(2-methylpropionitrile) [R $= Me_2(CN)C-$] in some less common solvents. A considerable range of solvent types and polarity is represented; thus, $\varepsilon = 4.9$ for dimethylaniline, whereas $\varepsilon = 64.9$ for propylene carbonate. These kinetic data are very precise, so the small rate differences in the several solvents are real; but the essential result is that the reaction is relatively insensitive to the solvent.

If two neutral reactant molecules yield a polar product, then presumably the transition state will be intermediate in polarity, and we anticipate an increase in rate as the solvent polarity is increased. A quantitative formulation of this case[63] is based on Kirkwood's expression[64] for the free energy of transfer of a dipole of

TABLE 8-4. Kinetics of Decomposition of 2,2'-Azobis-(2-methylpropionitrile) at 67°C

Solvent	$10^5 \ k/s^{-1}$
Diethylene glycol mono-*n*-butyl ether	2.72
Diethylene glycol di-*n*-butyl ether	2.44
Propylene carbonate	2.74
1,2-*bis*-(Benzyloxy)ethane	3.10
Diphenylmethane	2.89
N-Methylpropionamide	2.94
N-Methyl-*N*-benzylaniline	3.27
N,N-Dimethylaniline	3.39

Source: Reference 62.

moment μ from a medium having $\varepsilon = 1$ to a medium of dielectric constant ε,

$$\Delta G_{tr}^0 = -\frac{\mu^2}{r^3}\left(\frac{\varepsilon - 1}{2\varepsilon + 1}\right) \tag{8-20}$$

where r is the molecular radius. This equation accounts only for the electrostatic component, and it treats the solvent as a continuous isotropic medium. We can apply Kirkwood's equation to the reaction of two molecules A and B having dipole moments μ_A and μ_B and forming the transition state $M^{\ddagger}$ with dipole moment μ_M. Writing Eq. (8-3) as

$$\delta_M \Delta G^{\ddagger} = \Delta G_S^{\ddagger} - \Delta G_0^{\ddagger} = -RT \ln \frac{k_s}{k_0} \tag{8-21}$$

TABLE 8-5. Kinetic Data on the Menschutkin Reaction of Triethylamine and Ethyl Iodide at 25°C

No.	Solvent	ε	$10^6 \ k/M^{-1}s^{-1}$	$\Delta G^{\ddagger a}/kcal \ mol^{-1}$
1	Hexane	1.89	0.0135	26.99
2	Cyclohexane	2.02	0.0216	26.60
3	Diethyl ether	4.34	0.359	24.91
4	Carbon tetrachloride	2.23	0.422	24.77
5	1,1,1-Trichloroethane	7.25	3.14	23.60
6	Toluene	2.38	3.37	23.59
7	Cyclohexyl chloride	—	5.21	23.41
8	Cyclohexyl bromide	—	6.15	23.32
9	Benzene	2.28	5.37	23.21
10	Ethyl acetate	6.02	7.78	23.05
11	Dioxane	2.21	11.8	22.72
12	Tetrahydrofuran	7.32	11.7	22.70
13	Ethyl benzoate	6.02	23.8	22.61
14	Chlorobenzene	5.62	19.5	22.53
15	Bromobenzene	5.40	34.4	22.37
16	1,1-Dichloroethane	10.0	23.7	22.30
17	Chloroform	4.70	30.1	22.13
18	Methyl ethyl ketone	18.5	39.6	22.03
19	Iodobenzene	4.49	50.2	22.02
20	Acetone	20.7	65.4	21.62
21	1,2-Dichloroethane	10.4	94.1	21.47
22	Dichloromethane	8.9	79.9	21.42
23	Acetophenone	17.4	164.0	21.35
24	Benzonitrile	25.2	152.0	21.32
25	Propionitrile	28.9	118.0	21.25
26	Nitrobenzene	35	184.0	21.20
27	Dimethyl formamide	36.7	216.0	20.94
28	1,1,2,2-Tetrachloroethane	8.2	312	20.90
29	Acetonitrile	36.2	227	20.64
30	Nitromethane	38.6	333	20.47
31	Propylene carbonate	64.9	684	20.31
32	Dimethyl sulfoxide	49	873	20.06

Source: Reference 65.

$^a\Delta G^{\ddagger}$ is calculated on the mole fraction scale.

where k_0 is the rate constant in a hypothetical medium of $\varepsilon = 1$, and combining Eqs. (8-20) and (8-21) gives

$$\ln k_s = \ln k_0 + \frac{N}{RT}\left(\frac{\varepsilon - 1}{2\varepsilon + 1}\right)\left(\frac{\mu_M^2}{r_M^3} - \frac{\mu_A^2}{r_A^3} - \frac{\mu_B^2}{r_B^3}\right) \tag{8-22}$$

where N is Avogadro's number. Equation (8-22) predicts a linear relationship between $\log k_s$ and $(\varepsilon - 1)/(2\varepsilon + 1)$. Many tests of this prediction have been made. A favorite test reaction is the Menschutkin quaternization reaction:

$$R_3N + RX \rightarrow R_4N^+ + X^-$$

The neutral reactants possess permanent dipoles, the product is ionic, and the transition state must be intermediate in its charge separation, so an increase in solvent polarity should increase the rate. Except for selective solvation effects of the type cited in the preceding section, this qualitative prediction is correct.

Table 8-5 gives data for the Menschutkin reaction of triethylamine and ethyl

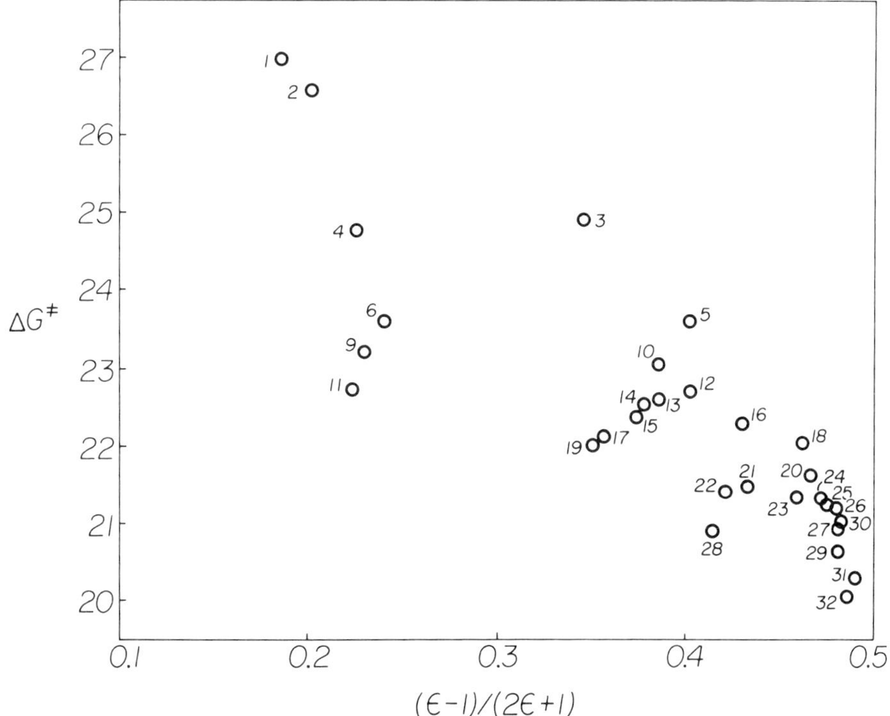

Figure 8-3. Plot of free energy of activation for the Menschutkin reaction $Et_3N + EtI \rightarrow Et_4N^+ + I^-$ against the Kirkwood dielectric constant function. Data are from Table 8-5, where the solvents are identified.

iodide in 32 solvents.[61] The second-order rate constant is expressed on the molar concentration scale, whereas the free energy of activation, $\Delta G^{\ddagger}$, is on the mole fraction scale. [This conversion (which results in some changes in the rank ordering) yields the unitary quantity discussed in Section 6.1; see Eq. (6.29).] A plot of $\Delta G^{\ddagger}$ against the Kirkwood dielectric constant function is shown in Fig. 8-3. Although the expected trend is seen, the adherence to Eq. (8-22) is obviously poor. Aside from the assumptions implicit in Eq. (8-22), selective solvation effects probably contribute to the scatter seen in Fig. 8-3. Certainly the dielectric constant alone is not capable of accounting for the solvent effect on this reaction. When the Menschutkin reaction is carried out in binary mixtures, much better adherence to a straight line is seen, and from the slope of the plot estimates of the dipole moment of the transition state can be made [see Eq. (8-22)]. We will return to the analysis of the data in Table 8-5 later in this section.

Some authors plot $\log k$ or $\Delta G^{\ddagger}$ against $1/\varepsilon$ rather than against the Kirkwood function.[66] Since $1/\varepsilon$ is nearly linearly related[67] to $(\varepsilon - 1)/(2\varepsilon + 1)$, within the assumptions of a theory in which the solvent is treated as a continuum this substitution of variable is not serious. Another approach is to interpret the solvent dependence of the Hammett reaction constant ρ on a dielectric constant function.[68]

Ion-Neutral Molecule Reactions

The quantitative theory of ionic reactions, within the limitations of a continuum model of the solvent, is based on the Born equation for the electrostatic free energy of transfer of an ion from a medium of $\varepsilon = 1$ to the solvent of dielectric constant ε,[69]

$$\Delta G_{\text{tr}}^0 = -\frac{Z_i^2 e^2}{2r_i}\left(1 - \frac{1}{\varepsilon}\right) \tag{8-23}$$

where Z_i is the charge of ion i. For the activation process

$$A + B \rightleftharpoons M^{\ddagger}$$

we apply Eq. (8-23), letting A and B both be ions, just as in the earlier development of Eq. (8-22). The result is[63, p. 427]

$$\ln k_s = \ln k_0 + \frac{Ne^2}{2RT}\left(1 - \frac{1}{\varepsilon}\right)\left(\frac{Z_M^2}{r_M} - \frac{Z_A^2}{r_A} - \frac{Z_B^2}{r_B}\right) \tag{8-24}$$

where $Z_M = Z_A + Z_B$. If A is an ion and B is a neutral molecule, this becomes

$$\ln k_s = \ln k_0 - \frac{NZ_A^2 e^2}{2RT}\left(1 - \frac{1}{\varepsilon}\right)\left(\frac{r_M - r_A}{r_A r_M}\right) \tag{8-25}$$

Since $r_M > r_A$, Eq. (8-25) predicts that k_s will decrease as ε increases. This is the same prediction that we would make with the Hughes–Ingold hypothesis, for the transition state is less polar than the initial state (the ionic charge being dispersed into a greater volume in the transition state).

S_N2 reactions with anionic nucleophiles fall into this class, and observations are generally in accord with the qualitative prediction. Unusual effects may be seen in solvents of low dielectric constant where ion pairing is extensive, and we have already commented on the enhanced nucleophilic reactivity of anionic nucleophiles in dipolar aprotic solvents owing to their relative desolvation in these solvents. Another important class of ion–molecule reaction is the hydroxide-catalyzed hydrolysis of neutral esters and amides. Because these reactions are carried out in hydroxylic solvents, the general medium effect is confounded with the acid–base equilibria of the mixed solvent lyate species. (This same problem occurs with S_N2 reactions in hydroxylic solvents.) This equilibrium is established in alcohol–water mixtures:

$$OH^- + ROH \rightleftharpoons RO^- + H_2O$$

The position of equilibrium depends upon the identity of the alcohol and the composition of the mixture. Figure 8-4, given by Burns and England,[70] shows the

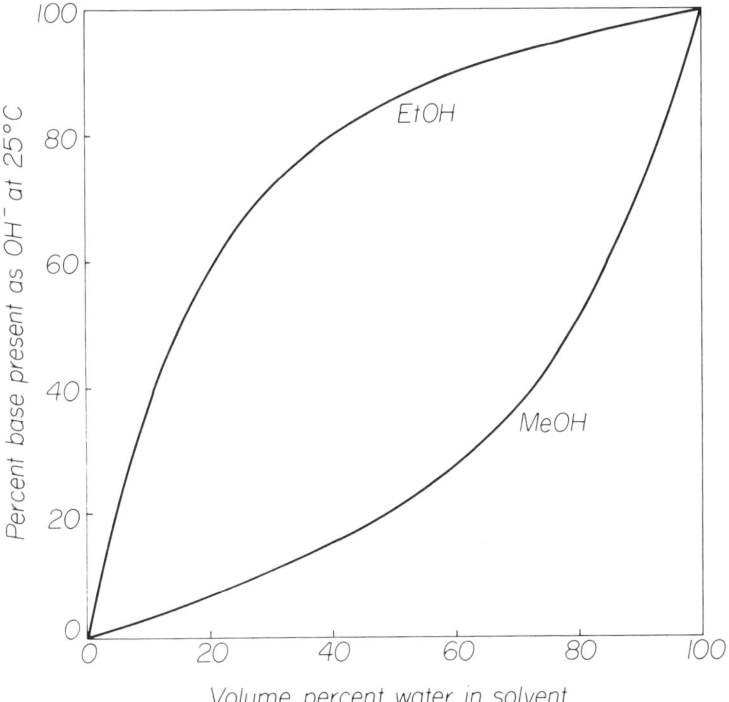

Figure 8-4. Fraction of total base present as hydroxide as a function of the percentage of water in the binary solvent, for ethanol–water and methanol–water mixtures.[70]

profound difference in this equilibrium for ethanol–water and methanol–water mixtures. For example, in a methanol solution containing 20% water, about 6% of total base present is hydroxide, the remainder being methoxide; in the corresponding ethanol–water solution, about 60% of the base is hydroxide. This difference has kinetic consequences because the nucleophilicity of these lyate ions varies in the order $C_2H_5O^- > CH_3O^- > OH^-$;[71] see Table 7-16 also.

Ion–Ion Reactions

Equation (8-24) is the result of the simple electrostatic model for the reaction of two ions A and B, combining to form the transition state $M^\ddagger$ in an isotropic continuum of dielectric constant ε. Because $Z_M = Z_A + Z_B$, if the two ions have opposite charge, the charge on the transition state will be zero or at least smaller than that of one of the ions, and Eq. (8-24) predicts that k_s will decrease as ε increases. If the ions have the same charge, Z_M^2 will be larger than $Z_A^2 + Z_B^2$, and the opposite solvent effect is predicted. The qualitative Hughes–Ingold hypothesis is in agreement with these predictions. Experiment shows reasonable agreement with the predicted $1/\varepsilon$ dependence for binary solvent mixtures.[67, pp. 390–3] Entelis and Tiger[3, Chap. 5] have given a detailed discussion of electrostatic models of medium effects.

Another way to examine ion–ion reactions is to study their response to ionic strength in aqueous electrolyte solutions. The usual formulation is to combine the transition state expression

$$k = k_0 \frac{\gamma_A \gamma_B}{\gamma^\ddagger} \tag{8-26}$$

with the Debye–Hückel equation for the dependence of activity coefficient on ionic strength I:

$$-\log \gamma_i = \frac{A Z_i^2 \sqrt{I}}{1 + aB\sqrt{I}} \tag{8-27}$$

It is common to write $a = a_A = a_B = a^\ddagger$ for the ion contact distance. With this condition, Eqs. (8-26) and (8-27) yield

$$\log k = \log k_0 + \frac{2 Z_A Z_B A \sqrt{I}}{1 + aB\sqrt{I}} \tag{8-28}$$

or, for very low ionic strengths,

$$\log k = \log k_0 + 2 Z_A Z_B A \sqrt{I} \tag{8-29}$$

Since $A = 0.509$ for water at 25°C, Eq. (8-29) predicts that the limiting slope of a plot of $\log k$ against $\sqrt{I}$ is approximately equal to the product $Z_A Z_B$. This prediction has been verified for many reactions.[74] Equation (8-28) constitutes a quantitative description of the primary salt effect. Note that the qualitative Hughes–Ingold hypothesis is in agreement with the direction of the salt effect, provided an increase in ionic strength is construed to be an increase in solvent polarity.

An important cautionary note must be inserted here. It may seem that the study of the salt effect on the reaction rate might provide a means for distinguishing between two kinetically equivalent rate terms such as $k[HA][B]$ and $k'[A^-][BH^+]$, for, according to the preceding development, the slope of log k vs. $\sqrt{I}$ should be 0, whereas that of log k' vs. $\sqrt{I}$ should be -1. This is completely illusory. These two rate terms are kinetically equivalent, which means that no kinetic experiment can distinguish between them. To show this, we write the rate equation in the two equivalent forms, making use of Eq. (8-26):

$$v = k[HA][B] = k_0 \frac{\gamma_{HA}\gamma_B}{\gamma^{\ddagger}} [HA][B] \tag{8-30}$$

$$v = k'[A^-][BH^+] = k_0' \frac{\gamma_{A^-}\gamma_{BH^+}}{\gamma^{\ddagger}} [A^-][BH^+] \tag{8-31}$$

where k_0 and k_0' are independent of ionic strength. Defining acid dissociation constants in the usual way

$$K_a^{HA} = \frac{[H^+][A^-]}{[HA]} \cdot \frac{\gamma_{H^+}\gamma_{A^-}}{\gamma_{HA}}$$

$$K_a^{BH^+} = \frac{[H^+][B]}{[BH^+]} \cdot \frac{\gamma_{H^+}\gamma_B}{\gamma_{BH^+}}$$

gives, upon dividing K_a^{HA} by $K_a^{BH^+}$ and rearranging,

$$[HA][B]\gamma_{HA}\gamma_B = \frac{K_a^{BH^+}}{K_a^{HA}} \gamma_{A^-}\gamma_{BH^+} [A^-][BH^+] \tag{8-32}$$

Substituting Eq. (8-32) into (8-30),

$$v = k_0 \frac{\gamma_{A^-}\gamma_{BH^+}}{\gamma^{\ddagger}} [A^-][BH^+] \frac{K_a^{BH^+}}{K_a^{HA}} \tag{8-33}$$

The dissociation constants are thermodynamic constants, independent of ionic strength. Equation (8-33), which was derived from (8-30), is, therefore, identical in its form, and its salt effect, with Eq. (8-31). Therefore, salt effects cannot be used to distinguish between Eqs. (8-30) and (8-31). Another way to express this is that if kinetically equivalent forms can be written, it is not possible to determine, on the basis of kinetic results, whether the salt effect is primary or secondary. Weil and Morris[73] encountered this problem in the chlorination of ammonia,

$$HOCl + NH_3 \rightarrow NH_2Cl + H_2O$$

the rate equation being expressible in the equivalent forms (as shown by a bell-shaped pH–rate profile):

$$v = k[HOCl][NH_3] = k'[OCl^-][NH_4^+]$$

The formation of urea from ammonium cyanate,

$$NH_4^+ + CNO^- \rightarrow (H_2N)_2C{=}0$$

is another example, the rate terms being[74]

$$v = k[NH_4^+][CNO^-] = k'[NH_3][HCNO]$$

The Cavity Model

In the models discussed thus far in this section, emphasis has been placed on electrostatic effects and solvent polarity. An alternative view that to some extent takes other forces into account begins with the idea that, in order to dissolve a solute molecule in a solvent, energy is required to create a cavity in the solvent; the solute is then inserted into this cavity. In Section 8.2 we saw that the energy to create a cavity can be expressed as a product of the surface area of the cavity and the surface tension of the solvent. An equivalent expression is obtained as the product of the volume of the cavity and the pressure exerted by the solvent, and we now explore this concept.

The pressure exerted by the solvent is called its *internal pressure* π, and it is defined by Eq. (8-34), where E is molar energy and V is molar volume.

$$\pi = \left(\frac{\partial E}{\partial V}\right)_T \tag{8-34}$$

The internal pressure is a differential quantity that measures some of the forces of interaction between solvent molecules. A related quantity, the *cohesive energy density* (ced), defined by Eq. (8-35), is an integral quantity that measures the total molecular cohesion per unit volume.[5, p. 56]

$$ced = \frac{\Delta E_{vap}}{V} = \frac{\Delta H_{vap} - RT}{V} \tag{8-35}$$

In Eq. (8-35), ΔE_{vap} is the molar energy of vaporization, and ΔH_{vap} is the molar heat of vaporization. In effect, π is a measure of the energy required to break *some* of the solvent–solvent forces, whereas ced is a measure of the energy required to

TABLE 8-6. Internal Pressure and Cohesive Energy Density (ced) of Solvents

Solvent	$\pi/cal\ cm^{-3}$	$ced/cal\ cm^{-3}$	ced/atm
Ethylene glycol	120	212	8 681
Dioxane	119.3	96	3 909
Acetophenone	109.3	109	4 466
Methyl iodide	89.5	99	4 047
Toluene	84.8	80.6	3 278
Ethyl acetate	84.5	83.0	3 374
Carbon tetrachloride	82.4	74.6	3 040
Acetone	80.5	95	3 853
Ethanol	69.5	168	6 871
Methanol	68.1	212	8 621
Diethyl ether	63.0	59.9	2 474
Pentane	54.8	50.2	2 038
Water	41.0	547.6	22 608

Source: Reference 75.

break *all* of the solvent–solvent forces. Consequently π and ced tend to have very similar values for nonpolar solvents, but ced is much larger than π for hydroxylic solvents where strong H bonds are not broken in the differential measurement but are broken in the integral measurement. Table 8-6 gives values of π and ced (in two units) for some solvents.[75] Barton[76] gives conversion factors for π and ced units.

This approach to solution chemistry was largely developed by Hildebrand[77] in his regular solution theory. A regular solution is one whose entropy of mixing is ideal and whose enthalpy of mixing is nonideal. Consider a binary solvent of components 1 and 2. Let n_1 and n_2 be numbers of moles of 1 and 2, ϕ_1 and ϕ_2 their volume fractions in the mixture, and V_1, V_2 their molar volumes. This treatment follows Shinoda.[78]

The total energy of condensation from the ideal gas to the liquid state (the reverse process of vaporization) as a consequence of 1–1 contacts (i.e., intermolecular interactions of component 1 with like molecules) is the product of the energy of condensation per unit volume, the volume of liquid, and the volume fraction of component 1 in the liquid, or

$$- \left(\frac{\Delta E_1}{V_1} \right)(n_1 V_1)\phi_1 \;=\; - \frac{\Delta E_1 n_1^{\,2} V_1}{n_1 V_1 + n_2 V_2} \tag{8-36}$$

where ΔE_1 is the molar energy of vaporization. A similar expression is written for 2–2 contacts.

We define the energy of condensation per cubic centimeter of 1–2 contacts as $- \Delta E_{12}/(V_1 V_2)^{1/2}$. Then the total energy of condensation in the mixture due to 1–2 contacts is

$$- \frac{\Delta E_{12}}{(V_1 V_2)^{1/2}} (n_1 V_1 \phi_2 + n_2 V_2 \phi_1) \;=\; - \frac{2\Delta E_{12} n_1 n_2 (V_1 V_2)^{1/2}}{n_1 V_1 + n_2 V_2} \tag{8-37}$$

Then to find the total energy of condensation of n_1 moles of 1 with n_2 moles of 2, we add the 1–1, 2–2, and 1–2 contributions to obtain

$$-E_M = -\frac{\Delta E_1 n_1^2 V_1 + 2\Delta E_{12} n_1 n_2 (V_1 V_2)^{1/2} + \Delta E_2 n_2^2 V_2}{n_1 V_1 + n_2 V_2} \tag{8-38}$$

For the condensation, without mixing, of the pure liquids, the energy change would be

$$-\Delta E_0 = -(n_1 \Delta E_1 + n_2 \Delta E_2) \tag{8-39}$$

Therefore, the energy of mixing is the difference between (8-38) and (8-39), or $\Delta E_M = -E_M - (-\Delta E_0) = \Delta E_0 - E_M$.

$$\Delta E_M = \frac{n_1 n_2 V_1 V_2}{n_1 V_1 + n_2 V_2} \left[\frac{\Delta E_1}{V_1} + \frac{\Delta E_2}{V_2} - \frac{2\Delta E_{12}}{(V_1 V_2)^{1/2}} \right] \tag{8-40}$$

Recall that regular solution theory deals with nonpolar solvents, for which the dispersion force is expected to be a major contributor to intermolecular interactions. The dispersion energy, from Eq. (8-15), is for 1–2 interactions

$$U_{12} = -\frac{3\alpha_1 \alpha_2}{2r^6} \cdot \frac{I_1 I_2}{I_1 + I_2}$$

where α is polarizability and I is ionization potential. For like molecules $\alpha_1 = \alpha_2$ and $I_1 = I_2$, giving

$$U_{11} = -3\alpha_1^2 I_1 / 4r^6; \qquad U_{22} = -3\alpha_2^2 I_2 / 4r^6$$

Combining these expressions yields

$$U_{12} = \frac{2(I_1 I_2)^{1/2}}{I_1 + I_2} (U_{11} U_{22})^{1/2}$$

Because molecules 1 and 2 are both nonpolar, they do not differ greatly, so we assume $I_1 \approx I_2$; therefore, $2(I_1 I_2)^{1/2}/(I_1 + I_2) \approx 1$. Hence,

$$U_{12} \approx (U_{11} U_{22})^{1/2} \tag{8-41}$$

Equation (8-41) is called the *geometric mean approximation*; it is often used in approximate theories. We apply this result to Eq.(8-40), defining $\Delta E_{12} = (\Delta E_1 \Delta E_2)^{1/2}$. This gives

$$\Delta E_M = \frac{n_1 n_2 V_1 V_2}{n_1 V_1 + n_2 V_2} \left[\left(\frac{\Delta E_1}{V_1}\right)^{1/2} - \left(\frac{\Delta E_2}{V_2}\right)^{1/2} \right]^2 \tag{8-42}$$

We encountered the quantity $\Delta E_{\text{vap}}/V$ in Eq. (8-35); it is the cohesive energy density. The square root of this quantity plays an important role in regular solution theory, and Hildebrand named it the *solubility parameter*, δ.

$$\delta = (\text{ced})^{1/2} = \left(\frac{\Delta E_{\text{vap}}}{V}\right)^{1/2} \tag{8-43}$$

Using Eq. (8-43) in (8-42),

$$\Delta E_{\text{M}} = \frac{n_1 n_2 V_1 V_2}{n_1 V_1 + n_2 V_2}(\delta_1 - \delta_2)^2 \tag{8-44}$$

Next we differentiate Eq. (8-44) with respect to n_2, obtaining the partial molar energy of mixing of 2 in 1:

$$\left(\frac{\partial \Delta E_{\text{M}}}{\partial n_2}\right)_{n_1} = \Delta \overline{E}_2 = V_2 \phi_1{}^2 (\delta_1 - \delta_2)^2 \tag{8-45}$$

For an ideal solution the entropy of mixing is $\Delta \overline{S}_2 = -R \ln x_2$, where x_2 is the mole fraction of 2. Writing $\Delta \overline{G}_2 = \Delta \overline{H}_2 - T\Delta \overline{S}_2$ and identifying $\Delta \overline{H}_2$ as ΔE_2 (which is equivalent to equating the Gibbs free energy and the Helmboltz free energy) yields

$$\Delta \overline{G}_2 = V_2 \phi_1{}^2 (\delta_1 - \delta_2)^2 + RT \ln x_2 \tag{8-46}$$

Because $\mu_2 - \mu_2^0 = \Delta \overline{G}_2 = RT \ln a_2$, where μ_2 is chemical potential and a_2 is activity, Eq. (8-46) becomes

$$\ln\frac{a_2}{x_2} = \ln \gamma_2 = \frac{V_2 \phi_1{}^2 (\delta_1 - \delta_2)^2}{RT} \tag{8-47}$$

where γ_2 is the activity coefficient of component 2. Equation (8-47) is a fundamental result of regular solution theory. The quantity x_2 is interpreted as the mole fraction equilibrium solubility, and a_2 is the activity of component 2, equal to the activity of the pure component. If the solution of component 2 is very dilute, $\phi_1 \approx 1$, and Eq. (8-47) becomes

$$RT \ln \gamma_2 = V_2 (\delta_1 - \delta_2)^2 \tag{8-48}$$

We now apply Eq. (8-48) to the kinetic problem expressed as Eq. (8-26), obtaining Eq. (8-49).

$$RT \ln k = \text{constant} + V_{\text{A}}(\delta_{\text{A}} - \delta_{\text{S}})^2 + V_{\text{B}}(\delta_{\text{B}} - \delta_{\text{S}})^2 - V^{\ddagger}(\delta_{\ddagger} - \delta_{\text{S}})^2 \tag{8-49}$$

which can be more succinctly written

$$RT \ln k = \text{constant} + \Sigma\, V_R(\delta_R - \delta_S)^2 - V^{\ddagger}(\delta_{\ddagger} - \delta_S)^2 \qquad (8\text{-}50)$$

where δ_S is the solubility parameter of the solvent. The molar volumes of reactants (V_R) and the transition state ($V^{\ddagger}$) are determined by their van der Waals radii and by the internal pressure of the solvent. Solvent–solvent cohesive interactions form a cavity about each solute species. In addition, solute–solvent interactions may act to decrease the cavity volume; this phenomenon is called *electrostriction*.

Introducing the activation volume $\Delta V^{\ddagger} = V^{\ddagger} - \Sigma\, V_R$ into Eq. (8-50) leads to

$$RT \ln k = \text{constant} + \Sigma\, V_R[(\delta_R - \delta_{\ddagger})^2 - 2\delta_S(\delta_R - \delta_{\ddagger})] - \Delta V^{\ddagger}(\delta_{\ddagger} - \delta_S)^2$$
$$(8\text{-}51)$$

For a given reaction studied in a series of solvents, $(\delta_R - \delta_{\ddagger})$ is essentially constant, and most of the change in $\ln k$ will come from the term $-\Delta V^{\ddagger}(\delta_{\ddagger} - \delta_S)^2$. If $\Delta V^{\ddagger}$ is positive, an increase in δ_S (increase in solvent internal pressure) results in a rate decrease. If $\Delta V^{\ddagger}$ is negative, the reverse effect is predicted. Thus reactivity is predicted by regular solution theory to respond to internal pressure just as it does to externally applied pressure (Section 6.2). This connection between reactivity and internal pressure was noted long ago,[79] and it has been systematized by Dack.[4,75]

It should be noted that the cohesive energy density, and, therefore, the solubility parameter, is a well-defined physical property, independently of any assumptions and approximations of regular solution theory. However, we must not expect regular solution theory to apply with exactness to polar liquids. A major application of the solubility parameter concept is to the solubility of nonvolatile solutes such as polymers. According to Eq. (8-46), the free energy of mixing is minimized when $\delta_1 = \delta_2$, that is, when the solubility parameters of solute and solvent are equal. This is why a blend of two solvents may dissolve a solute more effectively than either pure solvent. For example,[76] a mixture of ether ($\delta = 7.4$) and ethanol ($\delta = 12.7$) dissolves nitrocellulose ($\delta = 11.2$), although neither pure liquid is a good solvent for this polymer.

There have been many attempts to divide the overall solubility parameter into components corresponding to the several intermolecular forces. For example, a so-called three-dimensional solubility parameter concept is built on the assumption that the ced is an additive function of contributions from dispersion (d), polar (p), and H-bonding (h) forces. It follows that

$$\delta_0^2 = \delta_d^2 + \delta_p^2 + \delta_h^2 \qquad (8\text{-}52)$$

Evaluation of the individual contributions requires extrathermodynamic assumptions and analysis of a body of data so as to achieve a self-consistent set of numbers. For this reason, there may be small differences between δ_0 of Eq. (8-52) and the δ defined by Eq. (8-43). Table 8-7 gives value of δ, δ_0, δ_d, δ_p, δ_h for the solvents of Table 8-2.[76]

TABLE **8-7.** Solubility Parameters of Some Solvents[a]

No.[b]	Solvent	δ	δ_o	δ_d	δ_p	δ_h
		\multicolumn{5}{c}{$\delta / cal^{1/2}\ cm^{-3/2}$}
1	*n*-Pentane	7.0	7.1	7.1	0.0	0.0
2	*n*-Hexane	7.3	7.3	7.3	0.0	0.0
3	*n*-Heptane	7.4	7.5	7.5	0.0	0.0
4	Cyclohexane	8.2	8.2	8.2	0.0	0.0
5	1,4-Dioxane	10.0	10.0	9.3	0.9	3.6
6	Carbon tetrachloride	8.6	8.7	8.7	0.0	0.3
7	Benzene	9.2	9.1	9.0	0.0	1.0
8	Toluene	8.9	8.9	8.8	0.7	1.0
9	Diethyl ether	7.4	7.7	7.1	1.4	2.5
10	Chloroform	9.3	9.3	8.7	1.5	2.8
11	Bromobenzene	9.9	10.6	10.0	2.7	2.0
12	Chlorobenzene	9.5	9.6	9.3	2.1	1.0
13	Ethyl acetate	9.1	8.9	7.7	2.6	3.5
14	Acetic acid	10.1	10.5	7.1	3.9	6.6
15	Methyl acetate	9.6	9.2	7.6	3.5	3.7
16	Tetrahydrofuran	9.1	9.5	8.2	2.8	3.9
17	Dichloromethane	9.7	9.9	8.9	3.1	3.0
18	*t*-Butyl alcohol	—	—	—	—	—
19	Pyridine	10.7	10.7	9.3	4.3	2.9
20	Benzyl alcohol	12.1	11.6	9.0	3.1	6.7
21	2-Methoxyethanol	11.4	12.1	7.9	4.5	8.0
22	*n*-Butyl alcohol	11.4	11.3	7.8	2.8	7.7
23	*i*-Butyl alcohol	10.8	10.8	7.7	2.8	7.1
24	*i*-Propyl alcohol	11.5	11.5	7.7	3.0	8.0
25	*n*-Propyl alcohol	11.9	12.0	7.8	3.3	8.5
26	Acetone	9.9	9.8	7.6	5.1	3.4
27	Acetic anhydride	10.3	10.9	7.8	5.7	5.0
28	Ethanol	12.7	13.0	7.7	4.3	9.5
29	Methanol	14.5	14.5	7.4	6.0	10.9
30	Nitrobenzene	10.0	10.9	9.8	4.2	2.0
31	Acetonitrile	11.9	12.0	7.5	8.8	3.0
32	Dimethylformamide	12.1	12.1	8.5	6.7	5.5
33	Ethylene glycol	14.6	16.1	8.3	5.4	12.7
34	Dimethylacetamide	10.8	11.1	8.2	5.6	5.0
35	Nitromethane	12.7	12.3	7.7	9.2	2.5
36	Glycerol	16.5	17.7	8.5	5.9	14.3
37	Dimethylsulfoxide	12.0	13.0	9.0	8.0	5.0
38	Formic acid	12.1	12.2	7.0	5.8	8.1
39	Water[c]	23.4	23.4	7.6	7.8	20.7
40	Formamide	19.2	17.9	8.4	12.8	9.3

Source: Reference 76.

[a]To convert δ in $cal^{1/2}\ cm^{-3/2}$ to δ in $MPa^{1/2}$, multiply by 2.046.

[b]Numbered as in Table 8-2 and ranked in order of increasing dielectric constant.

[c]Values uncertain.

Strictly speaking Eq. (8-51) should be applied only to reacting systems whose molecular properties are consistent with the assumptions of regular solution theory. This essentially restricts the approach to the reactions of nonpolar species in nonpolar solvents. Even in these systems, which we recall do not exhibit a marked solvent dependence, correlations with δ_S^2 tend to be poor.[5, pp. 190-5] Nevertheless, the solubility parameter and its partitioning into dispersion, polar, and H-bonding components provide some insight into solvent behavior that is different from the information given by other properties such as those in Tables 8-2 and 8-3.

Separation of Initial and Transition State Solvent Effects

We have defined the solvent effect (medium effect) on a reaction by

$$\delta_M \Delta G^{\ddagger} = \delta G_S^{\ddagger} - \delta G_0^{\ddagger} = -RT \ln \frac{k_S}{k_0} \tag{8-53}$$

where k_S and k_0 are rate constants in solvents S and 0. Thus, $\delta_M \Delta G^{\ddagger}$ is accessible. However, this quantity is a composite of effects on the initial state and the transition state. We have, earlier in this chapter, made guesses about the relative importance of these contributions to the observed solvent effect, as when we applied the Hughes–Ingold hypothesis to the Menschutkin reaction, but we could not know if the guesses were correct. To achieve a better understanding of the effect of the solvent on reactivity it would be desirable to dissect $\delta_M \Delta G^{\ddagger}$ into initial and transition state effects.

Here is one way to attack the problem.[60,80,81] Equations (8-54) constitute definitions of $\Delta G^{\ddagger}$ in the two solvents:

$$\Delta G_0^{\ddagger} = G_0^{\ddagger} - G_0^R \tag{8-54a}$$

$$\Delta G_S^{\ddagger} = (G_S^{\ddagger} - G_S^R \tag{8-54b}$$

As usual the double dagger signifies the transition state and R the reactant or initial state. Subtracting these equations, with slight rearrangement of terms, gives

$$\delta_M \Delta G^{\ddagger} = (G_S^{\ddagger} - G_0^{\ddagger}) - (G_S^R - G_0^R) \tag{8-55}$$

We now define a quantity $\delta_M G_{tr}^i$ for state i by

$$\delta_M G_{tr}^i = G_S^i - G_0^i \tag{8-56}$$

Applying Eq. (8-56) to (8-55) yields

$$\delta_M \Delta G^{\ddagger} = \delta_M G_{tr}^{\ddagger} - \delta_M G_{tr}^R \tag{8-57}$$

The quantity $\delta_M G_{tr}^i$ is called the *transfer free energy*; it is the (standard) free energy change for the transfer of 1 mole of i from solvent 0 to solvent S. Equation (8-57) constitutes a formal separation of the observed solvent effect into effects on the reactant and transition states. (A similar analysis can be given for solvent effects on equilibria.)

To use Eq. (8-57) we require an experimental measure of the transfer free energy for the reactant; then Eq. (8-57) permits the transfer free energy for the transition state to be calculated. First, consider this equilibrium process:

$$\text{solute in solvent } 0 \overset{K_{eq}}{\rightleftharpoons} \text{solute in solvent S}$$

At equilibrium, the ratio of concentrations is an equilibrium constant, so we can write the standard free energy change for the process as

$$\Delta G^\circ = -RT \ln K_{eq} = -RT \ln \frac{c_S}{c_0}$$

However, ΔG° for this process is the transfer free energy, which we indicate by the symbolism in Eq. (8-58) to make clear the direction of transfer.

$$\delta_M G_{tr}(S\leftarrow 0) = -RT \ln \frac{c_S}{c_0} \tag{8-58}$$

This experiment requires that the solvents 0 and S be immiscible; then the ratio c_S/c_0 is the partition coefficient. The experiment should be carried out at sufficiently low concentrations that solute–solute interactions are negligible in both solvents.

If solvents 0 and S are miscible, the following scheme can be used. Consider this set of mutual equilibria:

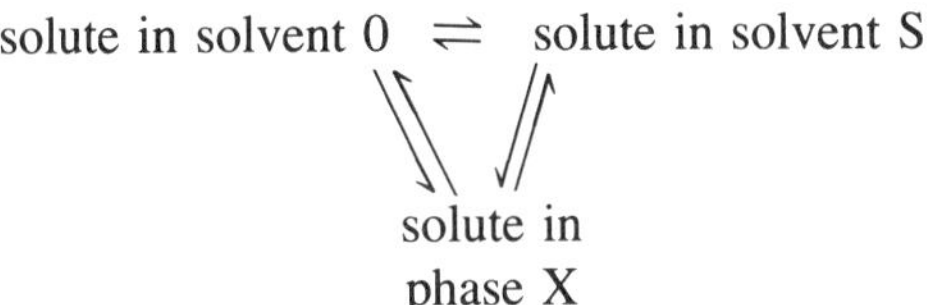

Defining equilibrium constants for each equilibrium shows that

$$K(S\leftarrow 0) = \frac{K(S\leftarrow X)}{K(0\leftarrow X)} = \frac{c_S}{c_0}$$

Now interpret phase X as pure solute; then c_S and c_0 become the equilibrium solubilities of the solute in solvents S and 0, respectively, and we can apply Eq. (8-58). Again the concentrations should be in the dilute range, but nonideality is not a great problem for nonelectrolytes. For volatile solutes vapor pressure measurements are suitable for this type of determination, and for electrolytes electrode potentials can be used.

Some authors[45,82,83] express the transfer free energy in the form of an activity coefficient defined by Eq. (8-59).

$$\delta_M G_{tr}(S\leftarrow 0) = -RT \ln \frac{1}{\gamma_{tr}(S\leftarrow 0)} = RT \ln \gamma_{tr}(S\leftarrow 0) \qquad (8\text{-}59)$$

$\gamma_{tr}(S\leftarrow 0)$ is variously called the *transfer activity coefficient, solvent* or *medium activity coefficient, degenerate activity coefficient*, or simply the *medium effect*. Several symbolisms are used. Because $\delta_M G_{tr}(S\leftarrow 0)$ describes solute–solvent interactions, so does $\gamma_{tr}(S\leftarrow 0)$; that is, this activity coefficient is a measure of solvation, not of solute–solute nonideality effects. For this reason the term *distribution coefficient* of Kolthoff and Bruckenstein,[50, p. 492] seems apt.

Because it is impossible to vary single ion concentrations independently, the activity coefficient of an electrolyte is a function of activity coefficients of the cation and anion of the electrolyte. For example, for 1:1 electrolytes the relationship is

$$\gamma_\pm = (\gamma_+ \gamma_-)^{1/2}$$

Without some additional relationship it is impossible to resolve $\gamma_\pm$ into γ_+ and γ_-. By introducing an extrathermodynamic assumption as this additional relationship, it becomes possible to estimate single ion transfer activity coefficients. A widely used assumption is that the transfer activity coefficients of the cation and anion of tetraphenylarsonium tetraphenylboride, $Ph_4As^+BPh_4^-$, are equal, i.e.,

$$\gamma_{tr}Ph_4As^+ (S\leftarrow 0) = \gamma_{tr}BPh_4^- (S\leftarrow 0)$$

By combining these ions with other counterions, single ion transfer activity coefficients are calculated. By these techniques transfer free energies or activity coefficients have been determined for many ions and nonelectrolytes in a wide variety of solvents.[45,48,81,84] Parker[85] has discussed the extrathermodynamic assumptions that lead to single ion quantities.

In Section 8.2 solvation energy ΔG_{solv} was defined as the free energy change upon transferring a solute from the gas into a solvent. We can now relate the transfer free energy to solvation energies:

$$\delta_M G_{tr}(S\leftarrow 0) = \Delta G_{solv}(S) - \Delta G_{solv}(0)$$

Usually $\delta_M G_{tr}$ is a small difference between two large numbers, so it is more accurate to measure $\delta_M G_{tr}$ directly by the techniques discussed above than to estimate it indirectly. Solvation is then usually considered in terms of transfer free energies or activity coefficients.

If a solute is better solvated in S than in 0, then $\delta_M G_{tr}(S\leftarrow 0)$ is negative and $\gamma_{tr}(S\leftarrow 0)$ is less than unity. Comparing Eqs. (8-58) and (8-59) shows that

$$\gamma_{tr}(S\leftarrow 0) = \frac{c_0}{c_S} \qquad (8\text{-}60)$$

TABLE 8-8.   Transfer Free Energies of Nonelectrolytes[a]

	$\delta_M G_{tr}(S \leftarrow MeOH)$ for		
Solvent, S	n-C_6H_{14}	C_6H_6	4-$NO_2C_6H_4CH_2Cl$
Methanol	(0)	(0)	(0)
Ethanol	-0.5	-0.1	-0.1
n-Propyl alcohol	-0.7	-0.3	-0.1
Dimethylsulfoxide	$+0.7$	-0.5	—
Nitromethane	$+0.3$	-0.5	-1.5
Acetonitrile	0	-0.6	-1.5
Nitrobenzene	-0.7	-1.0	-1.7
Acetophenone	-0.9	-1.1	-1.8
Acetone	-0.8	-0.9	-1.7
Benzene	-1.5	-1.2	-1.6
Carbon tetrachloride	-1.8	-1.1	-0.7
Hexane	-2.0	-0.9	$+0.5$

Source: From Reference 81.

[a]On the mole fraction scale in kilocalories per mole. Reference solvent is methanol.

so that $\gamma_{tr}(S \leftarrow 0)$ measures how strongly the solute is solvated in the reference solvent 0 compared with solvent S. The transfer activity coefficient for the transition state is accessible via Eq. (8-26)[85] or by combining Eqs. (8-57) and (8-59).

Table 8-8 gives some nonelectrolyte transfer free energies,[81] and Table 8-9 lists single ion transfer activity coefficients.[85] Note especially the remarkable values for anions in dipolar aprotic solvents, indicating extensive desolvation in these solvents relative to methanol. This is consistent with the enhanced nucleophilic reactivity of anions in dipolar aprotic solvents.[85] Parker[85] and Blandamer[86] have considered transfer activity coefficients for binary aqueous mixtures.

We are now in a position to apply these thermodynamic quantities to the kinetic

TABLE 8-9.   Single Ion Transfer Activity Coefficients[a]

	$log\ \gamma_{tr}(S \leftarrow MeOH)$				
Ion	H_2O	*DMF*	*DMSO*	*MeCN*	*MeNO$_2$*
Cl^-	-2.5	6.5	5.5	6.3	4.9
Br^-	-2.1	4.9	3.6	4.2	—
I^-	-1.5	2.6	1.3	2.4	2.6
N_3^-	-1.8	4.9	3.5	4.7	4.6
BPh_4^-	4.1	-2.7	-2.6	-1.6	—
OAc^-	-2.9	9.2	6.5	7.8	—
Ag^+	-0.8	-5.1	-8.2	-6.3	1.7
Na^+	—	-3.9	-3.6	$+1.4$	—
K^+	-1.5	-3.7	-4.5	-0.8	—
Ph_4As^+	4.1	-2.7	-2.6	-1.6	—

Source: Reference 85.

[a]At 25°C. Reference solvent is methanol.

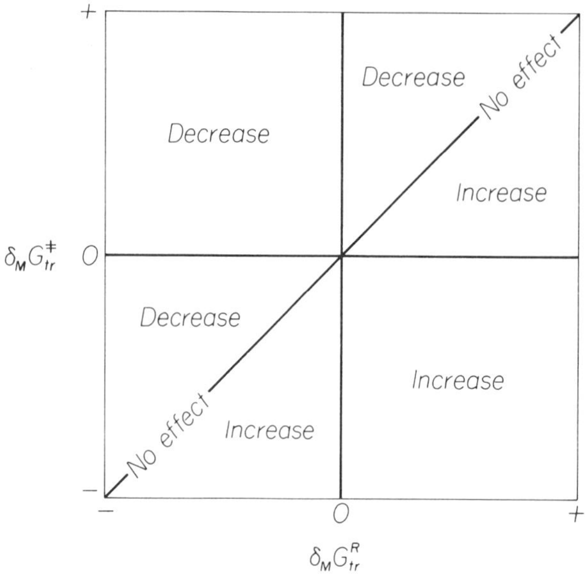

Figure 8-5. Two-dimensional representation of the possible combinations of transfer free energies of reactants (R) and transition state (‡). The words refer to rate increases or decreases relative to the reference solvent, which lies at the origin.

problem. We will work with the transfer free energies $\delta_M G_{tr}^{R}$ and $\delta_M G_{tr}^{\ddagger}$ for the reactant state and the transition state, respectively. For a given reaction transferred from reference solvent 0 to solvent S, each of these may be zero, positive, or negative, giving rise to nine possible combinations. Buncel and Wilson[60,80] have devised a classification scheme based on these combinations. Figure 8-5 is a graphical presentation of the possibilities. The reference solvent is at the origin. Any combination of $\delta_M G_{tr}^{R}$ and $\delta_M G_{tr}^{\ddagger}$ leading to placement on the 45° diagonal results in no rate change relative to the reference solvent, the solvent effects on the initial and transition states exactly compensating each other. Combinations to the left and above this line lead to rate decreases, those below the line to rate increases. To change the reference solvent it is only necessary to translate the origin to the desired reference solvent.

Table 8-10 gives pertinent data[65] for the Menschutkin reaction of triethylamine with ethyl iodide. These reactant molecules are volatile, so their transfer free energies were determined by a gas chromatographic variation of the vapor pressure method. For this reaction Eq. (8-57) is written

$$\delta_M \Delta G^{\ddagger} = \delta_M G_{tr}^{\ddagger} - \delta_M G_{tr}(Et_3N) - \delta_M G_{tr}(EtI)$$

A plot of $\delta_M G_{tr}^{\ddagger}$ against $\delta_m \Delta G^{\ddagger}$ (not shown here) reveals a reasonable correlation with a slope about unity, indicating that $\delta_M \Delta G^{\ddagger}$ is dominated by transition state

TABLE 8-10. Solvent Effects and Transfer Free Energies for the Menschutkin Reaction of Triethylamine and Ethyl Iodide[a,b]

No.	Solvent	$\delta_M \Delta G^{\ddagger}$	$\delta_M G_{tr}(Et_3N)$	$\delta_M G_{tr}(EtI)$	$\delta_M G_{tr}^{\ddagger}$
1	Hexane	6.05	-1.28	-0.01	4.76
2	Cyclohexane	5.66	-1.26	-0.21	4.19
3	Diethyl ether	3.97	-1.06	0.10	3.01
4	Carbon tetrachloride	3.83	-1.39	-0.26	2.18
5	1,1,1-Trichloroethane	2.66	-1.22	-0.26	1.18
6	Toluene	2.66	-1.29	-0.33	1.04
7	Cyclohexyl chloride	2.47	-0.99	-0.39	1.09
8	Cyclohexyl bromide	2.38	-1.02	-0.37	0.99
9	Benzene	2.27	-1.12	-0.35	0.80
10	Ethyl acetate	2.11	-0.74	-0.18	1.19
11	Dioxane	1.78	-0.56	-0.21	1.01
12	Tetrahydrofuran	1.76	-1.06	-0.43	0.27
13	Ethyl benzoate	1.67	-0.86	-0.30	0.51
14	Chlorobenzene	1.59	-1.16	-0.38	0.05
15	Bromobenzene	1.43	-1.09	-0.36	-0.02
16	1,1-Dichloroethane	1.36	-1.35	-0.17	-0.16
17	Chloroform	1.19	-1.92	-0.41	-1.14
18	Methyl ethyl ketone	1.10	-0.68	-0.18	0.24
19	Iodobenzene	1.08	-0.99	-0.46	-0.37
20	Acetone	0.68	-0.30	0.05	0.43
21	1,2-Dichloroethane	0.51	-0.83	-0.17	-0.49
22	Dichloromethane	0.48	-1.31	0.04	-0.79
23	Acetophenone	0.41	-0.61	-0.24	-0.44
24	Benzonitrile	0.38	-0.66	-0.10	-0.38
25	Propionitrile	0.31	-0.42	0.17	0.06
26	Nitrobenzene	0.26	-0.52	-0.11	-0.37
27	Dimethylformamide	(0)	(0)	(0)	(0)
28	1,1,2,2-Tetrachloroethane	-0.03	-2.51	-0.53	-3.07
29	Acetonitrile	-0.26	0.13	0.47	0.35
30	Nitromethane	-0.47	0.21	0.54	0.28
31	Propylene carbonate	-0.63	0.62	0.29	0.28
32	Dimethyl sulfoxide	-0.88	0.74	0.29	0.15

Source: Reference 65.

[a]All free energies are in kilocalories per mole on the mole fraction scale, relative to DMF.
[b]See Table 8-5 for rate data.

solvent effects for the series, as we would expect for this type of reaction. A plot of $\delta_M G_{tr}^{\ddagger}$ against $\delta_M G_{tr}^R$, in the manner of Fig. 8-5, is shown in Fig. 8-6. Several interesting results are apparent in this plot.

1. Dipolar aprotic solvents have similar effects on the transition state; any significant differences arise from variable effects on the reactants.
2. Polyhalogenated solvents all stabilize the initial state (relative to DMF), but have varied effects on the transition state. [Note particularly No. 28, tetrachloroethane, which shows no net solvent effect relative to DMF, yet profoundly

stabilizes both the initial and transition states. This is an excellent example of the additional information provided by this approach.]

3. Aromatics all stabilize the initial state, relative to DMF, and have small but varied effects on the transition state.

4. Saturated hydrocarbons show a slight stabilizing effect on the initial state but a very destabilizing effect on the transition state, consistent with arguments based on solvent polarity.

Demonstrations such as those embodied in Fig. 8-6 constitute the first and most important result of the separation of solvent effects into initial state and transition state components, but the analysis can be taken further by comparing the value of $\delta_M G_{tr}^{\ddagger}$ with the transfer free energy of a model of the transition state. In this way Abraham[65,87] has compared $\delta_M G_{tr}^{\ddagger}$ for the Menschutkin reaction of Table 8-10 with $\delta_M G_{tr}$ values for the ion pair $Et_4N^+I^-$ and the dissociated ions $Et_4N^+ + I^-$, reaching the conclusion that the transition state more closely resembles the ion pair than the free ions and that charge development in the transition state is about 0.4 unit charge.

Other examples of this "thermodynamic dissection" procedure are provided by

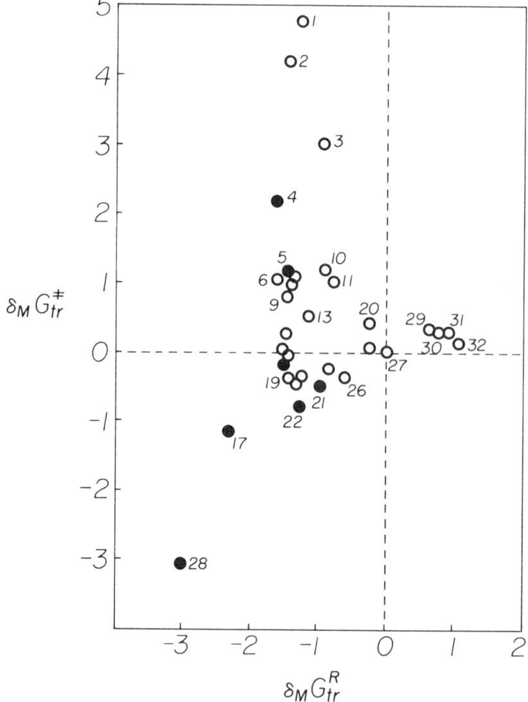

Figure 8-6. Plot according to Fig. 8-5 of transfer free energies of the transition state (ordinate) and reactant state (abscissa) for the Menschutkin reaction of triethylamine and ethyl iodide. The reference solvent is N, N-dimethylformamide (No. 27). Data are from Table 8-10, where the solvents are identified by number. Closed circles are polychlorinated solvents.

studies of Schiff base formation[88] and solvolysis of 2-chloroethyl methylsulfide.[89] An alternative approach to the problem of analyzing the solvent dependence of initial and transition states has been developed by Blandamer and co-workers,[90,91] who begin with the assumption that the standard chemical potentials are independent of solvent composition by definition and interpret initial state and transition state activity coefficients in terms of pairwise group interaction parameters. This procedure has been applied to water-rich cosolvent systems.

8.4 EMPIRICAL MEASURES AND CORRELATIONS

We have seen that physical properties fail to correlate rate data in any general way, although some limited relationships can be found. Many workers have, therefore, sought alternative measures of solvent behavior as means for correlating and understanding reactivity data. These alternative quantities are the empirical measures described in this section. The adjective *empirical* in this usage is synonymous with *model dependent*; this is, therefore, an extrathermodynamic approach, entirely analogous to the LFER methods of Chapter 7 with which structure–reactivity relationships can be studied.

Some of these model-dependent quantities were formulated as measures of a particular phenomenon, such as electron-pair donor ability; but many of them have been proposed as *empirical measures of solvent polarity*, with the goal, or hope, that they may embody a useful blend of solvent properties that quantitatively accounts for the solvent effect on reactivity. This section describes many, although not all, of these empirical measures. Reichardt[5, Chap. 7] has reviewed this subject.

Thermodynamic Measures

The chemistry of Lewis acid–base adducts (electron-pair donor–acceptor complexes) has stimulated the development of measures of the Lewis basicity of solvents. Jensen[92] and Persson[93] have reviewed these. Gutmann[94] defined the donor number (DN) as the negative of the enthalpy change (in kcal mol^{-1}) for the interaction of an electron-pair donor with $SbCl_5$ in a dilute solution in dichloroethane. DN has been widely used to correlate complexing data, but side reactions can lead to inaccurate DN values for some solvents.[95] Maria and Gal[95] measured the enthalpy change of this reaction

$$BF_3(gas) \; + \; :D(soln) \rightarrow D:BF_3(soln)$$

in dichloromethane, taking $-\Delta H^0$ as a measure of the Lewis basicity of donor D. A collection of these values is given in Table 8-11. $-\Delta H^0(BF_3)$ correlates fairly well with DN and with some other measures of donor ability.[93] Note that these measures do not characterize the neat solvents.

TABLE 8-11. Enthalpy Change for Boron Trifluoride Complex Formation with Donors at 25°C

Solvent	ΔH^0/kJ mol^{-1}	Solvent	ΔH^0/kJ mol^{-1}
Dichlormethane	10.0	Acetone	76.0
Nitrobenzene	35.8	Methyl ethyl ketone	76.1
Nitromethane	37.6	Cyclohexanone	76.4
Benzonitrile	55.4	Cyclopentanone	77.4
Methyl benzoate	59.4	Diethyl ether	78.8
Acetonitrile	60.4	Tetrahydrofuran	90.4
Propionitrile	61.0	Dimethylsulfoxide	105.3
Ethyl benzoate	61.2	Dimethyl formamide	110.5
Propylene carbonate	64.2	Dimethylacetamide	112.1
Dimethyl carbonate	67.6	Hexamethylphosphoramide	117.5
Acetaldehyde	69.6	Pyridine	128.1
Methyl formate	69.8	Triethylamine	135.9
Diethyl carbonate	71.0		
Ethyl formate	71.2		
Diethyl ketone	72.3		
Methyl acetate	72.8		
Dioxane	74.1		
Acetophenone	74.5		
Benzaldehyde	74.9		
Ethyl acetate	75.6		

Source: Reference 95.

Drago and co-workers[96] have correlated a large body of enthalpies of adduct formation in Lewis acid–base systems, including some solvents as reactants, with this four-parameter equation:

$$-\Delta H = E_A E_B + C_A C_B \tag{8-61}$$

The parameters E_A and C_A are associated with the Lewis acid, and E_B and C_B with the base. E_A and E_B are interpreted as measures of electrostatic interaction, and C_A and C_B as measures of covalent interaction. Drago has criticized the DN approach as being based upon a single model process, and this objection applies also to the $-\Delta H^0(\text{BF}_3)$ model. Drago's criticism is correct, yet we should be careful not to reject a simple concept provided its limits are appreciated. Indeed, many very useful chemical quantities are subject to this criticism; for example, pK_a values are measures of acid strength with reference to the base water.

Marcus[97] has defined a quantitative measure of softness, μ, by Eq. 8-62.

$$100\mu = \tfrac{1}{2}\left[\delta_M G_{tr}^{Na^+}(S \leftarrow W) + \delta_M G_{tr}^{K^+}(S \leftarrow W)\right] - \delta_M G_{tr}^{Ag^+}(S \leftarrow W) \tag{8-62}$$

That is, the softness of solvent S is measured by the difference in transfer free energies (in kJ mol^{-1}) from water to S of a hard solute (the mean of Na$^+$ and K$^+$)

and a soft solute, Ag^+. Water itself is a hard solvent. μ does not correlate with $-\Delta H°(BF_3)$.

In section 8.2 we described the solvophobic effect, which theory leads us to expect is related to the solvent surface tension. Abraham et al.[98] have developed a different measure of solvophobicity by relating the transfer free energy $\delta_M G_{tr}(S\leftarrow W)$ for several solutes from water to given solvent S (which may be an aqueous–organic mixture) to an empirical solute parameter R_T by

$$\delta_M G_{tr}(S\leftarrow W) = MR_T + D \tag{8-63}$$

R_T seems to be related to solute size, but it is obtained by an iterative curve-fitting procedure.[99] The parameters M and D are characteristic of the solvent. For water $M = 0$. A solvophobicity parameter Sp is then defined for the solvent having value M

$$Sp = 1 - M/M_{\text{hexadecane}} \tag{8-64}$$

where $M_{\text{hexadecane}} = -4.2024$. SP is a smooth but curved function of surface tension for mixed solvent systems.

Kinetic Measures

Most of the kinetic measures of solvent effects have been developed for the study of nucleophilic substitution (S_N) at saturated carbon, solvolytic reactions in particular. It may, therefore, be helpful to give a brief review of aliphatic nucleophilic substitution. Two mechanistic routes have been clearly identified. One of these is shown by

$$R\text{–}X \xrightarrow{\text{slow}} R^+ + X^-$$

$$R^+ + Y^- \xrightarrow{\text{fast}} R\text{–}Y$$

The first step, which is rate determining, is an ionization to a carbocation (carbonium ion in earlier terminology) intermediate, which reacts with the nucleophile in the second step. Because the transition state for the rate-determining step includes R–X but not Y^-, the reaction is unimolecular and is labeled S_N1. First-order kinetics are involved, with the rate being independent of the nucleophile identity and concentration.

The second S_N mechanism is the one-step concerted reaction,

$$Y^- + R\text{–}X \rightarrow [Y\cdots R\cdots X]^- \rightarrow R\text{–}Y + X^-$$

This bimolecular process is called the *S_N2 mechanism*. It yields overall second-order kinetics (unless the nucleophile is the solvent, in which case apparent first-order kinetics are seen).

The evidence supporting the duality of mechanisms is of several kinds. The kinetic behavior is an obvious feature. This is somewhat more complex than is implied by the preceding treatment. A quantitative description of the S_N1 mechanism requires recognition of the reversibility of the ionization step, thus

$$R-X \underset{k_{-1}}{\overset{k_1}{\rightleftharpoons}} R^+ + X^-$$

$$R^+ + Y^- \overset{k_2}{\rightarrow} R-Y$$

Because the cationic intermediate is unstable, it will be permissible to apply the steady-state approximation, leading to Eq. (8-65) for the reaction rate.

$$v = \frac{k_1[R-X]}{(k_{-1}[X^-]/k_2[Y^-]) + 1} \tag{8-65}$$

If the intermediate reacts with Y^- (which may be the solvent) to give product much faster than it does with X^- to revert to reactant, then Eq. (8-65) will tend to the simple first-order form, $v = k_1[R-X]$. In aqueous solvents *tert*-butyl bromide exhibits this kinetic behavior.

If $(k_{-1}[X^-]/k_2[Y^-])$ is not much smaller than unity, then as the substitution reaction proceeds, the increase in $[X^-]$ will increase the denominator of Eq. (8-65), slowing the reaction and causing deviation from simple first-order kinetics. This *mass-law* or *common-ion effect* is characteristic of an S_N1 process, although, as already seen, it is not a necessary condition. The common-ion effect (also called *external return*) occurs only with the common ion and must be distinguished from a general kinetic salt effect, which will operate with any ion. An example is provided by the hydrolysis of triphenylmethyl chloride (trityl chloride); the addition of 0.01 M NaCl decreased the rate by fourfold.[100] The solvolysis rate of diphenylmethyl chloride in 80% aqueous acetone was decreased by LiCl but increased by LiBr.[101] The S_N2 mechanism will also yield first-order kinetics in a solvolysis reaction, but it should not be susceptible to a common-ion rate inhibition.

The real world of S_N reactions is not quite as simple as the discussion has so far suggested. The preceding treatment in terms of two clearly distinct mechanisms, S_N1 and S_N2, implies that all substitution reactions will follow one or the other of these mechanisms. This is an oversimplification. The strength of the dual mechanism hypothesis and its limitations are revealed by these relative rates of solvolysis of alkyl bromides in 80% ethanol:methyl bromide, 2.51; ethyl bromide, 1.00; isopropyl bromide, 1.70; *tert*-butyl bromide, 8600. Addition of lyate ions increases the rate for the methyl, ethyl, and isopropyl bromides, whereas the *tert*-butyl bromide solvolysis rate is unchanged. The reaction with lyate ions is overall second-order for methyl and ethyl, first-order for *tert*-butyl, and first- or second-order for the isopropyl member, depending upon the concentrations. Similar results are found in other solvents. These data show that the methyl and ethyl bromides solvolyze by the S_N2 mechanism, and *tert*-butyl bromide by the S_N1 mechanism. The isopropyl bromide does not fit so neatly into one of these classes; depending upon the solvent and the concentrations, it can present features of either mechanistic class. Evidently a change in the alkyl substitution on the α-carbon is responsible for a change in mechanism, with primary alkyl substrates tending to react by the S_N2 mechanism and tertiary substrates by S_N1. At some point along this scale of substituent changes presumably the mechanistic switch occurs, and this appears to be, roughly, with the secondary substrate. This behavior has given rise to the concept of "borderline" reaction, which do not fit unambiguously into either the S_N1 or S_N2 classes.[7, pp. 320-24]

Several proposals have been made to fit the borderline reactions into a well-defined mechanistic scheme. Most of these adopt one of two viewpoints: either (1) borderline substrates undergo concurrent S_N1 and S_N2 processes, with the particular system determining which mechanism, if either, predominates or (2) all S_N reactions are related by essentially the same mechanism, which differs from case to case in the detailed disposition of electrons in the transition state. In this view pure S_N1 and S_N2 processes are merely the extreme limiting forms of a single mechanism, and the borderline mechanism is a merged process having some features of both.

The notion of concurrent S_N1 and S_N2 reactions has been invoked to account for kinetic observations in the presence of an added nucleophile[102] and for heat capacities of activation,[103] but the hypothesis is not strongly supported.[104] Interpretations of borderline reactions in terms of one mechanism rather than two have been more widely accepted. Winstein et al.[105] have proposed a classification of mechanisms according to the covalent participation by the solvent in the transition state of the rate-determining step. If such covalent interaction occurs, the reaction is assigned to the nucleophilic (N) class; if covalent interaction is absent, the reaction is in the limiting (Lim) class. At their extremes these categories become equivalent to $S_N{}^1$ and $S_N{}^2$, respectively, but the dividing line between $S_N{}^1$ and $S_N{}^2$ does not coincide with that between N and Lim. For example, a mass-law effect, which is evidence of an intermediate and therefore of the S_N1 mechanism, can be observed for some isopropyl compounds, but these appear to be in the N class in aqueous media.

The N–Lim classification does not eliminate the possibility of borderline cases between these two categories, but it leads to the suggestion that no sharp distinction can be made between the possible intermediates in these mechanisms and that perhaps all solvolyses proceed via an intermediate. The mechanistic category of a particular solvolysis then depends upon the relative weights of the canonical structures **3**, **4**, and **5** to the transition state resonance hybrid.

$$Y^-: R\text{–}X \; <\!\!-\!\!-\!\!-\!\!-\!\!> \; Y\text{–}R \; :X^- \; <\!\!-\!\!-\!\!-\!\!-\!\!> \; Y^-: R^+ \; :X^-$$

3   **4**   **5**

The greater the contribution of **4** to the transition state, the more firmly the system is placed in the N category; likewise a large contribution from **5** is characteristic of the Lim category. Bentley and Schleyer[104] state that the essential difference between the S_N1 and S_N2 mechanisms depends upon whether nucleophilic attack occurs before (S_N2) or after (S_N1) the transition state of the rate-determining step. This is equivalent to Winstein's N–Lim classification.

Now, it can be postulated[105] that solvolysis rate should be a function of two properties of the solvent: one is its "ionizing power," and the other is its nucleophilicity. An S_N1 process should be promoted by high ionizing power, and an S_N2 process by high solvent nucleophilicity. At this point, we are ready to bring the extrathermodynamic approach to bear on this problem. This was initiated by Grunwald and Winstein,[106] who defined a solvent ionizing power parameter Y by

$$Y = \log \frac{k^{t\text{-BuCl}}}{k_0^{t\text{-BuCl}}} \qquad (8\text{-}66)$$

where $k_0^{t\text{-BuCl}}$ is the rate constant for the solvolysis of t-butyl chloride in 80 vol% ethanol, and $k^{t\text{-BuCl}}$ is the corresponding rate constant in another solvent. This reaction was selected as the model process because it was believed to occur by an essentially pure S_N1 process. A linear free energy equation is written in the form familiar from Chapter 7:

$$\log \frac{k}{k_0} = mY \qquad (8\text{-}67)$$

Ideally the parameter m should be characteristic of only the substrate, whereas Y should be a function of the solvent. The equation is expected to apply to reactions very similar to the defining reaction, that is, S_N1 solvolyses. Table 8-12 gives Y

TABLE 8-12. The Solvent Ionizing Power Measure Y for Aqueous Mixed Solvents at 25°C

Volume % of organic solvent[a]	Organic solvent					
	Ethanol	Methanol	Dioxane	Acetone	Acetic acid[b]	Formic acid[b]
0	3.49	3.49	3.49	3.49	3.49	3.49
10	3.31	3.28	3.22	3.23	—	—
20	3.05	3.03	2.88	2.91	—	—
25	2.91	—	—	2.69	2.84	3.10
30	2.72	2.75	2.46	2.48	—	—
40	2.20	2.39	1.95	1.98	2.31	—
50	1.66	1.97	1.36	1.40	1.94	2.64
60	1.12	1.49	0.72	0.80	1.52	—
70	0.60	0.96	0.01	0.13	—	—
75	—	—	—	—	—	—
80	(0.00)	0.38	−0.83	−0.67	—	2.32
90	−0.75	−0.30	−2.03	−1.86	—	2.22
100	−2.03	−1.09	—	—	−1.64	2.05

Source: Reference 107.
[a]x vol % solution prepared by mixing x volumes of organic solvent with 100-x volumes of water.
[b]Containing about 0.07 M lithium salts.

values measured by Fainberg and Winstein.[107] Y is not simply related to solvent composition over wide ranges, although over limited composition ranges Y can be described as a linear function of composition. The great differences in Y for pure alcohols (methanol, -1.09; ethanol, -2.03; isopropyl alcohol, -2.73; *tert*-butyl alcohol, -3.26) seem surprising, but are removed by placing the Y's on a molar basis.[108]

It is found that m is solvent dependent.[109] The R part of substrate RX cannot be made drastically different from that in the model substrate without causing dispersion into separate lines for different binary solvents. The leaving group X introduces another type of specificity.

We noted above that solvent nucleophilicity is expected to play a kinetic role for reactions possessing S_N2 character. Winstein et al.[105] took this participation into account formally by writing Eq. (8-68),

$$d \log k = \left(\frac{\partial \log k}{\partial Y} \right)_N dY + \left(\frac{\partial \log k}{\partial N} \right)_Y dN \qquad (8\text{-}68)$$

where N is an (unspecified) measure of solvent nucleophilicity. This equation is usually written

$$\log \left(\frac{k}{k_0} \right) = mY + lN \qquad (8\text{-}69)$$

and much recent work in this field has been directed to the definition and measurement of different Y and N scales. We note that Eqs. (8-68) and (8-69) omit consideration of a possible "cross-interaction" term of the type we generated in Eq. (7-41) for structure–reactivity relationships.

If Y is to be a valid measure of solvent ionizing power, presumably the defining reaction should proceed via the Lim (pure S_N1) process. This was the basis for the original choice[106] of t-butyl chloride. It is now believed that t-butyl chloride solvolyzes with some solvent participation, and modern versions of Y are based on other compounds, of which 2-adamantyl tosylate (p-toluenesulfonate, OTs), **6**, is the most favored.[110]

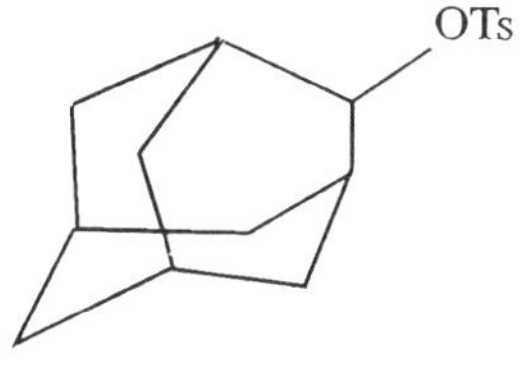

6

Although this is a secondary substrate, complete shielding from backside attack by nucleophiles leads to S_N1 solvolysis without solvent participation. The corresponding Y value is defined[111] by Eq. (8-70)

$$Y_{\text{OTs}} = \log \frac{k^{\text{2-AdOTs}}}{k_0^{\text{2-AdOTs}}} \qquad (8\text{-}70)$$

where the standard solvent is again 80 vol% ethanol. Many investigations have been made of the dependence of Y on the nature of the leaving group, giving rise to the quantities Y_{OClO_3}, Y_{Picrate}, Y_{OTf} (trifluoromethane sulfonate),[117,118] Y_{PFBS} (pentafluorobenzene sulfonate),[114] and Y_{OTr} (tresylate).[115] Kevill and Hawkinson[115] have concluded (on the basis of linear correlations among these scales) that sensitivity of the scale to the leaving group is not a matter of great importance. Table 8-13 gives some values of Y_{OTs}.[111,116]

Our next concern is the solvent nucleophilicity. Schadt et al.[111] chose the solvolysis of methyl tosylate, which should be an S_N2 process, as the defining process. For this standard reaction the parameter l in Eq. (8-69) was set at 1.00. An empirical estimate of m, describing the sensitivity of methyl tosylate to solvent ionizing power, was obtained as the slope of the two-point line for methyl tosylate solvolysis in

TABLE 8-13. Solvent Ionizing Power (Y_{OTs}) and Nucleophilicity (N_{OTs}) Based on 2-Adamantyl Tosylate

Pure solvents

Solvent	Y_{OTs}	N_{OTs}
H_2O	4.1	-0.44
$C_{H3}OH$	-0.92	-0.04
C_2H_5OH	-1.75	0.00
CH_3COOH	-0.61	-2.35
$HCOOH$	3.04	-2.35
CF_3COOH	4.57	-5.56
CF_3CH_2OH	1.80	-3.0

Solvent mixtures

Volume % organic component	Acetone Y_{OTs}	N_{OTs}	Ethanol Y_{OTs}	N_{OTs}	Methanol Y_{OTs}	N_{OTs}
10	3.58	-0.41	3.78	-0.41	3.78	-0.41
20	3.05	-0.38	3.32	-0.34	3.39	-0.35
30	2.50	-0.40	2.84	-0.35	2.97	-0.30
40	1.85	-0.38	2.08	-0.23	2.43	-0.21
50	1.26	-0.39	1.29	-0.09	2.00	-0.19
60	0.66	-0.41	0.92	-0.08	1.52	-0.13
70	—	—	0.47	-0.05	1.02	-0.08
80	—	—	(0.00)	(0.00)	0.47	-0.05
90	—	—	-0.58	0.01	-0.17	-0.01

Source: References 111 and 116.

formic and acetic acids, which are believed to be equally nucleophilic but to have different ionizing powers; this gave $m = 0.3$. Solving Eq. (8-69) for N and substituting these numerical values for l and m,

$$N_{OTs} = \log \left(\frac{k^{MeOTs}}{k_0^{MeOTs}} \right) - 0.3 \, Y_{OTs} \qquad (8\text{-}71)$$

Table 8-13 gives some N_{OTs} values.[111,116]

Kevill and co-workers[117–9] have developed nucleophilicity scales based on the solvolysis of cationic substrates $R-X^+$, the leaving group being neutral rather than anionic. Their $N_{Et_3O^+}$ scale is defined in Eq. (8-72).

$$N_{Et_3O^+} = \log \frac{k^{Et_3O^+}}{k_0^{Et_3O^+}} \qquad (8\text{-}72)$$

the mY^+ correction term corresponding to that of Eq. (8-71) being of negligible magnitude. [These workers believe that the factor 0.3 in Eq. (8-71) should be 0.55; this is an unresolved issue.[120]] One motivation for this work is the possibility that solvent participation may involve electrophilic assistance as well as a nucleophilic role, and Y scales based on neutral substrates will include the electrophilic participation together with the ionizing power.

Schadt et al.[111] have written Eq. (8-69) in the alternative form

$$\log \left(\frac{k}{k_0} \right) = (1 - Q) \log \left(\frac{k^{MeOTs}}{k_0^{MeOTs}} \right) + Q \log \frac{k^{2\text{-}AdOTs}}{k_0^{2\text{-}AdOTs}} \qquad (8\text{-}73)$$

which states that a solvolysis rate (k) is a linear combination of the S_N2 (k^{MeOts}) and S_N1 $(k^{2\text{-}AdOTs})$ models. Combination of Eqs. (8-70), (8-71), and (8-73) and comparison with Eq. (8-69) shows that $l = 1 - Q$; $m = 0.3 + 0.7 \, Q$; and $l = (1 - m)/0.7$. Note that parameters l and m are correlated.

These kinetic measures of solvent ionizing power and nucleophilicity can be used to assess the mechanism of a solvolysis reaction. The simplest procedure is to plot $\log k$ against Y or Y_{OTs}. A good linear correlation with a slope near unity suggests the S_N1 mechanism, that is, a mechanism similar to that of the defining reaction. A shallow slope, curvature, or a scatter diagram implicates solvent participation and calls for the multiple linear relationship Eq. (8-69). [Some solvolyses of S-alkylbenzothiophenium ions are well correlated with only the lN term of Eq. (8-69).[119]] Schadt et al.[111] show examples of the use of Eqs. (8-69) and (8-73) to fit solvolysis data for primary and secondary tosylates; the regression results are given in Table 8-14. In every instance use of the four-parameter Eq. (8-69) markedly improved the correlation relative to the two-parameter Eq. (8-67). The magnitudes of l and m can be interpreted by comparison with the model processes, namely, $l = 0$ and $m = 1$ for 2-AdOTs solvolysis and $l = 1$, $m = 0.3$ for MeOTs solvolysis. Evidently nucleophilic solvent participation is important in all of these reactions.

TABLE 8-14. Regression Parameters for Tosylate
Solvolysis Reactions[a]

| Tosylate | Eq. (8-69) | | Eq. (8-73) |
	l	m	Q
Ethyl	0.89	0.40	0.12
2-Propyl	0.49	0.62	0.48
2-Butyl	0.41	0.71	0.59
2-Pentyl	0.40	0.73	0.60
3-Pentyl	0.41	0.69	0.58
4-Heptyl	0.33	0.75	0.66
Cyclopentyl	0.37	0.70	0.60
Cyclohexyl	0.32	0.78	0.68
Benzyl	0.75	0.64	0.51

Source: Reference 111.
[a]At 25°C. The solvents were EtOH, 50 and 80 vol%
EtOH, HOAc, and HCOOH.

In 1955 Swain, et al.[121] proposed a four-parameter equation, Eq. (8-74), to
describe the solvent dependence of solvolytic reactions.

$$\log \left(\frac{k}{k_0} \right) = c_1 d_1 + c_2 d_2 \tag{8-74}$$

The parameters c_1, c_2 were postulated to be dependent only upon the substrate, and
d_1, d_2, upon the solvent. A large body of kinetic data, embodying many structural
types and leaving groups, was subjected to a statistical analysis. In order to achieve
a unique solution, these arbitrary conditions were imposed: $c_1 = 3.0\ c_2$ for MeBr;
$c_1 = c_2 = 1.0$ for t-BuCl; $3.0\ c_1 = c_2$ for Ph_3CF. Some remarkably successful
correlations [calculated vs. experimental $\log (k/k_0)$] were achieved, but the approach
appeared to lack physical significance and was not much used. Many years later
Peterson et al.[122] showed a correspondence between Eqs. (8-69) and (8-74); in
particular, the very simple result $d_1 + d_2 = Y$ was found.

The Menschutkin reaction of tripropylamine and methyl iodide was proposed by
Drougard and Decroocq[123] as a defining process for a kinetic measure of solvent
polarity. Drougard and Decroocq[123] and Abraham and Grellier[65] have shown that
solvent effects on pairs of Menschutkin reactions are linearly correlated, so any
well-behaved Menschutkin reaction could play the role of the standard process; for
example, $\delta_M \Delta G^{\ddagger}$ of Table 8-10, for the quaternization of triethylamine and ethyl
iodide, might serve as a polarity measure. Auriel and de Hoffmann[124] detected
specific solvation effects of protic solvents on a quaternization reaction by means
of deviations of these points from a linear correlation with the parameter of Drougard
and Decroocq.

Oshima and Nagai[125] used the 1,3-cycloaddition of diphenyldiazomethane to tetracyanoethylene as the standard reaction for the definition of a solvent parameter D_π,

$$D_\pi = \log \frac{k_S}{k_0}$$

where k_s and k_0 are second-order rate constants for the reaction in solvent s and in the reference solvent (benzene), respectively. D_π is considered to be a kinetic measure of solvent basicity relative to the Lewis acid tetracyanoethylene (TCNE). D_π correlates roughly with the equilibrium measure ΔH^0 of Table 8-11.

Spectroscopic Measures

The preceding empirical measures have taken chemical reactions as model processes. Now we consider a different class of model process, namely, a transition from one energy level to another within a molecule. The various forms of spectroscopy allow us to observe these transitions; thus, electronic transitions give rise to ultraviolet–visible absorption spectra and fluorescence spectra. Because of solute–solvent interactions, the electronic energy levels of a solute are influenced by the solvent in which it is dissolved; therefore, the absorption and fluorescence spectra contain information about the solute–solvent interactions. A change in electronic absorption spectrum caused by a change in the solvent is called *solvatochromism*.

The energy difference between the ground and excited states is given by

$$\Delta E = h\nu = hc/\lambda \tag{8-75}$$

where h is Planck's constant, c is the speed of light, ν is the frequency of the absorbed (or emitted) light, and λ is the wavelength of the light. A shift of the wavelength of the absorption maximum (caused, for example, by a change in solvent) to a longer wavelength is called a *bathochromic*, or red, shift; a shift to a lower wavelength is a *hypsochromic*, or blue, shift. Evidently a bathochromic shift indicates that ΔE has been decreased, whereas a hypsochromic shift means that ΔE has been increased, by the change in solvent.

Solvatochromic shifts are rationalized with the aid of the Franck–Condon principle, which states that during the electronic transition the nuclei are essentially immobile because of their relatively great masses. The solvation shell about the solute molecule minimizes the total energy of the ground state by means of dipole–dipole, dipole-induced dipole, and dispersion forces. Upon transition to the excited state, the solute has a different electronic configuration, yet it is still surrounded by a solvation shell optimized for the ground state. There are two possibilities to consider:

1. Let μ_{gr} and μ_{ex} be the dipole moments of the ground and excited states. Then if $\mu_{gr} > \mu_{ex}$, the less polar excited state is surrounded by a solvation shell

oriented for a polar solvent. Upon changing the medium to a solvent of greater polarity, the excited state is more destabilized relative to the ground state, ΔE increases, and a hypsochromic shift results.

2. If $\mu_{gr} < \mu_{ex}$, an increase in solvent polarity stabilizes the excited state relative to the ground state, producing a bathochromic shift.

Kosower[2,126] made the first use of this phenomenon for measuring solvent polarity. The model process is the absorption transition of 1-ethyl-4-carbomethoxy-pyridinium iodide, **7**:

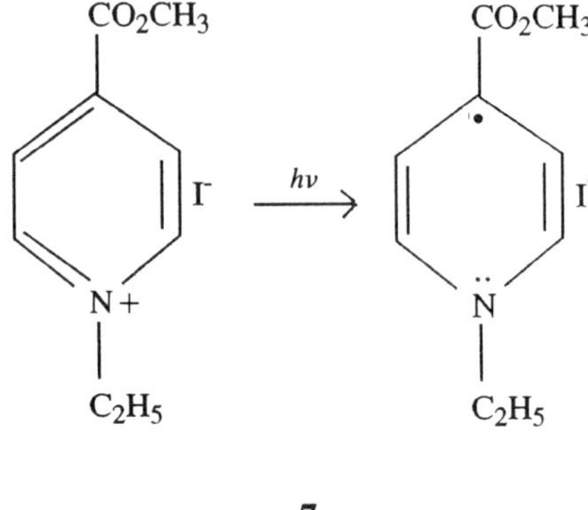

7

and the solvent polarity parameter is the transition energy (at the longest wavelength band maximum) in kilocalories per mole. This parameter, symbolized Z, is calculated with Eq. (8-76), where λ_{max} is in nanometers.

$$Z = 2.859 \times 10^4/\lambda_{max} \tag{8-76}$$

Figure 8-7 illustrates schematically the electronic natures of the polar ground state (an ion pair) and the less polar excited state. This is, therefore, a case of μ_{gr}

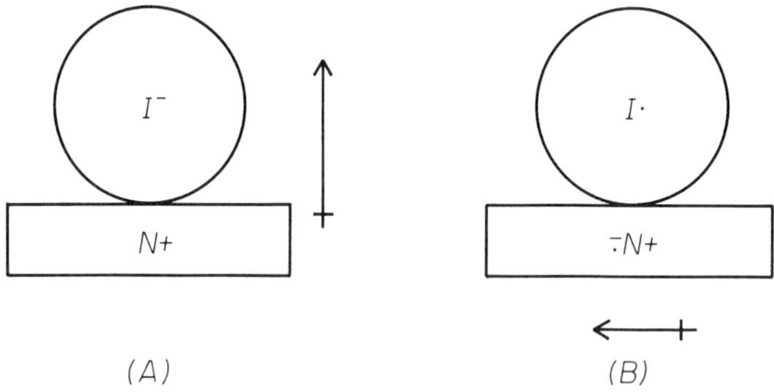

Figure 8-7. Model of the alkylpyridinium iodide complex: (*A*) ground state. (*B*) Excited state [after Kosower[126]].

$> \mu_{ex}$, so hypsochromic shifts are observed (upon changing to solvents of higher polarity), and Z is larger for more polar solvents.

Z values are obtained from Eq. (8-76) for solvents having Z in the approximate range 63–86. In more polar solvents the CT band is obscured by the pyridinium ion ring absorption, and in nonpolar solvents 1-ethyl-4-carbomethoxy-pyridinium iodide is insoluble. By using the more soluble pyridine-1-oxide as a secondary standard and obtaining an empirical equation between Z and the transition energy for pyridine-1-oxide, it is possible to measure the Z values of nonpolar solvents. The value for water must be estimated indirectly from correlations with other quantities. Table 8-15 gives Z values for numerous solvents.[126,127]

Another solvatochromic polarity measure, E_T (30), is the transition energy for compound **8**, which is 2,6-diphenyl-4-(2,4,6-triphenylpyridinio)phenolate, also referred to as *Dimroth–Reichardt's betaine*.

8 $(R = C_6H_5)$

Dimroth et al.[128] introduced **8** as a solvatochromic probe of solvent polarity having absorption in the visible region; it shows the largest solvatochromic shift of any substance yet reported. E_T (30) is calculated with Eq. (8-76), like Z. (The peculiar symbolism arose because compound **8** happened to be No. 30 on the list of substances studied by Dimroth et al.) The shift is hypsochromic as solvent polarity is increased. Table 8-16 gives some E_T (30) values.[5, Chap. 7] E_T (30) is linearly correlated with Z, and this correlation allows E_T (30) values to be indirectly estimated for carboxylic acid solvents, which protonate the phenolic oxygen of **8**. A secondary solvatochromic probe is also required for hydrocarbon solvents, in which **8** is not soluble.

Other solvatochromic probes have been proposed. Mukerjee et al.[51] used nitroxides for this purpose, finding that their transition energies correlate linearly with Z and E_T (30). Brooker et al.[129] prepared a polar merocyanine that shows a blue shift

TABLE 8-15. Z Values of Some Solvents

Solvent	Z/kcal mol^{-1}	Solvent	Z/kcal mol^{-1}
Benzene	54.0	Dimethylsulfoxide	70.2
Chlorobenzene	58.0	Nitromethane	71.2
Tetrahydrofuran	58.8	Acetonitrile	71.3
Bromobenzene	59.2	t-Butyl alcohol	71.3
Cyclohexane	60.1	i-Propyl alcohol	76.3
1,1-Dichloroethane	62.1	n-Butyl alcohol	77.7
Chloroform	63.2	i-Butyl alcohol	77.7
1,2-Dichloroethane	63.4	n-Propyl alcohol	78.3
Ethyl acetate	64.0	Benzyl alcohol	78.4
Methyl ethyl ketone	64.0	2-Methoxyethanol	78.5
Pyridine	64.0	Acetic acid	79.2
Dichloromethane	64.2	Ethanol	79.6
Dioxane	64.5	Glycerol	82.7
Benzonitrile	65.0	Formamide	83.3
Acetone	65.7	Methanol	83.6
Dimethylacetamide	66.9	Ethylene glycol	85.1
Dimethylformamide	68.5	Water	91.8

TABLE 8-16. $E_T(30)$ Values of Some Solvents[a]

Solvent	$E_T(30)$/kcal mol^{-1}	Solvent	$E_T(30)$/kcal mol^{-1}
Cyclohexane	30.9	Dimethylformamide	43.8
n-Hexane	31.0	Acetic anhydride	43.9
Carbon tetrachloride	32.4	Dimethyl sulfoxide	45.1
Toluene	33.9	Acetonitrile	45.6
Benzene	34.3	Nitromethane	46.3
Diethyl ether	34.5	2-Butanol	47.1
Dioxane	36.0	1-Octanol	48.3
Bromobenzene	36.6	i-Propyl alcohol	48.4
Chlorobenzene	36.8	i-Butyl alcohol	48.6
Tetrahydrofuran	37.4	1-Butanol	50.2
Ethyl acetate	38.1	Benzyl alcohol	50.4
Chloroform	39.1	1-Propanol	50.7
1,1-Dichloroethane	39.4	Acetic acid	51.7
Methyl acetate	40.0	Ethanol	51.9
Pyridine	40.5	2-Methoxyethanol	52.3
Dichloromethane	40.7	Formic acid	54.3
Nitrobenzene	41.2	Methanol	55.4
1,2-Dichloroethane	41.3	Ethylene glycol	56.3
Acetone	42.2	Formamide	56.6
t-Butyl alcohol	43.3	Glycerol	57.0
Dimethylacetamide	43.7	Water	63.1

Source: Reference 5 (Chap. 7).
[a]At 25° C.

(its transition energy being designated χ_B) and a nonpolar red-shifting merocyanine, whose transition energy is called χ_R. The χ_B and χ_R values do not correlate well with each other, although χ_B is linearly related to Z and E_T (30). Evidently χ_B provides essentially the same information as Z and E_T (30), whereas χ_R constitutes a different combination of responses to solvent properties; χ_R may be a suitable measure of solvent effects in nonpolar systems.[129]

Other forms of spectroscopy can be used. Gutmann[94] defined the acceptor number (AN) of a solvent as the ^{31}P-NMR chemical shift of triethylphosphine oxide in the solvent relative to the chemical shift of the $Et_3PO{\rightarrow}SbCl_5$ adduct, which is taken as 100. AN is considered to be a measure of the Lewis acid electron-pair acceptor ability of the solvent and, therefore, related to electrophilicity. Since E_T (30) and Z are closely related to AN, it was concluded that these other quantities do not measure solely polarity. Nicolet et al.[130] measured the infrared stretching frequency of the carbonyl groups in Cl_3CCOOH and $Cl_3CCOOCH_3$, finding a good linear correlation for non-H-bond acceptor solvents. Deviations from this line were observed for H-bond acceptor solvents, and the difference $\Delta\nu$ between the observed frequency and that expected in the absence of H bonding is a measure of the H-bond base strength of the solvent. We will subsequently see other examples of this comparison method. Dong and Winnik[131] introduced the polarity parameter Py, which is the ratio of intensities of two vibrational bands in the fluorescence spectrum of pyrene. Accurate Py determinations require the elimination of instrumental artifacts[132] and careful temperature control.[133] Py is linearly related to Z and E_T,[30] with protic and aprotic solvents giving different correlation lines. Values of Z, E_T,[30] and Py have been measured in mixed solvents.[5, Chap. 7; 126, 134-7]

Each of these spectroscopic measures of solvent character has been constructed from a single model process, chosen because it might usefully reflect a combination of solute–solvent and solvent–solvent interactions applicable to many other chemically interesting phenomena. We have seen that some of these measures are mutually correlated, indicating that they are reporting on the same interactions. However, any attempt to express, in a single number derived from a single model process, all of the possibilities of solvent effects in a general way seems now to be improbable, and as a consequence most current efforts in this field are directed to the dissection of net effects into component parts, such as polarity, polarizability, H bonding, etc. These approaches tend to make use of many model processes in an averaging manner designed to produce measures of very wide applicability. A long series of papers by Kamlet, Taft, and their co-workers[138,139] constitutes a major development of this type of approach, and we will outline their solvatochromic method.

The basic premise of Kamlet and Taft is that attractive solute–solvent interactions can be represented as a linear combination of a nonspecific dipolarity/polarizability effect and a specific H-bond formation effect, this latter being divisible into solute H-bond donor (HBD)–solvent H-bond acceptor (HBA) interactions and the converse possibility. To establish the dipolarity/polarizability scale, a solvent set was chosen with neither HBD nor HBA properties, and the spectral shifts of numerous solvatochromic dyes in these solvents were measured. These shifts, $\Delta\nu$, were related to a dipolarity/polarizability parameter π^* by $\Delta\nu = s\pi^*$. The quantity π^* was scaled to $\pi^* = 0.000$ for cyclohexane and 1.000 for DMSO. The resulting π^* values were quantities averaged over all the solvatochromic dyes.[140,141]

A solvent HBD parameter α (acidity) is determined by a solvatochromic comparison method[142] based on these conditions:

1. A linear correlation is observed between solvatochromic shifts for two solutes, one capable of, one incapable of, H bonding, in a series of solvents for which H bonding is not possible.
2. Points in this correlation representing solvents in which H bonding is possible should show significant deviations from the linear "reference" line.
3. These deviations should possess magnitudes and directions that are chemically reasonable.

This procedure was applied to numerous solvatochromic scales [Z, χ_R, E_T (30)], using 4-nitroanisole as the solute incapable of H bonding. As in the development of the π^* scale, the α scale is averaged over several processes. A β scale of HBA ability (basicity) was established in a similar manner. Table 8-17 lists some π^*, α, and β values.[46]

It is then proposed that any solvent-dependent energy property, such as $\log k$ and ΔG^0, can be expressed as the sum

$$\log \left(\frac{k}{k_0} \right) = s\pi^* + a\alpha + b\beta \tag{8-77}$$

Since π^*, α, and β are approximately normalized scales, the coefficients s, a, and b are measures of the relative weights of the dipolarity/polarizability, HBD ability, and HBA ability of the solvent. Equation (8-77) has been extended to take account of the cavity effect and certain anomalies, as we will see later in this section.

This approach to separating the different types of interactions contributing to a net solvent effect has elicited much interest. Tests of the π^*, α, and β scales on other solvatochromic or related processes have been made,[143] an alternative π^* scale based on chemically different solvatochromic dyes has been proposed,[144] and the contribution of solvent polarizability to π^* has been studied.[145,146] Opinion is not unanimous, however, that the Kamlet–Taft system constitutes the best or ultimate extrathermodynamic approach to the study of solvent effects. There are two objections: One of these is to the averaging process by which many model phenomena are combined to yield a single best-fit value. We encountered this problem in Section 7.2 when we considered alternative definitions of the Hammett substituent constant, and similar comments apply here; Reichardt has discussed this in the context of the Kamlet–Taft parameters.[5, p. 379] The second objection is to the claim of generality for the parameters and the correlation equation[147]; we will return to this controversy later.

Many linear correlation equations have been described between pairs of empirical parameters, especially the solvatochromic parameters.[5, Chap. 7; 6, 124] Aside from their fundamental meaning that linearly correlated scales are responding similarly to

TABLE 8-17. Solvatochromic π^*, α, and β Values

Solvent	π^*	α	β
n-Hexane	-0.08	0.00	0.00
Diethyl ether	0.27	0.00	0.47
Carbon tetrachloride	0.28	0.00	0.00
t-Butyl alcohol	0.41	0.68	$(1.01)^a$
n-Butyl alcohol	0.47	0.79	(0.88)
i-Propyl alcohol	0.48	0.76	(0.95)
n-Propyl alcohol	0.52	0.78	—
Toluene	0.54	0.00	0.11
Ethanol	0.54	0.83	0.77
Dioxane	0.55	0.00	0.37
Ethyl acetate	0.55	0.00	0.45
Chloroform	0.58	(0.44)	0.00
Tetrahydrofuran	0.58	0.00	0.55
Benzene	0.59	0.00	0.10
Methyl acetate	0.60	0.00	0.42
Methanol	0.60	0.93	0.62
Acetic acid	0.64	1.12	—
Chlorobenzene	0.71	0.00	0.07
Acetone	0.71	0.08	0.48
Acetonitrile	0.75	0.19	0.31
Acetic anhydride	0.76	0.00	—
Bromobenzene	0.79	0.00	0.06
Dichloromethane	0.82	(0.30)	0.00
Nitromethane	0.85	0.22	—
Pyridine	0.87	0.00	0.64
Dimethylformamide	0.88	0.00	0.69
Dimethylacetamide	0.88	0.00	0.76
Ethylene glycol	0.92	0.90	0.52
Formamide	0.97	0.71	—
Benzyl alcohol	0.98	—	(0.50)
Dimethylsulfoxide	1.00	0.00	0.76
Nitrobenzene	1.01	0.00	0.39
Water	1.09	1.17	(0.18)

Source: Reference 46.

^{a}Values in parentheses are less certain.

changes in solvent properties, these correlations can be useful for estimating missing data and for obtaining values of parameters that may be experimentally inaccessible for reasons of insolubility, spectral properties, or instability. Related to this issue of correlations between solvent parameters is the independent development of scales of solvent polarity by chromatographers, who have need to characterize the solute–solvent interaction capabilities of the stationary phase in gas–liquid chromatography and the mobile phase in liquid–liquid chromatography. Several ingenious measures have been developed.[148–50] These polarity scales are now being reevaluated, largely with the aid of correlations with solvatochromic measures.[151–4] It may happen that the solvatochromic parameters themselves will serve to characterize chromatographic liquid phases.

Linear Solvent Effect Relationships

The empirical solvent parameters are energy quantities or are proportional to energies, so a correlation of reactivity as $\log k$ or $\log (k/k_0)$ against one or more of these parameters is a correlation of energy quantities. Some authors call these *linear solvation energy relationships* (LSERs); this designation is perhaps too precise to be generally accurate, so we will use the term *linear solvent effect relationship*. LSERs may be univariate or multivariate correlations; the parameters of the equation and their uncertainties are obtained by least-squares regression analysis (Section 2.3 and Reference 155).

Several uses may be made of such correlations. Very precise correlations constitute a means for estimating (i.e., predicting) the reactivity in an unstudied solvent. Such correlations constitute normal or expected behavior for the data set, and significant (i.e., statistically and chemically significant) deviations may be indications of special effects, such as selective solvation. At a deeper level, inferences may be drawn about the molecular level interactions responsible for the solvent effect. This last application is a fundamental one, yet it is also subject to great uncertainty, in part because our knowledge of the "constitution" of the empirical parameters is imperfect, being merely hypothetical and because the correlation may be complicated by particular issues, such as a limited solvent set or colinearity between parameters. Brady and Carr[156] have cautioned against oversimplified molecular interpretations of LSERs.

Univariate LSERs may possess the conventional LFER form, as exemplified by Eq. (8-67), the Grunwald–Winstein equation, or they may simply be plots of $\log k$ against a solvent parameter such as Z, E_T (30), or π^*. Brownstein[157] developed an LFER form for the latter type of correlation, writing

$$\log \left(\frac{k}{k_0} \right) = RS$$

where R is characteristic of the reaction and S of the solvent. R was defined to be unity for Kosower's model process, and S was defined to be zero for ethanol, the standard solvent; then from Z values a scale of S values could be developed.

It is expected that the excellence of a correlation should diminish as the correlated process and the model process are made increasingly different in their mechanistic character. Thus, we should not expect great generality from univariate correlations; on the other hand, from the quality of the correlation we may be able to learn something about the correlated process. It is not surprising that the rate of one Menschutkin reaction is well correlated with the rate of another Menschutkin reaction, for their mechanisms should be very similar. The solvent effect data ($\delta_M \Delta G^{\ddagger}$) in Table 8-10 are poorly correlated with the Kirkwood dielectric constant function, but a plot against E_T (30) shows some improvement.[5, pp. 199, 391] In this case the microscopic quantity E_T (30) appears to mimic the solvation effects on the reaction rate better than does the bulk property ε. Any mechanistic inference from the correlation must be very limited, but there is a sense that the solvatochromic correlation has expanded our ability to comprehend this reaction by relating it to a quite different phenomenon. This modest achievement is typical of many LFERs and LSERs.

Langhals[158] has described a remarkable relationship of most of the empirical solvent parameters [Z, E_T (30), Y, etc.] to composition in binary solvent mixtures:

$$P = E_D \ln \left(\frac{c}{c^*} + 1 \right) + P^0 \tag{8-78}$$

where P is the solvent parameter in the solvent mixture, P^0 is its value in the pure, less polar component, c is the molar concentration of the more polar component, and E_D and c^* are adjustable parameters. For many binary mixtures a plot of P against $\ln[(c/c^*) + 1]$ is linear over the entire composition range; both aqueous–organic and organic–organic mixtures show this behavior. For some aqueous–organic mixtures, such as dioxane–water and ethanol–water, discontinuities occur at compositions independent of the polarity measure and, therefore, characteristic of the solvent. Some organic–organic mixtures show a maximum in the plot. Rezende[159,160] has applied Eq. (8-78) to salt solutions, treating the salt as the more polar component. If the solvent polarity parameter is correlated by this function of concentration, then a log k that is linear in P will also be correlated.

Numerous authors have devised multiple linear regression approaches to the correlation of solvent effects, the intent being to widen the applicability of the correlation and to develop insight into the molecular factors controlling the correlated process.[161–3] For example, Zilian[163] treated polarity as a combination of effects measured by molar refraction, AN, and DN. Koppel and Palm[162] write

$$\log \left(\frac{k}{k_0} \right) = yY_\varepsilon + pP_n + eE + bB \tag{8-79}$$

where Y_ε is the Kirkwood dielectric constant function (considered a measure of polarity), P_n is the molar refraction refractive index function (measuring polarizability), E is E_T (30) (said to be a measure of electrophilic solvating power), and B, a measure of nucleophilic solvating power, is the infrared frequency of the oxygen–deuterium band. The correlation coefficients y, p, e, b are interpreted as measures of the role played by the respective factors. The $yY + pP$ portion is considered to describe the nonspecific solvent effect, and the $eE + bB$ describes specific effects.

The most familiar multivariate LSER is that of the Kamlet–Taft group, which in its fully rigged form is now written

$$\log \left(\frac{k}{k_0} \right) = s(\pi^* + d\delta) + a\alpha + b\beta + h\delta_H^2 + e\xi \tag{8-80}$$

where

π^* is the dipolarity/polarizability parameter.
α is the solvent HBD (acidity) parameter.

β is the solvent HBA (basicity) parameter.

δ_H is the Hildebrand solubility parameter.

δ is an empirical polarizability correction term[140,141] having the values

   $\delta = 0.0$ (nonchlorinated aliphatic solvents).

   $\delta = 0.5$ (polychlorinated aliphatic solvents).

   $\delta = 1.0$ (aromatic solvents).

ξ is a coordinate covalency index.

We have already encountered the π^*, α, and β quantities. The δ_H^2 term is inserted to account for the cavity effect. Equation (8-80) is a 12-parameter equation for which considerable generality is claimed, in that it is said to be applicable to chemical rates and equilibria, spectra, solubilities, partition coefficients, and even biological responses. Usually, of course, by judicious selection of solvents, it is possible to reduce the number of parameters by ensuring that some terms are negligible.[46] An example requiring most of the parameters in Eq. (8-80) is the solvolysis/dehydrohalogenation of t-butyl chloride in 21 HBD and non-HBD solvents, for which this correlation was found:[164]

$$\log k = -14.60 + 5.10\pi^* + 4.17\alpha + 0.73\beta + 0.0048\delta_H^2$$

It is claimed that a measures electrophilic assistance by the solvent, b measures nucleophilic assistance, and that at least three, and sometimes four, parameters are required to perform a dissection into these separate effects. These workers also decomposed Y into π^* and α contributions, and N into π^* and β contributions. Abraham et al.[165] have compared the performance of Eqs. (8-79) and (8-80) and find that they are about equally successful in correlating data.

Equations (8-79) and (8-80) constitute attempts to account for all types of solvent effects on all types of processes, and this goal of a general quantitative description provides an interesting contrast with the situation in structure–reactivity studies, where more limited objectives are established. Thus, for example, we are content to deal with aromatic and aliphatic substrates by means of separate empirical parameters and correlations. An analogy with solvent effects might divide these into effects of protic and aprotic solvents or into finer subdivisions. The search for generality has subjected authors of these multiple regression equations (particularly the Kamlet–Taft group) to criticism that the generality does not actually exist and that apparent successes are a consequence of limited or selected data sets that explore only a small part of the universe of possibilities.[147]

Statistical Approaches

In the above paragraphs we saw that multiple linear regression analysis on equations of the form

$$\log \left(\frac{k}{k_0} \right) = \Sigma a_i x_i \tag{8-81}$$

can be used to describe solvent effects; in Eq. (8-81) the x_i are properties of the solvents i, and the a_i are coefficients expressing sensitivity to the solvent properties. The x_i are given, and least-squares regression provides estimates of the a_i. All LFERs and LSERs are of the form of Eq. (8-81).

We now consider a type of analysis in which the data (which may consist of solvent properties or of solvent effects on rates, equilibria, and spectra) again are expressed as a linear combination of products as in Eq. (8-81), but now the statistical treatment yields estimates of *both* a_i and x_i. This method is called *principal component analysis* or *factor analysis*.[166] A key difference between multiple linear regression analysis and principal component analysis (in the chemical setting) is that regression analysis adopts chemical models a priori, whereas in factor analysis the chemical significance of the factors emerges (if desired) as a result of the analysis. We will not explore the statistical procedure, but will cite some results. We have already encountered examples in Section 8.2 on the classification of solvents[47] and in the present section in the form of the Swain et al.[121] treatment leading to Eq. (8-74).

Cramer[167] applied factor analysis to six physical properties (aqueous solvation energy, partition coefficient, boiling point, molar refraction, volume, and vaporization enthalpy) of 114 solvents, finding that about 96% of the variance in these properties is expressible in terms of two parameters identified as *bulk* and *cohesiveness* of the solvent molecules. This reduction in number of variables was suggested as the basis for the apparent simplicity of many LSERs. Maria et al.[168] applied factor analysis to 10 measures of basicity for 22 nonprotogenic solvents, isolating two factors that accounted for 95% of the total variance. One of these factors is a blend of electrostatic and CT character, the other is essentially electrostatic. Chastrette and Carretto[169] studied 57 aprotic and 24 protic solvents, the data base consisting of π^*, E_T (30), Z, δ_H, μ, ε, R_M, etc. It was concluded that, for the protic solvents, E_T (30) measures polarity, polarizability, and cohesion to the extent of 43, 39, and 18%, respectively; for π^* these figures were 53, 18, and 29%. One must admire the potential of a technique that can lead to such conclusions even while one remains somewhat skeptical of the particular result.

Swain et al.[170] analyzed solvent effects on 1080 pieces of rate and equilibrium data, showing that more than 98% of the effects could be correlated by the four-parameter equation

$$\log \left(\frac{k}{k_0} \right) = aA + bB \tag{8-82}$$

where factor A is identified with anion-solvating ability (called *acity*) and B with cation-solvating ability (*basity*). It was not found necessary to add terms for H bonding; indeed, A is said to account for anion solvation, H-bond acidity, and electrophilicity, and likewise B incorporates cation solvation, H-bond basicity, and nucleophilicity. Swain et al.[170] suggest that the sum $A + B$ is a measure of solvent polarity. Taft et al.[171] responded to this two-factor analysis by pointing out that of the 77 reactions and properties studied by Swain, 71 involved non-HBD solvents and the other 6 were very weak or non-HBA solvents, so conditions for the applicability of β were not met. Because B correlates well with π^*, Taft et al. conclude that B measures solvent dipolarity/polarizability rather than basity. Swain[172] then cautioned against overinterpreting small deviations in correlations, namely, 0.3 unit or less in log k.

A different sort of controversy has developed between the Kamlet–Taft group and Sjöström and Wold, who in an exchange of papers[147,173–6] debate the respective strengths and weaknesses of the chemical model approach (Kamlet–Taft) and the factor analysis approach (Sjöström–Wold). Sjöström and Wold argue that the generality claimed by Kamlet and Taft for their LSER as embodied in Eq. (8-80) is illusory and that LFERs and LSERs are merely empirical representations of similarity, being most effective when the correlated process is closely similar to the model process. Thus, a failure to appreciate this "local" applicability of LFERs leads to confusion of the type evidenced by the profusion of σ substituent constant scales. Kamlet and Taft reply that the strength of the chemical model approach is that it leads to interpretations having chemical meaning, which the factors of principal component analysis do not.

This controversy appears to confound two separate issues. One of these is the *general* (i.e., the fundamental and valid in all chemical systems) versus the *particular* (the locally valid in the sense in which a Taylor's series expansion of any function leads to a locally valid linear representation). The other issue is the philosophical distinction between *understanding* (i.e., understanding chemically) and *describing* (with its statistical capability of predicting). These are important issues in the context of LFERs and LSERs. They raise the question of the relevance of chemical insight and scientific model building in an era when correlation functions can be arrived at without the introduction of chemical concepts. One possible outcome may be that chemical understanding is best achieved in the context of the particular solutions, whereas a general statement will tend to be merely a quantitative description, whose chemical meaning remains obscure because of the inherent complexity of the subject.

8.5 STRONGLY ACID SOLUTIONS

Acid catalysis is an important kinetic phenomenon, and its study often requires the use of concentrated acid solutions, in which the conventional pH scale is not applicable. In such solutions (e.g., sulfuric acid–water mixtures covering the full range of compositions) the acid component simultaneously functions both as an acid and as a solvent; thus, a medium effect is superimposed on the acidity effect. In this section we briefly describe the acidity function approach to coping with this problem. (A comparable approach can be taken to the study of highly basic solutions.) For more detailed reviews see Hammett,[177] Rochester,[178] and Stewart.[179]

Acidity Functions

Most organic compounds are bases, that is, they are capable of accepting a proton. The best-studied organic bases are the moderately strong ones, which will receive a proton in dilute aqueous solutions; amines are the most important examples. The pK_a value of the protonated base, referred to the infinitely dilute aqueous solution, is the usual measure of base strength, and the pH of the solution is a quantitative measure of solvent acidity, or ability to transfer a proton.

Many functional groups are too weakly basic to be appreciably protonated in dilute aqueous solution. Nearly all of the oxygen functional groups belong to this category, for example. In order to measure the basicity of such substances in terms of a pK_a value, it is necessary to devise a measure of solution acidity. In strongly acid solutions the pH scale is inapplicable; the problem is that any operationally significant change in acidity can only be accomplished with a concomitant change in the medium.

Consider a neutral base B of such strength that it can be protonated in dilute aqueous solution in the acidic range, say pH 1–2. In the conventional manner the acid dissociation constant K_{BH^+} is defined,

$$K_{BH^+} = \frac{a_{H^+}a_B}{a_{BH^+}} = \frac{a_{H^+}f_B c_B}{f_{BH^+}c_{BH^+}} \tag{8-83}$$

where the activity coefficients become unity in the infinitely dilute aqueous solution. If B is an indicator base, the concentration ratio c_B/c_{BH^+} can be measured spectrophotometrically. Thus, K_{BH^+} can be evaluated.

Select now a second neutral indicator base C that is weaker than B by roughly an order of magnitude; thus, a solvent can be found of such acidity that a significant fraction of both B and C will be protonated, but this will no longer be a dilute aqueous solution, so the individual activity coefficients will in general deviate from unity. For this solution containing low concentrations of both B and C,

$$pK_{CH^+} - pK_{BH^+} = -\log \frac{c_C}{c_{CH^+}} + \log \frac{c_B}{c_{BH^+}} - \log \frac{f_C f_{BH^+}}{f_{CH^+}f_B} \tag{8-84}$$

If B and C are not only of the same charge type but also of the same structural type, it is reasonable to postulate that the ratio $f_C f_{BH^+}/f_{CH^+}f_B$ will not be markedly different from unity. Let us make this *cancellation assumption;* then Eq. (8-84) becomes

$$pK_{CH^+} - pK_{BH^+} = -\log \frac{c_C}{c_{CH^+}} + \log \frac{c_B}{c_{CH^+}} \tag{8-85}$$

Because the concentration ratios can be measured spectrophotometrically (in separate solutions usually) and pK_{BH^+} is known, the unknown pK_{CH^+} is obtained.

This procedure can now be repeated with a base D that is slightly weaker than C, using C as the reference. In this stepwise manner, a series of pK_a determinations can be made over the acidity range from dilute aqueous solution to highly concentrated mineral acids. Table 8-18 gives pK_{BH^+} values determined in this way for nitroaniline bases in sulfuric and perchloric acid solutions.[177] This technique of determining weak base acidity constants is called the *overlap method,* and the series of pK_{BH^+} values is said to be anchored to the first member of the series, which means that all of the members of the series possess the same standard state, namely, the hypothetical ideal 1 M solution in water.

Writing Eq. (8-83) in logarithmic form,

$$pK_{BH^+} = -\log \frac{c_B}{c_{BH^+}} - \log \frac{a_{H^+} f_B}{f_{BH^+}} \tag{8-86}$$

The last term is conveniently designated according to

$$h_0 = \frac{a_{H^+} f_B}{f_{BH^+}} \tag{8-87}$$

$$H_0 = -\log h_0 \tag{8-88}$$

Thus Eq. (8-86) becomes

$$H_0 = pK_{BH^+} + \log \frac{c_B}{c_{BH^+}} \tag{8-89}$$

H_0, the *acidity function* introduced by Hammett, is a measure of the ability of the solvent to transfer a proton to a base of neutral charge.[180] In dilute aqueous solution h_0 becomes equal to a_{H^+} and H_0 is equal to pH, but in strongly acid solutions H_0 will differ from both pH and $-\log c_{H^+}$. The determination of H_0 is accomplished with the aid of Eq. (8-89) and a series of neutral indicator bases (the nitroanilines in Table 8-18) whose pK_{BH^+} values have been measured by the overlap method. Table 8-19 lists H_0 values for some aqueous solutions of common mineral acids.[181]

Analogous acidity functions have been defined for bases of other structural and charge types, such as H_A for amides and H_R for bases that ionize with the production of a carbocation:

$$ROH + H^+ \rightleftharpoons R^+ + H_2O$$

O'Connor[182] and Cox and Yates[183] have reviewed the many acidity function scales. A major use of acidity functions is for the measurement of the strengths of very weak bases. The procedure utilizes spectrophotometric measurements of the concentration ratio c_B/c_{BH^+} in solutions of known acidity function and application of Eq. (8-89). One problem is the estimation of the spectra of the pure forms (protonated and unprotonated) of the base, for the spectra are subject to the medium effect, and corrections must be applied.[177,184] Another problem is that the base

TABLE 8-18. pK_{BH^+} Values of Nitroanilines

	Acid	
Base	H_2SO_4	$HClO_4$
m-Nitroaniline	$+2.50$	
p-Nitroaniline	$+0.99$	
o-Nitroaniline	-0.29	-0.29
4-Chloro-2-nitroaniline	-1.03	-1.07
2,5-Dichloro-4-nitroaniline	-1.78	-1.79
2-Chloro-6-nitroaniline	-2.43	-2.41
2,6-Dichloro-4-nitroaniline	-3.27	-3.20
2,4-Dichloro-6-nitroaniline	-3.27	
2,4-Dinitroaniline	-4.53^a	-4.26
2,6-Dinitroaniline	-5.54	-5.25
4-Chloro-2,6-dinitroaniline	-6.14	-6.12
2-Bromo-4,6-dinitroaniline	-6.68	-6.69
3-Methyl-2,4,6-trinitroaniline	-8.22	-8.56
3-Bromo-2,4,6-trinitroaniline	-9.46	-9.77
3-Chloro-2,4,6-trinitroaniline	-9.71	
2,4,6-Trinitroaniline	-10.10	

Source: Reference 177.
^{a}Uncertain.

TABLE 8-19. H_0 Values at 25°C

Acid concentration/M	HCl	$HClO_4$	H_2SO_4
0.1	$+0.98$	—	$+0.83$
0.25	$+0.55$	—	$+0.44$
0.5	$+0.20$	—	$+0.13$
0.75	-0.03	-0.04	-0.07
1.0	-0.20	-0.22	-0.26
1.5	-0.47	-0.53	-0.56
2.0	-0.69	-0.78	-0.84
2.5	-0.87	-1.01	-1.12
3.0	-1.05	-1.23	-1.38
3.5	-1.23	-1.47	-1.62
4.0	-1.40	-1.72	-1.85
4.5	-1.58	-1.97	-2.06
5.0	-1.76	-2.23	-2.28
5.5	-1.93	-2.52	-2.51
6.0	-2.12	-2.84	-2.76
7.0	-2.56	-3.61	-3.32
8.0	-3.00	-4.33	-3.87
9.0	-3.39	-5.05	-4.40
10.0	-3.68	-5.79	-4.89

Source: Reference 181.

being studied should possess the same response to acidity changes as do the indicator bases used to establish the acidity function scale used in the calculation. It is a corollary of this requirement that the relative base strengths of two bases of different structural type may vary with the identity and concentration of the acid used to study them.[177] Arnett et al.[185] have summarized the limitations of the acidity function concept and have proposed that the enthalpy of solution of the weak base in a strong acid be used as a quantitative measure of basicity. Heats of protonation were determined calorimetrically for the dissolution of bases in concentrated sulfuric acid and in fluorosulfuric acid, FSO_3H. A good linear correlation was obtained between ΔH and pK_a, covering 40 kcal mol^{-1} in the enthalpies and 22 units in pK_a.

Bell[186] has calculated H_0 values with fair accuracy by assuming that the increase in acidity in strongly acid solutions is due to hydration of hydrogen ions and that the hydration number is 4. The addition of "neutral" salts to acid solutions produces a marked increase in acidity,[181] and this too is probably a hydration effect in the main. Critchfield and Johnson[187] have made use of this salt effect to titrate very weak bases in concentrated aqueous salt solutions. The addition of DMSO to aqueous solutions of strong bases increases the alkalinity of the solutions.

The proliferation of acidity functions is a consequence of the activity coefficient cancellation assumption. According to Eq. (8-89), a plot of $\log(c_B/c_{BH^+})$ against H_0 should be linear with unit slope. Such plots are usually linear (for bases of closely related structure), but the slopes often differ from unity.[184,188] This behavior is an indication that the cancellation assumption (also called the *zero-order approximation*) is not valid, and several groups[189–91] have devised alternatives. We will use the symbolism of Cox and Yates.[191]

Writing Eq. (8-83) in logarithmic form

$$pK_{BH^+} = -\log c_{H^+} - \log \frac{c_B}{c_{BH^+}} - \log \frac{f_B f_{H^+}}{f_{BH^+}} \qquad (8\text{-}90)$$

and similarly for base C shows that the zero-order approximation is stated by Eq. (8-91):

$$\log \frac{f_B f_{H^+}}{f_{BH^+}} = \log \frac{f_C f_{H^+}}{f_{CH^+}} \qquad (8\text{-}91)$$

In the linear or first-order approximation, it is postulated that these activity coefficient terms are directly proportional, as in Eq. (8-92):

$$\log \frac{f_B f_{H^+}}{f_{BH^+}} = m^* \log \frac{f_C f_{H^+}}{f_{CH^+}} \qquad (8\text{-}92)$$

where m^* is not necessarily equal to unity. Now let base C be defined as a standard base[191] or even a hypothetical base,[191] and write

$$\log \frac{f_C f_{H^+}}{f_{CH^+}} = X \qquad (8\text{-}93)$$

Then Eq. (8-90) becomes

$$pK_{BH^+} = -\log c_{H^+} - \log \frac{c_B}{c_{BH^+}} - m^*X \qquad (8\text{-}94)$$

X is an acidity function based on the first-order approximation, Eq. (8-92). Values of X have been assigned[191] by an iterative procedure. The data consist of values of c_B/c_{BH^+} as functions of c_{H^+} for a large number of indicators. For each indicator an initial estimate of pK_{BH^+} and m^* is made and X is calculated with Eq. (8-94). This yields a large body of X values, which are fitted to a polynomial in acid concentration. From this fitted curve smoothed X values are obtained, and Eq. (8-94), a linear function in X, allows refined values of pK_{BH^+} and m^* to be obtained. This procedure continues until the parameters undergo no further change. Table 8-20 gives X values for sulfuric and perchloric acid solutions.[191]

Let us compare the expressions for pK_{BH^+} under ideal (Eq. 8-95), nonideal zero-order approximation (Eq. 8-96), and nonideal first-order approximation (Eq. 8-97) conditions:

$$pK_{BH^+} = -\log c_{H^+} - \log \frac{c_B}{c_{BH^+}} \qquad (8\text{-}95)$$

$$pK_{BH^+} = H_0 - \log \frac{c_B}{c_{BH^+}} \qquad (8\text{-}96)$$

$$pK_{BH^+} = -(\log c_{H^+} + m^*X) - \log \frac{c_B}{c_{BH^+}} \qquad (8\text{-}97)$$

Comparing Eqs. (8-95) and (8-97) shows that the quantity m^*X is a measure of solution acidity in excess of the ideal state; this gives rise to the term *excess acidity method* for the X acidity function approach. We also have, from Eqs. (8-87) and (8-88),

$$H_0 = -\log c_{H^+} - \log \frac{f_B f_{H^+}}{f_{BH^+}} \qquad (8\text{-}98)$$

Comparing Eqs. (8-92) and (8-98) shows that

$$H_0 + \log c_{H^+} = -m^*X \qquad (8\text{-}99)$$

Observed m^* values cluster about 1, but considerable variation is seen among structural types.[191] For some systems, however, $H_0 + \log c_{H^+} \approx -X$; in these cases the excess acidity method is equivalent to Bunnett and Olsen's method,[189] in which the sum $H_0 + \log c_{H^+}$ plays the acidity function role.

Equations (8-96) and (8-97) provide alternate routes to the estimation of pK_{BH^+} values for weak bases. Because we have no independent knowledge of a true value, it is difficult to judge which method is superior. Johnson and Stratton[193]

TABLE 18-20. Values of X for Aqueous Sulfuric and Perchloric
Acid Solutions

	$H_2SO_4{}^a$		$HClO_4$
Weight % acid	$Log\ c_{H^+}$	X	X
5	-0.205	0.103	0.091
10	0.117	0.231	0.212
15	0.315	0.387	0.371
20	0.461	0.573	0.571
25	0.577	0.790	0.819
30	0.674	1.038	1.116
35	0.757	1.317	1.468
40	0.828	1.628	1.879
45	0.891	1.969	2.354
50	0.945	2.345	2.908
52.5	0.970	2.548	3.220
55	0.992	2.763	3.561
57.5	1.014	2.992	3.932
60	1.033	3.238	4.337
62.5	1.052	3.505	4.775
65	1.069	3.795	5.241
67.5	1.084	4.112	5.727
70	1.097	4.459	6.220
72	1.108	4.759	6.607
74	1.118	5.080	6.983
76	1.128	5.421	7.356
78	1.136	5.779	7.766
80	1.143	6.150	
82	1.143	6.528	
84	1.133	6.906	
86	1.109	7.277	
88	1.066	7.637	
90	0.996	7.985	
92	0.894	8.340	
94	0.749	8.743	
95	0.654	8.989	
96	0.539	9.285	
97	0.392	9.656	
98	0.187	10.132	
99	-0.153	10.754	
99.5	-0.504	11.136	

Source: Reference 191.
$^a c_{H^+}$ calculated from the dissociation constant of the bisulfate ion.

correlated pK_{BH^+} values calculated by each method against other measures such
as σ or ΔH^0, seeking the more chemically reasonable relationship, and concluded
that Eq. (8-96) provided slightly better results. Wojcik[194] compared the zero-order,
first-order, and second-order (i.e., quadratic) approximations and reached pessi-
mistic conclusions about the potential accuracy of any pK_{BH^+} measured by an
acidity function method.

Mechanisms of Acid Catalysis

The principal use of acidity functions has been for the study of reaction mechanisms in acid-catalyzed reactions.[181] We consider acid-catalyzed reactions in which a nucleophile, often water, may be a reactant. Three mechanisms are commonly considered:

$A1$. A fast preequilibrium protonation of substrate followed by a slow rate-determining reaction of the protonated substrate. Subsequent steps (such as attack by water) are fast.

$$S + H^+ \underset{}{\overset{fast}{\rightleftharpoons}} SH^+$$

$$SH^+ \underset{slow}{\overset{k_1}{\rightarrow}} I^+$$

$$I^+ + H_2O \underset{fast}{\rightarrow} products$$

$A2$. A fast preequilibrium protonation followed by a slow rate-determining attack by nucleophile.

$$S + H^+ \overset{fast}{\rightleftharpoons} SH^+$$

$$SH^+ + H_2O \underset{slow}{\overset{k_2}{\rightarrow}} I^+$$

$$I^+ \underset{fast}{\rightarrow} products$$

$A\!-\!S_E2$. A slow protonation of substrate followed by fast steps.

$$S + H^+ \overset{k_3}{\rightarrow} SH^+$$

$$SH^+ \underset{fast}{\rightarrow} products$$

Most acid-catalyzed hydrolyses of carboxylic acid derivatives proceed by the A2 mechanism, as shown for ester hydrolysis:

$$RCOOR + H^+ \rightleftharpoons R\overset{\displaystyle \overset{+OH}{\|}}{C}\!-\!OR$$

$$\overset{\overset{\displaystyle +OH}{\parallel}}{R-C-OR} + H_2O \rightleftharpoons \underset{\underset{\displaystyle OH_2}{\mid}}{\overset{\overset{\displaystyle +OH}{\mid}}{R-C-OR}}$$

$$\underset{\underset{\displaystyle +OH_2}{\mid}}{\overset{\overset{\displaystyle OH}{\mid}}{R-C-OR}} \rightleftharpoons RCOOH + ROH + H^+$$

In highly concentrated acids, however, with leaving groups that facilitate bond cleavage, the A1 route may prevail, the intermediate being the acylium ion:

$$RCOX + H^+ \rightleftharpoons \overset{\overset{\displaystyle +OH}{\parallel}}{R-C-X}$$

$$\overset{\overset{\displaystyle +OH}{\parallel}}{R-C-X} \rightarrow R-\overset{+}{C}{=}O + HX$$

$$R-\overset{+}{C}{=}O + H_2O \rightarrow RCOOH + H^+$$

The A–S_E2 mechanism is exemplified by some aromatic electrophilic substitutions:

The kinetic problem is to distinguish among these mechanisms and particularly between the A1 and A2 routes. The first effective solution to this problem was provided by Zucker and Hammett.[195] The key difference between these mechanisms is the presence (A2) or absence (A1) of a molecule of water in the transition state of the rate-determining step.

Consider the A1 scheme.[192] The observed first-order rate constant is defined by the experimental rate equation

$$v = k_{obs}S_t = k_{obs}(c_S + c_{SH^+}) \tag{8-100}$$

and according to transition state theory the rate equation can be written

$$v = \frac{k_1 c_{SH^+} f_{SH^+}}{f_\ddagger} \tag{8-101}$$

where $f_\ddagger$ is the transition state activity coefficient. The acidity constant of the substrate is

$$K_{SH^+} = \frac{a_{H^+} c_S f_S}{c_{SH^+} f_{SH^+}} \tag{8-102}$$

Making use of Eq. (8-87), which defines h_0, gives, upon combination with these equations,

$$k_{obs} \left(\frac{c_S + c_{SH^+}}{c_S} \right) = \frac{k_1 h_0 f_{BH^+} f_S}{K_{SH^+} f_B f_\ddagger} \tag{8-103}$$

where B is a Hammett base, i.e., a base appropriate to the H_0 acidity function. In logarithmic form,

$$\log k_{obs} - \log \left(\frac{c_S}{c_S + c_{SH^+}} \right) = -H_0 + \log \frac{k_1}{K_{SH^+}} + \log \frac{f_{BH^+} f_S}{f_B f_\ddagger} \tag{8-104}$$

The quantity on the left side of the equation is referred to as the *logarithm of the rate constant corrected for protonation;* often the correction term is negligible. If the activity coefficient term on the right side is negligible, Eq. (8-104) predicts a linear relationship between the corrected $\log k_{obs}$ and $-H_0$, the slope being unity. A similar treatment of the A–S_E2 mechanism also predicts a linear plot of $\log k_{obs}$ against $-H_0$.

The result for the A2 route is somewhat different; Eq. (8-105) is obtained for this mechanism,

$$\log k_{obs} - \log \left(\frac{c_S}{c_S + c_{SH^+}} \right) = -H_0 + \log a_w + \log \frac{k_2}{K_{SH^+}} + \log \frac{f_{BH^+} f_S}{f_B f_\ddagger} \tag{8-105}$$

where a_w is the activity of water (or other nucleophile participating in the reaction). Because a_w is a function of the acid concentration, the plot of the left side against $-H_0$ will be nonlinear. This distinguishes between the A1 and A2 mechanisms.

As originally proposed, the Zucker–Hammett hypothesis[195] states that for A1 reactions $\log k_{obs}$ is linear in $-H_0$, whereas for A2 reactions $\log k_{obs}$ is linear in $\log c_{H^+}$. This latter statement is now known not to be generally correct. Moreover, the slopes of plots against $-H_0$ often differ from unity.[177] Bunnett and Olsen[189] showed that plots of $\log k_{obs} + H_0$ against $H_0 + \log c_{H^+}$ were linear; as we saw earlier, this approach is similar to the excess acidity method.

In 1961 Bunnett[196] reported that plots of $\log k_{obs} + H_0$ against $\log a_w$ are often linear [see Eq. (8-105)]. The slope of such a plot is labeled w. In some cases a plot of $\log k_{obs} - \log c_{H^+}$ is linear, with a slope w^*. The protonation correction may have to be applied. [O'Connor[182] has tabulated a_w for solutions of HCl, H_2SO_4, $HClO_4$, and H_3PO_4.] A range of w values is observed, and by utilizing independent mechanistic evidence, Bunnett proposed the empirical mechanistic criteria in Table 8-21 to replace the Zucker–Hammett criteria. The w and w^* values have been interpreted as functions of the difference in number of water molecules between the initial and transition states. This interpretation is certainly not fully correct, for other effects seem to contribute to these parameters, but the hydration hypothesis probably is qualitatively valid. Thus, this hypothesis predicts $\Delta S^{\ddagger}$ will decrease as w increases, as a consequence of the increasing hydration of transition states, and a rough correlation is observed.

The excess acidity treatment makes use of X rather than H_0. For the A1 mechanism Eqs. (8-100)–(8-102) apply, but now we replace a_{H^+} with $c_{H^+} f_{H^+}$ and rearrange to obtain

$$\log k_{obs} - \log \left(\frac{c_S}{c_S + c_{SH^+}} \right) - \log c_{H^+} = \log \frac{k_1}{K_{SH^+}} + \log \frac{f_S f_{H^+}}{f_{\ddagger}} \tag{8-106}$$

Now by analogy with Eq. (8-92) we postulate[192]

$$\log \frac{f_S f_{H^+}}{f_{\ddagger}} = m^{\ddagger} \log \frac{f_S f_{H^+}}{f_{SH^+}} = m^{\ddagger} m^* X \tag{8-107}$$

where m^* pertains to base S and may be independently determinable by means of Eq. (8-94). We, thus, obtain

$$\log k_{obs} - \log \left(\frac{c_S}{c_S + c_{SH^+}} \right) - \log c_{H^+} = m^{\ddagger} m^* X + \log \frac{k_1}{K_{SH^+}} \tag{8-108}$$

The analogous equation for the A2 mechanism includes $\log a_w$ on the right side.

A plot of the left side of Eq. (8-108) against X is linear for A1 and nonlinear for A2 reactions. The curved A2 plot can be transformed to a linear plot by

TABLE 8-21. Mechanistic Criteria Based on w and w^*

Substrate type	w	w^*	Role of H_2O in rate-determining step
Protonated on O or N	-2.5 to 0.0		Is not involved
Protonated on O or N	$+1.2$ to $+3.3$	<-2	Acts as a nucleophile
Protonated on O or N	$>+3.3$	>-2	Acts as general acid–base
Hydrocarbonlike bases	About 0		Acts as general acid–base

subtracting log a_w from the ordinate quantity. Other nucleophiles can be dealt with similarly; for example, bisulfate ion may function as a nucleophile in concentrated solutions of sulfuric acid. It is observed that $m^{\ddagger}$ is greater than 1 for A1, about 1 for A2, and less than 1 for A–S_E2 reactions.[192]

REFERENCES

1. Amis, E.S. "Solvent Effects on Reaction Rates and Mechanisms"; Academic Press: New York, 1966.
2. Kosower, E.M. "An Introduction to Physical Organic Chemistry"; Wiley: New York, 1968; Part 2.
3. Entelis, S.G.; Tiger, R.P. "Reaction Kinetics in the Liquid Phase"; Wiley (Halsted): New York, 1976.
4. Dack, M.R.J. "Solutions and Solubilities"; Dack, M.R.J., Ed., Wiley-Interscience: New York, 1975; Vol VIII, Part II, Chapter XI.
5. Reichardt, C. "Solvents and Solvent Effects in Organic Chemistry"; VCH: Weinheim, 1988; Chapters 5, 7.
6. Hughes, E.D.; Ingold, C.K. *J. Chem. Soc.* **1935**, 244, 252.
7. Ingold, C.K. "Structure and Mechanism in Organic Chemistry"; Cornell University Press: Ithaca, N.Y., 1953; pp 345–55.
8. McClellan, A.L. "Tables of Experimental Dipole Moments"; Rahara Enterprises: El Cerrito, Calif., 1974; Vol. 2, p 2.
9. Debye, P. "Polar Molecules"; Chemical Catalog Co., 1929. (Dover, New York, reprint.)
10. Harned, H.S.; Owen, B.B. "The Physical Chemistry of Electrolytic Solutions"; 3rd ed., Reinhold: New York, 1958; pp 161, 713.
11. Moreau, C.; Douhéret, G. *J. Chem. Thermodyn.* **1976**, 8, 403.
12. Connors, K.A.; Wright, J.L. *Anal. Chem.* **1989**, 61, 194.
13. Hirschfelder, J.O.; Curtiss, C.F.; Bird, R.B. "Molecular Theory of Gases and Liquids"; Wiley: New York, 1954; Chapter 1.
14. Mulliken, R.S. *J. Am. Chem. Soc.* **1952**, 74, 811.
15. Mulliken, R.S.; Person, W.B. "Molecular Complexes"; Wiley-Interscience: New York; 1969.
16. Benesi, H.A.; Hildebrand, J.H. *J. Am. Chem. Soc.* **1949**, 71, 2703.
17. Andrews, L.J; Keefer, R.M. "Molecular Complexes in Organic Chemistry"; Holden-Day: San Francisco, 1964.
18. Gur'yanova, E.N.; Gol'dshtein, I.P.; Romm, I.P. "Donor-Acceptor Bond"; Wiley (Halsted): New York, 1975.
19. Pimentel, G.C.; McClellan, A.L. "The Hydrogen Bond"; Freeman: San Francisco, 1960; p 195.
20. Frank, H.S.; Evans, M.W. *J. Chem. Phys.* **1945**, 13, 507.
21. Kauzmann, W. *Adv. Protein Chem.*, **1959**, 14, 1.
22. Richards, F.M. *Ann. Rev. Biochem.* **1963**, 32, 269.
23. Tanford, C. "The Hydrophobic Effect", 2nd ed.; Wiley-Interscience: New York, 1980.
24. Israelachvili, J.N. "Intermolecular and Surface Forces"; Academic Press: London, 1985; pp 103–6.
25. Herman, R.B. *J. Phys. Chem.* **1975**, 79, 163.
26. Sinanoglu, O.; Abdulnur, S. *Photochem. Photobiol* **1964**, 3, 333.
27. Sinanoglu, O. In "Molecular Associations in Biology"; Pullman, B., Ed.; Academic Press: New York, 1968; p 427.
28. Halicioglu, T.; Sinanoglu, O. *Ann. N.Y. Acad. Sci.* **1969**, 158, 308.
29. Sinanoglu, O.; Fernandez, A. *Biophys. Chem.* **1985**, 21, 157, 167.
30. Crothers, D.M.; Ratner, D.I. *Biochemistry* **1968**, 7, 1823.
31. Mukerjee, P.; Ghosh, A.K. *J. Am. Chem. Soc.* **1970**, 92, 6419.
32. Cecil, R. *Nature* **1967**, 214, 369.

33. Albergo, D.D.; Turner, D.H. *Biochemistry,* **1981,** *20,* 1413.

34. Fornili, S.L.; Sgroi, G.; Izzo, V. *J. Chem. Soc., Faraday Trans. 2* **1983,** *79,* 1085.

35. Granot, J. *J. Am. Chem. Soc.* **1978,** *100,* 6745.

36. Jencks, W.P. "Catalysis in Chemistry and Enzymology"; McGraw-Hill: New York, 1969; pp 417–36.

37. Roseman, M.; Jencks, W.P. *J. Am. Chem. Soc.* **1975,** *97,* 631.

38. Howarth, O.W. *J. Chem. Soc., Faraday Trans. 1* **1975,** *71,* 2303.

39. Cramer, R.D. *J. Am. Chem. Soc.* **1977,** *99,* 5408.

40. Lee, B. *Biopolymers,* **1985,** *24,* 813.

41. Mirejovsky, D.; Arnett, E.M. *J. Am. Chem. Soc.* **1983,** *105,* 1112.

42. Ben-Naim, A. "Water and Aqueous Solutions"; Plenum: New York, 1974; pp 428–36.

43. Ben-Naim, A. *J. Phys. Chem.* **1978,** *82,* 874.

44. Grunwald, E. *J. Am. Chem. Soc.* **1986,** *108,* 5726.

45. Parker, A.J. *Advan. Phys. Org. Chem.* **1967,** *5,* 173.

46. Kamlet, M.J.; Abboud, J.-L.M.; Abraham, M.H.; Taft, R.W. *J. Org. Chem.* **1983,** *48,* 2877.

47. Chastrette, M.; Rajzmann, M.; Chanon, M.; Purcell, K.F. *J. Am. Chem. Soc.* **1985,** *107,* 1.

48. Gordon, J.E. "The Organic Chemistry of Electrolyte Solutions"; Wiley-Interscience: New York, 1975; p 160.

49. Leo, A.; Hansch, C.; Elkins, D. *Chem. Revs.* **1971,** *71,* 525.

50. Funasaki, N.; Hada, S.; Neya, S. *J. Phys. Chem.* **1985,** *89,* 3046.

51. Mukerjee, P.; Ramachandran, C.; Pyter, R.A. *J. Phys. Chem.* **1982,** *86,* 3189.

52. Coetzee, J.F.; Ritchie, C.D., Eds. "Solute-Solvent Interactions"; Dekker: New York, 1976; Vol. 2.

53. Kolthoff, I.M., Bruckenstein, S. In "Treatise on Analytical Chemistry"; Kolthoff, I.M., Elving, P.J., Eds.; Interscience: New York, 1959; Part I, Vol I, Chapter 13.

54. Denison, J.T.; Ramsey, J.B. *J. Am. Chem. Soc.* **1955,** *77,* 2615.

55. Winstein, S.; Clippinger, E.; Fainberg, A.H.; Heck, R.; Robinson, G.C. *J. Am. Chem. Soc.* **1956,** *78,* 328.

56. Gurney, R.W. "Ionic Processes in Solution"; McGraw-Hill: New York, 1953; pp 4, 28–30, 248–51. (Dover, New York, reprint, 1962.)

57. Kosower, E.M. "An Introduction to Physical Organic Chemistry"; Wiley: New York, 1968; Part 2.

58. Stepukovich, A.D.; Lapshova, N.I.; Efimova, T.D. *Zh. Fiz.. Khim.* **1961,** *35,* 2532.

59. Hart, H.; Cassis, F.A.: Bordeaux, J.J. *J. Am. Chem. Soc.* **1954,** *76,* 1639.

60. Buncel, E.; Wilson, H. *J. Chem. Educ.* **1980,** *57,* 629.

61. Gould, E.S. "Mechanism and Structure in Organic Chemistry"; Holt, Rinehart & Winston: New York, 1959; pp 724–7.

62. Petersen, R.C.; Markgraf, J.H.; Ross, S.D. *J. Am. Chem. Soc.* **1961,** *83,* 3819.

63. Glasstone, S.; Laidler, K.J.; Eyring, H. "The Theory of Rate Processes"; McGraw-Hill: New York, 1941; pp 419–23.

64. Kirkwood, J.G. *J. Chem. Phys.* **1934,** *2,* 351.

65. Abraham, M.H.; Grellier, P.L. *J. Chem. Soc., Perkin Trans. 2* **1976,** 1735.

66. Moelwyn-Hughes, E.A. "The Kinetics of Reactions in Solution"; 2nd ed.; Oxford University Press: London, 1947; pp 212–3.

67. Wiberg, K.B. "Physical Organic Chemistry"; Wiley: New York, 1964; pp 381–2.

68. Poh, B. *Aust. J. Chem.* **1980,** *33,* 1175.

69. Marcus, Y. "Introduction to Liquid State Chemistry"; Wiley-Interscience: New York, 1977; p 220.

70. Burns, R.G.; England, B.D. *Tetrahedron Lett.* **1960** (24), 1.

71. Bernasconi, C.F. *J. Am. Chem. Soc.* **1970,** *92,* 4682.

72. LaMer, V.K. *Chem. Revs.* **1932,** *10,* 179.

73. Weil, I.; Morris, J.C. *J. Am. Chem. Soc.* **1949,** *71,* 1664.

74. Frost, A.A.; Pearson, R.G. "Kinetics and Mechanism"; Wiley: New York, 1953; pp 257–65.

75. Dack, M.R.J. *J. Chem. Educ.* **1974,** *51,* 231.

76. Barton, A.F.M. *Chem. Revs.* **1975,** *75,* 731.

77. Hildebrand, J.H.; Prausnitz, J.M.; Scott, R.L. "Regular and Related Solutions"; Van Nostrand Reinhold: New York, 1970.

78. Shinoda, K. "Principles of Solution and Solubility"; Dekker: New York, 1978; Chapter 4.

79. Richardson, M.; Soper, F.G. *J. Chem. Soc.* **1929**, 1873.

80. Buncel, E.; Wilson, H. *Acc. Chem. Res.* **1979**, *12*, 42.

81. Abraham, M.H. *Prog. Phys. Org. Chem.* **1974**, *11*, 1.

82. Parker, A.J. *J. Chem. Soc.* **1966A**, 220.

83. Bates, R.G. In "Solute-Solvent Interactions"; Coetzee, J.F., Ritchie, C.D., Eds.; Dekker: New York, 1969; p 57.

84. Buncel, E.; Wilson, H. *Advan. Phys. Org. Chem.* **1977**, *14*, 133.

85. Parker, A.J. *Chem. Revs.* **1969**, *69*, 1.

86. Blandamer, M.J. *Advan. Phys. Org. Chem.* **1977**, *14*, 203.

87. Abraham, M.H. *Pure Appl. Chem.* **1985**, *57*, 1055.

88. Arcoria, A.; Cipria, A.; Longo, M.L.; Maccarone, E.; Tomaselli, G.A. *Gazz. Chim. Ital.* **1987**, *117*, 723.

89. Sedaghat-Herati, M.R.; Harris, J.M.; McManus, S.P. *Tetradhedron* **1988**, *44*, 7479.

90. Blokzijl, W.; Engberts, J.B.F.N.; Jager, J.; Blandamer, M.J. *J. Phys. Chem.* **1987**, *91*, 6022.

91. Galema, S.A.; Blandamer, M.J.; Engbergs, J.B.F.N. *J. Org. Chem.* **1989**, *54*, 1227.

92. Jensen, W.B. *Chem. Revs.*, **1978**, *78*, 1.

93. Persson, I. *Pure Appl. Chem.* **1986**, *58*, 1153.

94. Gutmann, V. *Coord. Chem. Rev.* **1976**, *18*, 225.

95. Maria, P.-C.; Gal, J.-F. *J. Phys. Chem.* **1985**, *89*, 1296.

96. Drago, R.S.; Vogel, G.C.; Needham, T.E. *J. Am. Chem. Soc.* **1971**, *93*, 6014.

97. Marcus, Y. *J. Phys. Chem.* **1987**, *91*, 4422.

98. Abraham, M.H.; Grellier, P.L.; McGill, R.A. *J. Chem. Soc., Perkin Trans. 2* **1988**, 339.

99. Abraham, M.H. *J. Am. Chem. Soc.* **1979**, *101*, 5477; **1982**, *104*, 2085.

100. Swain, C.G.; Scott, C.B.; Lohmann, K.H. *J. Am. Chem. Soc.* **1953**, *75*, 136.

101. Spieth, F., quoted by Streitwieser, A., Jr. "Solvolytic Displacement Reactions"; McGraw-Hill: New York, 1962; p 53.

102. Bunton, C.A. "Nucleophilic Substitution at a Saturated Carbon Atom"; Elsevier: Amsterdam, 1963; pp 21, 164.

103. Kohnstam, G. *Advan. Phys. Org. Chem.* **1967**, *5*, 121.

104. Bentley, T.W.; Schleyer, P. von R. *Advan. Phys. Org. Chem.* **1978**, *14*, 2.

105. Winstein, S.; Grunwald, E.; Jones, H.W. *J. Am. Chem. Soc.* **1951**, *73*, 2700.

106. Grunwald, E.; Winstein, S. *J. Am. Chem. Soc.* **1948**, *70*, 846.

107. Fainberg, A.H.; Winstein, S. *J. Am. Chem. Soc.* **1956**, *78*, 2770.

108. Wells, P.R. "Linear Free Energy Relationships"; Academic Press: New York, 1968; Chapter 4.

109. Winstein, S.; Fainberg, A.H.; Grunwald, E. *J. Am. Chem. Soc.* **1957**, *79*, 4146.

110. Bentley, T.W.; Schleyer, P.v.R. *J. Am. Chem. Soc.* **1976**, *98*, 7658.

111. Schadt, F.L.; Bentley, T.W.; Schleyer, P.v.R. *J. Am. Chem. Soc.* **1976**, *98*, 7667.

112. Bentley, T.W.; Roberts, K. *J. Org. Chem.* **1985**, *50*, 4821.

113. Kevill, D.N.; Anderson, S.W. *J. Org. Chem.* **1985**, *50*, 3330.

114. Hawkinson, D.C.; Kevill, D.N. *J. Org. Chem.* **1988**, *53*, 3857.

115. Kevill, D.N.; Hawkinson, D.C. *J. Org. Chem.* **1989**, *54*, 154.

116. Bentley, T.W.; Carter, G.E. *J. Org. Chem.* **1983**, *48*, 579.

117. Kevill, D.N.; Lin, G.M.L. *J. Am. Chem. Soc.* **1976**, *101*, 3916.

118. Kevill, D.N.; Rissmann, T.J. *J. Org. Chem.* **1985**, *50*, 3062.

119. Kevill, D.N.; Anderson, S.W.; Fujimoto, E.K. In "Nucleophilicity"; Harris, J.M.; McManus, S.P. Eds.; Adv. Chem. Ser. 215; American Chemical Society: Washington, D.C., 1987; Chapter 19.

120. Bentley, T.W. In "Nucleophilicity"; Harris, J.M.; McManus, S.P., Eds.; Adv. Chem. Ser. 215; American Chemical Society: Washington, D.C., 1987; Chapter 18.

121. Swain, C.G.; Moseley, R.B.; Bown, D.E. *J. Am. Chem. Soc.* **1955**, *77*, 3731.

122. Peterson, P.E.; Vidrine, D.W.; Waller, F.J.; Henrichs, P.M.; Magaha, S.; Stevens, B. *J. Am. Chem. Soc.* **1977**, *99*, 7968.

123. Drougard, Y.; Decroocq, D. *Bull. Soc. Chim. Fr.* **1969**, 2972.

124. Auriel, M.; de Hoffmann, E. *J. Am. Chem. Soc.* **1975**, *97*, 7433.

125. Oshima, T.; Nagai, T. *Bull. Chem. Soc. Jpn.* **1982**, *55*, 555.

126. Kosower, E.M. *J. Am. Chem. Soc.* **1958**, *80*, 3253, 3261, 3267.

127. Griffiths, T.R.; Pugh, D.C. *Coord. Chem. Rev.* **1979**, *29*, 129.

128. Dimroth, K.; Reichardt, C.; Sipemann, T.; Bohlmann, F. *Liebig's Ann. Chem.* **1963**, *661*, 1.

129. Brooker, L.G.S.; Craig, A.C.; Heseltine, D.W.; Jenkins, P.W.; Lincoln, L.L. *J. Am. Chem. Soc.* **1965**, *87*, 2443.

130. Nicolet, P.; Laurence, C.; Lucon, M. *J. Chem. Soc., Perkin Trans.* 2 **1987**, 483.

131. Dong, D.C.; Winnik, M.A. *Can. J. Chem.* **1984**, *62*, 2560.

133. Street, K.W. Jr.; Acree, W.E., Jr.; Street, K.W. Jr.; *Analyst* **1986**, *111*, 1197.

133. Waris, R.; Acree, W.E. Jr.; *Analyst* **1988**, *113*, 1465.

134. Gowland, J.A.; Schmid, G.H. *Can. J. Chem.* **1969**, *47*, 2953.

135. Street, K.W., Jr.; Acree, W.E., Jr. *J. Liq. Chromatogr.* **1986**, *9*, 2799.

136. De Vijlder, M. *Bull. Soc. Chim. Belg.* **1982**, *91*, 947.

137. Johnson, B.P.; Khaledi, M.G.; Dorsey, J.G. *Anal. Chem.* **1986**, *58*, 2354.

138. Kamlet, M.J.; Abboud, J.L.M.; Taft, R.W. *Prog. Phys. Org. Chem.* **1981**, *13*, 485.

139. Taft, R.W.; Abboud, J.L.M.; Kamlet, M.J.; Abraham, M.H. *J. Solution Chem.* **1985**, *14*, 153.

140. Kamlet, M.J. Abboud, J.L.; Taft, R.W. *J. Am. Chem. Soc.* **1977**, *99*, 6027.

141. Kamlet, M.J.; Hall, T.N.; Boykin, J.; Taft, R.W. *J. Org. Chem.* **1979**, *44*, 2599.

142. Taft, R.W.; Kamlet, M.J. *J. Am. Chem. Soc.* **1976**, *98*, 2886.

143. Kolling, O.W. *Anal. Chem.* **1981**, *53*, 54; **1983**, *55*, 143; **1984**, *56*, 2988.

144. Buncel, E.; Rajagopal, S. *J. Org. Chem.* **1989**, *54*, 798.

145. Brady, J.E.; Carr, P.W. *J. Phys. Chem.* **1982**, **86**, 3053.

146. Bekarek, V. *J. Chem. Soc., Perkin Trans.* **1986**, 2, 1425.

147. Eliasson, B.; Johnels, D.; Wold, S.; Edlund, U.; Sjöström, M. *Acta Chem. Scand.* **1987**, *B41*, 291.

148. Snyder, L.R. In "Separation and Purification", 3rd ed.; Perry, E.S.; Weissberger, A., Eds.; Wiley-Interscience: New York, 1978; Chapter 2.

149. Skoog, D.A. "Principles of Instrumental Analysis", 3rd ed.; Saunders College Publishing, Holt, Rinehart & Winston: Philadelphia, 1985; Chapters 26 and 27.

150. Miller, J.M. "Chromatography: Concepts and Contrasts"; Wiley-Interscience: New York, 1988.

151. Dorsey, J.G.; Johnson, B.P. *J. Liq. Chromatogr.* **1987**, *10*, 2695.

152. Poole, C.F.; Poole, S.K. *Chem. Revs.* **1989**, *89*, 377.

153. Park, J.H.; Carr, P.W. *J. Chromatogr.* **1989**, *465*, 123.

154. Rutan, S.C.; Carr, P.W.; Cheong, W.J.; Park, J.H.; Snyder, L.R. *J. Chromatogr.* **1989**, *463*, 21.

155. Shorter, J. "Correlation Analysis of Organic Reactivity"; Research Studies Press (Wiley): Chicester, 1982.

156. Brady, J.E.; Carr, P.W. *J. Phys. Chem.* **1984**, *88*, 5796.

157. Brownstein, S. *Can J. Chem.* **1960**, *38*, 1590.

158. Langhals, H. *Angew. Chem. Intern. Ed.* **1982**, *21*, 724.

159. Rezende, M.C.; Zanette, D.; Zucco, C. *Tetrahedron Lett.* **1984**, *25*, 3423.

160. Rezende, M.C. *Tetrahedron Lett.* **1988**, *44*, 3513.

161. Fowler, F.W.; Katritzky, A.R.; Rutherford, R.J.D. *J. Chem. Soc. B* **1971**, 460.

162. Koppel, I.A.; Palm, V.A. "Advances in LFER"; Chapman, N.B.; Shorter, J. Eds.; Plenum: London, 1973; Chapter 5.

163. Zilian, U. *Chem. Ztg.* **1984**, *108*, 381.

164. Abraham, M.H.; Doherty, R.M.; Kamlet, M.J.; Harris, J.M.; Taft, R.W. *J. Chem. Soc., Perkin Trans.* 2 **1987**, 913, 1097.

165. Abraham, M.H.; Grellier, P.L.; Abboud, J.L.M.; Doherty, R.M.; Taft, R.W. *Can. J. Chem.* **1988**, *66*, 2673.

166. Weiner, P.H. *CHEMTECH* **1977**, *7*, 321.

167. Cramer, R.D. *J. Am. Chem. Soc.* **1980**, *102*, 1837, 1849.

168. Maria, P.C.; Gal, J.-F.; de Franceschi, J.; Fargin, E. *J. Am. Chem. Soc.* **1987**, *109*, 483.

169. Chastrette, M.; Carretto, J. *Can. J. Chem.*, **1985**, *63*, 3492.

170. Swain, C.G.; Swain, M.S.; Powell, A.L.; Alunni, S. *J. Am. Chem. Soc.* **1983**, *105*, 502.

171. Taft, R.W.; Abboud, J.L.M.; Kamlet, M.J. *J. Org. Chem.* **1984**, *49*, 2001.

172. Swain, C.G. *J. Org. Chem.* **1984,** *49,* 2005.

173. Sjöström, M.; Wold, S. *Acta Chem. Scand.* **1981** *B35,* 537.

174. Wold, S.; Sjöström, M. *Acta Chem. Scand.* **1986,** *B40,* 270.

175. Kamlet, M.J.; Taft, R.W. *Acta Chem. Scand.* **1985,** *B39,* 611.

176. Kamlet, M.J.; Doherty, R.M.; Famini, G.R.; Taft, R.W. *Acta Chem. Scand.* **1987,** *B41,* 589.

177. Hammett, L.P. "Physical Organic Chemistry", 2nd ed.; McGraw-Hill: New York, 1970; Chapter 9.

178. Rochester, C.H. "Acidity Functions"; Academic Press: London, 1970.

179. Stewart, R. "The Proton: Applications to Organic Chemistry"; Academic Press: Orlando, Fla., 1985; Chapters 2, 3, and 7.

180. Hammett, L.P. Deyrup, A.J. *J. Am. Chem. Soc.* **1932,** *54,* 2721.

181. Paul, M.A.; Long, F.A. *Chem. Revs.* **1957,** *57,* 1.

182. O'Connor, C.J. *J. Chem. Educ.* **1969,** *46,* 686.

183. Cox, R.A.; Yates, K. *Can. J. Chem.* **1983,** *61,* 2225.

184. Cookson, R.F. *Chem. Revs.* **1974,** *74,* 5.

185. Arnett, E.M.; Quirk, R.P.; Burke, J.J. *J. Am. Chem. Soc.* **1970,** *92,* 1260.

186. Bell, R.P. "The Proton in Chemistry"; Cornell University Press: Ithaca, N.Y., 1959; pp 81–4.

187. Critchfield, F.E.; Johnson, J.B. *Anal. Chem.* **1958,** *30,* 1247; **1959,** *31,* 570.

188. Arnett, E.M. *Prog. Phys. Org. Chem.* **1963,** *1,* 223.

189. Bunnett, J.F.; Olsen, F.P. *Can. J. Chem.* **1966,** *44,* 1899.

180. Marziano, N.C.; Cimino, G.M.; Passerini, R.C. *J. Chem. Soc., Perkin Trans.* 2 **1973,** 1915.

191. Cox, R.A.; Yates, K. *J. Am. Chem. Soc.* **1978,** *100,* 3861.

192. Cox, R.A. *Acc. Chem. Res.* **1987,** *20,* 27.

193. Johnson, C.D.; Stratton, B. *J. Org. Chem.* **1986,** *51,* 4100; **1987,** *52,* 4798.

194. Wojcik, J.F. *J. Phys. Chem.* **1982,** *86,* 145; **1985,** *89,* 1748.

195. Zucker, L.; Hammett, L.P. *J. Am. Chem. Soc.* **1939,** *61,* 2791.

196. Bunnett, J.F.; *J. Am. Chem. Soc.* **1961,** *83,* 4956, 4968, 4973, 4978.

197. Riddick, J.A.; Bunger, W.B. "Organic Solvents", 3rd ed.; Wiley-Interscience: New York, 1970.

198. Gordon, A.J.; Ford, R.A. "The Chemist's Companion"; Wiley-Interscience: New York, 1972.

PROBLEMS

1. These are data for this reaction:

$$N_2(COO)_2^{2-} + H^+ \rightarrow \text{products}$$

Volume % dioxane	log k	$\Delta S^{\ddagger}/eu$	$\Delta H^{\ddagger}/kcal\ mol^{-1}$
0	10.34	12.9	10.2
10	10.60	14.1	10.2
20	10.89	14.9	10.1
30	11.22	15.3	9.8
40	11.64	15.7	9.3
50	12.20	20.2	9.9
60	12.95	25.3	10.4

Calculate $\delta_M \Delta S^{\ddagger}$ and $\delta_M \Delta H^{\ddagger}$ from these data (relative to pure water).

2. (a) Derive an equation giving $\delta_M \Delta S^{\ddagger}$ as a function of k_M and k_0, where k_M and k_0 are rate constants in medium M and a reference medium, respectively, for the case in which $\Delta H^{\ddagger}$ is the same in the two solvents.

(b) Calculate k_M/k_0, assuming $\delta_M \Delta H^{\ddagger} = 0$, at the following values of $\delta_M \Delta S^{\ddagger}$: 1, 2, 10, 20, 30 eu.

3. Calculate Mukerjee's H value for pure methanol.

4. For the isotropic continum model of the reaction of an ion with a neutral molecule [see Eq. (8-25)], obtain an expression for the electrostatic entropy of activation.

5. Sketch qualitative reaction coordinate diagrams corresponding to the several areas and lines in Fig. 8-5.

6. Acidity functions other than H_0 have been defined based on indicators other than neutral bases:

$$H_R \quad ROH + H^+ \rightleftharpoons R^+ + H_2O$$

$$H_- \quad HA \rightleftharpoons H^+ + A^-$$

$$H_+ \quad BH^{2+} \rightleftharpoons B^+ + H^+$$

Give formal definitions of H_R, H_-, and H_+ analogous to the definition of H_0.

7. Amides undergo acid-catalyzed hydrolysis, the rate constant typically passing through a maximum at 2–6 M acid. Develop a hypothesis to account for the maximum.

APPENDIX A

Answers to Selected Problems

CHAPTER 1

4. $v = -\dfrac{1}{2}\dfrac{d[KMnO_4]}{dt} = -\dfrac{1}{3}\dfrac{d[H_2SO_4]}{dt} = -\dfrac{1}{5}\dfrac{d[H_2O_2]}{dt} = \dfrac{1}{2}\dfrac{d[MnSO_4]}{dt}$

$$= \dfrac{1}{8}\dfrac{d[H_2O]}{dt} = \dfrac{1}{5}\dfrac{d[O_2]}{dt} = \dfrac{d[K_2SO_4]}{dt}.$$

5. $v = k[I^-][C_2H_5Br]$.
6. First-order: time^{-1} (usually s^{-1}).
   Second-order: concentration^{-1} time^{-1} (e.g., M^{-1} s^{-1}).
   Third-order: concentration^{-2} time^{-1} (M^{-2} s^{-1}).
7. $k_{30}/k_{20} = 3.10$.
8. $v = \dfrac{1}{v_i}\dfrac{dc_i}{dt} + \dfrac{c_i}{v_i V}\dfrac{dV}{dt}$.

CHAPTER 2

2. $t_{1/2} = 0.693\tau$.
3. $v_0' = -(\varepsilon_R - \varepsilon_p)bv_0$.
4. $1/c_A^2 = 1/(c_A^0)^2 + 2kt$.
5. (a) $k = 5.84 \times 10^{-3}$ s^{-1}.
   (b) ε (ester) $= 4.20 \times 10^3$; ε (phenol) $= 1.59 \times 10^4$.
6. (a) $k = 0.539$ M^{-1} s^{-1}.
   (b) 1.29×10^3 s.
7. First-order.

8. (a) Using the Guggenheim method, 10^3 k/s^{-1} = 1.96 (pH 11.59), 2.52 (pH 11.69), 3.11 (pH 11.79).
   (b) $k = 0.356$ M^{-1} s^{-1}.
   (c) $k = 0.503$ M^{-1} s^{-1}.
9. (a) $k = 0.253$ M^{-1} min^{-1} = 4.22 $\times$ 10^{-3} M^{-1} min^{-1}.
   (b) $k = 1.49$ M^{-2} s^{-1}.
10. $(V_\infty = V_t)/V_\infty = \exp(-kt)$.
11. Cf. Connors, K.A. *Anal. Chem.* **1975,** *47,* 2066.
12. $k = v_0/(c^0)^n$.
13. (a) t_{90} (zero-order) = 0.1 c^0/v_0.
    (b) t_{90} (first-order) = 0.105 c^0/v_0.
15. $w_i = \alpha/\sigma^2_{1/c}$, where $\sigma^2_{1/c} = \sigma^2_c/c^4$.
16. (a) $\sigma = 0.00283$.
    (b) % RSD = 0.63%.

CHAPTER 3

1. $k_{obs} = k_1[OMe^-] + k_{-1}$.
2. $k_1 [H_2O] = 5.53 \times 10^{-5}$ s^{-1}; $k_{-1} = 1.27 \times 10^{-6}$ s^{-1}.
   Using $[H_2O] = 55.5$ M, $k_1 = 9.96 \times 10^{-7}$ M^{-1} s^{-1}.
3. $k' = 0.316$ M^{-2} s^{-1}.
4. $c_B^{max} = c_A^0(k_2/k_1)^{k_2/(k_1 - k_2)}$
5. $100F/2c_A^0$.
7. s^{-1}.
9. $k_{OH} = k_1k_2/(k_{-1} + k_2)$.
10. $\lambda^3 - p\lambda^2 + q\lambda = 0$, where $p = k_1 + k_2 + k_3 + k_{-1} + k_{-2} + k_{-3}$, and
    $q = k_1k_2 + k_1k_3 + k_1k_{-2} + k_{-1}k_{-3} + k_{-1}k_{-2} + k_{-1}k_3 + k_2k_3 + k_2k_{-3} + k_{-2}k_{-3}$.
    The roots are
    $\lambda_1 = 0$.
    $\lambda_2 = [p + (p^2 - 4q)^{1/2}]/2$.
    $\lambda_3 = [p - (p^2 - 4q)^{1/2}]/2$.
15. (a) $v/[S] = V_m/K_m - v/K_m$
    (b) $[S]/v = [S]/V_m + K_m/V_m$

CHAPTER 4

1. Five relaxation times.
2. $\tau^{-1} = (k_1 + k_{-1})c_C$.
3. $\tau^{-1} = k_1(\bar{c}_A + \bar{c}_B) + 4k_{-1}\bar{c}_C$.
4. $\tau^{-1} = 4k_1\bar{c}_A + k_{-1}$.
5. $\tau^{-1} = k_{12}\bar{c}_B + k_{21}(\bar{c}_C + \bar{c}_D)$.
7. The experimental results are $k = 5 \times 10^{10}$ M^{-1} s^{-1} and $k_{-1} = 8.6 \times 10^6$ s^{-1}.

8. See Table 4-1.
9. $k_{\text{MeOH}\frac{1}{2}} = 3.6 \times 10^9$ M^{-1} s^{-1}.
   $k_{\text{MeO}^-} = 7.5 \times 10^8$ M^{-1} s^{-1}.
12. (a) 0.32 Hz.
    (b) 3.50 Hz.

CHAPTER 5

1. $K = k_1 k_2 / k_{-1} k_{-2}$
3. $\Delta S^{\ddagger} = \Delta H^{\ddagger}/T - 4.576 \log (T/k) - 47.22$, where the time unit is seconds.
4. $E_a = E + RT/2$.
6. $\Delta G^{\ddagger} = \Delta G^0/2 + 2\Delta G_0^{\ddagger}$.
9. (b) Substitute the coordinates at the transition state, substitute the coordinates at the initial state, and subtract.

CHAPTER 6

1. $E_a = 123$ kJ mol^{-1}.
2. $E_a = 83.5$ kJ mol^{-1}; $A = 5 \times 10^{11}$ M^{-1} s^{-1};
   $k_{36} = 3.8 \times 10^{-3}$ M^{-1} s^{-1}.
3. $k_{50} = 1.79 \times 10^{-2}$ s^{-1}.
4. $\Delta H^{\ddagger} = 29.1$ kcal mol^{-1}; $\Delta S^{\ddagger} = 7.2$ eu.
5. $\Delta H^{\ddagger} = 15.6$ kcal mol^{-1}; $\Delta S^{\ddagger} = -45.6$ eu.
7. Let $y = \ln k$; then ^{2} $^2/k^2 = $ constant, and $w = \alpha/^2 = $ constant.
8. le Noble[33] reports $\Delta V^{\ddagger} = 16 \pm 1$ cm^3 mol^{-1}.
9. See Fig. 6-5.
10. $k_{\text{OH}} = 169$ M^{-1} s^{-1}.
11. k_c/M^{-1} $s^{-1} = 2.70$ (pyridine); 13.7 (NMIM); 2000 (DMAP).
14. $k_{\text{obs}} = \dfrac{k_1 k_2 [\text{B}][\text{X}_2]}{k_{-1}[\text{BH}^+] + k_2[\text{X}_2]}$.
15. $S^- + H_2O \rightleftharpoons I^-$.
    $I^- + H^+ \rightleftharpoons IH$.
    $IH \rightarrow$ products.
16. $\Delta E_a = 1.15$ kcal mol^{-1}.

CHAPTER 7

1. (a) ρ (58.25°C) = +1.57.
   ρ (79.65°C) = +1.25. Note that benzoic anhydrides have two identical sites of reaction.

3. (a) $k = 5.9 \times 10^{-3}$ M^{-1} s^{-1}.
   (b) $k/k_0 = 2.15 \times 10^7$.
   (c) $k/k_0 = 0.769$.
6. Slope $= -\dfrac{1}{2.3R}\left(\dfrac{1}{T} - \dfrac{1}{\beta}\right)$.

CHAPTER 8

3. $H = 0.456$.
4. $-\Delta S_{el} = (\partial \Delta G/\partial T)_P = $ constant x $d(1/\varepsilon)/dT$.
7. The increase at low acid concentrations is due to catalysis; the decrease at high acid concentrations is due to reduction in the water activity.

Physical Constants

Quantity	Symbol	Value
Avogadro's number	N_A	6.023×10^{23} mol^{-1}
Boltzmann constant	k	1.380×10^{-23} J K^{-1}
		1.380×10^{-16} erg K^{-1}
Gas constant	R	8.314 J K^{-1} mol^{-1}
		8.314×10^{7} ergs K^{-1} mol^{-1}
		1.987 cal K^{-1} mol^{-1}
Faraday	F	9.6487×10^{4} C mol^{-1}
Elementary charge	e	1.602×10^{-19} C
Electron volt	eV	1.602×10^{-19} J
Planck's constant	h	6.626×10^{-34} J s
		6.626×10^{-27} erg s
Speed of light in vacuum	c	2.9979×10^{8} m s^{-1}

Index

Absolute activity, 202
Absolute reaction rate theory, 200
Absorbance, 34
Absorption coefficient, 145
Absorption mode, 164
Absorption phase, 68
Absorption spectroscopy, 71
Absorptivity, 34
Acceptor ability, 439
Acceptor number, 439
Accuracy, 51
Acetylation kinetics, 32
Acetylation of alcohols, 117
Acid
    hard, 8, 361
    soft, 8, 361
Acid-base catalysis, general, 344
Acid-base reactions, fast, 149
Acid catalysis, 446
    general, 265, 268
    mechanisms of, 453
    specific, 264, 268
Acidity
    excess, 451
    Lewis, 426
Acidity function, 448
Acity, 445
Activated complex, 200
Activation energy, 14, 188, 208, 246
    determination of, 246, 256
    diffusion-limited, 136
    experimental, 14
    intrinsic, 256
    typical values of, 260
Activation parameters, 207, 246
    standard states and, 253
    uses of, 259
Activation volume, 261, 416
Activity, 255
    absolute, 202

Activity coefficient, 255
    degenerate, 420
    ionic strength and, 410
    regular solution theory and, 415
    single ion transfer, 420
    transfer, 420
    transition state, 209
Acyl-oxygen fission, 7
Acyl transfer, nucleophilic, 349
Addition, 8
Adiabatic assumption, 194
    failure of, 229
Alkyl-oxygen fission, 7, 10
Alpha effect, 355
Analog simulation, 114
Angular momentum, 160
    nuclear, 153
Angular velocity, 155
Anion nucleophilicity, 421
Anion solvating ability, 445
Anti-Hammond behavior, 232
Apolar solvents, 398
Approximation, catalysis by, 263
Approximation effect, 365
Aprotic solvents, 398
Area as a variable, 81
Area differential, 81
Arrhenius equation, 14, 187, 208, 245,
    259
    extensions of, 252
Arrhenius plot, 188, 246, 254
    curvature in, 251
Autocatalysis, 22
Avoided crossing, 233

Barrier
    energy, 3
    intrinsic, 226, 239
Base
    hard, 8, 361

Base (*continued*)
  indicator, 447
  soft, 8, 361
Base catalysis
  general, 265, 268, 271
  specific, 264, 268
Base strength, measurement of, 447
Basicity, 8
  Lewis, 425
Basicity and nucleophilicity, 345, 350, 360
Basity, 445
Batch mixing, 176
Bathochromic shift, 435
Beer's law, 34
Bell-shaped pH curve, 285
Bloch equations, 160, 163
Blue shift, 435
Bodenstein approximation, 101
Boltzmann constant, 202
Boltzmann distribution, 157, 201
Bond dissociation curve, 191, 293
Bond dissociation energy, 196
Bond order, 223
Borderline reactions, 429
Born equation, 408
Born–Oppenheimer approximation, 193
Brønsted acid catalysis, 265
Brønsted base catalysis, 265
Brønsted coefficient, 225, 345, 347
Brønsted plot, 346
Brønsted relationships, 345
Brønsted-type plot, 350
  curved, 351
Buffer, 24
Buffer catalysis, 269
Burst effect, 118

Cage, solvent, 134
Cancellation assumption, 447
Catalysis, 263
  acid, 453
  buffer, 269
  definitions of, 263
  electrophilic, 265
  general acid, 265, 268
  general base, 265, 268, 271
  intermolecular, 266
  intramolecular, 266
  nucleophilic, 266, 268, 271

by pyridine, 7
  specific acid, 264, 268
  specific base, 264, 268
Cation-anion recombination, 358
Cation solvating ability, 445
Cavity theory, 395, 412
Characteristic equation, 92
Charge-transfer, 394
Chemical exchange, 166, 173
Chemical flux, 60
Chemical potential, 202, 254
  standard, 254
Classification of solvents, 397
  acid-base, 397
  hydrogen-bonding, 397
  statistical, 399
  structural, 397
Clausius–Mosotti equation, 389
Closed systems, 10
Coalescence, 169
Cohesive energy density, 412
Collision, bimolecular, 188
Collision theory, 188
Common-ion effect, 428
Common-ion inhibition, 183
Compensation effect, 369
Competitive reactions, 59
Complex
  activated, 200
  charge-transfer, 394
  electron donor-acceptor, 394
Complexation rates, 150, 152
Complex reaction, 4, 12, 59, 139
Complicated rate equation, 59
Composite reaction, 4
Concentration jump, 179
Concentration scales, 254
Concentration-time curve, 120
  area under, 81
Concentration units, 11
Concerted reaction, 230
Concurrent reactions, 62
Confidence interval, 48, 49
Conformational change, 175
Consecutive reactions, 59, 66
Continuous flow, 178
Continuous wave, 170
Coordinate system, rotating, 168, 170
Coordination number, 134, 403

Correlation time, 165
  rotational, 165
Cosphere, 402
Coulombic integral, 194
Coulomb's law, 392
Covariance, 40, 47
Cross-interaction constant, 332
Cross-reaction, 229
Curve-fitting, 88
Cybotactic region, 403

Data analysis, 31
Dative form, 394
Dead time, 177, 179
Debye–Hückel equation, 410
Degenerate states, 153
Degrees of freedom, 47
Desolvation, 362, 421
Detailed balance, 125
Determinant, 92
Deterministic rate equations, 114
Deuterium oxide, properties of, 300
Dielectric constant, 387, 389
  Kirkwood function of, 406
  microscopic, 403
Differential equations, solution of, 86
Differential rate equation, 13, 59, 77
Diffusion, 134
Diffusion coefficient, 134
Diffusion-controlled rate, 124, 134
Diffusion-controlled reactions, 149
Dimroth-Reichardt betaine, 437
Dipolar aprotic solvents, 398
  effect on anion nucleophilicity, 404, 421
Dipolarity-polarizability measure, 439
Dipole-dipole interaction, 392
  magnetic, 159, 165
Dipole moment, 389
  excited state, 435
Disparity reaction, 237
Dispersion force, 392
Dispersion mode, 164
Dissociating power, solvent, 402
Dissociation, 401
Dissociation constant, 402
Dissolution rate, 24
Distribution coefficient, 420
Donor ability, 425
Donor number, 425

Double dagger symbol, 205
Drug absorption, 68
Drug elimination, 68

Effective molarity, 365
Eigenvalue, 90, 92
Electric field discharge, 144
Electric field jump, 144
Electrochemical reactions, 181
Electrode reactions, 182
Electrofuge, 9
Electron affinity, 361
Electron donor-acceptor complex, 394
Electroneutrality principle, 147
Electronic effects,
  aliphatic, 337
  aromatic, 316
  hydrogen and deuterium, 300
  separation of, 325
  susceptibility to, 328
  transmission of, 328
Electronic transition energy, 435
Electron paramagnetic resonance, 153
Electrophile, 8
Electrophilic catalysis, 265
Electrophilic substituent constant, 322
Electrostatic factor, 398
Electrostatic free energy, 406, 408
Electrostatic interactions, 391
Electrostriction, 262, 416
Elementary reaction, 3, 7, 11, 59, 201
  rate equation of, 12, 17
Elimination, 8
  unimolecular, 9
Elimination phase, 68
Empirical polarity measures, 425
Encounter, 134
Endergonic reaction, 223
Energy barrier, 3
  height of, 220
  position of, 220
Enthalpy-entropy relationship, 369
Enthalpy of activation, 207
  determination of, 246
Entropy control, intramolecular reactivity
    and, 367
Entropy of activation, 207
  determination of, 246
  as mechanistic criterion, 220

Entropy of activation (*continued*)
  sign of, 256
Entropy unit, 242
Enzyme catalysis, 102
Enzyme-substrate complex, 102
Equilibrium, 60, 97, 99, 105, 125, 136
  condition for, 205
  displacement from, 62, 78
  in transition state theory, 201, 205
Equilibrium assumption, 96
Equilibrium constant, 61, 138
  complexation, 152
  dissociation, 402
  ionization, 402
  kinetic determination of, 279
  partition functions in, 204
  pressure dependence of, 144
  temperature dependence of, 143, 257
  transition state, 207
Equivalence, kinetic, 123
Error analysis, 40
Error propagation, 40
Ester hydrolysis, 4
Euler's method, 106
Excess acidity method, 451
Exchange
  chemical, 166, 173
  fast, 168
  isotopic, 300
  slow, 168
Exchange frequency, 167
Exchange integral, 194
Exergonic reaction, 223
Exponential function, Laplace transform
  of, 83
Extent of reaction, 10
External return, 428
Extrapolation, 64
Extrathermodynamic relationships, 311,
  425
Eyring plot, 246

Factor analysis, 445
Fast preequilibrium, 97
Fast reactions, 63, 133
Feathering technique, 73
Fick's first law, 134
Field effect, 336, 338
Final state, 3

First-order
  apparent, 23
  pseudo, 23
First-order approximation, 450
First-order decay, 18
First-order plot, 18, 35
First-order rate constant, 18, 31, 61
First-order rate equation, 18, 31, 34
First-order reaction, 18, 60
Flip-flop problem, 68
Flow methods, 177
Fluorescence quenching, 180
Flux, 134
  chemical, 60
Force constant, 294
Force of interaction, intermolecular,
  391
Forcing function, 143
  periodic, 144
  transient, 143
Fourier transform, 170
Fractional time, 29
Fractionation factor, 301
Fraction theorem, general partial, 85
Frame, rotating, 170
Franck–Condon principle, 435
Free energy, 211
  transfer, 418
Free energy of activation, 207, 210
Free induction decay, 170
Free radical reactions, 8
Frequency, 153, 155
  exchange, 167
  precessional, 155, 165
  relaxation, 167
  resonance, 164
Frequency domain, 170
Frequency factor, 188

Gas-phase reactivity, 385
General acid-base catalysis, 344
General acid catalysis, 265, 268
General base catalysis, 265, 268, 271
General partial fraction theorem, 85
Geometric mean approximation, 414
Graphical analysis, 52
Group transfer, nucleophilic, 357
Growth curve, 23
Guggenheim method, 36

Half-life, 29
concentration dependence of, 29
first-order, 18
Hammett acidity function, 448
Hammett equation, 315
Hammett plot, 318
nonlinear, 333
Hammett substituent constant, 316
Hammond behavior, 232
Hammond postulate, 220
geological analogy to, 220
optical analogy to, 223
Hardness, 8, 361
Hard-soft acid-base concept, 360
Harmonic mean, 370
Harmonic oscillator, 221, 294
Heat capacity of activation, 251
Heating time constant, 144
Heat of ionization, 257
of water, 257
Heisenberg uncertainty principle, 158
Henderson-Hasselbalch equation, 281
Heterogeneous reactions, 10
Heterolytic reactions, 8
Hindered base, nucleophilicity of, 272
Homogeneous reactions, 10
Homolytic reactions, 8
Hughes–Ingold rule, 387
Hydration sphere, 152
Hydrogen bond acidity, 440
Hydrogen bond basicity, 440
Hydrogen bond formation, effect of, 439
Hydrogen bonding, 395
Hydrolysis, 7, 147, 148
ester, 4
Hydrophobic effect, 395
Hydrophobic interaction, 395
Hydrophobicity, 395
Hydroxamic acid, 118
Hydroxylic solvents
ionization of, 409
polarity of, 401
Hypsochromic shift, 435

Identity reaction, 229, 358
Induced dipole, 392
Induction forces, 392
Induction period, 75, 120
Inductive effect, 323, 325, 338

Infinity time, 34, 36
Infinity value, unknown, 36
Inhibition, common ion, 183
Initial rate, 28, 103
Initial state, 3
Instrumental data, 34
Integrated rate equation, 24, 60, 77
deviation from, 25
Integrating factor, 66
Integration
by Laplace transforms, 84
numerical, 105
Runge–Kutta, 107
Intermediate, 4–7, 67, 79, 116
anionic, 230
cationic, 230
common, 119
detection of, 117, 188, 271, 272
ionization of, 292
trapping of, 272
Intermolecular catalysis, 266
Intermolecular forces, 391
Internal pressure, 412
Intramolecular assistance, 334
Intramolecular catalysis, 266
Intramolecular reactivity, 363
Intrinsic acidity, 402
Intrinsic barrier, 226, 239
Intrinsic well, 239
Inverse transform, 83
Ion contact distance, 410
Ionic reactions, 8
Ionic strength, 386, 410
Ionization, 401
heat of, 257
Ionization constant, 402
Ionization potential, 361, 414
Ionizing power, solvent, 402, 430
Ion pair, 401
contact, 402
external, 402
internal, 402
intimate, 402
solvent-separated, 402
Ion solvation, 401
Irreversibility, 116
Isergonic reaction, 223, 225
Isoenthalpic reaction, 369
Isoentropic reaction, 369

Isokinetic relationship, 261, 368
Isokinetic temperature, 368
Isolation technique, 26, 78
Isotope effects
    inverse, 299
    kinetic, 292
    primary, 293, 295
    secondary, 298
    solvent, 272, 300
Isotopic exchange, 300
Isotopic substitution, 6

Joule heating, 144

Kilocalories, 14
    conversion of, 246
Kilojoules, 14
    conversion of, 246
Kinetically equivalent terms, 123
Kinetically indistinguishable terms, 123
Kinetic equivalence, 123, 136, 217, 349
    in intramolecular catalysis, 267
    salt effect and, 411
Kinetic isotope effects, 292
    primary, 293
    secondary, 298
    solvent, 300
Kinetic scheme, 4, 7, 59
    cyclic, 99
    development of, 115
    rate equations in, 122
Kirkwood dielectric constant function, 406

Lagtime, 75
Laplace transform, 82
Larmor precessional frequency, 155, 165
Laser pulse absorption, 144
Lattice energy, 403
Law of mass action, 60, 125
Least-squares analysis
    linear, 41
    nonlinear, 49
    univariate, 44
    unweighted, 44, 51
    weighted, 46, 51, 247
Leaving group, 9, 340, 349, 357
Lennard–Jones potential, 393
Lewis acid-base adduct, 425
Lewis acid catalysis, 265
Lewis acidity, 426

Lewis basicity, 425
Life-time, 18
    fluorescence, 181
    state, 153, 158, 167
Lifetime broadening, 158, 165
Linear free energy relationships, 312
    and reactivity-selectivity principle, 375
Linear functions, 41
Linearization, 137, 139
Linear operator, 83
Linear solvent effect relationships, 442
Line broadening, 158, 159
Lineweaver–Burk plot, 103
Line width, 167
Lipophilicity, 396
London force, 392
Longitudinal relaxation, 161
Lorentzian line shape, 164, 170
Lorenz-Lorentz equation, 389
Lyate ion catalysis, 264
Lyonium ion catalysis, 264

Magnetic moment, 153, 155, 160
Magnetic quantum number, 153
Magnetization, 160
Magnetogyric ratio, 153, 160
Main reaction, 237
Marcus equation, 227, 238, 314
Marcus plot, slope of, 227, 354
Marcus theory, applicability of, 358
    reactivity-selectivity principle and, 375
Mass, reduced, 189, 294
Mass action law, 11, 60, 125, 428
Mass balance relationships, 19, 21, 34, 60,
    64, 67, 89, 103, 140, 147
Maximum velocity, enzyme-catalyzed, 103
Mean, harmonic, 370
Mechanism
    classification of, 8
    definition of, 3
    study of, 6, 115
Medium effects, 385, 418, 420
    physical theories of, 405
Meisenheimer complex, 129
Menschutkin reaction, 404, 407, 422
Mesomerism, 323
Method of residuals, 73
Michaelis constant, 103
Michaelis–Menten equation, 103
Microscopic reversibility, 125

Microwave heating, 144
Mixing
  batch, 176
  complete, 178
  flow, 177
  rapid, 176
Mixing time, 133
Model building, 115
Model function, 42
Molarity, effective, 365
Molarity scale, 254
Molar polarization, 389
Molar refraction, 389
Molar volume, 261
Molecularity, 9, 24
  reaction order and, 24
Mole fraction scale, 254
Monte Carlo simulation, 109
  fluctuations in, 114
Morse equation, 196
Multiple linear regression, solvent effects
  and, 443

Naked anions, 360
Neighboring group participation, 334
Neutralization, 147
No-bond form, 394
Nonisothermal kinetics, 250
Nonlinear function, 41
Nonlinear least-squares regression,
  49
Nonpolar solvents, 398
Normal equations, 43
  nonlinear, 49
Nuclear magnetic resonance, 153,
  155
Nuclear relaxation, 160
Nuclear spin, 153
Nuclear Zeeman splitting, 154
Nucleofuge, 9
Nucleophile, 8, 349, 357
Nucleophilic attack, 266
Nucleophilic catalysis, 266, 268, 271
Nucleophilicity, 8, 349, 357
  anion, 404
  solvent, 431
Nucleophilic parameter, 358, 361
Nucleophilic substitution, solvent effects
  on, 427
Numerical integration, 105

Objective function, least-squares, 42
Occam's razor, 115
Operator, delta, 312, 386
Orbital steering, 366
Order
  determination of, 24, 28, 275
  overall, 13
  reaction, 13, 24
  with respect to concentration, 28, 29
  with respect to time, 24
Orientation effect, 366
Ortho effect, 334
Overlap integral, 194
Overlap method, 448
Oxidation and nucleophilicity, 360

Parabolic energy barrier, 221
Parallel displacement, 232
Parallel reactions, 59, 62
Paramagnetism, 166
Parameter, 41, 42
  variance of, 46
Partial fraction theorem, 85
Partial molar free energy, 254
Partial rate factor, 373
Partition coefficient, 400
  transfer free energy and, 419
Partition function, 201
  factorization of, 203
  translational, 204
  vibrational, 204, 206
Peak width, 164
Perpendicular displacement, 232
pH, temperature dependence of, 258
Pharmacokinetics, 68, 73
pH buffer, 24
pH effects, 273
pH-rate profiles, 273
  bell-shaped, 285
  compilation of, 292
  sigmoid, 277
P-jump, 144
Polar effect, 323, 337, 339
Polarity, solvent, 387, 399
Polarity measures, empirical, 425
Polarizability, 338, 389, 392, 414
  nucleophilicity and, 361
Polarization, molar, 389
Polarization forces, 392
Polar solvents, 398

Potential, chemical, 202
Potential energy, 211
  diatomic molecule, 191
Potential energy functions, 391
Potential energy surfaces, 191
Precession, 155
Precessional frequency, 160, 165
Preequilibrium, fast, 142
Preequilibrium assumption, 96, 105, 116
Preexponential factor, 14, 188, 246
  collision theory and, 190
  temperature dependence of, 190, 207,
    251
Pressure
  effect on rates, 261
  internal, 412
Pressure jump, 144
Pressure units, 262
Pre-steady-state, 104
Primary isotope effect, 293, 295
Primary salt effect, 386, 411
Principal component analysis, 445
Principle of detailed balance, 125
Principle of microscopic reversibility, 125
Product state, 3
Propagation of errors, 40, 48, 248
Propinquity effect, 263, 365
Protolysis, 147, 148
Proton inventory technique, 302
Proton transfer, 166
  direct, 148
  extent of, 346
  fast, 97, 146, 173
  isotope effect in, 296
  partial, 395
Proximity effect, 365
Pseudo-first-order rate constant, 23
Pseudo-first-order reaction, 61
Pseudo-order rate constant, 23
Pseudo-order reaction, 23
Pseudo-order technique, 26, 78
Pulse NMR, 170

Quadrupole moment, 155, 166, 391
Quasi-steady-state approximation, 101
Quenching, fluorescence, 180

Rapid mixing, 176
Rate, diffusion-controlled, 134

Rate coefficient, 13
Rate constant, 13
  catalytic, 268
  determination of, 31
  diffusion-limited, 135
  first-order, 18, 31, 61
  pressure dependence of, 261
  pseudo-order, 23
  second-order, 20
  temperature dependence of, 13, 245
  transition state theory, 206
  zero-order, 17
Rate-determining step, 213
  change in, 352
Rate equation, 11, 17
  complicated, 13, 59
  deterministic, 114
  differential, 59, 77
  elementary reaction, 17
  first-order, 18
  integrated, 17, 60, 77
  simple, 13, 17, 59
  transition state, 205
  zero-order, 17
Rate-equilibrium relationship, 224
Rate-limiting step, 213
Rate measurement, 77
Rate of reaction, 10, 11
Reactant state, 3
Reaction
  acyl transfer, 349
  borderline, 429
  complex, 4, 12, 59
  composite, 4
  concerted, 230
  diffusion-controlled, 134
  disparity, 237
  displacement, 427
  electrode, 182
  electrophile-nucleophile, 358
  elementary, 3, 11, 59, 201
  endergonic, 223
  enzyme-catalyzed, 102
  exergonic, 223
  fast, 133
  group transfer, 358
  identity, 229, 358
  ion-ion, 410
  ion-neutral molecule, 408

isergonic, 223
isoenthalpic, 369
isoentropic, 369
main, 237
Menschutkin, 404, 407, 422
neutral–neutral molecule, 405
spontaneous, 268, 277
stepwise, 230
uncatalyzed, 268, 277
water, 268, 277
Reaction cone, 366
Reaction constant, 328
Reaction coordinate, 192
  motion along, 206
Reaction coordinate diagram, 192, 209
Reaction mechanism, 3
Reaction order, 13
  determination of, 24
Reaction parameter, 314
Reaction path, 4
Reaction rate, 11
Reactions
  competitive, 59
  concurrent, 62
  consecutive, 59, 66
  parallel, 59, 62
  reversible, 59, 60
  series, 59, 66
Reaction series, 312
Reaction site, 313
Reaction variable, 19, 21, 33, 137, 138,
    140, 147
Reactivity, 371
Reactivity-selectivity principle, 371
  Marcus theory and, 375
Reagent, 8, 315
Red shift, 435
Reduced mass, 189, 294
Reference frame, rotating, 168
Reference state, 255
Regression analysis, 41
Regression equations, normal, 43, 44
Regular solution theory, 413
Relaxation, 136, 157
  longitudinal, 161
  nuclear, 160
  spin-lattice, 157, 161, 164
  spin-spin, 159, 163, 165, 167
  transverse, 163

Relaxation amplitude, 143
Relaxation kinetics, 62, 136
Relaxation spectrum, 140
Relaxation time, 138, 146
  spin-lattice, 158, 161, 164
  spin-spin, 159, 163, 165, 167
Repulsive force, 393
Residuals, method of, 73
Resonance
  nuclear magnetic, 155
  push–pull, 328
  through, 320
Resonance effect, 323, 325, 338
Resonance integral, 194
Reversibility, 116
  microscopic, 125
Reversible reactions, 59
  simulation of, 113
Rotating frame, 168, 170
Rotation, single bond, 174
Rule of six, 344
Runge–Kutta method, 107

Salt effect, 124, 386
  primary, 386, 411
  secondary, 386
Saturation factor, 159, 164
Secondary isotope effect, 298
Secondary salt effect, 386
Second-order rate constant, 20
  determination of, 32
Second-order rate equation, 20, 34
Second-order reaction, 20
  simulation of, 113
Secular equation, 90, 92
Selectivity, 371
Selectivity factor, 373
Sensitivity coefficient, 225
Separability postulate, 314
Series reactions, 59, 66
Shelf-life, 18
Sigmoid curve, 23
Sigmoid pH-rate profile, 277
Simple rate equation, 59
Simulation, analog, 114
  Monte Carlo, 109
Six number, 344
Slow-fast ambiguity, 68, 72
Slow passage, 163

Softness, 8, 361, 426
Solubility, aqueous, 400
Solubility and transfer free energy, 419
Solubility equilibrium, 24
Solubility parameter, 415
Solvation, 401
  change in, 354
  preferential, 403
  selective, 403
Solvation and polarity, 399
Solvation energy, 403, 420
Solvation of anions, 360
Solvation shell, 403
Solvatochromic comparison method, 439
Solvatochromic polarity measures, 436
Solvatochromism, 435
Solvent cage, 134
Solvent coordination number, 134, 403
Solvent effects, 385, 418
  initial and transition state, 418
  kinetic measures of, 427
Solvent ionizing power parameter, 430
Solvent isotope effects, 272, 300
Solvent nucleophilicity, 431
Solvent participation, covalent, 429
Solvent polarity, 399, 425
Solvent polarity parameter, 436
Solvent properties, 389
Solvent-separated complex, 152
Solvent sorting, 404
Solvent structure, 402
Solvophobic interaction, 395
Solvophobicity parameter, 427
Sound absorption
  chemical, 145
  classical, 145
Spatiotemporal hypothesis, 368
Specific acid catalysis, 264, 268, 446
Specific base catalysis, 264, 268
Specific rate, 13
Spectroscopy, 153
Speed, molecular, 189
Spin, nuclear, 153
Spin echo technique, 172
Spin-lattice relaxation, 157, 161, 164, 172

Spin-lattice relaxation time, 158, 161, 164
Spin quantum number, 153
Spin–spin relaxation, 159, 163, 165, 167
Spin–spin relaxation time, 159, 163–165, 167
Spin system, 157, 159
Splitting, Zeeman, 154
Stability study, accelerated, 260
Standard potential, 254, 359
Standard reaction, 315
Standard state, 253
  selection of, 208
Stationary-state hypothesis, 101
Statistical analysis, 40
Statistical analysis of solvent effects, 444
Steady state, 100
Steady-state approximation, 100, 105, 116
Stepwise reaction, 230
Stereopopulation control, 367
Steric compression, 367
Steric constant, 335, 342
Steric effect
  aliphatic, 340, 342
  aromatic, 335
Steric requirements, hydrogen and deuterium, 299
Stern–Volmer plot, 181
Stiff differential equations, 109
Stochastic simulation, 109
Stoichiometric coefficients, 11
Stokes–Einstein equation, 135
Stopped flow, 179
Structured water, 395
Structure-reactivity relationships, 311
Sublimation energy, 403
Substituent, 313
Substituent constant, 323
  alkyl group, 341
  electrophilic, 322
  Hammett, 316
  inductive, 325, 338
  normal, 324
  polar, 339
  primary, 324
  resonance, 325

Substituent effects
  aliphatic, 338
  aromatic, 315
  multiple, 332
  reagent, 344
Substituent parameter, 314
Substitution, 8
  electrophilic, 9
  nucleophilic, 8
Substrate, 8, 315
Surface area, cavity, 395
Surface tension, 391, 395
Swain-Scott equation, 359
Symbiotic effect, 360

Taft equation, 339
Taylor's series expansion, 40, 49,
  332
Temperature effects, 13, 245
Temperature jump, 143
Third-order reaction, 54
Time
  elimination of, 79
  replacement with area, 81
Time domain, 170
Time-ratio method, 69, 76
Time scales, 99, 173
Titrimetric analysis, 32, 69
T-jump, 143
Torque, 160
Tunneling, 136, 197, 295
Transfer activity coefficient, 420
Transfer free energy, 418
Transform
  inverse, 83
  Laplace, 82
  Fourier, 170
Transition, electronic, 435
  nuclear, 154, 157
Transition state, 3, 5, 193, 201, 205
  attractive, 199
  composition of, 216
  early, 199, 234
  late, 197, 234
  location of, 220, 224, 235
  loose, 236
  probe of, 351
  productlike, 197
  reactantlike, 199

  repulsive, 197
  tight, 237
Transition state, microscopic reversibility
  and, 126
Transition state theory, 200
  thermodynamics of, 207
Transmission coefficient, 207
Transverse relaxation, 163

Ultrasonic absorption, 144
Uncertainty principle, 158
Unitary quantities, 255
Units, 14
  activation energy, 246
  dipole moment, 389
  energy, 202
  first-order rate constant, 18
  molar, 189
  pressure, 262
  rate constant, 212
  second-order rate constant, 20
  solubility parameter, 417
  third-order rate constant, 27
  zero-order rate constant, 17
Univariate least-squares regression, 44
Univariate linear model, 44, 48
Unweighted regression equations, 44

Valence bond description, 233
van der Waals forces, 393
van der Waals radius, 343
van't Hoff equation, 143, 257
Variable, 44
  calculated, 42
  dependent, 42
  independent, 42
Variable, transformation of, 45
Variance, 40, 45, 47
Variance of observations, 47
Variance of parameters, 46
Variate, 44
Variational method, 195
Velocity of reaction, 11
Vibrational energy, zero-point, 294,
  299
Vibrational quantum number, 294
Volume, activation, 261, 416
  molar, 261

Water, structure of, 395
Weight, 45, 49
Weighting, 44, 45, 51, 248

Zero-order, pseudo, 23
Zero-order approximation, 450

Zero-order rate equation, 34
Zero-order reaction, 17
Zero-point vibrational energy, 294, 299
Zero time, 176
Zucker-Hammett hypothesis, 455